Lecture Notes in Artificial Intelligence 7123

Subseries of Lecture Notes in Computer Science

LNAI Series Editors

Randy Goebel
  University of Alberta, Edmonton, Canada
Yuzuru Tanaka
  Hokkaido University, Sapporo, Japan
Wolfgang Wahlster
  DFKI and Saarland University, Saarbrücken, Germany

LNAI Founding Series Editor

Joerg Siekmann
  DFKI and Saarland University, Saarbrücken, Germany

Fernando Bobillo   Paulo C.G. Costa
Claudia d'Amato   Nicola Fanizzi
Kathryn B. Laskey   Kenneth J. Laskey
Thomas Lukasiewicz   Matthias Nickles
Michael Pool (Eds.)

Uncertainty Reasoning for the Semantic Web II

International Workshops
URSW 2008-2010 Held at ISWC
and UniDL 2010 Held at FLoC
Revised Selected Papers

 Springer

ISSN 0302-9743   e-ISSN 1611-3349
ISBN 978-3-642-35974-3   e-ISBN 978-3-642-35975-0
DOI 10.1007/978-3-642-35975-0
Springer Heidelberg Dordrecht London New York

Library of Congress Control Number: 2012954778

CR Subject Classification (1998): I.2.3-4, I.2.6, H.3.5, H.5.3, F.4.1, G.3

LNCS Sublibrary: SL 7 – Artificial Intelligence

Typesetting: Camera-ready by author, data conversion by Scientific Publishing Services, Chennai, India

Printed on acid-free paper

Springer is part of Springer Science+Business Media (www.springer.com)

Preface

This is the second volume on "Uncertainty Reasoning for the Semantic Web," containing revised and significantly extended versions of selected workshop papers presented at three workshops on Uncertainty Reasoning for the Semantic Web (URSW), held at the International Semantic Web Conferences (ISWC) in 2008, 2009, and 2010, or presented at the First International Workshop on Uncertainty in Description Logics (UniDL) in 2010. The first volume contained the proceedings of the first three workshops on URSW at ISWC in 2005, 2006, and 2007.

The two volumes together represent a comprehensive compilation of state-of-the-art research approaches to uncertainty reasoning in the context of the Semantic Web, capturing different models of uncertainty and approaches to deductive as well as inductive reasoning with uncertain formal knowledge.

The World Wide Web community envisions effortless interaction between humans and computers, seamless interoperability and information exchange among Web applications, and rapid and accurate identification and invocation of appropriate Web services. As work with semantics and services grows more ambitious, there is increasing appreciation of the need for principled approaches to the formal representation of and reasoning under uncertainty. The term *uncertainty* is intended here to encompass a variety of forms of incomplete knowledge, including incompleteness, inconclusiveness, vagueness, ambiguity, and others. The term *uncertainty reasoning* is meant to denote the full range of methods designed for representing and reasoning with knowledge when Boolean truth values are unknown, unknowable, or inapplicable. Commonly applied approaches to uncertainty reasoning include probability theory, Dempster-Shafer theory, fuzzy logic and possibility theory, and numerous other methodologies.

A few Web-relevant challenges that are addressed by reasoning under uncertainty include:

Uncertainty of available information: Much information on the World Wide Web is uncertain. Examples include weather forecasts or gambling odds. Canonical methods for representing and integrating such information are necessary for communicating it in a seamless fashion.

Information incompleteness: Information extracted from large information networks such as the World Wide Web is typically incomplete. The ability to exploit partial information is very useful for identifying sources of service or information. For example, that an online service deals with greeting cards may be evidence that it also sells stationery. It is clear that search effectiveness could be improved by appropriate use of technologies for handling uncertainty.

Information incorrectness: Web information is also often incorrect or only partially correct, raising issues related to trust or credibility. Uncertainty representation and reasoning helps to resolve tension among information sources having different confidence and trust levels, and can facilitate the merging of controversial information obtained from multiple sources.

Uncertain ontology mappings: The Semantic Web vision implies that numerous distinct but conceptually overlapping ontologies will co-exist and interoperate. It is likely that in such scenarios, ontology mapping will benefit from the ability to represent degrees of membership and/or likelihoods of membership in categories of a target ontology, given information about class membership in the source ontologies.

Indefinite information about Web services: Dynamic composability of Web services will require runtime identification of processing and data resources and resolution of policy objectives. Uncertainty reasoning techniques may be necessary to resolve situations in which existing information is not definitive.

Uncertainty is thus an intrinsic feature of many important tasks on the Web and the Semantic Web, and a full realization of the World Wide Web as a source of processable data and services demands formalisms capable of representing and reasoning under uncertainty. Unfortunately, none of these needs can be addressed in a principled way by current Web standards. Although it is to some degree possible to use semantic mark-up languages such as OWL or RDF(S) to represent qualitative and quantitative information about uncertainty, there is no established foundation for doing so, and feasible approaches are severely limited. Furthermore, there are ancillary issues such as how to balance representational power vs. simplicity of uncertainty representations, which uncertainty representation techniques address uses such as the examples listed above, how to ensure the consistency of representational formalisms and ontologies, etc.

In response to these pressing demands, in recent years, several promising approaches to uncertainty reasoning on the Semantic Web have been proposed. The present volume covers a representative cross section of these approaches, from extensions to existing Web-related logics for the representation of uncertainty to approaches to inductive reasoning under uncertainty on the Web.

In order to reflect the diversity of the presented approaches and to relate them to their underlying models of uncertainty, the contributions to this volume are grouped as follows:

Probabilistic and Dempster-Shafer Models

Probability theory provides a mathematically sound representation language and formal calculus for rational degrees of belief, which gives different agents the freedom to have different beliefs about a given hypothesis. As this provides a compelling framework for representing uncertain, imperfect knowledge that can come from diverse agents, there are many distinct approaches using probability in the context of the Semantic Web. Classes of probabilistic models covered with the present volume are Bayesian networks, probabilistic extensions to description and first-order logics, and models based on the Dempster-Shafer theory (a generalization of the classic Bayesian approach).

Fuzzy and Possibilistic Models

Fuzzy formalisms allow for the representing and processing of degrees of truth about vague (or imprecise) pieces of information. In fuzzy description logics and ontology languages, concept assertions, role assertions, concept inclusions, and role inclusions have a degree of truth rather than a binary truth value. The present volume presents various approaches that exploit fuzzy logic and possibility theory in the context of the Semantic Web.

Inductive Reasoning and Machine Learning

Machine learning is supposed to play an increasingly important role in the context of the Semantic Web by providing various tasks, such as the learning of ontologies from incomplete data or the (semi-)automatic annotation of data on the Web. Results obtained by machine learning approaches are typically uncertain. As a logic-based approach to machine learning, inductive reasoning provides means for inducing general propositions from observations (example facts). Papers in this volume exploit the power of inductive reasoning for the purpose of ontology learning, and project future directions for the use of machine learning on the Semantic Web.

Hybrid Approaches

This volume segment contains papers that either combine approaches from two or more of the previous segments, or that do not rely on any specific classic approach to uncertainty reasoning.

We would like to express our gratitude to the authors of this volume for their contributions and to the workshop participants for inspiring discussions, as well as to the members of the workshop Program Committees and the additional reviewers for their reviews and for their overall support.

October 2012

Fernando Bobillo
Paulo C.G. Costa
Claudia d'Amato
Nicola Fanizzi
Kathryn B. Laskey
Kenneth J. Laskey
Thomas Lukasiewicz
Matthias Nickles
Michael Pool

Organization

Reviewers

Saminda Abeyruwan	University of Miami, USA
Fernando Bobillo	University of Zaragoza, Spain
Silvia Calegari	University of Milano-Bicocca, Italy
Rommel Carvalho	George Mason University, USA
Paulo C.G. Costa	George Mason University, USA
Fabio G. Cozman	Universidade de São Paulo, Brazil
Claudia d'Amato	University of Bari, Italy
Nicola Fanizzi	University of Bari, Italy
Bart Gajderowicz	Ryerson University, Canada
Juan Gómez-Romero	Carlos III University of Madrid, Spain
Pavel Klinov	University of Manchester, UK
Marcelo Ladeira	University of Brasilia, Brazil
Kathryn B. Laskey	George Mason University, USA
Anders L. Madsen	Hugin Expert A/S, Denmark
Trevor Martin	University of Bristol, UK
Pasquale Minervini	University of Bari, Italy
Pierluigi Miraglia	Goldman Sachs, USA
Matthias Nickles	Technische Universität München, Germany
Rafael Peñaloza	Technische Universität Dresden, Germany
Michael Pool	Convera Technologies, Inc., USA
Livia Predoiu	Universität Magdeburg, Germany
Guilin Qi	Southeast University, China
David Robertson	University of Edinburgh, UK
Daniel Sánchez	University of Granada, Spain
Lutz Schröder	Universität Bremen, Germany
Nikolaos Simou	National Technical University of Athens, Greece
Pavel Smrž	Brno University of Technology, Czech Republic
Giorgos Stoilos	University of Oxford, UK
Umberto Straccia	ISTI-CNR, Italy
Matthias Thimm	Universität Koblenz-Landau, Germany
Andreas Tolk	Old Dominion University, USA
Ubbo Visser	University of Miami, USA
Yining Wu	University of Luxembourg, Luxembourg

4th International Workshop on Uncertainty Reasoning for the Semantic Web (URSW 2008)

Organizing Committee

Fernando Bobillo	University of Zaragoza, Spain
Paulo C.G. Costa	George Mason University, USA
Claudia d'Amato	University of Bari, Italy
Nicola Fanizzi	University of Bari, Italy
Kathryn B. Laskey	George Mason University, USA
Kenneth J. Laskey	MITRE Corporation, USA
Thomas Lukasiewicz	University of Oxford, UK
Trevor Martin	University of Bristol, UK
Matthias Nickles	University of Bath, UK
Michael Pool	Convera Technologies, Inc., USA
Pavel Smrž	Brno University of Technology, Czech Republic

Program Committee

Ameen Abu-Hanna	Universiteit van Amsterdam, The Netherlands
Fernando Bobillo	University of Zaragoza, Spain
Silvia Calegari	University of Milano-Bicocca, Italy
Paulo C.G. Costa	George Mason University, USA
Fabio G. Cozman	Universidade de São Paulo, Brazil
Claudia d'Amato	University of Bari, Italy
Ernesto Damiani	University of Milan, Italy
Nicola Fanizzi	University of Bari, Italy
Francis Fung	Eduworks, Inc., USA
Kathryn B. Laskey	George Mason University, USA
Kenneth J. Laskey	MITRE Corporation, USA
Thomas Lukasiewicz	University of Oxford, UK
Anders L. Madsen	Hugin Expert A/S, Denmark
M. Scott Marshall	Universiteit van Amsterdam, The Netherlands
Trevor Martin	University of Bristol, UK
Matthias Nickles	University of Bath, UK
Yung Peng	University of Maryland, USA
Michael Pool	Convera, Inc., USA
Livia Predoiu	Universität Mannheim, Germany
David Robertson	University of Edinburgh, UK
Elie Sanchez	Université de La Méditerranée Aix-Marseille II, France
Daniel Sánchez	University of Granada, Spain
Nematollaah Shiri	Concordia University, Canada
Oreste Signore	ISTI-CNR, Italy
Sergej Sizov	University of Koblenz-Landau, Germany
Pavel Smrž	Brno University of Technology, Czech Republic
Umberto Straccia	ISTI-CNR, Italy

Heiner Stuckenschmidt   Universität Mannheim, Germany
Masami Takikawa   Cleverset, Inc., USA
Peter Vojtáš   Charles University Prague, Czech Republic

Additional Reviewers

Ahmed Alasoud
Paolo Ceravolo

5th International Workshop on Uncertainty Reasoning for the Semantic Web (URSW 2009)

Organizing Committee

Fernando Bobillo   University of Zaragoza, Spain
Paulo C.G. Costa   George Mason University, USA
Claudia d'Amato   University of Bari, Italy
Nicola Fanizzi   University of Bari, Italy
Kathryn B. Laskey   George Mason University, USA
Kenneth J. Laskey   MITRE Corporation, USA
Thomas Lukasiewicz   University of Oxford, UK
Trevor Martin   University of Bristol, UK
Matthias Nickles   University of Bath, UK
Michael Pool   Convera Technologies, Inc., USA
Pavel Smrž   Brno University of Technology, Czech Republic

Program Committee

Fernando Bobillo   University of Zaragoza, Spain
Silvia Calegari   University of Milano-Bicocca, Italy
Rommel N. Carvalho   George Mason University, USA
Paulo C.G. Costa   George Mason University, USA
Fabio G. Cozman   Universidade de São Paulo, Brazil
Claudia d'Amato   University of Bari, Italy
Nicola Fanizzi   University of Bari, Italy
Marcelo Ladeira   University of Brasilia, Brazil
Kathryn B. Laskey   George Mason University, USA
Kenneth J. Laskey   MITRE Corporation, USA
Thomas Lukasiewicz   University of Oxford, UK
Anders L. Madsen   Hugin Expert A/S, Denmark
Trevor Martin   University of Bristol, UK
Matthias Nickles   University of Bath, UK
Jeff Pan   University of Aberdeen, UK
Yung Peng   University of Maryland, USA
Michael Pool   Convera, Inc., USA

Livia Predoiu   Universität Mannheim, Germany
Guilin Qi   University of Karlsruhe, Germany
Carlos H. Ribeiro   Instituto Tecnológico de Aeronáutica, Brazil
David Robertson   University of Edinburgh, UK
Daniel Sánchez   University of Granada, Spain
Sergej Sizov   University of Koblenz-Landau, Germany
Pavel Smrž   Brno University of Technology, Czech Republic
Giorgos Stoilos   National Technical University of Athens, Greece
Umberto Straccia   ISTI-CNR, Italy
Andreas Tolk   Old Dominion University, USA
Peter Vojtáš   Charles University Prague, Czech Republic
Johanna Völker   University of Karlsruhe, Germany

Additional Reviewer

Zhiqiang Gao

6th International Workshop on Uncertainty Reasoning for the Semantic Web (URSW 2010)

Organizing Committee

Fernando Bobillo   University of Zaragoza, Spain
Rommel N. Carvalho   George Mason University, USA
Paulo C.G. Costa   George Mason University, USA
Claudia d'Amato   University of Bari, Italy
Nicola Fanizzi   University of Bari, Italy
Kathryn B. Laskey   George Mason University, USA
Kenneth J. Laskey   MITRE Corporation, USA
Thomas Lukasiewicz   University of Oxford, UK
Trevor Martin   University of Bristol, UK
Matthias Nickles   University of Bath, UK
Michael Pool   Vertical Search Works, Inc., USA

Program Committee

Fernando Bobillo   University of Zaragoza, Spain
Rommel N. Carvalho   George Mason University, USA
Paulo C.G. Costa   George Mason University, USA
Fabio G. Cozman   Universidade de São Paulo, Brazil
Claudia d'Amato   University of Bari, Italy
Nicola Fanizzi   University of Bari, Italy
Marcelo Ladeira   University of Brasilia, Brazil
Kathryn B. Laskey   George Mason University, USA
Kenneth J. Laskey   MITRE Corporation, USA
Thomas Lukasiewicz   University of Oxford, UK
Trevor Martin   University of Bristol, UK

Matthias Nickles	University of Bath, UK
Jeff Pan	University of Aberdeen, UK
Michael Pool	Convera, Inc., USA
Livia Predoiu	Universität Mannheim, Germany
Guilin Qi	University of Karlsruhe, Germany
David Robertson	University of Edinburgh, UK
Daniel Sánchez	University of Granada, Spain
Sergej Sizov	University of Koblenz-Landau, Germany
Giorgos Stoilos	University of Oxford, UK
Umberto Straccia	ISTI-CNR, Italy
Andreas Tolk	Old Dominion University, USA
Johanna Völker	University of Karlsruhe, Germany
Peter Vojtáš	Charles University Prague, Czech Republic

Additional Reviewers

Edward Thomas
Xiaowang Zhang
Yuan Ren

First International Workshop on Uncertainty in Description Logics (UniDL 2010)

Organizing Committee

Thomas Lukasiewicz	University of Oxford, UK
Rafael Peñaloza	Technische Universität Dresden, Germany
Anni-Yasmin Turhan	Technische Universität Dresden, Germany

Program Committee

Eyal Amir	University of Illinois, Urbana-Champaign, USA
Fernando Bobillo	University of Zaragoza, Spain
Simona Colucci	Technical University of Bari, Italy
Fabio G. Cozman	Universidade de São Paulo, Brazil
Manfred Jaeger	Aalborg University, Denmark
Pavel Klinov	University of Manchester, UK
Ralf Möller	Hamburg University of Technology, Germany
Mathias Niepert	Universität Mannheim, Germany
Stefan Schlobach	Vrije Universiteit Amsterdam, The Netherlands
Luciano Serafini	IRS Trento, Italy
Giorgos Stoilos	University of Oxford, UK
Guilin Qi	Southeast University, China
Umberto Straccia	ISTI-CNR, Italy

Additional Reviewer

Abner Guzman-Rivera

Table of Contents

Probabilistic and Dempster-Shafer Models

Fuzzy and Possibilistic Models

Inductive Reasoning and Machine Learning

Hybrid Approaches

PR-OWL 2.0 – Bridging the Gap
to OWL Semantics

Rommel N. Carvalho, Kathryn B. Laskey, and Paulo C.G. Costa

Department of SEOR / Center of Excellence in C4I,
George Mason University, USA
rommel.carvalho@gmail.com, {klaskey,pcosta}@gmu.edu
http://www.gmu.edu

Abstract. The past few years have witnessed an increasingly mature body of research on the Semantic Web (SW), with new standards being developed and more complex use cases being proposed and explored. As complexity increases in SW applications, so does the need for principled means to represent and reason with uncertainty in SW applications. One candidate representation for uncertainty representation is PR-OWL, which provides OWL constructs for representing Multi-Entity Bayesian Network (MEBN) theories. This paper reviews some shortcomings of PR-OWL 1.0 and describes how they are addressed in PR-OWL 2. A method is presented for mapping back and forth between OWL properties and MEBN random variables (RV). The method applies to properties representing both predicates and functions.

Keywords: uncertainty reasoning, OWL, PR-OWL, MEBN, probabilistic ontology, Semantic Web, compatibility.

1 Introduction

Appreciation is growing within the Semantic Web community of the need to represent and reason with uncertainty. Several approaches have emerged to treating uncertainty in Semantic Web applications (e.g., [5,6,8,9,12,14,17,18]). In 2007, the World Wide Web Consortium (W3C) created the Uncertainty Reasoning for the World Wide Web Incubator Group (URW3-XG) to identify requirements for reasoning with and representing uncertain information in the World Wide Web. The URW3-XG concluded that standardized representations are needed to express uncertainty in Web-based information [11]. A candidate representation is Probabilistic OWL (PR-OWL) [5], an OWL upper ontology for representing probabilistic ontologies based on Multi-Entity Bayesian Networks (MEBN) [10].

Compatibility with OWL was a major design goal for PR-OWL [5]. However, there are several ways in which the initial release of PR-OWL falls short of complete compatibility. First, there is no mapping in PR-OWL to properties of OWL. Second, although PR-OWL has the concept of meta-entities, which allows the definition of complex types, it lacks compatibility with existing types already present in OWL.

These problems have been noted in the literature [16]:

F. Bobillo et al. (Eds.): URSW 2008-2010/UniDL 2010, LNAI 7123, pp. 1–18, 2013.

PR-OWL does not provide a proper integration of the formalism of MEBN and the logical basis of OWL on the meta level. More specifically, as the connection between a statement in PR-OWL and a statement in OWL is not formalized, it is unclear how to perform the integration of ontologies that contain statements of both formalisms.

This Chapter describes the need for a formal mapping between random variables defined in PR-OWL and properties defined in OWL, and proposes an approach to such a mapping. We then explain why PR-OWL 1.0 does not support such a mapping. Next, we present an approach to overcome the limitations in PR-OWL 1.0 by introducing new relationships created in PR-OWL 2.0. Finally, we present a scheme for the mapping back and forth from triples into random variables.

2   PR-OWL - An OWL Upper Ontology for Defining MEBN Models

PR-OWL was proposed as an extension to the OWL language to define probabilistic ontologies expressed in MEBN [10], a first-order probabilistic language (FOPL) [13]. Before delving into the details of PR-OWL, we provide a brief overview of MEBN.

As a running example, we consider an OWL ontology for the public procurement domain. A fuller treatment of the procurement ontology can be found in [4]. The ontology defines concepts such as procurement, winner of a procurement, members of a committee responsible for a procurement, etc. Figure 1 presents an OWL ontology with a few of the concepts that would be present in this domain. In the figure we can see that a front man is defined as a person who is a front for some organization (as shown in the equivalent class expression Person and isFrontFor some Organization for the FrontMan class in Figure 1).

Although there is great interest in finding people acting as fronts, it is in general unknown whether a given person meets this definition. This is a typical case where we would benefit from reasoning with uncertainty. For example, if an enterprise wins a procurement for millions of dollars, but the responsible person for this enterprise makes less than 5 thousand dollars a year or if that person has only a middle school education, then it is likely that this responsible person is a front for that enterprise. That is, we can identify potential fronts by examining the value of the procurement, the income of the responsible person, and his/her education level. Although we are not certain that this person is in fact a FrontMan, we would like to at least use the available information to draw an inference that the person is likely to be a front. This strategy is preferable to ignoring the evidence supporting this hypothesis and saying that we simply do not know whether or not this person is a front. It is also preferable to creating an arbitrary rule declaring that certain combinations of education level and income imply with certainty that a person is a FrontMan.

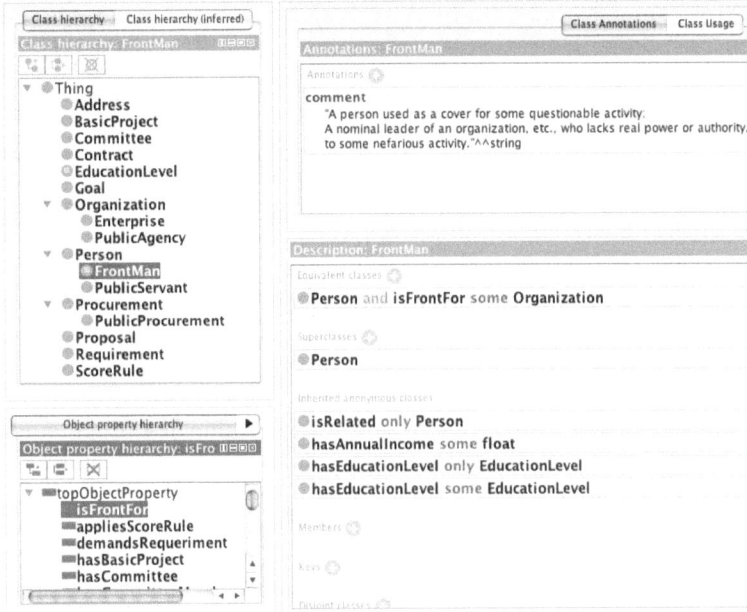

Fig. 1. OWL ontology for the public procurement domain

Figure 2 shows a formalization of this uncertain relationship in MEBN logic. MEBN represents knowledge as a collection of MEBN Fragments (MFrags), which are organized into MEBN Theories (MTheories).

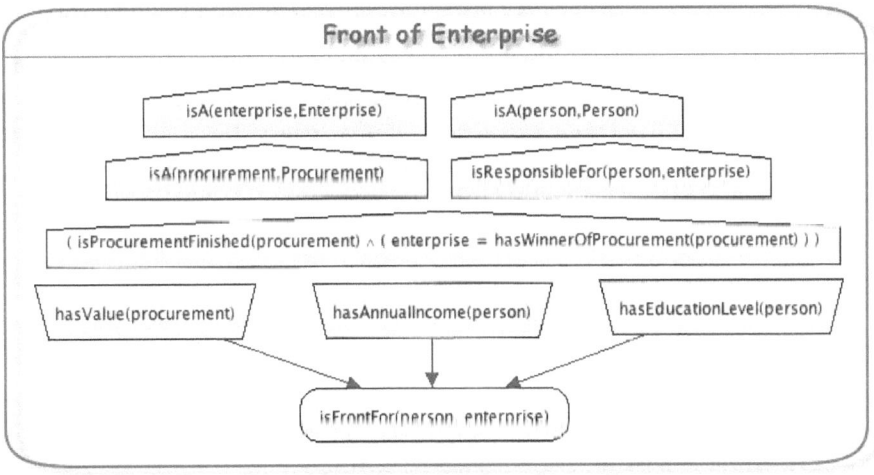

Fig. 2. Front of an Enterprise MFrag

An MFrag contains random variables (RVs) and a fragment graph representing dependencies among these RVs. It represents a repeatable pattern of knowledge that can be instantiated as many times as needed to form a BN addressing a specific situation, and thus can be seen as a template for building and combining fragments of a Bayesian network. It is instantiated by binding its arguments to domain entity identifiers to create instances of its RVs. There are three kinds of nodes: context, resident and input. Context nodes represent conditions that must be satisfied for the distributions represented in the MFrag to apply. Input nodes may influence the distributions of other nodes in an MFrag, but their distributions are defined in their home MFrags. Distributions for resident nodes are defined within the MFrag by specifying local distributions conditioned on the values of the instances of their parents in the fragment graph.

Figure 2 presents a MEBN Fragment, where we see that the education level and annual income of a responsible person and the value of a procurement influence whether the person is front for the procurement. However, in order for the probabilistic relations described to hold, some conditions have to be satisfied, namely that the person we are considering as a possible front must be responsible for the enterprise we are examining, which is the winner of the procurement that is already finished. In other words, if the person is not responsible for the enterprise, there is no reason for this person to be considered a front for this enterprise. The same principle holds if the enterprise did not win that procurement, *i.e.*, the value of a procurement that was not won by that enterprise will not affect the likelihood of having a front for that enterprise. These conditions that must be satisfied for the probabilistic relationship to hold are depicted inside the green pentagonal shapes in the figure.

Figure 2 shows only the structure of our reasoning just described. In order to be complete, we also need to define the conditional probability distribution, also called local probability distribution (LPD), for the random variable being defined. Listing 1.1[1] presents the LPD for the random variable isFrontFor(person, enterprise). When a random variable has its LPD defined within the MFrag where it appears, we call it a resident node. Nodes that are not resident nodes, but influence the distribution of resident nodes, are called input nodes and they have their LPDs defined in another MFrag. A collection of MFrags that guarantees a joint probability distribution over instances of random variables form a MEBN Theory (MTheory).

Listing 1.1. LPD for isFrontFor(person, enterprise)

```
1   if any procurement have ( hasValue = From100kTo500k ) [
2       if any person have ( hasAnnualIncome = Lower10k |
           hasEducationLevel = NoEducation ) [
3           true = .9 ,
4           false = .1
5       ] else if any person have ( hasAnnualIncome = From10kTo30k |
           hasEducationLevel = MiddleSchool ) [
```

[1] This LPD is notional only. No real data or statistics was used.

```
 6                  true = .6,
 7                  false = .4
 8          ] else [
 9                  true = .00001,
10                  false = .99999
11          ]
12  ] else if any procurement have ( hasValue = From500kTo1000k ) [
13          if any person have ( hasAnnualIncome = Lower10k |
                hasEducationLevel = NoEducation ) [
14                  true = .95,
15                  false = .05
16          ] else if any person have ( hasAnnualIncome = From10kTo30k |
                hasEducationLevel = MiddleSchool ) [
17                  true = .8,
18                  false = .2
19          ] else if any person have ( hasAnnualIncome = From30kTo60k |
                hasEducationLevel = HighSchool ) [
20                  true = .6,
21                  false = .4
22          ] else [
23                  true = .00001,
24                  false = .99999
25          ]
26  ] else if any procurement have ( hasValue = Greater1000k ) [
27          if any person have ( hasAnnualIncome = Lower10k |
                hasEducationLevel = NoEducation ) [
28                  true = .99,
29                  false = .01
30          ] else if any person have ( hasAnnualIncome = From10kTo30k |
                hasEducationLevel = MiddleSchool ) [
31                  true = .9,
32                  false = .1
33          ] else if any person have ( hasAnnualIncome = From30kTo60k |
                hasEducationLevel = HighSchool ) [
34                  true = .8,
35                  false = .2
36          ] else if any person have ( hasAnnualIncome = From60kTo100k |
                hasEducationLevel = Undergraduate ) [
37                  true = .6,
38                  false = .4
39          ] else [
40                  true = .00001,
41                  false = .99999
42          ]
43  ] else [
44          true = .00001,
45          false = .99999
46  ]
```

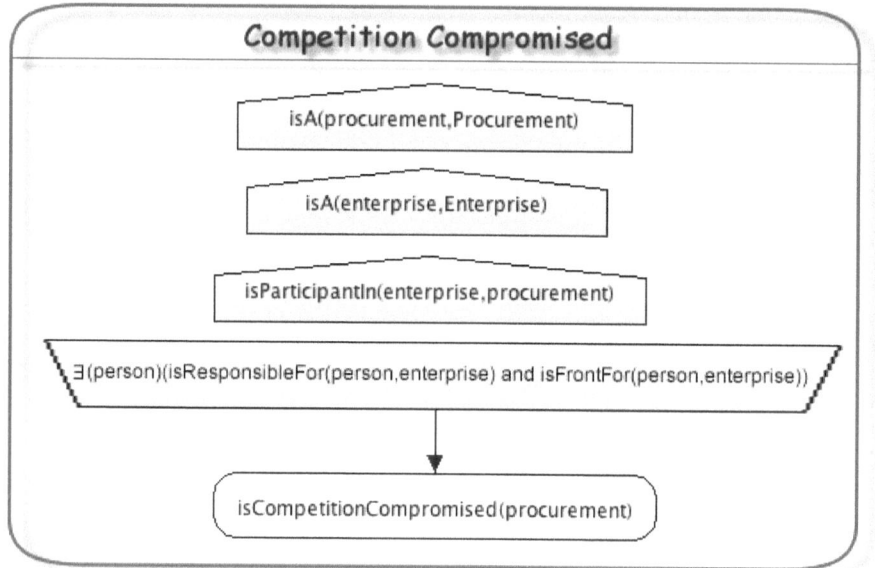

Fig. 3. Competition Compromised MFrag

The major advantage of MEBN when compared to Bayesian networks (BN), is the ability to represent repeated structure. In order to understand what kind of repetition can be represented and why it is important, let's introduce a new MFrag called Competition Compromised. In this MFrag (see Figure 3) we have the probabilistic rule that says that if any participant enterprise has at least one responsible person as a front, then it is more likely that the competition is going to be compromised in this procurement.

Now we can see how we benefit from the MEBN model. Figures 4 and 5 present two different BNs generated from our MEBN model given the information available in two different scenarios. In one we only have two participant enterprises (`enterprise2` and `enterprise3`), while in the other we have four participant enterprises (`enterprise1`, `enterprise2`, `enterprise3`, and `enterprise4`). Depending on the number of enterprises participating in a given procurement and the information available about the person responsible for those enterprises, we would have a different BN being used. In other words, with just one MEBN representation we can instantiate many different BN models representing problems with different numbers of individuals, as appropriate for each query of interest.

Probabilistic OWL (PR-OWL) is an upper ontology defined in OWL for representing MEBN theories. In other words, PR-OWL defines classes and properties for MEBN terms, like `MTheory`, `MFrag`, `hasMFrag`, and `ResidentNode` and restrictions on those terms (*e.g.*, MTheory is a collection of MFrags - *i.e.*, `hasMFrag some MFrag`) in order to allow the definition of MEBN models. In PR-OWL, a probabilistic ontology (PO) has to have at least one individual of class `MTheory`, which is basically a label linking a group of MFrags that collectively form a valid

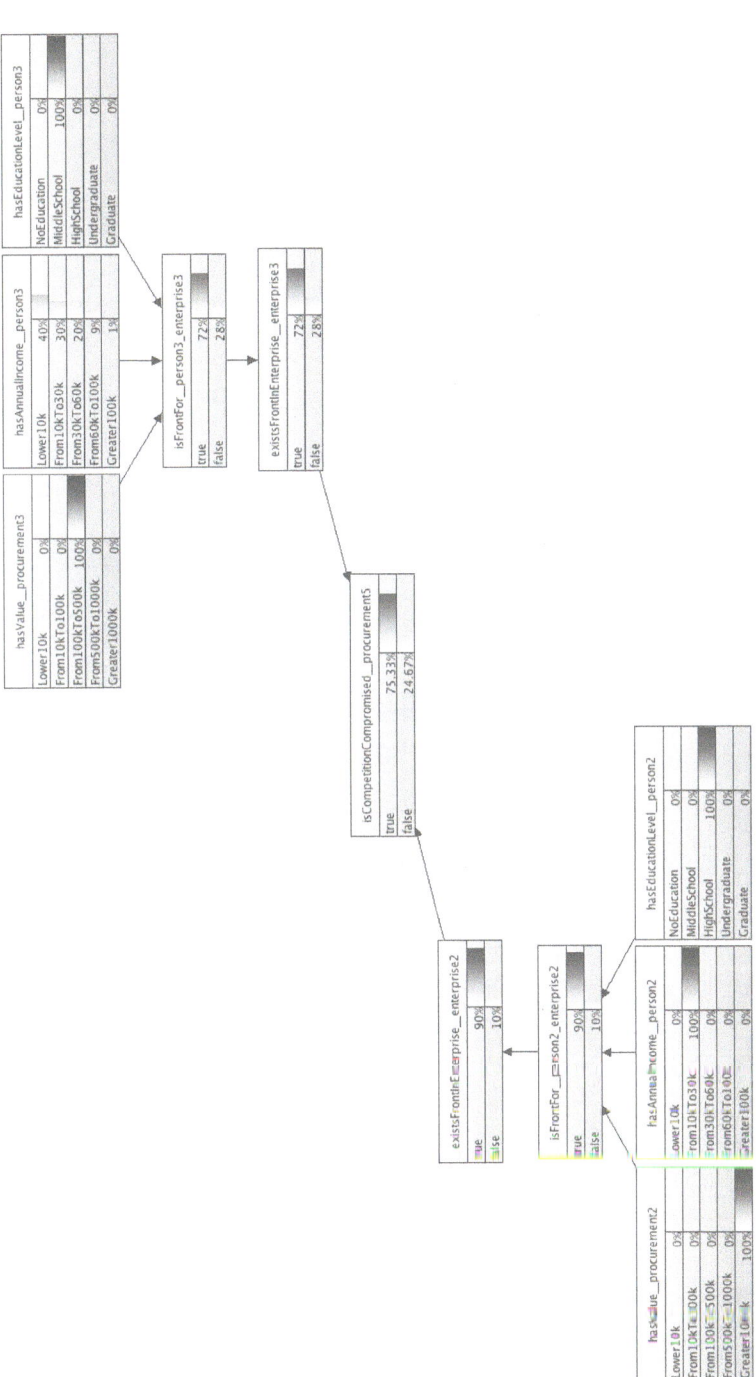

Fig. 4. Situation specific BN (SSBN) generated for the query isCompetitionCompromised(procurement5) given that there are two enterprises participating in this procurement

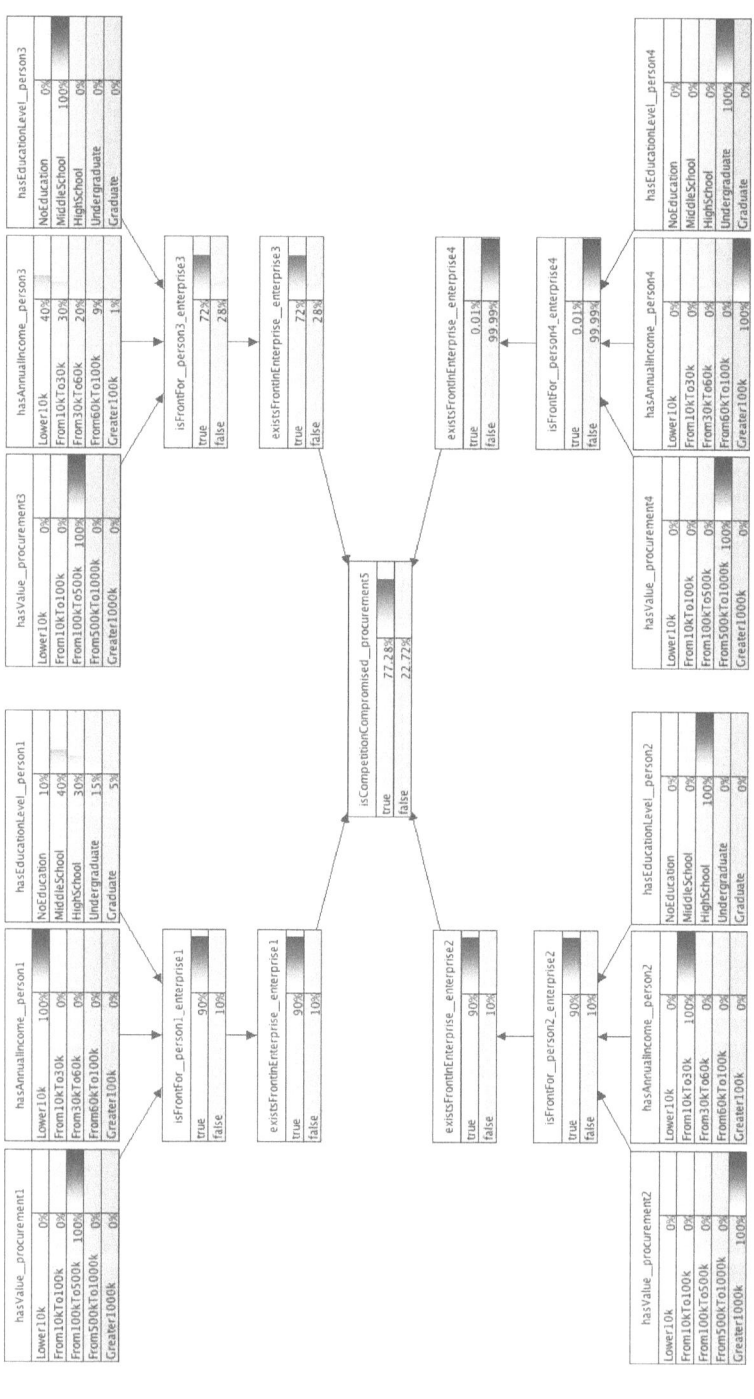

Fig. 5. SSBN generated for the query isCompetitionCompromised(procurement5) given that there are four enterprises participating in this procurement

MTheory. In actual PR-OWL syntax, that link is expressed via the object property `hasMFrag` (which is the inverse of object property `isMFragIn`). Individuals of class `MFrag` are comprised of nodes. Each individual of class `Node` is a random variable (RV) and thus has a mutually exclusive, collectively exhaustive set of possible states. In PR-OWL, the object property `hasPossibleValues` links each node with its possible states, which are individuals of class `Entity`. Finally, random variables (represented by the class `Node` in PR-OWL) have unconditional or conditional probability distributions, which are represented by class `ProbabilityDistribution` and linked to their respective nodes via the object property `hasProbDist`. This property would define a LPD like the one presented in Listing 1.1. Figure 6 presents an example of a PO for the procurement domain using the PR-OWL language.

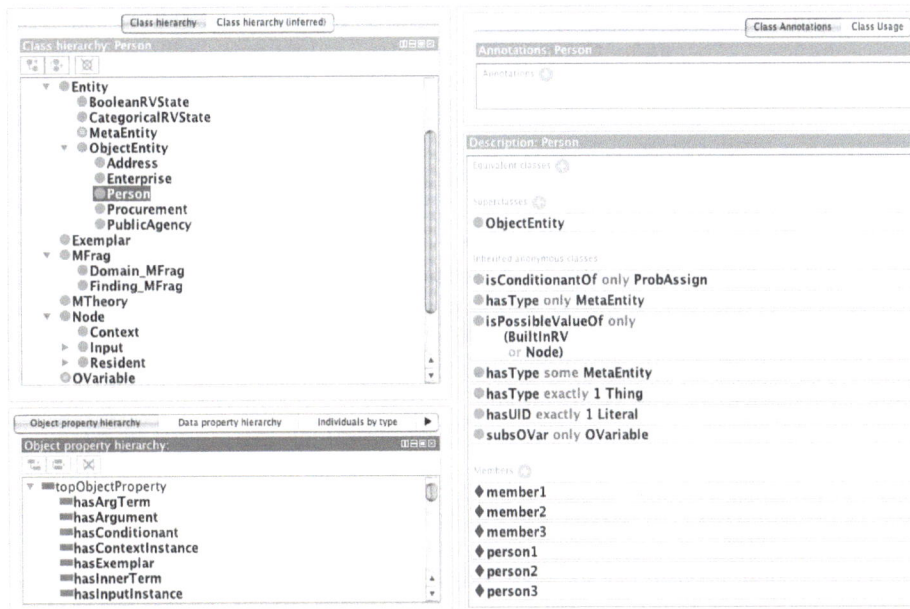

Fig. 6. Example of a PO for the procurement domain using PR-OWL language

3   Why Map PR-OWL Random Variables to OWL Properties?

Ideally, it should be possible to use PR-OWL to reason probabilistically about uncertain aspects of an existing ontology represented in OWL. For instance, Figure 7 presents some information we might have available in an OWL ontology for the procurement domain. Specifically, the ontology represents a person called Joe who has a middle school education level and an income of $5,000. As shown in the figure, we might want to generate a BN in order to draw inferences about whether Joe is a front for a procurement for which he is responsible. Although we cannot say for certain that John_Doe is a FrontMan, the likelihood is high given his low annual income, his low education level, and the high value of the procurement won by his enterprise. In order to construct the BN allowing us to draw this inference, we need to relate the knowledge expressed in the OWL ontology to PR-OWL random variables.

The problem with PR-OWL 1.0 is that it has no mapping between the random variables used in PR-OWL and the properties used in OWL. In other words, there is nothing in the language that tells us that the RV hasEducationLevel(person) defines the uncertainty of the OWL property hasEducationLevel. So, even if we have information about the education level of a specific person, we cannot connect that information In other words, even if we we have the triple John_Doe hasEducationLevel middleSchool, we would not be able to instantiate the random variable hasEducationLevel(person) for John_Doe. Although the OWL property hasEducation and the RV hasEducationLevel(person) have similar syntax, there is no formal representation of this link (as depicted in Figure 8). In other words, we cannot use the information available in an OWL ontology (the triples with information about individuals) to perform probabilistic reasoning. Full compatibility between PR-OWL and OWL requires this ability.

In fact, Poole *et al.* [15] states that it is not clear how to match the formalization of random variables from probabilistic theories with the concepts of individuals, classes and properties from current ontological languages like OWL. However, Poole *et al.* [15] suggests, "We can reconcile these views by having properties of individuals correspond to random variables." This is exactly the approach used in this work to integrate MEBN logic and the OWL language. This integration is a major feature of PR-OWL 2.0 [1].

4   The Bridge Joining OWL and PR-OWL

The key to building the bridge that connects the deterministic ontology defined in OWL and its probabilistic extension defined in PR-OWL is to understand how to translate one to the other. On the one hand, given a concept defined in OWL, how should its uncertainty be defined in PR-OWL in a way that maintains its semantics defined in OWL? On the other hand, given a random variable defined in PR-OWL, how should it be represented in OWL in a way that respects its uncertainty already defined in PR-OWL?

Imagine we are trying to define the RV hasEducationLevel_RV[2], which represents the MEBN RV hasEducationLevel(person) used in Figure 2. Let's also assume that we have an OWL property called hasEducationLevel, which is a functional property with domain Person and range EducationLevel,

---

[2] This is the OWL syntax for this RV. In MEBN we represent a RV by its name followed by the arguments in parentheses. In OWL the arguments are defined by the property hasArgument.

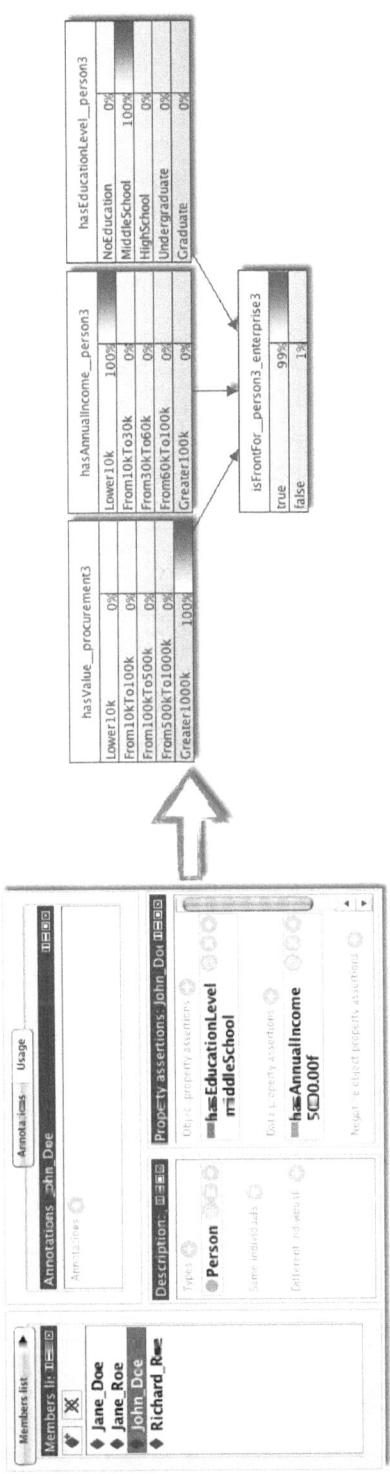

Fig. 7. Using triples for probabilistic reasoning

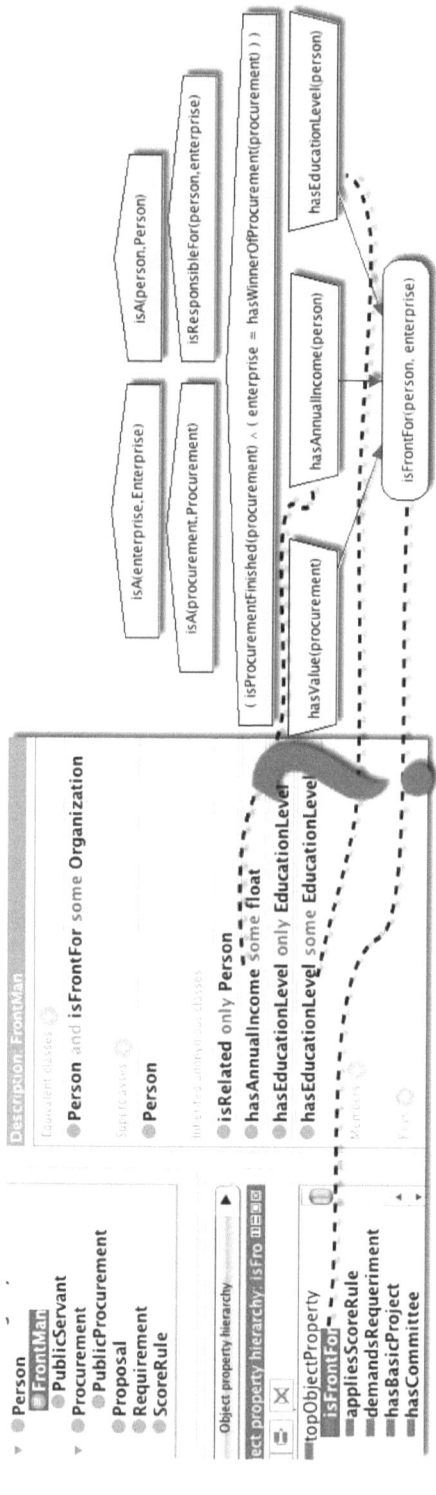

Fig. 8. Unknown mapping between PR-OWL 1.0 RVs and OWL properties

and an OWL property called `aspiresEducationLevel`, which is also a functional property with domain `Person` and range `EducationLevel`. As shown in Figure 9, in PR-OWL 1.0 it is not possible to distinguish whether the `hasEducationLevel_RV` is defining the uncertainty of the OWL property `hasEducationLevel` or `aspiresEducationLevel`. To clarify this problem, imagine that John Doe has only middle school (`John_Doe hasEducationLevel middleSchool`), but he aspires to have a graduate degree (`John_Doe aspiresEducationLevel graduate`). If we do not explicitly say which OWL property should be used to instantiate the `hasEducationLevel_RV`, we might end up saying that `hasEducationLevel(John_Doe) = graduate`, instead of saying that `hasEducationLevel(John_Doe) = middleSchool`, which is the intended semantics.

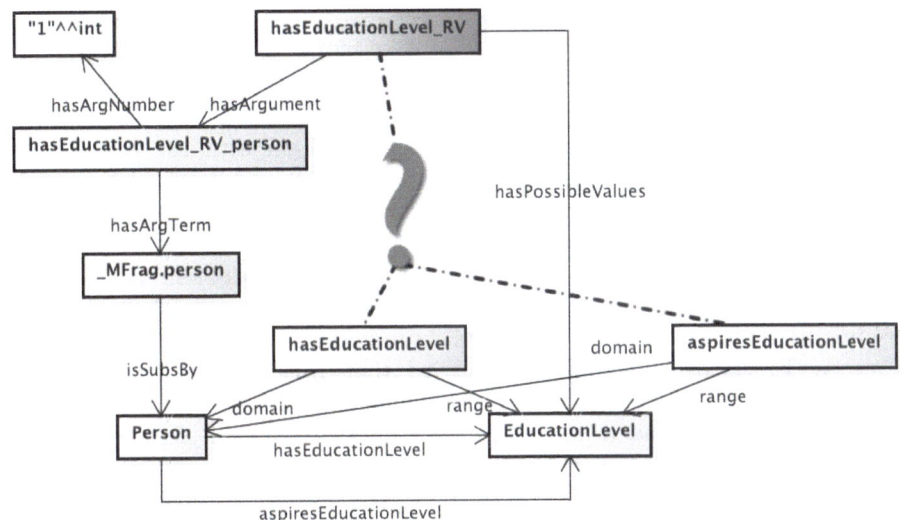

Fig. 9. PR-OWL 1.0 lack of mapping from RVs to OWL properties

A simple solution is to add a relation between a PR-OWL RV and the OWL property that this RV defines the uncertainty of, as suggested by Poole et al. [15]. In PR-OWL 2.0 this relation is called `definesUncertaintyOf` [1][3]. However, this is not enough to have a complete mapping between RVs and OWL properties. Another problem appears when we try to define n-ary RVs. This mapping is not as straight forward as the previous one because OWL only supports binary properties (for details on suggested work arounds to define n-ary relations in OWL see [7]).

---

[3] We make the distinction between property and property_RV with the `definesUncertaintyOf` relation, instead of just using the property as the RV, in order to stay within OWL DL. For more information see Section 4.4.2 in [1].

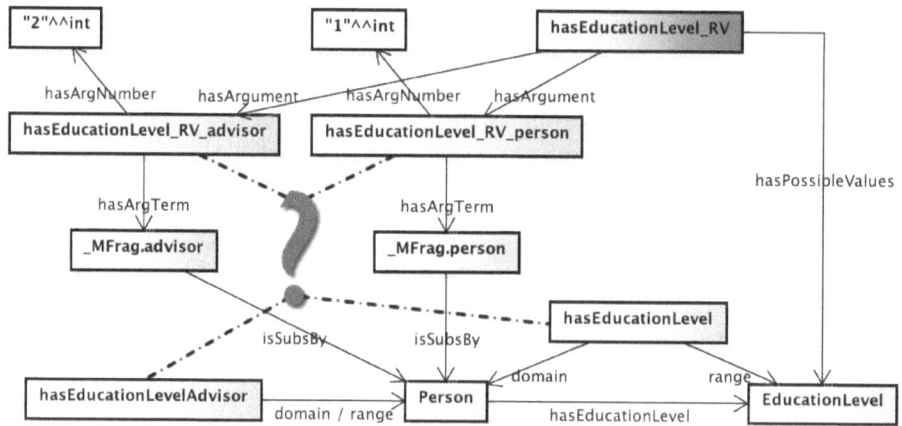

Fig. 10. PR-OWL 1.0 lack of mapping from arguments to OWL properties

Imagine we now want to represent not only the education level a person has, but also who was the advisor when this person attained that education level. So now, besides having the property `hasEducationLevel`, we also have the property `hasEducationLevelAdvisor`, which has `Person` as both domain and range. Thus, our RV now is `hasEducationLevel(person,advisor)`. With this new scenario, we can see that a similar problem occurs with the mapping of arguments. As it can be seen in Figure 10, there is nothing in PR-OWL 1.0 that tells which argument is associated with which property. To clarify the problem, imagine that Richard Roe has graduate education level (`Richard_Roe hasEducationLevel graduate`) and that his advisor was J. Pearl (`Richard_Roe hasEducationLevelAdvisor J_Pearl`). When instatiating the `hasEducationLevel(person,advisor)` RV, machines would not know who is the student and who is the advisor. Although this mapping is obvious for a human being, without an explicit mapping of the arguments, machines could end up using Richard Doe as the advisor and J. Pearl as the student (`hasEducationLevel(J_Pearl,Richard_Roe)`), instead of using J. Pearl as the advisor and Richard Doe as the student (`hasEducationLevel(Richard_Roe,J_Pearl)`).

As expected, to a similar problem we apply a similar solution. In PR-OWL 2 we have a relation between an argument to a RV and the OWL property it refers to. However, unlike the RV mapping, the argument mapping refers to either the domain or the range of a property, not to the property itself. For instance, in the `hasEducationLevel(person,advisor)` RV, the `person` argument refers to the domain of the OWL property `hasEducationLevel`, which is a `Person`. The `advisor` argument, on the other hand, refers to the range of the OWL property `hasEducationLevelAdvisor`, which is also a `Person`, but a different one (a person cannot be his/her own advisor). Therefore, in order to differentiate when the argument refers to the domain or to the range of a property, we add to PR-OWL 2.0 the relations `isSubjectIn` and `isObjectIn`. More examples of

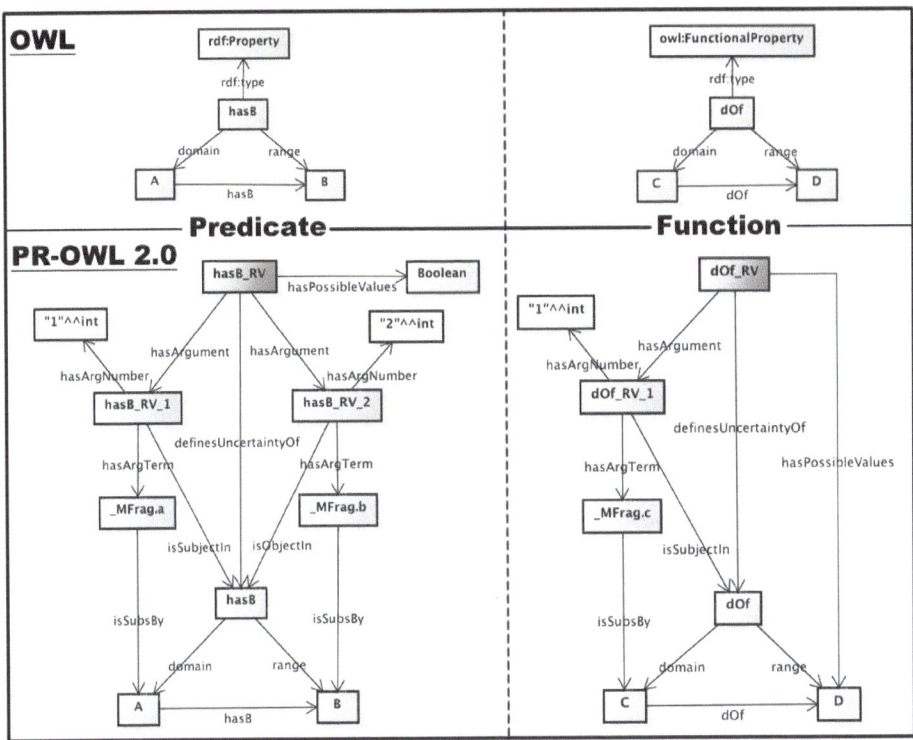

Fig. 11. The bridge joining OWL and PR-OWL

random variables in this new format can be found in [1, 3]. Here, a scheme is given in Figure 11 for the 2-way mapping between triples and random variables. Functions and predicates are considered as separate cases.

If a property (`hasB` or `dOf`) is defined in OWL, then its domain and range are already represented (`A` and `B`; `C` and `D`, respectively). The first thing to be done is to create the corresponding RV in PR-OWL (`hasB_RV` and `dOf_RV`, respectively) and link it to this OWL property through the property `definesUncertaintyOf`.

For binary relations, the domain of the property (`A` and `C`, respectively) will usually be the type (`isSubsBy`) of the variable (`_MFrag.a` and `_MFrag.c`, respectively) used in the first argument (`hasB_RV_1` and `dOf_RV_1`, respectively) of the RV. For n-ary relations see example given earlier in this Section on the RV `hasEducationalLevel(person,advisor)` and also [1,3].

If the property is non-functional (`hasB`), then it represents a predicate that may be true or false. Thus, instead of having the possible values of the RV in PR-OWL (`hasB_RV`) being the range of the OWL property (`B`), it must be Boolean. So, its range (`B`) has to be mapped to the second argument (`hasB_RV_2`) of the RV, the same way the domain (`A`) was mapped to the first argument (`hasB_RV_1`) of the RV. On the other hand, if the the property is functional (`dOf`), the possible values of its RV (`dOf_RV`) must be the same as its range (`D`).

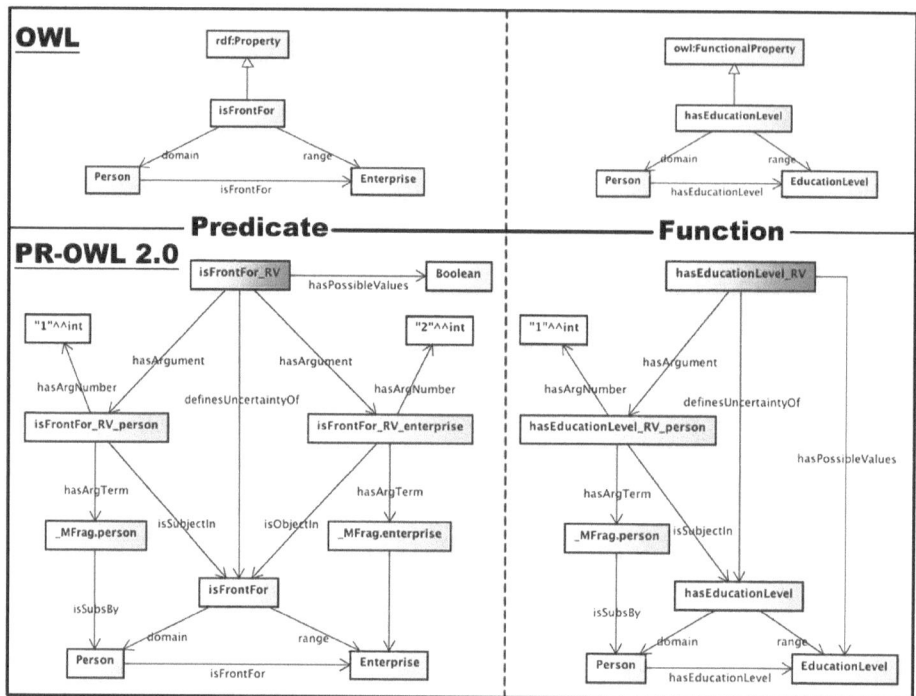

Fig. 12. Example of binary RVs mapping to OWL properties for both predicate and function

It is important to note that not only is the RV linked to the OWL property by `definesUncertaintyOf`, but also its arguments are linked to their respective OWL properties by either `isSubjectIn` or `isObjectIn`, depending on what they refer to (domain or range of the OWL property, respectively). This feature is especially important when dealing with n-ary relations, where each variable will be associated with a different OWL property (see explanation of Figure 10 earlier in this Section for details).

Finally, if the RV is already defined in PR-OWL with all its arguments and its possible values, the only thing that needs to be done is to create the corresponding OWL property, link the RV to it using `definesUncertaintyOf`, create the OWL properties for the arguments, if necessary, link them using either `isSubjectIn` or `isObjectIn`, depending on what they refer to (domain or range of the OWL property, respectively), and make sure that the domain and range of the property matches the RV definition, as explained previously.

Figure 12 presents examples of instantiations of the scheme just presented. In it we have the mapping of the RV `isFrontFor(person,enterprise)` to the OWL property `isFront`, which is a predicate, and the mapping of the RV `hasEducationLevel(person)` to the OWL property `hasEducationLevel`, which is a function.

The mapping described in this Section provides the basis for a formal definition of consistency between a PR-OWL probabilistic ontology and an OWL ontology, in which rules in the OWL ontology correspond to probability one assertions in the PR-OWL ontology. A formal notion of consistency can lead to development of consistency checking algorithms. For details on PR-OWL 2.0 abstract syntax and semantics see Carvalho [1].

5   Conclusion

With this mapping it is possible to not only resuse existing OWL semantics, but also automatically retrive available information from the mapped OWL ontology to use as evidence for probabilistic reasoning. This was not possible in PR-OWL 1.0.

Moreover, a scheme was given for how to do the mapping back and forth between PR-OWL random variables and OWL triples (both predicates and functions). Besides providing the scheme, a few examples were presented to illustrate how it works.

For full description of PR-OWL 2.0 abstract syntax and semantics, see Carvalho [1]. In it Carvalho also addresses other issues with PR-OWL 1.0 presented in [2].

Acknowledgments. The authors would like to thank the Brazilian Office of the Comptroller General (CGU) for their active support since 2008 and for providing the human resources necessary to conduct this research.

References

1. Carvalho, R.N.: Probabilistic Ontology: Representation and Modeling Methodology. PhD, George Mason University, Fairfax, VA, USA (2011)
2. Carvalho, R.N., Laskey, K.B., Costa, P.C.G.: Compatibility formalization between PR-OWL and OWL. In: Proceedings of the First International Workshop on Uncertainty in Description Logics (UniDL) on Federated Logic Conference (FLoC) 2010, Edinburgh, UK (July 2010)
3. Carvalho, R.N., Laskey, K.B., Costa, P.C.G.: PR-OWL 2.0 - Bridging the Gap to OWL Semantics. In: Bobillo, F., Costa, P.C.G., d'Amato, C., Fanizzi, N., Laskey, K.B., Laskey, K.J., Lukasiewicz, T., Nickles, M., Pool, M. (eds.) URSW 2008-2010/UniDL 2010. LNCS (LNAI), vol. 7123, pp. 1–18. Springer, Heidelberg (2013)
4. Carvalho, R.N., Matsumoto, S., Laskey, K.B., Costa, P.C.G., Ladeira, M., Santos, L.L.: Probabilistic Ontology and Knowledge Fusion for Procurement Fraud Detection in Brazil. In: Bobillo, F., Costa, P.C.G., d'Amato, C., Fanizzi, N., Laskey, K.B., Laskey, K.J., Lukasiewicz, T., Nickles, M., Pool, M. (eds.) URSW 2008-2010/UniDL 2010. LNCS (LNAI), vol. 7123, pp. 19–40. Springer, Heidelberg (2013)
5. Costa, P.C.G.: Bayesian Semantics for the Semantic Web. PhD, George Mason University, Fairfax, VA, USA (July 2005)
6. Ding, Z., Peng, Y., Pan, R.: BayesOWL: Uncertainty Modeling in Semantic Web Ontologies. In: Ma, Z. (ed.) Soft Computing in Ontologies and Semantic Web. STUDFUZZ, vol. 204, pp. 3–29. Springer, Heidelberg (2006)

7. Hayes, P., Rector, A.: Defining n-ary relations on the semantic web (2006),
   http://www.w3.org/TR/swbp-n-aryRelations/
8. Heinsohn, J.: Probabilistic description logics. In: Proceedings of the 10th Annual
   Conference on Uncertainty in Artificial Intelligence, UAI 1994, Seattle, Washington, USA, pp. 311–318. Morgan Kaufmann (1994)
9. Koller, D., Levy, A., Pfeffer, A.: P-CLASSIC: a tractable probabilistic description
   logic. In: Proceedings of AAAI 1997, pp. 390–397 (1997)
10. Laskey, K.B.: MEBN: a language for First-Order Bayesian knowledge bases. Arti-
    ficial Intelligence 172(2-3), 140–178 (2008)
11. Laskey, K., Laskey, K.B.: Uncertainty reasoning for the World Wide Web: Report
    on the URW3-XG incubator group. URW3-XG, W3C (2008)
12. Lukasiewicz, T.: Expressive probabilistic description logics. Artificial Intelli-
    gence 172(6-7), 852–883 (2008)
13. Milch, B., Russell, S.: First-Order Probabilistic Languages: Into the Unknown.
    In: Muggleton, S.H., Otero, R., Tamaddoni-Nezhad, A. (eds.) ILP 2006. LNCS
    (LNAI), vol. 4455, pp. 10–24. Springer, Heidelberg (2007)
14. Pan, J.Z., Stoilos, G., Stamou, G., Tzouvaras, V., Horrocks, I.: f-SWRL: A Fuzzy
    Extension of SWRL. In: Spaccapietra, S., Aberer, K., Cudré-Mauroux, P. (eds.)
    Journal on Data Semantics VI. LNCS, vol. 4090, pp. 28–46. Springer, Heidelberg
    (2006)
15. Poole, D., Smyth, C., Sharma, R.: Semantic Science: Ontologies, Data and Prob-
    abilistic Theories. In: da Costa, P.C.G., d'Amato, C., Fanizzi, N., Laskey, K.B.,
    Laskey, K.J., Lukasiewicz, T., Nickles, M., Pool, M. (eds.) URSW 2005-2007. LNCS
    (LNAI), vol. 5327, pp. 26–40. Springer, Heidelberg (2008)
16. Predoiu, L., Stuckenschmidt, H.: Probabilistic extensions of semantic web lan-
    guages - a survey. In: The Semantic Web for Knowledge and Data Management:
    Technologies and Practices. Idea Group Inc. (2008)
17. Straccia, U.: A fuzzy description logic for the semantic web. In: Fuzzy Logic and
    the Semantic Web, Capturing Intelligence, pp. 167–181. Elsevier (2005)
18. Tao, J., Wen, Z., Hanpin, W., Lifu, W.: PrDLs: A New Kind of Probabilistic
    Description Logics About Belief. In: Okuno, H.G., Ali, M. (eds.) IEA/AIE 2007.
    LNCS (LNAI), vol. 4570, pp. 644–654. Springer, Heidelberg (2007)

Probabilistic Ontology and Knowledge Fusion for Procurement Fraud Detection in Brazil

Rommel N. Carvalho[1], Shou Matsumoto[1], Kathryn B. Laskey[1],
Paulo C.G. Costa[1], Marcelo Ladeira[2], and Laécio L. Santos[2]

[1] Department of SEOR / Center of Excellence in C4I
George Mason University
4400 University Drive
Fairfax, VA 22030-4400 USA
{rommel.carvalho,cardialfly}@gmail.com, {klaskey,pcosta}@gmu.edu
http://www.gmu.edu
[2] Department of Computer Science
University of Brasilia
Campus Universitario Darcy Ribeiro
Brasilia DF 70910-900 Brazil
mladeira@unb.br, laecio@gmail.com
http://www.unb.br

Abstract. To cope with citizens' demand for transparency and corruption prevention, the Brazilian Office of the Comptroller General (CGU) has carried out a number of actions, including: awareness campaigns aimed at the private sector; campaigns to educate the public; research initiatives; and regular inspections and audits of municipalities and states. Although CGU has collected information from hundreds of different sources - Revenue Agency, Federal Police, and others - the process of fusing all this data has not been efficient enough to meet the needs of CGU's decision makers. Therefore, it is natural to change the focus from data fusion to knowledge fusion. As a consequence, traditional syntactic methods should be augmented with techniques that represent and reason with the semantics of databases. However, commonly used approaches, such as Semantic Web technologies, fail to deal with uncertainty, a dominant characteristic in corruption prevention. This paper presents the use of probabilistic ontologies built with Probabilistic OWL (PR-OWL) to design and test a model that performs information fusion to detect possible frauds in procurements involving Federal money in Brazil. To design this model, a recently developed tool for creating PR-OWL ontologies was used with support from PR-OWL specialists and careful guidance from a fraud detection specialist from CGU. At present, the task of procurement fraud detection is done manually by an auditor. The number of suspicious cases that can be analyzed by a single person is small. The experimental results obtained with the presented approach are preliminary, but show the viability of developing a tool based on PR-OWL ontologies to automatize this task. This paper also examplifies how to use PR-OWL 2.0 to provide a link between the deterministic and probabilistic parts of the ontology.

F. Bobillo et al. (Eds.): URSW 2008-2010/UniDL 2010, LNAI 7123, pp. 19–40, 2013.
© Springer-Verlag Berlin Heidelberg 2013

Keywords: Probabilistic Ontology, PR-OWL, Ontology, Procurement, Fraud Detection, Fraud Prevention, Knowledge Fusion, MEBN, UnBBayes.

1   Introduction

A primary responsibility of the Brazilian Office of the Comptroller General (CGU) is to prevent and detect government corruption. To carry out this mission, CGU must gather information from a variety of sources and combine it to evaluate whether further action, such as an investigation, is required. One of the most difficult challenges is the information explosion. Auditors must fuse vast quantities of information from a variety of sources in a way that highlights its relevance to decision makers and helps them focus their efforts on the most critical cases. This is no trivial task. Brazil's Growing Acceleration Program (PAC) alone has a budget greater than 250 billion dollars with more than one thousand projects in the state of Sao Paulo alone[1]. Each of these projects must be audited and inspected by CGU – yet CGU has only three thousand employees. Therefore, CGU must optimize its processes in order to carry out its mission.

The Semantic Web (SW), like the document-based web that preceded it, is based on radical notions of information sharing. These ideas [1] include: (i) the Anyone can say Anything about Any topic (AAA) slogan; (ii) the open world assumption, in which we assume there is always more information that could be known, and (iii) non-unique naming, which acknowledges that different authors on the Web might use different names to define the same entity. In a fundamental departure from assumptions of traditional information systems architectures, the Semantic Web is intended to provide an environment in which information sharing can thrive and a network effect of knowledge synergy is possible. Although a powerful concept, this style of information gathering can generate a chaotic landscape rife with confusion, disagreement, and conflict.

We call an environment characterized by the above assumptions a Radical Information Sharing (RIS) environment. The challenge facing SW architects is therefore to avoid the natural chaos to which RIS environments are prone, and move to a state characterized by information sharing, cooperation, and collaboration. According to [1], one solution to this challenge lies in modeling, and this is where ontology languages such as Web Ontology Language (OWL) come in.

As noted in Section 4 below, procurement fraud detection is carried out within a RIS environment. The ability to deal with uncertainty is especially important in applications such as fraud detection, in which perpetrators seek to conceal illicit intentions and activities, making crisp assertions problematic. In such environments, partial or approximate information is more the rule than the exception.

Bayesian networks (BNs) have been widely applied to information and knowledge fusion in the presence of uncertainty. However, BNs are not expressive enough for many important applications [11]. Specifically, BNs assume a simple attribute-value representation – that is, each problem instance involves reasoning

---

[1] http://www.brasil.gov.br/pac/

about the same fixed number of attributes, with only the evidence values changing from problem instance to problem instance. Complex problems on the scale of the Semantic Web often involve intricate relationships among many variables. The limited representational power of BNs is insufficient for models in which the variables and relationships are not fixed in advance.

To address this weakness of BNs it is common to extend this formalism with approaches based on first-order logic (FOL). FOL is highly expressive but has no built-in capability to reason with uncertainty. To combine the strengths of both approaches, researchers have used FOL expressions to specify relationships among fragments of BNs. The resulting model specifies a probability distribution over many different "ground models" obtained by instantiating the fragments as many times as needed for the given situation and combining into a Bayesian network. Multi-Entity Bayesian Network (MEBN) is an example of this style of language. The ground BN generated after instantiating the variables with domain objects has been called a Situation-Specific Bayesian Network (SSBN). Inference in the SSBN can be performed with a standard belief updating algorithm.

Multi-Entity Bayesian Network (MEBN) logic can represent and reason with uncertainty about any propositions that can be expressed in first-order logic [19]. Probabilistic OWL (PR-OWL), an OWL upper ontology for expressing MEBN theories, is a language for expressing probabilistic ontologies (PO) [21]. The ability to represent and compute with probabilistic ontologies represents a major step towards semantically aware, probabilistic knowledge fusion systems. Although compatibility with OWL was a major design goal for PR-OWL [8], there are several ways in which the initial release of PR-OWL fell short of complete compatibility [4,5]. These shortcomings were addressed in PR-OWL 2.0, which extends PR-OWL by formalizing the relationship between the probabilistic and deterministic parts of a probabilistic ontology [2]. Therefore, PR-OWL 2.0 provides better integration between the probabilistic and deterministic parts of an ontology.

As a result of its focus on the probabilistic aspecs of the ontology, previous literature on PR-OWL did not discuss its relationship to the deterministic part of the ontology as defined by OWL semantics. (e.g., [17,18,22,21,10,9]). This paper uses PR-OWL 2.0 to design and test a model for fusing knowledge to detect possible frauds in procurements involving Federal funds. Unlike previous literature, this paper explicitly addresses the use of PR-OWL 2.0 to provide the link between the deterministic and probabilistic parts of the ontology. We discuss how the new features of PR-OWL 2.0 enable a more natural fusion of information available from multiple sources (see Section 4 for details).

The major contribution of this paper is to clarify how to map properties of entities of a deterministic ontology into random variables of a probabilistic ontology and how to perform hybrid ontological and probabilistic reasoning using PR-OWL 2.0. This approach can be used for the task of information fusion based on reuse of available deterministic ontologies.

This paper is organized as follows. Section 2 introduces MEBN, an expressive Bayesian logic, and PR-OWL, an extension of the OWL language that can represent probabilistic ontologies having MEBN as its underlying logic. Section 3 presents a case study from CGU to demonstrate the power of PR-OWL ontologies for knowledge representation and inferring rare events like fraud. Then, Section 4 describes how to extend this PO to gather information from other sources and perform knowledge fusion in order to improve the likelihood of finding frauds. Finally, Section 5 presents some concluding remarks.

2   MEBN and PR-OWL

Multi-Entity Bayesian Networks (MEBN) [17,20] extend BNs to achieve first-order expressive power. MEBN represents knowledge as a collection of MEBN Fragments (MFrags), which are organized into MEBN Theories (MTheories).

An MFrag contains random variables (RVs) and a fragment graph representing dependencies among these RVs. An MFrag is a template for a fragment of a Bayesian network. It is instantiated by binding its arguments to domain entity identifiers to create instances of its RVs. There are three kinds of RV: context, resident and input. Context RVs represent conditions that must be satisfied for the distributions represented in the MFrag to apply. Input nodes represent RVs that may influence the distributions defined in the MFrag, but whose distributions are defined in other MFrags. Distributions for resident RV instances are defined in the MFrag. Distributions for resident RVs are defined by specifying local distributions conditioned on the values of the instances of their parents in the fragment graph.

A set of MFrags represents a joint distribution over instances of its random variables. MEBN provides a compact way to represent repeated structures, which can then be instantiated as many times as needed to build an actual BN tailored for the specific situation at hand. An important advantage of MEBN is that there is no fixed limit on the number of RV instances, and the random variable instances are dynamically instantiated as needed.

An MTheory is a set of MFrags that satisfies conditions of consistency ensuring the existence of a unique joint probability distribution over its random variable instances.

To apply an MTheory to reason about particular scenarios, one needs to provide the system with specific information about the individual entity instances involved in the scenario. On receipt of this information, Bayesian inference can be used both to answer specific questions of interest (*e.g.*, how likely is it that a particular procurement is being directed to a specific enterprise?) and to refine the MTheory (*e.g.*, each new situation includes additional data about the likelihood of fraud for that set of circumstances). Bayesian inference is used to perform both problem specific inference and learning in a sound, logically coherent manner (for more details see [20,24]).

State-of-the-art systems are increasingly adopting ontologies as a means to ensure formal semantic support for knowledge sharing [6,7,12,3,13,15,28]. Representing and reasoning with uncertainty is becoming recognized as an essential capability in many domains. In fact, the W3C created the Uncertainty Reasoning for the World Wide Web Incubator Group (URW3-XG) to research the use of uncertainty in semantic technologies. The group was created in 2007 and, one year later, presented its conclusion that standardized representations were needed to express uncertainty in Web-based information [23].

A candidate representation for uncertainty reasoning in the Semantic Web is Probabilistic OWL (PR-OWL) [8], an OWL upper ontology for representing probabilistic ontologies based on Multi-Entity Bayesian Networks (MEBN) [20]. More specifically, PR-OWL is an upper ontology (*i.e.* an ontology that represents fundamental concepts common to various disciplines and applications) for probabilistic systems. It consists of a set of classes, subclasses and properties that collectively form a framework for building probabilistic ontologies.

There are several ways in which the initial release of PR-OWL fell short of fully integrating the deterministic and probabilistic parts of an ontology [4,5]. In fact, Poole *et al.* [27] emphasizes that it is not clear how to match the formalization of random variables from probabilistic theories with the concepts of individuals, classes and properties from current ontological languages like OWL. However, Poole *et al.* [27] says "We can reconcile these views by having properties of individuals correspond to random variables." This is the approach used in PR-OWL 2.0 [2] to integrate MEBN and OWL.

Matsumoto [25] describes a Java implementation of PR-OWL 2.0, including a GUI, API and inference engine in the UnBBayes framework [26]. With this tool it is possible to drag and drop OWL properties into MFrags. The MFrag designer can define and edit the probabilistic definition for that property as shown in Figure 1. This action creates a RV that represents a probability distribution for the OWL property being mapped into the MFrag. Of course, an OWL property with no uncertainty can be mapped with a probability distribution with probability 0 or 1.

With this tool it is possible to drag-and-drop OWL properties into MFrags, which will automatically create a RV, allowing the definition of the probabilistic definition for that property within the context of the MFrag it is defined in as shown in Figure 1. The tool has proven to be a simple, yet powerful, asset for designing probabilistic ontologies and for uncertain reasoning in complex situations such as procurement fraud detection.

3   Procurement Fraud Detection

A major source of corruption is the procurement process. Although laws were enacted to ensure a competitive and fair process, perpetrators find ways to turn the process to their advantage while appearing to be legitimate. To better understand these many, usually creative ways of circumventing the laws, a specialist from CGU has didactically structured the different kinds of procurement frauds that CGU has dealt with in past years. Those different kinds of procurement frauds have resulted in several MFrags built by the authors with the use of the tool UnBBayes.

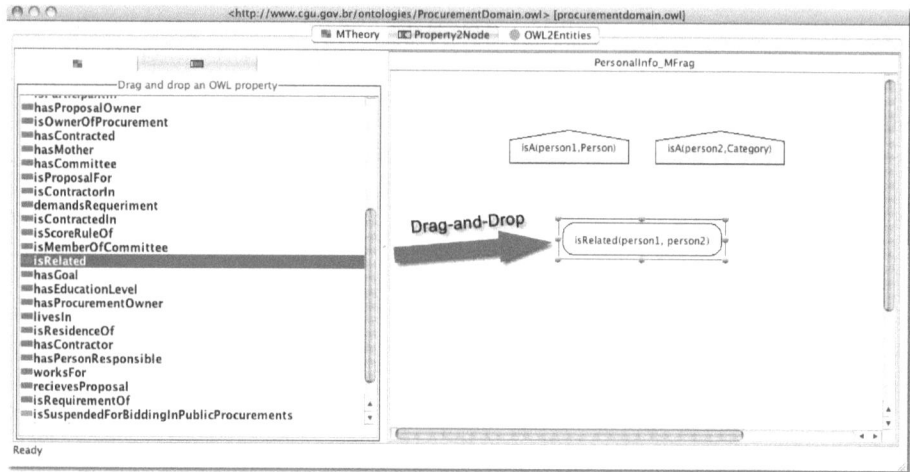

Fig. 1. Drag-and-drop of an OWL property for defining its probabilistic semantics

Different fraud types are characterized by criteria, such as business owners who work as a front for the company, use of accounting indices that are not common practice, among others. Indicators have been established by the CGU specialist to help identify cases of each of these fraud types. For instance, one principle that must be followed in public procurement is that of competition. Every public procurement should establish minimum requisites necessary to guarantee the execution of the contract in order to maximize the number of participating bidders. Nevertheless, it is common to have a fake competition when different bidders are, in fact, owned by the same person. This is usually done by having someone as a front for the enterprise, which is often someone with little or no education. Instead of calling this person a front, a common word used in Brazil is "laranja" (Portuguese for orange)[2].

Computerized support for procurement fraud detection must represent and reason about this kind of domain knowledge. The goal of this case study is to show how to structure the specialist's knowledge in a way that an automated system can reason with the evidence in a manner similar to the specialist. Such an automated fraud detection system is intended to be a decision support system to support specialists in carrying out their tasks. The system could also be used to help train new specialists. The case study focuses on a few selected criteria as a proof of concept. It is shown that the model can be incrementally updated to incorporate new criteria. In this process, it becomes clear that a number of different sources should be consulted to come up with the necessary indicators to create new and useful knowledge for decision makers about procurements.

---

[2] After a large chain letters hoax that happened in the late seventies in Brazil. People at the losing end were called the "laranjas," while the perpretators were called the "limões" (Portuguese for limes).

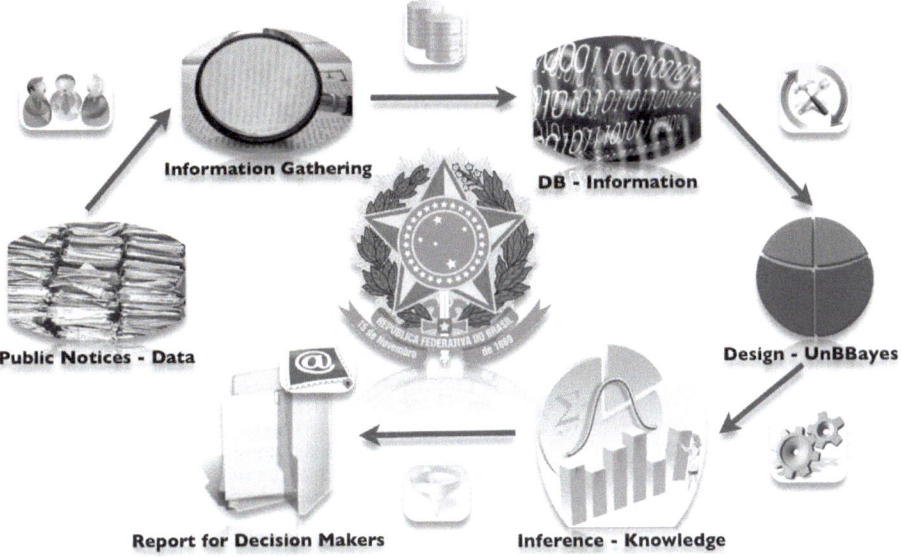

Fig. 2. Procurement fraud detection and prevention overview

Figure 2 presents an overview of the procurement fraud detection process. The data for our case study represent several requests for proposal and electronic auctions that are issued by the Federal, State, and Municipal Offices (Public Notices – Data). Our focus is on representing the specialist's knowledge and reasoning through probabilistic ontologies. We assume that analysts collect information (Information Gathering) through questionnaires specifically designed to capture indicators of the selected criteria. These questionnaires can be created using a system that is already in production at CGU. The questionnaire results provide the necessary information (DB – Information). UnBBayes, using the probabilistic ontology designed by experts (Design – UnBBayes), will collect these millions of items of information and transform them into dozens or hundreds of items of knowledge. This will be achieved through logic and probabilistic inference. For instance, procurement announcements, contracts, reports, etc. - an enormous amount of data - are analyzed to obtain relevant relations and properties - a large amount of information. Then, these relevant relations and properties are used to draw conclusions about possible irregularities - a smaller number of items of knowledge (Inference – Knowledge). This knowledge can be filtered so that only the procurements that show a probability higher than a threshold, *e.g.* 50%, are automatically forwarded to the responsible department along with the inferences about potential fraud and the supporting evidence (Report for Decision Makers).

For this proof of concept, the criteria selected by the specialist were the use of accounting indices and the demand for experience in just one contract. There are four common types of indices (acronyms in Portuguese) that are usually used as requirements in procurements (ILC for current ratio, ILG for general liquidity index, ISG for general solvency index, and IE for indebtedness index). Any other type could indicate a made-up index specifically designed to direct the procurement to some specific company. As the number of uncommon accounting indices used in a procurement increases, the chance of fraud increases. In addition, a procurement specifies a minimum value for these accounting indices. The minimum value that is usually required is 1.0. The higher this minimum value, the more the competition is narrowed, and therefore the higher the chance the procurement is being directed to some enterprise.

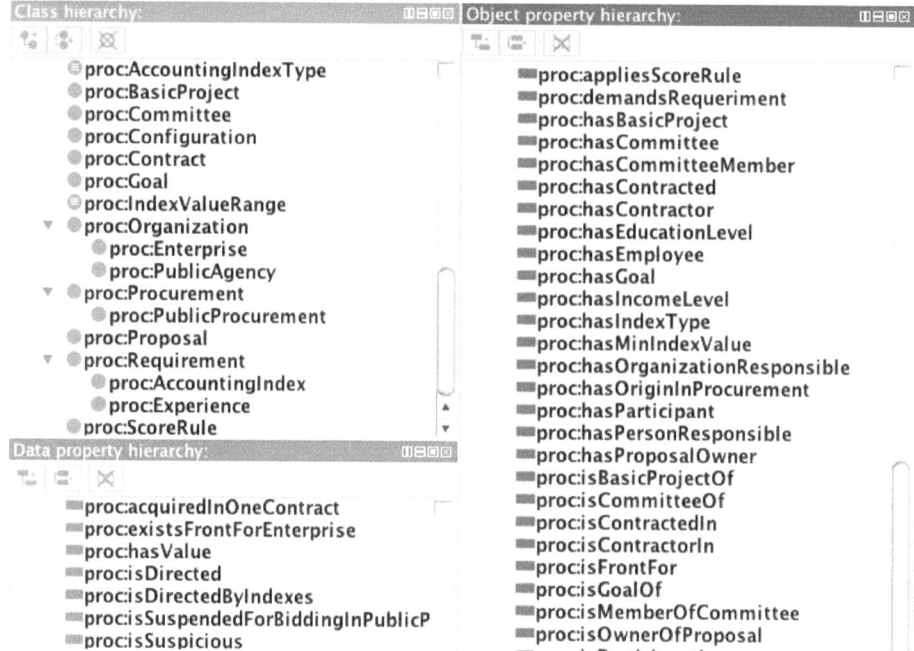

Fig. 3. A few classes, object and data properties of the OWL ontology for the procurement domain

The other criterion, demanding proof of experience in only one contract, is suspect because in almost every case, competence is attained not from a specific contract, but by repeatedly performing a given kind of work. It does not matter whether one has built 1,000 ft^2 of wall in just one contract or 100 ft^2 in 10 different contracts. The experience gained is basically the same.

Before implementing the probabilistic rules described above, we start by looking for an existing ontology that describes the procurement domain. The focus of this chapter is on how to model probabilistic ontologies and not OWL ontologies. Therefore we assume that the ontology depicted in Figure 3 is an existing ontology created by CGU and available at http://www.cgu.gov.br/ontologies/ProcurementDomain.owl.

Using UnBBayes PR-OWL 2.0 plugin [25] we are able to drag and drop OWL properties from a Protégé OWL ontology [16] to create corresponding RVs in our MEBN model. This provides a means to define the probabilistic rules described above. These rules were implemented in three different MFrags, built under the supervision of the CGU specialist.

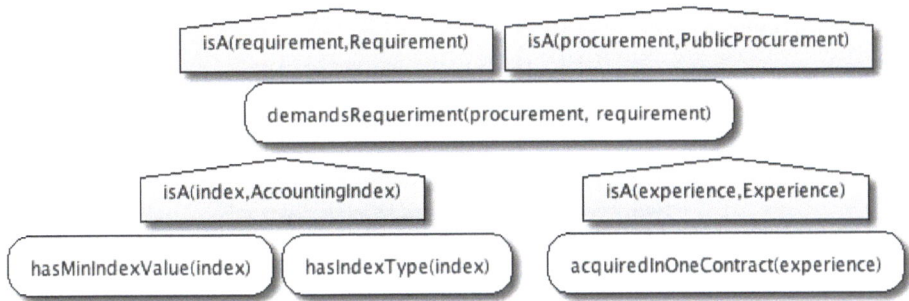

Fig. 4. Procurement Requirement MFrag

The first, Figure 4, represents the criteria required from an enterprise to participate in the procurement. The MFrag represents the type of accounting index (defined by the RV `hasIndexType(index)`, which has individuals of class `AccountingIndexType` as its possible values), as well as the minimum required value (defined by the RV `hasMinIndexValue(index)`, which has individuals of class `IndexValueRange` as its possible values). This MFrag also represents the type of requirement demanded by the procurement (defined by the RV `demandsRequirement(procurement, requirement)`, which has the datatype Boolean as its possible value), as well as whether the procurement demands experience in only one contract (defined by the RV `acquiredInOneContract(experience)`, which has the datatype Boolean as its possible value). As presented in Section 2, there are three kinds of RV: context, resident and input. Context nodes are depicted as green pentagons and represent conditions that should be satisfied for the distributions represented in the MFrag to apply. Resident nodes are depicted as yellow rounded rectangles. Probability distributions for the resident RVs are defined in the MFrag conditioned on the values of the instances of their parents in the fragment graph. Input nodes are depicted as gray trapezoids. They point to RVs that are resident in another MFrag but influence the distribution of RVs resident in this MFrag.

Both `AccountingIndexType` and `IndexValueRange` are nominal classes defined in OWL. The first has `ILC`, `ILG`, `ISG`, `IE`, and `other` as its possible individuals and the second has `between0And1`, `between1And2`, `between2And3`, and `greaterThan3` as its possible individuals.

The second MFrag, shown in Figure 5, represents whether the procurement is being directed to a specific enterprise by the use of unusual accounting indices (defined by the RV `isDirectedByIndexes(procurement)`, which has the

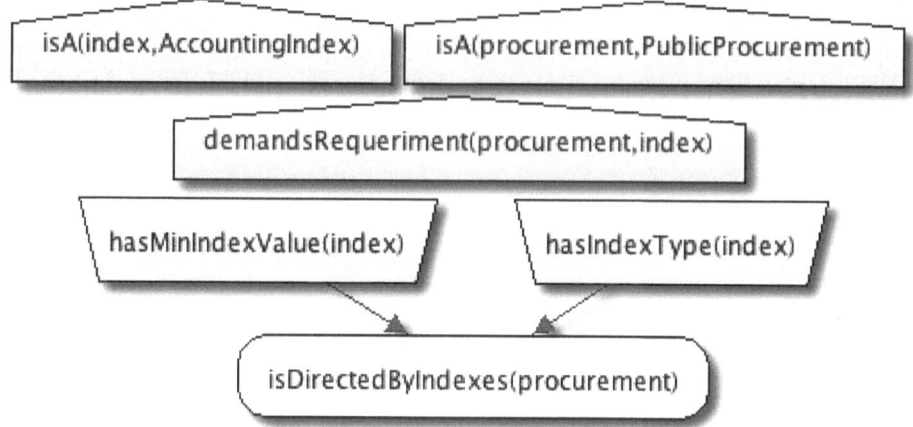

Fig. 5. Directing Procurement by Indexes MFrag

datatype Boolean as its possible value). As explained before, this analysis is based on the type of the index and the minimum value it requires (represented by the `hasIndexType(index)` and `hasMinIndexValue(index)` input nodes, respectively). This evaluation takes into consideration only the indices demanded as requirements for this specific procurement. This is represented by the context node `demandsRequirement(procurement, index)`. Notice that this RV is defined in Figure 4 as `demandsRequirement(procurement, requirement)`, where the second argument is a `Requirement`. However, in Figure 5 the second argument is an `AccountingIndex`. This is a new feature of UnBBayes PR-OWL 2.0 plugin, which allows the use of subtypes in our probabilistic ontology (in our OWL ontology in Figure 3 `AccountingIndex` is defined as a subtype of `Requirement`, and this semantics is inherited in PR-OWL 2.0).

The last MFrag, Figure 6, represents the overall possibility that the procurement is being directed to a specific enterprise (defined by the RV `isDirected(procurement)`, which has the datatype Boolean as its possible value) based on the result of it being directed by the use of unusual indices (represented by the input node `isDirectedByIndexes(procurement)`) and by the requirement of experience in only one contract (represented by the input node `acquiredInOneContract(experience)`), as explained before. Notice that we also make use of subtyping in this MFrag by considering only the experiences demanded as requirements for this specific procurement (in our OWL ontology in Figure 3 `Experience` is defined as a subtype of `Requirement`).

These three MFrags represent knowledge fragments for the domain of procurement fraud detection. The goal is to quantify the probability distribution of the resident RV `isDirected(procurement)` in order to use it to make a decision about whether a procurement is or not suspicious. The next step is to join those MFrags, respecting the logical conditions defined by the context nodes, to generate a Situation-Specific Bayesian Network (SSBN). The algorithm to generate

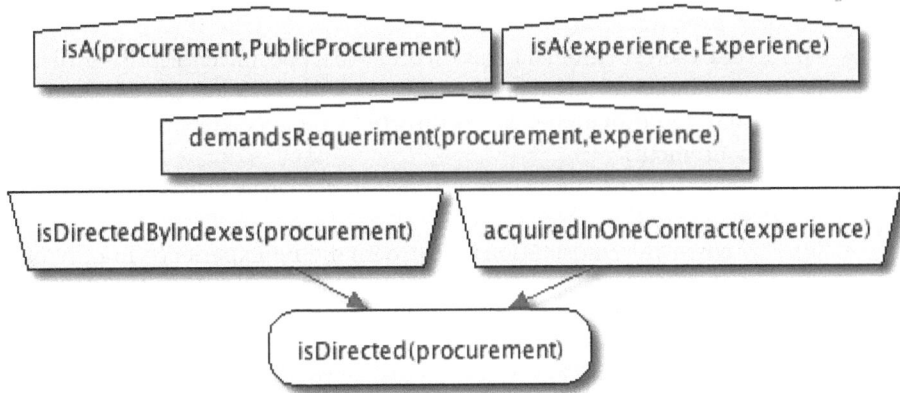

Fig. 6. Directing Procurement MFrag

SSBN proposed by Laskey [20] was implemented in UnBBayes. An SSBN is also a BN, so one can use a belief updating algorithm to do probabilistic inference after entering all available findings. The UnBBayes belief updating is exact and is performed through the strong junction tree algorithm [14].

The probability distributions for resident RVs were estimated with the CGU specialist support based on his knowledge of real cases registered at CGU.

To test the model, two scenarios, that represent the two groups of suspect and non suspect procurements, were chosen from a set of real cases, as shown:

- Suspect procurement (`procurement1`):
  - `index1 = ILC >= 2.0`;
  - `index2 = ILG >= 1.5`;
  - `index3 = other >= 3.0`.
  - It demands experience in only one contract.
- Non suspect procurement (`procurement2`):
  - `index4 = IE >= 1.0`;
  - `index5 = ILG >= 1.0`;
  - `index6 - ILC >- 1.0`;
  - It does not demand experience in only one contract.

The information above was introduced in our model as known individuals and evidence (as simple RDF triples defined in our OWL ontology). After that we queried the system to give us information about the node `isDirected(procurement)` for both `procurement1` and `procurement2`. UnBBayes PR-OWL 2.0 plugin then executed the SSBN algorithm and generated the same node structure as shown in Figure 7, because both procurements have three accounting indices and information about whether the demanding experience is in only one contract or not. However, as expected, the parameters and findings are different giving different results for the query, as shown below:

- Non suspect procurement:
  - 0.01% that the procurement was directed to a specific enterprise by using accounting indices;
  - 0.10% that the procurement was directed to a specific enterprise.
- Suspect procurement:
  - 55.00% that the procurement was directed to a specific enterprise by using accounting indices;
  - 29.77%, when the information about demanding experience in only one contract was omitted, and 72.00%, when it was given, that the procurement was directed to a specific enterprise.

The specialist from CGU analyzed and agreed with the knowledge generated by the probabilistic ontology developed using PR-OWL/MEBN in UnBBayes. By interpreting the resulting probabilities as high, medium, and low chances of something being true, he was able to state that the probabilities represented what he would think when analyzing the same individuals and evidence.

The SSBNs generated for this proof of concept model have the same structure. In practice, the context commonly varies from procurement to procurement in a way that would require SSBNs with different structures. For instance, we have come across several procurements that, in addition to the four common indices, include other indices as well. In this case, if there are two additional indices (index5 and index6), then the resulting SSBN would have two more copies for nodes hasIndexType(index) and hasMinIndexValue(index). Standard BNs cannot be used for such problems with varying structures. The ability to make multiple copies of nodes based on a context is only available in a more expressive formalism, such as MEBN.

4   Probabilistic Ontology Knowledge Fusion

From the criteria presented and modeled in Section 3, we can clearly see the need for a principled way of dealing with uncertainty. But what is the role of Semantic Web in this domain? Well, it is easy to see that our domain of fraud detection is a RIS environment. The data CGU has available does not come only from its audits and inspections. In fact, much complementary information can be retrieved from other Federal Agencies, including Federal Revenue Agency, Federal Police, and others. Imagine we have information about the enterprise that won the procurement, and we want to know information about its owners, such as their personal data and annual income. This type of information is not available at CGU's Data Base (DB), but should be retrieved from the Federal Revenue Agencys DB. Once the information about the owners is available, it might be useful to check their criminal history. For that (see Figure 8), information from the Federal Police (Polícia Federal) must be used. In this example, we have different sources saying different things about the same person: thus, the AAA slogan applies. Moreover, there might be other Agencies with crucial information related to our person of interest; in other words, we are operating in an

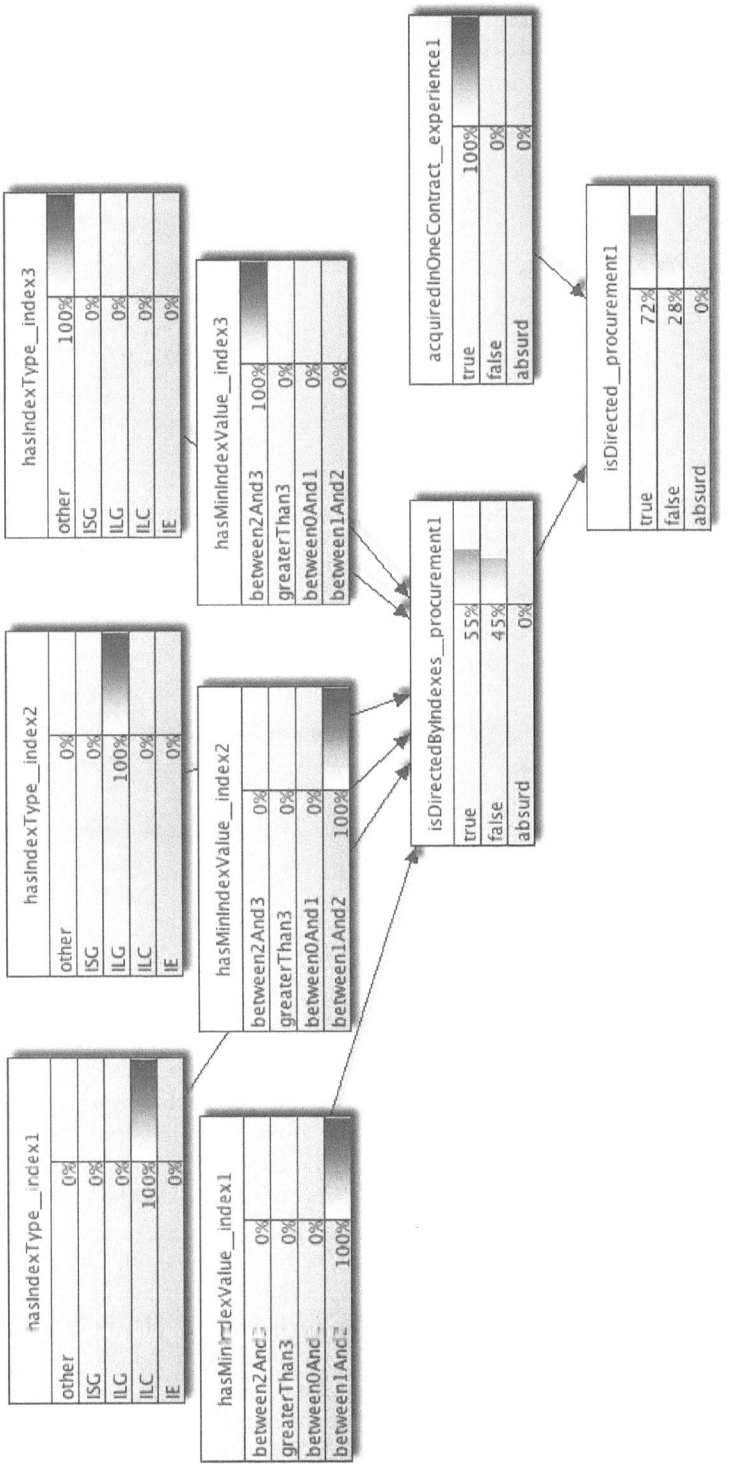

Fig. 7. SSBN generated for query isDirected(procurement1)

open-world enviroment which introduces uncertainty. Finally, to make this sharing and integration process possible, we have to make sure we are talking about the same person, who may (especially in case of fraud) be known by different names in different contexts.

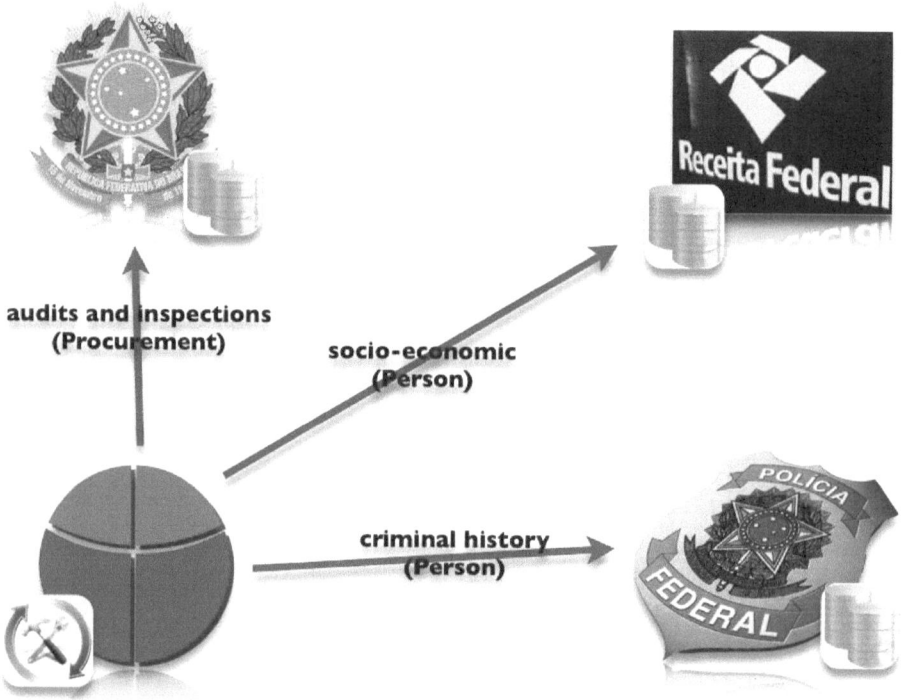

Fig. 8. Knowledge fusion from different Government Agencies DBs

We illustrate the need to fuse knowledge from different sources through the introduction of a new probabilistic reasoning rule. This rule, mentioned in Section 3, addresses the question of whether a person is front for some enterprise. Typically, a person acting as a front has a low annual income, and therefore is unlikely to not own properties such cars and houses. In fact, fronts are often gardeners or maids who work for the person who really makes the decisions for the enterprise.

So, by looking at a person's education level, annual income, and lack of properties (*e.g.*, whether the person has a car) we can determine whether this person is more likely to be a front for the enterprise for which he/she is listed as responsible. However, CGU does not have information about a person's education level, annual income, and property ownership. This information is available, but it is collected by other Federal Agencies. Information about education level can be retrieved from the Education Ministry (MEC). Information about annual income can be retrieved from the Federal Revenue Agency (Receita Federal). Finally, information about property ownership can be retrieved from the relevant agencies, such as the Department of Motor Vehicles (DENATRAN) for the case of motor vehicles.

CGU has been engaging in collaborations with different Agencies for some years now in order to gather more information that might help identify and prevent frauds in public procurements. In this Section we show how CGU can exploit SW technologies in order to add a new probabilistic rule to our probabilistic ontology and reason with the information provided from other Agencies.

In our proof of concept architecture we assume each Agency has its own ontology with focus on its domain of application. Furthermore, we assume that all Government Agencies use a common ontology with basic concepts for people (name, address, relationship, etc), which is the ontology created by the Federal Government and available at `http://www.brasil.gov.br/ontologies/People.owl`[3]. The Education Ministry (MEC) provides an ontology for education available at `http://www.mec.gov.br/ontologies/Education.owl`. The Department of Motor Vehicles (DENATRAN) provides an ontology for motor vehicle information (*e.g.*, ownership and license) available at `http://www.denatran.gov.br/ontologies/MotorVehicle.owl`. The Federal Revenue Agency (Receita Federal) provides an ontology for internal revenue services available at `http://www.receita.fazenda.gov.br/ontologies/InternalRevenue.owl`.

Because we need to use concepts from all these ontologies to define our new probabilistic rule for identifying a front, we need to import them into our probabilistic ontology. Once they have been imported, we can start creating our MFrags. Below we describe a set of MFrags representing information imported from other ontologies, rules for using this information to determine the likelihood that a person is a front, and using this information to reason about whether a procurement is fraudulent.

Figure 9 depicts an MFrag representing information associated with an enterprise. For our proof of concept, this MFrag contains a single RV for identifying the responsible person for an organization, which is the RV `isResponsibleForOrganization(person, enterprise)`. Although the range of the OWL ontology for the property `isResponsibleForOrganization` is an `Organization`, we define it here as an `Enterprise`, which is a subtype of `Organization`, because our procurement rule concerns enterprises. The ability to reason with subtypes in probabilistic ontologies is a new feature in UnBBayes PR-OWL 2.0 plugin.

Figure 10 shows an MFrag representing information associated with a procurement. The MFrag defines a RV `hasParticipant(procurement, enterprise)` for identifying whether an enterprise is participating in a procurement. Again,

---

[3] The ontologies presented in this chapter were created by the authors. In order to illustrate the idea of information fusion we will present them as being ontologies created and distributed by different agencies of the Brazilian Government. Thus, the URI provided here is for illustration only.

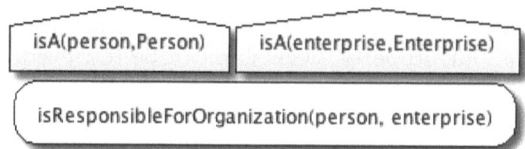

Fig. 9. Enterprise Information MFrag

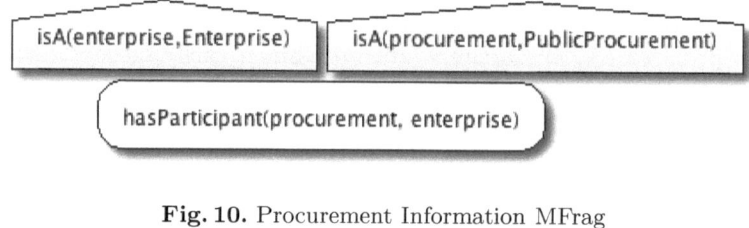

Fig. 10. Procurement Information MFrag

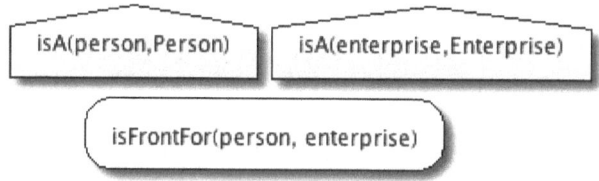

Fig. 11. Front for Enterprise MFrag

we make use of subtypes instead of the more general type defined in our OWL ontology.

The MFrag of Figure 11 defines the RV for verifying whether a person is a front for a specific enterprise, which is the RV isFrontFor(person, enterprise). Here we also make use of subtyping by using **Enterprise** instead of the more general class **Organization**.

Figure 12 presents the main rule for defining whether a person is front for an enterprise. As discussed above, the idea is that if a person is front for an enterprise, then this person is more likely to have a low annual income (defined by the RV hasIncomeLevel(person)), no motor vehicle (defined by the RV hasMotorVehicle(person)), and little or no education level (defined by the RV hasEducationLevel(person)). These RVs are mapped to OWL properties from the Federal Revenue Agency (Receita Federal), Department of Motor Vehicle (DENATRAN), and Education Ministry (MEC) ontologies, respectively. Notice that the only persons analyzed are the ones responsible for that enterprise (constrained by the context node isResponsibleForOrganization(person, enterprise)).

Figure 13 defines a RV collecting all information about potential fronts to assess whether there is a front for a given enterprise. This *existential* assertion is a built-in RV in MEBN, but had to be defined manually in UnBBayes because

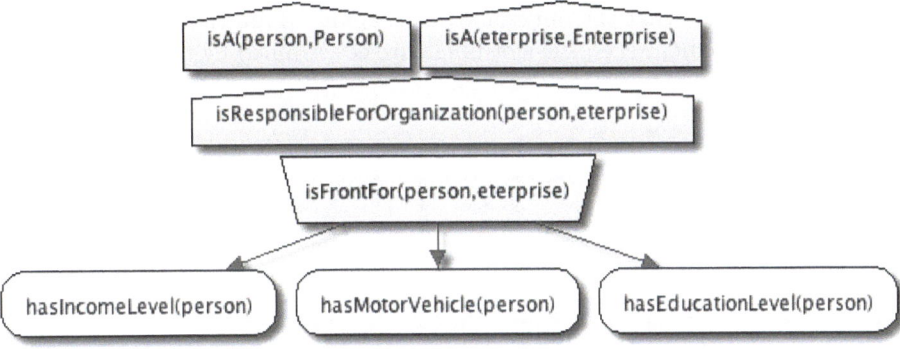

Fig. 12. Personal Information MFrag

this feature of MEBN has not yet been implemented there. The logic of this RV has the same logic as the built-in RV as defined in PR-OWL and MEBN [2]. That is, at least one of the potential fronts for an enterprise actually is a front, then there exists a front for the enterprise. Notice that the only persons included in the existential assertion are those responsible for that enterprise (i.e., the slot fillers are constrained by the context node `isResponsibleForOrganization(person, enterprise)`).

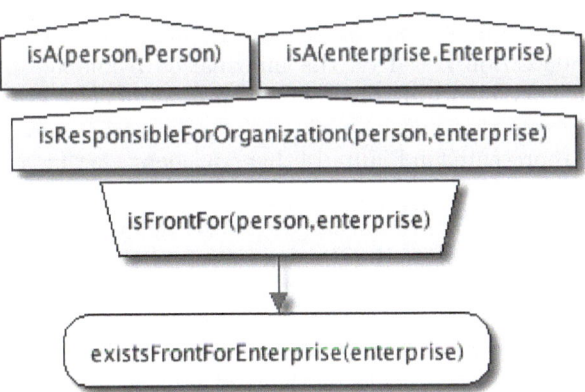

Fig. 13. Exists Front for Enterprise MFrag

Finally, Figure 14 integrates the two major probabilistic rules we have in our probabilistic ontology, namely identifying whether a procurement is being directed for a specific enterprise (represented by the input node `isDirected(procurement)`) and whether an enterprise has a front (represented by the input node `existsFrontForEnterprise(enterprise)`). The resident RV `isSuspicious(procurement)` of this MFrag represents whether the procurement is suspicious. Notice that the only enterprises analyzed are the ones participating in this procurement (i.e., the slot fillers are constrained by the context node hasParticipant(procurement, enterprise)).

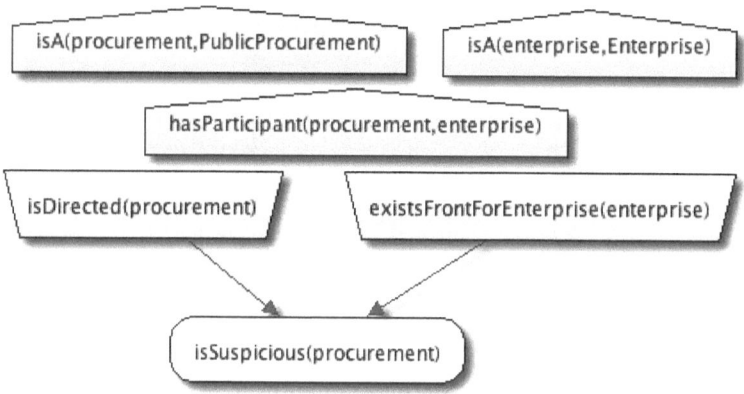

Fig. 14. Suspicious Procurement MFrag

The above MFrags represent modular pieces of knowledge that can be reused when necessary. This fact allows incremental enhancement of the model being designed by the fraud detection analyst. So, Figures 4 to 6 represent the first model of fraud detection based on the information available at CGU records only. The MFrags depicted in Figures 9 to 13 present general probabilistic rules representing concepts that can be aggregated to form the more complex fraud detection model represented in Figure 14 that considers whether the procurement is being directed or whether one of the participants is a front for an enterprise. In either of these cases, the procurement is considered suspicious.

This more complex model was built using both probabilistic models of the procurement being directed (Figure 6) and the existence of a front for an enterprise (Figure 11) which is a participant of the procurement (Figure 10).

To validate our knowledge fusion architecture we published each ontology on a different computer, but accessible via the network. We then have a user enter a query to our fraud detection and prevention ontology, which gathers information from the external ontologies for use in probabilistic inference. Unlike the example described in Section 3, the evidence collected in this Section is fictitious.

Figure 15 presents the SSBN generated with information about John_Doe who is responsible for ITBusiness and Jane_Doe who is responsible for TechBusiness. Both enterprises are participating in procurement1 from Section 3. It can be seen that the information about Jane_Doe favors the hypothesis of procurement1 being suspicious, since she seems to be a front for TechBusiness. Information about John_Doe, on the other hand, does not favor this hypothesis, since he does not seem to be a front for ITBusiness.

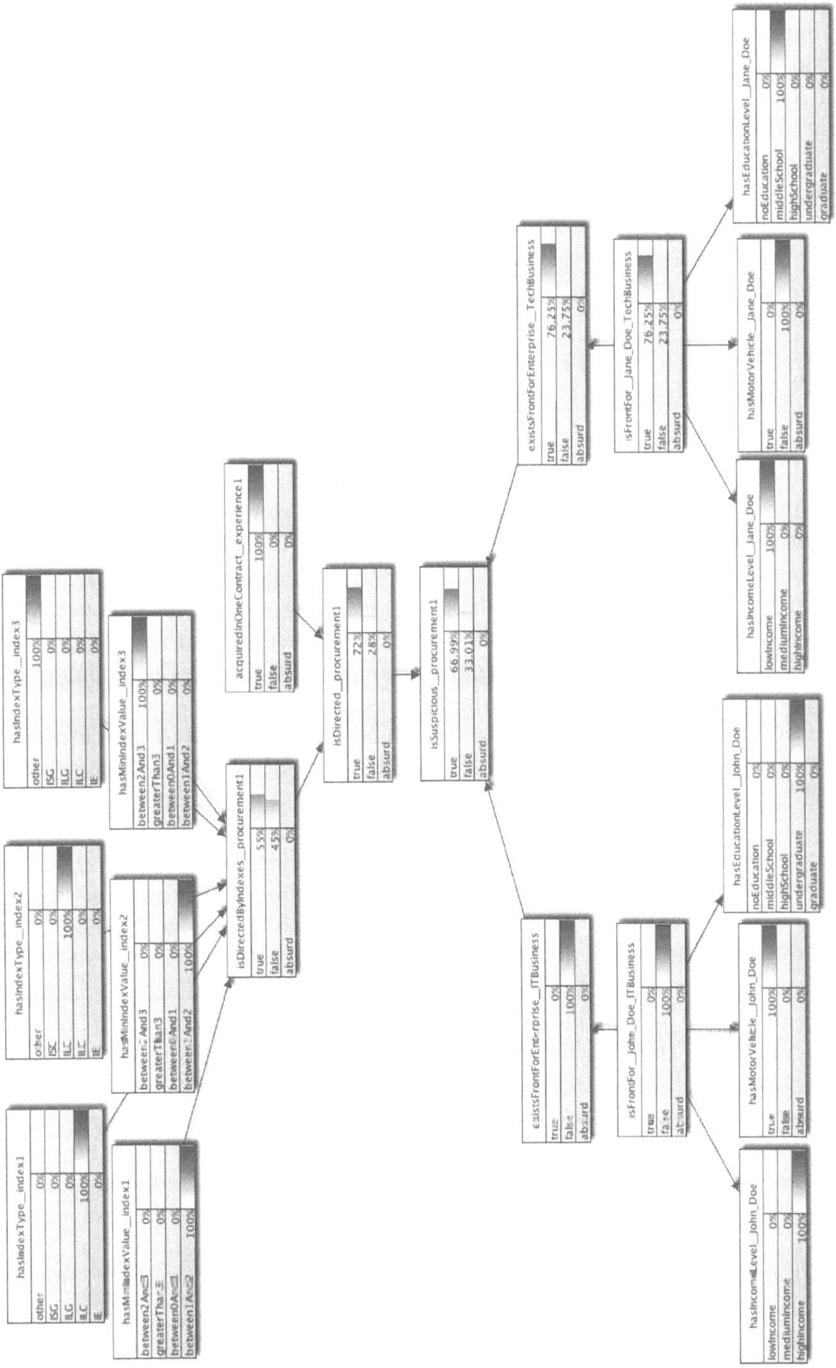

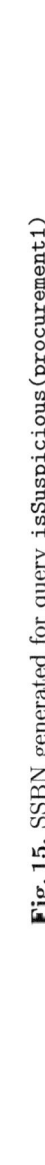

Fig. 15. SSBN generated for query isSuspicious(procurement1)

The importance of knowledge fusion is noticed when we compare the results of fusion with what we would be able to infer by considering each source separately. Having only the annual income of Jane Doe gives us a posterior probability of 8.26% that Jane Doe is a front. Considering only the information that Jane Doe does not have a motor vehicle gives us a posterior probability of 0.05% that she is a front. Finally, considering only the information about her education level gives us a posterior probability of 0.07% that she is a front. It is easy to see that separately these items of evidence do not provide strong evidence that Jane Doe is a front. However, if we fuse all the information, we now have strong evidence that Jane Doe is a front. This is shown in Figure 15 by the posterior probability of 76.25% that Jane Doe is a front for `TechBusiness`.

5   Conclusion

The problem that CGU and many other Agencies have faced of processing all the available data into useful knowledge is starting to be solved with the use of probabilistic ontologies, as the procurement fraud detection model showed. In addition to enabling fusion of available information from multiple external sources, the designed model was able to represent the specialist's knowledge for the two real cases we evaluated. UnBBayes reasoning given the evidence and using the designed model were accurate both in suspicious and non suspicious scenarios. These results are encouraging, suggesting that a fuller development of our proof of concept system is promising.

In addition, it is straightforward to introduce new criteria and indicators in the model in an incremental manner. That is, new rules for identifying fraud can be added without rework. After a new rule is incorporated into the model, a set of new tests can be added to the previous one with the objective of always validating the new model proposed, without doing everything from scratch. This was shown in Section 4 where we added a new rule that uses knowledge fusion to identify whether a person is a front for an enterprise. The new rule could be added, without making any changes to the existing MFrags.

Furthermore, the use of this formalism through UnBBayes allows advantages such as impartiality in the judgment of irregularities in procurements (given the same conditions the system will always deliver the same result), scalability (automatization implies expanding the capacity of the specialist to analyze more procurements in a short period of time) and a joint analysis of large volumes of indicators (the higher the number of indicators to examine jointly the more difficult it is for the specialist's analysis to be objective and consistent). The results described here are preliminary, but show that the development of a tool based on PR-OWL ontologies to automatize this task on CGU is viable. This paper also illustrates how to use PR-OWL 2.0 to provide a link between the deterministic and probabilistic parts of the ontology.

As a next step, CGU is choosing new criteria to be incorporated into the designed probabilistic ontology. This next set of criteria will require information from different Brazilian Agencies databases, as shown in Section 4. Therefore, combining the semantic power of ontologies with the uncertainty handling capability of PR-OWL will be extremely useful for fusing information from different sources.

Acknowledgments. Rommel Carvalho gratefully acknowledges full support from the Brazilian Office of the Comptroller General (CGU) for the research reported in this paper, and its employees involved in this research, especially Mário Vinícius Claussen Spinelli, the domain expert.

References

1. Allemang, D., Hendler, J.A.: Semantic Web for the Working Ontologist. Morgan Kaufmann (2008)
2. Carvalho, R.N.: Probabilistic Ontology: Representation and Modeling Methodology. PhD, George Mason University, Fairfax, VA, USA (2011)
3. Carvalho, R.N., Haberlin, R., Costa, P.C.G., Laskey, K.B., Chang, K.-C.: Modeling a probabilistic ontology for maritime domain awareness. In: Proceedings of the 14th International Conference on Information Fusion, Chicago, USA (July 2011)
4. Carvalho, R.N., Laskey, K.B., Costa, P.C.G.: Compatibility formalization between PR-OWL and OWL. In: Proceedings of the First International Workshop on Uncertainty in Description Logics (UniDL) on Federated Logic Conference (FLoC) 2010, Edinburgh, UK (July 2010)
5. Carvalho, R.N., Laskey, K.B., Costa, P.C.G.: PR-OWL 2.0 - Bridging the Gap to OWL Semantics. In: Bobillo, F., Costa, P.C.G., d'Amato, C., Fanizzi, N., Laskey, K.B., Laskey, K.J., Lukasiewicz, T., Nickles, M., Pool, M. (eds.) URSW 2008-2010/UniDL 2010. LNCS (LNAI), vol. 7123, pp. 1–18. Springer, Heidelberg (2013)
6. Chen, H., Wu, Z.: On Case-Based knowledge sharing in semantic web. In: Proceedings of the 15th IEEE International Conference on Tools with Artificial Intelligence, ICTAI 2003, pp. 200–207. IEEE Computer Society, Los Alamitos (2003)
7. Chen, H., Wu, Z., Xu, J.: KB-Grid: enabling knowledge sharing on the semantic web. In: International Workshop on Challenges of Large Applications in Distributed Environments, p. 70. IEEE Computer Society, Los Alamitos (2003)
8. Costa, P.C.G.: Bayesian Semantics for the Semantic Web. PhD, George Mason University, Fairfax, VA, USA (July 2005)
9. Costa, P.C.G., Chang, K.-C., Laskey, K.B., Carvalho, R.N.: High level fusion and predictive situational awareness with probabilistic ontologies. In: Proceedings of the AFCEA-GMU C4I Center Symposium, George Mason University, Fairfax, VA, USA (May 2010)
10. Costa, P.C.G., Laskey, K.B., Laskey, K.J.: Probabilistic ontologies for efficient resource sharing in semantic web services. In: Proceedings of the Second Workshop on Uncertainty Reasoning for the Semantic Web, URSW 2006, Athens, GA, USA (November 2006)
11. Costa, P.C.G., Laskey, K.B., Takikawa, M., Pool, M., Fung, F., Wright, E.J.: MEBN logic: A key enabler for network centric warfare. In: Proceedings of the 10th International Command and Control Research and Technology Symposium, 10th ICCRTS. CCRP Publications, McLean (2005)
12. Costa, P.C.G., Chang, K.-C., Laskey, K.B., Carvalho, R.N.: A Multi-Disciplinary approach to high level fusion in predictive situational awareness. In: Proceedings of the 12th International Conference on Information Fusion, Seattle, Washington, USA, pp. 248–255 (July 2009)

13. Dadzie, A.-S., Bhagdev, R., Chakravarthy, A., Chapman, S., Iria, J., Lanfranchi, V., Magalhães, J., Petrelli, D., Ciravegna, F.: Applying semantic web technologies to knowledge sharing in aerospace engineering. Journal of Intelligent Manufacturing 20(5), 611–623 (2008)
14. Jensen, F., Jensen, F.V., Dittmer, S.L.: From influence diagrams to junction trees. In: Proceedings of the Tenth Conference on Uncertainty in Artificial Intelligence (1994)
15. Kings, N.J., Davies, J.: Semantic web for knowledge sharing. In: Semantic Knowledge Management, pp. 103–111. Springer, Heidelberg (2009), doi:10.1007/978-3-540-88845-1_8
16. Knublauch, H., Fergerson, R.W., Noy, N.F., Musen, M.A.: The Protégé OWL Plugin: An Open Development Environment for Semantic Web Applications. In: McIlraith, S.A., Plexousakis, D., van Harmelen, F. (eds.) ISWC 2004. LNCS, vol. 3298, pp. 229–243. Springer, Heidelberg (2004)
17. Laskey, K.B., Costa, P.C.G.: Of starships and klingons: Bayesian logic for the 23rd century. In: Proceedings of the 21st Annual Conference on Uncertainty in Artificial Intelligence, UAI 2005. AUAI Press, Arlington (2005)
18. Laskey, K.B., Costa, P.C.G., Wright, E.J., Laskey, K.J.: Probabilistic ontology for Net-Centric fusion. In: Proceedings of the 10th International Conference on Information Fusion, pp. 1–8 (2007)
19. Laskey, K.B., Mahoney, S.M., Wright, E.: Hypothesis management in Situation-Specific network construction. In: Proceedings of the 17th Conference in Uncertainty in Artificial Intelligence, UAI 2001, pp. 301–309. Morgan Kaufmann Publishers Inc., San Francisco (2001), ACM ID: 720228
20. Laskey, K.B.: MEBN: a language for First-Order bayesian knowledge bases. Artificial Intelligence 172(2-3), 140–178 (2008)
21. Laskey, K.B., da Costa, P.C.G., Janssen, T.: Probabilistic ontologies for knowledge fusion. In: Proceedings of the 11th International Conference on Information Fusion, pp. 1–8 (2008)
22. Laskey, K.B., da Costa, P.C.G., Janssen, T.: Probabilistic ontologies for Multi-INT fusion. Technical report, George Mason University C4I Center (May 2008)
23. Laskey, K., Laskey, K.B.: Uncertainty reasoning for the world wide web: Report on the URW3-XG incubator group. URW3-XG, W3C (2008)
24. Mahoney, S., Laskey, K.B.: Constructing situation specific belief networks. In: Proceedings of the 14th Annual Conference on Uncertainty in Artificial Intelligence, UAI 1998. Morgan Kaufmann, San Francisco (1998)
25. Matsumoto, S.: Framework Based in Plug-ins for Reasoning with Probabilistic Ontologies. M.Sc., University of Brasília, Brasilia, Brazil (forthcoming)
26. Matsumoto, S., Carvalho, R.N., Ladeira, M., da Costa, P.C.G., Santos, L.L., Silva, D., Onishi, M., Machado, E., Cai, K.: UnBBayes: a java framework for probabilistic models in AI. In: Java in Academia and Research. iConcept Press (2011)
27. Poole, D., Smyth, C., Sharma, R.: Semantic Science: Ontologies, Data and Probabilistic Theories. In: da Costa, P.C.G., d'Amato, C., Fanizzi, N., Laskey, K.B., Laskey, K.J., Lukasiewicz, T., Nickles, M., Pool, M. (eds.) URSW 2005-2007. LNCS (LNAI), vol. 5327, pp. 26–40. Springer, Heidelberg (2008)
28. Veres, G.V., Huynh, T.D., Nixon, M.S., Smart, P.R., Shadbolt, N.R.: The military knowledge information fusion via semantic web technologies. Technical report, School of Electronics and Computer Science, University of Southampton (2006)

Understanding a Probabilistic Description Logic via Connections to First-Order Logic of Probability*

Pavel Klinov[1] and Bijan Parsia[2]

[1] Institute of Artificial Intelligence
University of Ulm, Germany
pavel.klinov@uni-ulm.de
[2] School of Computer Science
University of Manchester, United Kingdom
bparsia@cs.man.ac.uk

Abstract. This paper analyzes the probabilistic description logic P-$\mathcal{SROIQ}$ as a fragment of well-known first-order probabilistic logic (FOPL). P-$\mathcal{SROIQ}$ was suggested as a language that is capable of representing and reasoning about different kinds of uncertainty in ontologies, namely generic probabilistic relationships between concepts and probabilistic facts about individuals. However, some semantic properties of P-$\mathcal{SROIQ}$ have been unclear which raised concerns regarding whether it could be used for representing probabilistic ontologies. In this paper we provide an insight into its semantics by translating P-$\mathcal{SROIQ}$ into FOPL with a specific subjective semantics based on possible worlds. We prove faithfulness of the translation and demonstrate the fundamental nature of some limitations of P-$\mathcal{SROIQ}$. Finally, we briefly discuss the implications of the exposed semantic properties of the logic on probabilistic modeling.

1 Introduction

One common complaint about Description Logic (DL) based ontology languages, such as the Web Ontology Language (OWL), is that they fail to support non-classical uncertainty, in particular, probability. One answer to this complaint is the P-$\mathcal{S}$ family of logics which allows for the incorporation of probabilistic formulae as an extension of the familiar and widely used $\mathcal{S}$ DLs, such as $\mathcal{SHOIN}$ or $\mathcal{SROIQ}$ [1]. Unlike Bayesian or Markov extensions to DLs and OWL, the P-$\mathcal{S}$ logics do not require any machinery based on graphical models to answer probabilistic queries; their syntax, semantics, and inference services are extended in a purely logical way. These logics are also decidable, generally of the same worst case complexity as the base logic, and can be implemented on top of existing DL reasoners. Work on practical implementations is ongoing and produces some encouraging results [2].

---

* This work has been carried out when the first author was a doctoral student at the University of Manchester, UK.

F. Bobillo et al. (Eds.): URSW 2008-2010/UniDL 2010, LNAI 7123, pp. 41–58, 2013.
© Springer-Verlag Berlin Heidelberg 2013

However, there are several issues with the P-$\mathcal{S}$ family both from an expressivity and from a theoretical point of view. First, it has not been fully clear how, or whether, it actually combines statistical and subjective probabilities. Second, probabilistic ABoxes have a number of strong restrictions including no support of roles assertions between probabilistic individuals and only one probabilistic individual per ABox. Finally, the semantics of the P-$\mathcal{S}$ family (in terms of possible worlds) is not particularly familiar in the DL setting.

Often, insight into a DL (and associated extensions and reasoning techniques) has followed by considering its standard first-order translation, that is, as a fragment of first order-logics. In this paper, we attempt to apply this methodology to the P-$\mathcal{S}$ family by considering them as fragments of a first-order logic extended with various forms of probability (FOPL). We show that we can understand P-$\mathcal{S}$ logics as fragments of FOPL and explain its limitations on the basis of the known properties of FOPL with semantics based on possible worlds.

2   Background

This section presents necessary preliminaries on P-$\mathcal{SROIQ}$ and first-order probabilistic logic by Bacchus [3] and Halpern [4].

2.1   P-$\mathcal{SROIQ}$

P-$\mathcal{SROIQ}$ [1] is a probabilistic generalization of the DL $\mathcal{SROIQ}$ [5]. It provides means for expressing probabilistic relationships between arbitrary $\mathcal{SROIQ}$ concepts and a certain class of probabilistic relationships between classes and individuals. Any $\mathcal{SROIQ}$, and thus OWL 2 DL, ontology can be used as a basis for a P-$\mathcal{SROIQ}$ ontology, which facilitates transition from classical to probabilistic ontologies.

The additional syntactic feature in P-$\mathcal{SROIQ}$ is the conditional constraint.

Definition 1 (Conditional Constraint). *A conditional constraint is an expression of the form $(D|C)[l, u]$, where C and D are concept expressions in $\mathcal{SRIQ}$ (i.e., $\mathcal{SROIQ}$ without nominals) called **evidence** and **conclusion**, respectively, and $[l, u] \subseteq [0, 1]$ is a closed real-valued interval. In the case where C is $\top$ the constraint is called **unconditional**.*

Ontologies in P-$\mathcal{SROIQ}$ are separated into a classical and a probabilistic part. It is assumed that the set of individual names N_I is partitioned onto two sets: classical individuals N_{CI} and probabilistic individuals N_{PI}.

Definition 2 (PTBox, PABox, and Probabilistic Knowledge Base). *A **probabilistic TBox** (PTBox) is a pair $PT = (\mathcal{T}, \mathcal{P})$ where $\mathcal{T}$ is a classical $\mathcal{SROIQ}$ TBox and $\mathcal{P}$ is a finite set of conditional constraints. A **probabilistic ABox** (PABox) is a finite set of conditional constraints associated with a probabilistic individual $o_p \in N_{PI}$. A **probabilistic knowledge base** (or a probabilistic ontology) is a triple $PO = (\mathcal{T}, \mathcal{P}, \{\mathcal{P}_{o_p}\}_{o_p \in N_{PI}})$, where the first two components define a PTBox and the last is a a set of PABoxes.*

Informally, a PTBox constraint $(D|C)[l, u]$ expresses a conditional statement of the form "if a *randomly* chosen individual is an instance of C, the probability of it being an instance of D is in $[l, u]$". A PABox constraint, which we write as $(D|C)_o[l, u]$ where o is a probabilistic individual, states that "if a *specific* individual (that is, o) is an instance of C, the probability of it being an instance of D is in $[l, u]$". That distinction is important for default reasoning in P-$\mathcal{SROIQ}$.

Definition 3 (Probabilistic Signature). *Given a probabilistic knowledge base PO let $\mathcal{CE}(PO)$ be the set of all concept expressions that appear either as evidence or conclusion in some conditional concept (in the PTBox or in a PABox). Then **probabilistic signature** of PO, denoted as $\Phi(PO)$, is the smallest set of concept expressions such that i) no expression is syntactically a union, intersection, or complement of other expressions and ii) its closure under union, intersection, and complementation is a superset of $\mathcal{CE}(PO)$.*

$\Phi(PO)$ (or simply Φ when the ontology is clear from context) is a finite set because $\mathcal{CE}(PO)$ is finite. It can be computed by starting from $\Phi = \mathcal{CE}(PO)$ and exhaustively applying the following rules, where C and D are concept expressions:

- If $C \sqcup D \in \Phi$, then $\Phi \leftarrow (\Phi \cup \{C, D\}) \setminus \{C \sqcup D\}$;
- If $C \sqcap D \in \Phi$, then $\Phi \leftarrow (\Phi \cup \{C, D\}) \setminus \{C \sqcap D\}$;
- If $\neg C \in \Phi$, then $\Phi \leftarrow (\Phi \cup \{C\}) \setminus \{\neg C\}$;

The process of applying the rules will terminate since every rule reduces the syntactic length of expressions in Φ.

We conclude with an example which shows a small ontology which, first, defines breast cancer (BRC), duct cancer, and lobular cancer using DL, second, expresses generic knowledge that 10%–11% of cancer incidence among women is breast cancer, and third, unconditionally states that *Mary* has $\geq 90\%$ chance of having duct cancer.[1]

Example 1 (Fragment of a probabilistic ontology about breast cancer)

$$\mathcal{T} = \{BRC \equiv Cancer \sqcap \exists occursIn.\exists partOf.Breast$$
$$Duct \sqsubseteq \exists partOf.Breast, Lobule \sqsubseteq \exists partOf.Breast\}$$
$$\mathcal{P} = \{(Woman \sqcap \exists disease.BRC \mid Woman \sqcap \exists disease.Cancer)[0.1, 0.11]\}$$
$$\mathcal{P}_{Mary} = \{(Woman \sqcap \exists hasDisease.(Cancer \sqcap \exists occursIn.Duct))[0.9, 1]\}$$

According to Definition 3, $\mathcal{CE}(PO)$ and $\Phi(PO)$, where $PO = (\mathcal{T}, \mathcal{P}, \mathcal{P}_{Mary})$, are the following sets.

---

[1] In many cases it is convenient to introduce new concept names, such as $WomanWithBreastCancer$, to avoid repetition of complex expressions in conditional constraints.

$$\mathcal{CE}(PO) = \{Woman \sqcap \exists disease.BRC, Woman \sqcap \exists disease.Cancer,$$
$$Woman \sqcap \exists hasDisease.(Cancer \sqcap \exists occursIn.Duct)\}$$
$$\Phi(PO) = \{Woman, \exists disease.BRC, \exists disease.Cancer,$$
$$\exists hasDisease.(Cancer \sqcap \exists occursIn.Duct)\}$$

The semantics of P-$\mathcal{SROIQ}$ is based on the notion of a *world*.

Definition 4 (World). *Given the probabilistic signature Φ a **world** W is a subset of Φ. A concept $C \in \Phi$ occurs **positively**, or is **satisfied**, in a world W if $C \in W$, otherwise it is said to occur **negatively**. The satisfaction relation is extended recursively to Boolean expressions over Φ in a standard way (e.g., W satisfies $A \sqcup B$ if W satisfies A or W satisfies B).*

Finally, we extend the definition of satisfaction in a world to $\mathcal{SROIQ}$ TBoxes.

Definition 5 (Possible World, Index Set). *A world W is a **possible world** with respect to a $\mathcal{SROIQ}$ TBox $\mathcal{T}$, written as $W \models \mathcal{T}$, if $\mathcal{T} \cup \{\{o\} \sqsubseteq C | C \in W\} \cup \{\{o\} \sqsubseteq \neg C | C \notin W, C \in \Phi\}$ is satisfiable, where o is an individual name not occurring in $\mathcal{T}$. The set of all possible worlds over Φ with respect to $\mathcal{T}$, also called the **index set**, is denoted as $\mathcal{W}_\Phi(\mathcal{T})$.[2]*

Possible worlds correspond to what is commonly known as realizable concept types in the DL literature [6]. Each world W can be thought of as a conjunctive concept expression $X \equiv (\sqcap_{C \in W} C) \sqcap (\sqcap_{C \notin W, C \in \Phi} \neg C)$ so that the world is possible iff X is satisfiable (i.e., there is a realization of the concept type given a TBox).

In what follows we assume a *linear order* of basic concepts in Φ. Since Φ is a finite set we can denote the i-th basic concept in Φ by C_i. For a given possible world W we also use the notation W_i to denote either C_i if C_i occurs positively in W or $\neg C_i$ if it occurs negatively. For a given PTBox the order of basic concepts is fixed across all possible worlds.

Definition 6 (Probabilistic Interpretation, Probability of a Concept). *A **probabilistic interpretation** Pr of a PTBox $(\mathcal{T}, \mathcal{P})$ is a function $Pr : \mathcal{W}_\Phi(\mathcal{T}) \to [0,1]$ such that $\sum_{W \in \mathcal{W}_\Phi(\mathcal{T})} Pr(W) = 1$. The **probability of a concept** C, denoted as $Pr(C)$, is defined as $\sum_{W \models C} Pr(W)$. $Pr(D|C)$ is an abbreviation for $Pr(C \sqcap D)/Pr(C)$ if $Pr(C) > 0$ and undefined otherwise.*

In other words, a probabilistic interpretation is a probability distribution over possible worlds. It can be thought of as a function which maps each concept type Ct over Φ to the probability that a randomly chosen named individual is a realization of Ct.

Definition 7 (Satisfaction by Probabilistic Interpretation). *A probabilistic interpretation Pr satisfies (is a model of) a conditional constraint*

---

[2] We omit "w.r.t. $\mathcal{T}$" and simply write "possible world" (or $\mathcal{W}_\Phi$ instead of $\mathcal{W}_\Phi(\mathcal{T})$) when $\mathcal{T}$ is clear from context.

$(D|C)[l, u]$, written as $Pr \models (D|C)[l, u]$, if $Pr(C) = 0$ or $Pr(D|C) \in [l, u]$. Pr satisfies (is a model of) a set of conditional constraints $\mathcal{F}$ if it satisfies all constraints in $\mathcal{F}$. A PTBox $PT = (\mathcal{T}, \mathcal{P})$ is called **satisfiable** if there exists a probabilistic interpretation that satisfies $\mathcal{P}$.

Observe that a conditional constraint is satisfied by all probabilistic interpretations that assign zero probability to the evidence. Given the definition of satisfaction we formulate logical consequence in a standard way.

Definition 8 (Logical Consequence). *A conditional constraint $(D|C)[l, u]$ is a **logical consequence** of a PTBox $(\mathcal{T}, \mathcal{P})$, written as $(\mathcal{T}, \mathcal{P}) \models (D|C)[l, u]$, if all models of $(\mathcal{T}, \mathcal{P})$ also satisfy $(D|C)[l, u]$. $(D|C)[l, u]$ is a **tight logical consequence** of $(\mathcal{T}, \mathcal{P})$, written as $(\mathcal{T}, \mathcal{P}) \models_{tight} (D|C)[l, u]$ if l (resp. u) is the minimum (resp. the maximum) of $Pr(D|C)$ over all models Pr of $(\mathcal{T}, \mathcal{P})$ such that $Pr(C) > 0$.*

In this paper we consider only the monotonic problems of satisfiability (PSAT) and tight logical entailment (TLogEnt) formulated in the standard way. Ignored for space reasons are non-monotonic problems of default consistency and tight lexicographic entailment (TLexEnt) which are reducible to PSAT and TLogEnt. Details could be found in [1,7].

Probabilistic Satisfiability (PSAT): Given a PTBox PT decide if it is satisfiable.

Tight Logical Entailment (TLogEnt : Given a PTBox PT and $\mathcal{SROIQ}$ concepts C and D, compute rational numbers $l, u \in [0, 1]$ such that $PT \models_{tight} (D|C)[l, u]$.

Both problems are reducible to classical reasoning in $\mathcal{SROIQ}$ and Linear Programming (LP), as explained in [1,7].

We conclude the presentation of P-$\mathcal{SROIQ}$ with an example of what *cannot* be naturally expressed. Assume Example 1 is extended with another probabilistic individual, *Rebecca*:

$$\mathcal{P}_{Rebecca} = \{(Woman \sqcap \exists hasDisease.(Cancer \sqcap occursIn.Lobule))[0.9, 1]\}$$

Then one cannot express the uncertainty that *Rebecca* is a relative of *Mary* in a natural way, i.e., by asserting that probability of $relativeOf(Rebecca, Mary) = 0.9$. Informally, each probabilistic individual in P-$\mathcal{SROIQ}$ belongs to its own PABox which may not be "connected" via role assertions. Our translation to FOPL will demonstrate why this is the case.

2.2 First Order Logics of Probability

Next we describe a family of first-order logics of probability (FOPL) as defined and analyzed by Bacchus, Halpern, and Abadi [3,4,8]. To our knowledge, these logics represent the most general formalisms allowing fusion of classical first-order and probabilistic knowledge. They treat probabilities in a natural way on both syntactic and semantic levels, do not make commitments to point-valued, or even quantitative probabilities, and are well suited for representing probabilistic statements of different natures. Finally, the logics can serve as bases for designing systems of default reasoning. These features make them a perfect framework for studying probabilistic extensions of description logics.

The FOPL family of logics consists of three members: FOPL_I, FOPL_II, and FOPL_III. The main distinction lies in the semantics of probabilistic formulas which are interpreted, respectively, via probability distributions over the domain (Type I), possible worlds (Type II), or both (Type III). Type I is a frequentist semantics since it relates probability of a formula to the proportion of domain elements which satisfy it. Type II is a subjective semantics which presumes alternative states (worlds) such that truth of a non-probabilistic formula can be world-dependent. Finally, Type III is an attempt to capture both interpretations of probability in a single semantic theory. It has been noted that choosing an inappropriate semantics, e.g., Type I for beliefs or Type II for statistics, leads to unsatisfactory results [3,4]. Since P-$\mathcal{SROIQ}$'s semantics is clearly not combined, it is important to understand which, Type I, Type II, or neither, it corresponds to. We will show that it is a fragment of FOPL_II, which we describe next (others are described in [4]).

The syntax of FOPL_II is based on a two-sorted first-order vocabulary. The first sort consists of predicates and function names of different arity Φ and a countable set $\mathcal{X}^o$ of *object variables* $x^o, y^o, \ldots$. Intuitively, these variables range over the abstract domain as in standard FOL. The second sort is composed of constants 0 and 1, the binary function names $+$ and $\times$, the binary predicate names $>$ and $=$, and a countable set $\mathcal{X}^f$ of *field variables* $x^f, y^f, \ldots$. Intuitively, these variables range over field elements, i.e. numbers. Object terms are simply the closure of $\mathcal{X}^o$ under function applications. Formulas of FOPL_II and field terms are defined simultaneously in the following way:

- $0, 1$ and all field variables are field terms.
- If P is an n-ary predicate name in Φ and $t_1, \ldots, t_n$ are object terms, then $P(t_1, \ldots, t_n)$ is an atomic formula.
- If t and s are field terms then so are $t + s$ and $t \times s$.
- If t and s are field terms then $t = s$ and $t > s$ are atomic formulas.[3]
- If ϕ is a formula then $w(\phi)$ is a field term.
- Formulas of the form $w(\phi|\psi) = t$ and $w(\phi|\psi) > t$ are abbreviations for $w(\phi \wedge \psi) = t \times w(\psi)$ and $w(\phi \wedge \psi) > t \times w(\psi)$ respectively.
- The set of formulas is closed under conjunction, negation, and universal quantification. Both object and field variables can be bound by the universal quantifier.
- Logical symbols $\vee, \rightarrow$, and $\exists$ are standard abbreviations defined in terms of $\wedge, \neg$, and $\forall$. Field predicates $\leq$ and $\geq$ are defined in a similar way.

Terms of the form $w(\phi)$ are informally interpreted as "the probability of ϕ". Since the Type II logic does not deal with probability distributions over the

---

[3] If object equality is considered part of the language then $t = s$ is also an atomic formula whenever t and s are object terms.

domain, there is no notion of a random choice of domain objects that will satisfy ϕ with some probability. Instead, the logic allows talking about the probability of (typically closed) formulas which is defined with respect to *possible worlds*.

The notion of a "possible world" is made precise in the following way. A Type II probability structure is a tuple $M = (D, S, \pi, \mu)$, where D is the domain, S is a set of possible worlds (or states), π is a world-specific first-order interpretation function (i.e. it may interpret function and predicate names differently in different worlds), and μ is a discrete probability distribution over S. The key difference between the Type I and Type II semantics is that now the probability distributions are taken over the set of worlds S and not over the domain D.

A Type II structure M, a world $s \in S$, and a valuation v (a function which assigns domain objects to variables) collectively associate every object and field term with an element of D and $\mathbb{R}$ respectively, and every formula ϕ with a truth value. As before we write $(M, s, v) \models \phi$ if the tuple (M, s, v) maps ϕ to true. Next we present few important clauses to define the relation $\models$ for FOPL_{II} (see [3] for the complete list):

- $(M, s, v) \models P(x)$ iff $v(x) \in \pi(s)(P)$. Note that each world can be regarded as a first-order structure with its own interpretation function. All worlds are assumed to share the same domain but this restriction can be lifted [4].
- $(M, s, v) \models t_1 = t_2$ iff $[t_1]_{(M,s,v)} = [t_2]_{(M,s,v)}$;
- $(M, s, v) \models \forall x^o \phi$ iff $(M, s, v[x^o/d]) \models \phi$ for all $d \in D$;
- $[w(\phi)]_{(M,v)} = \mu(\{s \in S | (M, s, v) \models \phi\})$. Here the interpretation of field terms of the form $w(\phi)$ does not depend on a world since it is defined as a probability of all worlds in which ϕ is true.

The semantics of FOPL_{II} is generic in the sense that it allows for any (non-empty) set to be used as a set of possible worlds. However, it is common to take S as the set of all interpretations of symbols in Φ over D (see, for example, [9]). In what follows we will refer to such choice of worlds as "natural" and omit π in the structure (since every state s is by itself an interpretation). Also, if all formulas appearing in terms $w(\phi)$ are closed,[4] the components v and s of a Type II structure become fixed so we can write $M \models w(\phi) \leq t$ instead of $(M, v) \models w(\phi) \leq t$.

FOPL_{II} is capable of representing a wide range of belief statements, as the following examples demonstrate:

Example 2 (Belief Formulas in FOPL_{II})

- *Universally quantified beliefs:* $w(\forall x \; bird(x) \rightarrow hasWings(x)) \leq 0.1$. It is believed with probability no more than 10% that all birds have wings.
- *Ground beliefs:* $w(loves(Mary, Fido)) \geq 0.9$. It is believed with probability more than 90% that Mary loves Fido.
- *Conditional and qualitative probabilities:*
  $w(loves(Mary, Fido)|dog(Fido)) \geq w(loves(Mary, Fido)|cat(Fido))$. It is more likely that Mary loves Fido if it is a dog than if it is a cat.

---

[4] Our translation does not require free variables in terms of the form $w(\phi)$, as will be explained in the next section.

3   P-$\mathcal{SROIQ}$ as a Fragment of FOPL$_{\text{II}}$

This section presents a translation between P-$\mathcal{SROIQ}$ and FOPL$_{\text{II}}$. For brevity we will limit our attention to $\mathcal{ALC}$ concepts (calling the resulting logic P-$\mathcal{ALC}$) as the translation can be extended to more expressive DLs in a straightforward, but technically involved way. We will show that the translation preserves entailments so that P-$\mathcal{ALC}$ (and, consequently, P-$\mathcal{SROIQ}$) can be viewed as a fragment of FOPL$_{\text{II}}$.

3.1   Translation of PTBoxes into FOPL

We define the injective function κ to be the mapping of P-$\mathcal{ALC}$ formulas to FOPL$_{\text{II}}$. It is a superset of the standard translation of $\mathcal{ALC}$ axioms into the formulas of FOL [10]. In the Table 1 A, B stand for concept names, R for role names, C, D for concept expressions, r for a fresh constant, $var \in \{x, y\}$; $var' = x$ if $var = y$ and y if $var = x$. We use $\kappa(C)$ instead of $\kappa(C, var)$ when the free variables is substituted by a constant, e.g., r.

Table 1. Translation of P-$\mathcal{ALC}$ formulae into FOPL$_{\text{II}}$

P-$\mathcal{ALC}$	FOPL$_{\text{II}}$
$\kappa(A, var)$	$A(var)$
$\kappa(\neg C, var)$	$\neg(\kappa(C, var))$
$\kappa(R, var, var')$	$R(var, var')$
$\kappa(C \sqcap D, var)$	$\kappa(C, var) \wedge \kappa(D, var)$
$\kappa(C \sqcup D, var)$	$\kappa(C, var) \vee \kappa(D, var)$
$\kappa(\forall R.C, var)$	$\forall(var')(R(var, var') \rightarrow \kappa(C, var'))$
$\kappa(\exists R.C, var)$	$\exists(var')(R(var, var') \wedge \kappa(C, var'))$
$\kappa(a : C)$	$\kappa(C, x)[a/x]$
$\kappa((a, b) : R)$	$R(a, b)$
$\kappa(C \sqsubseteq D, x)$	$\forall(x)(\kappa(C, x) \rightarrow \kappa(D, x))$
$\kappa((D|C)[l, u], x)$	$l \leq w(\kappa(D)(r)|\kappa(C)(r)) \leq u$

For a possible world W we use the notation $\kappa(W)$ to denote the following conjunctive formula with a single free variable: $\bigwedge\{\kappa(C)\}_{C \in W} \wedge \bigwedge\{\kappa(\neg C)\}_{C \notin W}$ (since each C is a concept, each $\kappa(C)$ is a monadic predicate).

This function transforms a P-$\mathcal{ALC}$ PTBox into a FOPL$_{\text{II}}$ theory. The most important thing is that it translates *generic* PTBox constraints into *ground* probabilistic formulas for a fresh constant r, the same for all constraints. This explicates the fact that PTBox constraints are not (sort of) universally quantified statements which naturally apply to all probabilistic individuals but rather statements about a single object (the implications are discussed in the next section). Next we will show that this translation is faithful (i.e., it preserves satisfiability and entailments) and then generalize it to multiple PABoxes.

Faithfulness can be shown by establishing a correspondence between models in P-$\mathcal{ALC}$ and FOPL$_{\mathrm{II}}$. Observe that in contrast to [11] we consider the natural choice of states in Type II probability structures in which they correspond to first-order models of the knowledge base.

Theorem 1. *Let $PT = (\mathcal{T}, \mathcal{P})$ be a PTBox in P-$\mathcal{ALC}$, where $\Sigma_\mathcal{T}$ and Φ stand for the signature of $\mathcal{T}$ and the probabilistic signature of PT respectively, and let $F = \{\kappa(\phi) | \phi \in \mathcal{T} \cup \mathcal{P}\}$ be the translation according to Table 1. Then for every P-$\mathcal{ALC}$ model Pr of PT there exists a corresponding Type II structure $M = (D, S, \mu)$ such that:*

1. *for any axiom ϕ over $\Sigma_\mathcal{T}$, $(M, s) \models \kappa(\phi)$ for each $s \in S$ iff $\mathcal{T} \models \phi$,*
2. *for any Boolean concept expression X over Φ, $[w(\kappa(X)(r))]_M = Pr(X)$.*

and vice versa.

The first claim says that the translation preserves classical entailments over the signature of $\mathcal{T}$. The second claim implies that the translation preserves probabilities of concepts (they correspond to probabilities of ground formulas with the new constant r). The latter means that conditional probabilities are also preserved and, therefore, so are probabilistic entailments over Φ.

Proof. We first prove ($\Rightarrow$). Let $Pr : \mathcal{W}_\Phi \to [0, 1]$ be a model of PT (recall that $\mathcal{W}_\Phi$ is the set of all possible worlds over Φ, i.e., concept types which are realizable w.r.t. $\mathcal{T}$). Pr is a probability distribution so $\mathcal{W}_\Phi$ must be non-empty, which means that $\mathcal{T}$ is satisfiable (a concept type cannot be realizable w.r.t. an unsatisfiable TBox). We first define an extension of $\mathcal{T}$, called $\mathcal{T}'$, as follows: $\mathcal{T}' = \mathcal{T} \cup \bigcup_{W \in \mathcal{W}_\Phi} \{W(o_w) | W \in \mathcal{W}_\Phi\}$, where $W(o_w) = \{\{o_w : C\}_{C \in W} \cup \{\neg o_w : C\}_{C \notin W}\}$ and for each world o_w is a new individual name.[5] $\mathcal{T}'$ has exactly the same set of entailments w.r.t. $\Sigma_\mathcal{T}$ as $\mathcal{T}$ (we simply ruled out all models of $\mathcal{T}$ which do not realize some world[6]). Therefore, it is sufficient to prove the claims w.r.t. $\mathcal{T}'$.

Now we use $\Im$ for the set of all models of $\mathcal{T}'$ over a finite (or countable if generalized to $\mathcal{SROIQ}$) domain and define $M = (D, S, \mu)$ as follows:

- $D = \bigcup \{\Delta^\mathcal{I} | \mathcal{I} \in \Im\}$,
- For every formula Z which is a translation of a concept or a role over Σ_I let $s(\mathcal{I}, \cdot, Z) = \kappa^{-1}(Z)^\mathcal{I}$,
- For every world $W \in \mathcal{W}_\Phi$ let $s(\mathcal{I}, W, r) = d \in \Delta^\mathcal{I}$, where d is chosen such that $d \in C^\mathcal{I}$, if $C \in W$, and $d \in (\neg C)^\mathcal{I}$ otherwise,
- $S = \{s(\mathcal{I}, W, \cdot) | \mathcal{I} \in \Im, W \in \mathcal{W}_\Phi\}$,
- $\mu(\sigma(W)) = Pr(W)$ for each $W \in \mathcal{W}_\Phi$, where:
  $\sigma(W) = \{s(\mathcal{I}, W, \cdot) \in S | s(\mathcal{I}, W, r) \in s(\mathcal{I}, \cdot, \kappa(W))\}$,

---

[5] So in this case $\mathcal{T}'$ is a combination of a TBox and an ABox. In contrast to $\mathcal{SROIQ}$, assertions of the form $C(w)$ are not expressible via TBox axioms in $\mathcal{ALC}$.

[6] This step will be more awkward for logics with nominals, e.g., $\mathcal{SROIQ}$, since they can force an upper bound on the cardinality of the domain.

We take the domain D to be the domain union for all models of $\mathcal{T}'$. Next we define the interpretation function $s(\mathcal{I}, W, \cdot)$ which interprets each predicate Z (i.e., a translation of either a concept or a role) in the same way as $\cdot^{\mathcal{I}}$ interprets $\kappa^{-1}(Z)$. In addition, it interprets the new constant r as some realization of the world W in $\mathcal{I}$ (here we use the fact that all worlds are realized in models of $\mathcal{T}'$). Then we take the set of states as all possible interpretations s over $\mathfrak{I}$ and $\mathcal{W}_\Phi$. On the last step we define the probability distribution μ over S. For that we first define a function σ which maps each world $W = \{C_1, \ldots, C_k\}$ to a subset of states $\sigma(W) \subseteq S$ as follows: $\sigma(W) = \{s \in S | s \models \kappa(W)(r)\}$ ($s \models \kappa(W)(r)$ is equivalent to $s(I, W, r) \in s(I, \cdot, \kappa(W))$. Intuitively, $\sigma(W)$ is a set of first-order interpretations which satisfy $\kappa(C_i)(r)$ iff $C_i \in W$, so there is a one-to-one correspondence between worlds on the P-$\mathcal{ALC}$ side and interpretations of r on the FOPL$_{\mathrm{II}}$ side. Finally, we take $\mu(\sigma(W)) = Pr(W)$ for each possible world.

The first claim follows from the second bullet above because κ encompasses a standard and faithful translation from $\mathcal{ALC}$ to FOL and r is a fresh constant (so can be ignored for entailments over $\Sigma_{\mathcal{T}}$ which does not include it). The second claim is more complicated. First, observe that μ is a probability distribution (i.e. non-negative and countably additive) because it mimics the probability distribution Pr. The probability of a concept expression X over Φ, i.e., $Pr(X)$, is defined as $\sum_{W \models X} Pr(W)$, which is equal to $\sum_{W \models X} \mu(\sigma(W))$ or, using the definition of σ, equals to $\sum_{W \models X} \mu(\{s | s \models \kappa(W)(r)\})$, which is exactly $\mu(\{s | s \models \kappa(X)(r)\})$ or $[w(\kappa(X)(r))]_M$.

We now sketch the proof of ($\Leftarrow$). Let $M = (D, S, \mu)$ be a Type II model of F. We construct an $\mathcal{ALC}$ interpretation $\mathcal{I} = (\Delta^{\mathcal{I}}, \cdot^{\mathcal{I}})$ as follows: $\Delta^{\mathcal{I}} = D$ and $(\kappa^{-1}(\phi))^{\mathcal{I}} = s(\phi)$ for an arbitrarily chosen state $s \in S$ and an arbitrary closed classical formula ϕ (the choise of s does not matter since ϕ, being a closed non-probabilistic formula, is true at every state). $\mathcal{I} \models \phi$ due to the faithfulness of the $\mathcal{ALC}$ to FOL translation, so the first claim holds.

To construct Pr we first construct the set of possible worlds $\mathcal{W}_\Phi$. We take $\Phi = \bigcup_{(l \leq w(\psi(r)|\phi(r)) \leq u) \in F} \{\kappa^{-1}(\psi), \kappa^{-1}(\phi)\}$ as the probabilistic signature of PT. The key is that for every world $W \subseteq \Phi$, it satisfies $\mathcal{T}$ iff $\kappa(W)(r)$ is satisfiable w.r.t. F (since $\Delta^{\mathcal{I}} = D$, W could be a concept type of $s(r)$ for some s which satisfies $\kappa(W)(r)$). Let $\mathcal{W}_\Phi$ be the set of worlds which satisfy $\mathcal{T}$. Now we take $Pr(W) = \mu(\sigma(W))$, where $\sigma(W) = \{s \in S \mid s \models \kappa(W)(r)\}$. The second claim now follows in the same way is in ($\Rightarrow$).

The following result is a straightforward corollary of the above theorem.

Corollary 1. *Let $PT = (\mathcal{T}, \mathcal{P})$ be a PTBox in P-$\mathcal{ALC}$, where $\Sigma_{\mathcal{T}}$ and Φ stand for the signature of $\mathcal{T}$ and the probabilistic signature of PT respectively, and $F = \{\kappa(\phi) | \phi \in \mathcal{T} \cup \mathcal{P}\}$ be the translation according to Table 1. Then the following is true:*

1. *PT is satisfiable iff F is satisfiable,*
2. *$PT \models (D|C)[l, u]$ iff $F \models l \leq w(\kappa(D)(r)|\kappa(C)(r)) \leq u$.*

Next, we extend the translation from PTBoxes to probabilistic KBs.

3.2 Translation of PABoxes into FOPL

One particularly odd characteristic of P-$\mathcal{SROIQ}$ is that PABoxes cannot be combined into a single set of formulas. The separation between PABoxes and the PTBox can partly be justified because they are meant to contain different kinds of probabilistic knowledge, i.e., generic relationships and information about particular individuals respectively. However, the same argument does not hold in the case of separated PABoxes. Their separation has purely technical foundations: PABox constraints are modeled as generic constraints and the information about the individual is present only on a meta-level (as a label of the PABox). Therefore, to extend our translation to PABoxes we either have to translate them into a corresponding disjoint set of FOPL$_\mathrm{II}$ theories (with similar meta-labels) or make special arrangements to faithfully translate them into a combined FOPL$_\mathrm{II}$ theory. We opt for the latter because it will let us get rid of any meta-logical aspects and view a P-$\mathcal{SROIQ}$ ontology as a single, standard theory in FOPL$_\mathrm{II}$.

Since PABoxes in P-$\mathcal{SROIQ}$ are isolated from each other, the translation should preserve that isolation. The most obvious way to prevent any interaction between sets of formulas in a single logical theory is to make their signatures disjoint. That is, PABoxes can be translated into FOPL$_\mathrm{II}$ sub-theories with disjoint signatures. However, the translation should not only respect disjointness of PABoxes but also preserve their interaction with PTBox and the classical part of the ontology. We give an example to illustrate the issue.

Example 3. Consider the PTBox: $PT = \{\emptyset, \{(FlyingObject|Bird)[0.9,1],$ $(FlyingObject|\neg Bird)[0,0.5]\}$ and two PABoxes: $\mathcal{P}_{Tweety} = \{(Bird|\top)[1,1]\}$, $\mathcal{P}_{Sam} = \{(\neg Bird|\top)[1,1]\}$. If these sets of axioms are translated and combined into a single FOPL$_\mathrm{II}$ theory then it will contain a conflicting pair of formulas $\{w(Bird(r)) = 0.9, w(\neg Bird(r)) = 1\} \subseteq F$.

This inconsistency can be avoided by introducing fresh first-order predicates for every PABox: $\{w(Bird_{Tweety}(r)) = 1, w(\neg Bird_{Sam}(r)) = 1\}$. However, this would break any connection between PTBox and PABox axioms, for example, prevent the expected entailments $w(FlyingObject_{Tweety}(r)) \geq 0.9$ and $w(FlyingObject_{Sam}(r)) \leq 0.5$.

Another way to faithfully extend the translation to PABoxes is to introduce fresh concept names to *relativize* each TBox and PTBox axiom for every probabilistic individual and thus avoid inconsistencies. More formally, the transformation consists of the following steps:

- Firstly, we transform a P-$\mathcal{ALC}$ ontology $PO = (\mathcal{T}, \mathcal{P}, \{\mathcal{P}_o\}_{o \in N_{PI}})$ into a set of PTBoxes $\{(\mathcal{T}, \mathcal{P})\} \cup \{(\mathcal{T}, \mathcal{P} \cup \mathcal{P}_o)\}_{o \in N_{PI}}$. Informally, we create a copy PTBox for every probabilistic individual (PT_o) and make them isolated from each other. Now, instead of one PTBox and a set of PABoxes we have just a set of PTBoxes. This step preserves probabilistic entailments in the following sense: $PO \models (B|A)[l,u]$ iff $(\mathcal{T}, \mathcal{P}) \models (B|A)[l,u]$ and $PO \models (B|A)[l,u]$ for o iff $PT_o \models (B|A)[l,u]$ (classical entailments are trivially preserved).

- Secondly, we transform every PTBox PT_o into PT_o' by renaming every concept name C into C_o in all TBox axioms and conditional constraints. It is easy to see that $PT_o \models C \sqsubseteq D$ iff $PT_o' \models C_o \sqsubseteq D_o$ and $PT_o \models (B|A)[l,u]$ iff $PT_o' \models (B_o|A_o)[l,u]$. Intuitively, we have created a fresh copy of each PTBox to guard against possible conflicts between PABox constraints for different probabilistic individuals. Signatures of PT_o' are pairwise disjoint and denoted as Σ_o.
- Next, we union all PT_o' with disjoint signatures (including the original $PT = (\mathcal{T}, \mathcal{P})$) into a single unified PTBox $PT_U = \bigcup_{o \in I_p} PT_o \cup PT$ with signature $\Sigma_U = \bigcup_{o \in I_p} \Sigma_o \cup \Sigma$.
- Finally we can apply the previously presented faithful translation to PT_U and obtain a single FOPL_{II} theory which corresponds to the original P-$\mathcal{ALC}$ ontology.

A necessary condition for faithfulness of this transformation is that the original isolation of PABoxes is preserved by creating fresh copies of PTBoxes. In particular, this means that the unified PTBox cannot entail any subsumption relation between concept expressions C_{o_1} and C_{o_2} defined over disjoint signatures except of the case when one of them is either $\top$ or $\bot$. If this is false, for example, if $PT_U \models C_{o_1} \sqsubseteq C_{o_2}$ then the following PABox constraints represented as $(C_{o_1}|\top)[1,1]$ and $(C_{o_2}|\top)[0,0]$ will be contradictory in PT_U (but they were consistent in the original P-$\mathcal{ALC}$ because they belonged to different PABoxes isolated from each other). This condition is formalized in the following lemma:

Lemma 1. *Let $\mathcal{T}_1$ and $\mathcal{T}_2$ be copies of a satisfiable $\mathcal{ALC}$ ontology $\mathcal{T}$ with disjoint signatures Σ_1 and Σ_2, and $\mathcal{T}_U$ be the union of $\mathcal{T}_1$ and $\mathcal{T}_2$. Then for any concept expressions C_1, C_2 over Σ_1 and Σ_2 respectively such that $\mathcal{T}_1 \nvDash C_1 \sqsubseteq \bot$ and $\mathcal{T}_1 \nvDash \top \sqsubseteq C_2$, $\mathcal{T}_U \nvDash C_1 \sqsubseteq C_2$.*

Proof. Let $\mathcal{I}_1 = (\Delta^{\mathcal{I}_1}, \cdot^{\mathcal{I}_1})$ and $\mathcal{I}_2 = (\Delta^{\mathcal{I}_2}, \cdot^{\mathcal{I}_2})$ be models of $\mathcal{T}_1$ and $\mathcal{T}_2$ respectively, $x \in C_1^{\mathcal{I}_1}, y \in \Delta^{\mathcal{I}_2} \setminus C_2^{\mathcal{I}_2}$. We can assume that $\Delta^{\mathcal{I}_1}$ and $\Delta^{\mathcal{I}_2}$ are countably infinite because $\mathcal{ALC}$ models are closed under disjoint union. Next, we choose two linear orderings $p_i : \Delta^{\mathcal{I}_i} \to \mathbb{N}$ ($i \in \{1,2\}$) such that $p_1(x) = p_2(y) = 1$ and pick a new countable domain $\Delta^{\mathcal{I}_U} = \{d_{U_1}, d_{U_2}, \dots\}$. Finally, we construct an interpretation function $\cdot^{\mathcal{I}_U}$ such that for any concept name C_i (resp. role name R_i), $C_i^{\mathcal{I}_U} = \{d_{U_j} | p_i^{-1}(j) \in C_i^{\mathcal{I}_i}\}$ (resp. $R_i^{\mathcal{I}_U} = \{(d_{U_j}, d_{U_k}) | (p_i^{-1}(j), p_i^{-1}(k)) \in R_i^{\mathcal{I}_i}\}$).

Informally, we order both domains such that x and y are in the first position each. Then the domains are *aligned* such that elements at the same position, e.g., x and y, coincide. This induces a model $\mathcal{I}_U = (\Delta^{\mathcal{I}_U}, \cdot^{\mathcal{I}_U})$ of $\mathcal{T}_U$ which interpretation function agrees with $\mathcal{I}_i$ on all concepts and roles from Σ_i and which does *not* satisfy $C_1 \sqsubseteq C_2$.

Now we can obtain the main result:

Theorem 2. *Let $PO = (\mathcal{T}, \mathcal{P}, \{\mathcal{P}_o\}_{o \in N_{PI}})$ be a P-$\mathcal{ALC}$ ontology, where $\Sigma_{\mathcal{T}}$ and Φ_o stand for the signature of $\mathcal{T}$ and the probabilistic signature of $(\mathcal{T}, \mathcal{P} \cup \mathcal{P}_o)$ for $o \in N_{PI}$ respectively. Let F be the FOPL_{II} theory obtained by combining the*

PABoxes and translating the resulting PTBox into $\mathrm{FOPL_{II}}$. *If all PTBoxes of the form* $PT_o = (\mathcal{T}, \mathcal{P} \cup \mathcal{P}_o)$ *are satisfiable, then for every P-$\mathcal{ALC}$ model* Pr_o *of every* PT_o *there exists a corresponding Type II structure* $M = (D, S, \mu)$ *such that:*

1. *for any axiom* ϕ *over* $\Sigma_{\mathcal{T}}$, $(M, s) \models \kappa(\phi)$ *for each* $s \in S$ *iff* $\mathcal{T} \models \phi$,
2. *for any Boolean concept expression* X *over* Φ_o, $[w(\kappa(X)(r))]_M = Pr(X)$.

and vice versa. Furthermore, $\mathcal{F}$ *is unsatisfiable iff* $(\mathcal{T}, \mathcal{P} \cup \mathcal{P}_o)$ *is unsatisfiable for some* $o \in N_{PI}$.

Proof. Due to Theorem 1 it suffices to show that the steps 1-3 of the transformation preserve probabilistic models. This can be done by establishing a correspondence between possible worlds of each PT_o and PT_U. Since there are no subsumptions between concept expressions over signatures of different PTBoxes (see Lemma 1), each possible world W_o in PT_o corresponds to a finite set of possible worlds of PT_U defined as: $\sigma(W_o) = \{W_U \mid C_{i_o} \in W_U \text{ iff } C_i \in W_o\}$ (each C_{i_o} is a new concept name for C_i introduced on step 2). Then, a probability distribution over all possible worlds in PT_U can be defined as $Pr_U(W_U) = Pr_o(W_o)/|\sigma(W_o)|$. It follows that for any concept C over Σ_o, $Pr_o(C)$ is equal to $Pr_U(C_o)$ where C_o is the correspondingly renamed concept. Therefore, $Pr_U \models (B_o|A_o)[l, u]$ if $Pr_o \models (B|A)[l, u]$. The reverse direction can be proved along the same lines (i.e., $Pr_o(W_o)$ can be defined as $\sum_{W_U \in \sigma(W_o)} Pr_U(W_U)$).

It is worth stressing that if PABox for some probabilistic individual o contradicts the PTBox $(\mathcal{T}, \mathcal{P})$ then the entire $\mathrm{FOPL_{II}}$ theory is unsatisfiable. Therefore, for practical considerations it might be important to work with P-$\mathcal{SROIQ}$ ontologies as with stratified theories. Such separation between general knowledge and knowledge about particular individuals had been known before P-$\mathcal{SROIQ}$, for example, it was used in the default reasoning system developed by Geffner and Pearl [12].

4   Properties and Limitations of P-$\mathcal{SROIQ}$

The translation highlights two major properties of P-$\mathcal{SROIQ}$:

PI. P-$\mathcal{SROIQ}$ has a subjective, interpretation-based (Type II) semantics.
PII. Only a single constant is required to translate all probabilistic knowledge in a P-$\mathcal{SROIQ}$ ontology into a $\mathrm{FOPL_{II}}$ theory.[7]

PI implies that any claims that P-$\mathcal{SROIQ}$ handles different kinds of probabilities, especially statistics, require a careful examination. PII, which is the basis of P-$\mathcal{SROIQ}$'s direct inference mechanism, explains issues with handling degrees of belief since, intuitively, a single constant cannot be sufficient for modeling probability distributions over relational structures. We will discuss these issues in 4.2 and 4.3 but before we briefly discuss why some other, perhaps more natural-looking ways of translating P-$\mathcal{SROIQ}$ into $\mathrm{FOPL_{II}}$ are incorrect.

---

[7] Here we mean a "probabilistic" constant since all nominals occurring in the *classical* part of the ontology will be translated into corresponding constants in $\mathrm{FOPL_{II}}$.

4.1   Interpretation of Probabilistic Statements

According to the translation, all probabilistic statements in P-$\mathcal{SROIQ}$ express *degrees of belief* about a single, yet unnamed, individual (denoted as r). This is not an easily expected outcome because the variable-free syntax may give a misleading impression that PTBox constraints correspond to universally quantified formulas in FOPL$_{II}$, similarly to how TBox axioms in $\mathcal{SROIQ}$ correspond to universally quantified implications formulas in FOL. One may wonder whether a more natural translation is possible. We consider two such candidate translations: probabilistic implications and universally quantified conditional formulas.

An interpretation of conditional constraints $(D|C)[l, u]$ as formulas of the form $l \leq w(\forall x[c(x) \rightarrow d(x)]) \leq u$ lets us view them as probabilistic generalizations of TBox axioms $C \sqsubseteq D$ (which are translated into $\forall x[c(x) \rightarrow d(x)]$). It is easy to see how their semantics is different from P-$\mathcal{SROIQ}$'s. Such formulas are unconditional so, for example, the pair of formulas $w(\forall x[c_1(x) \rightarrow d(x)]) \geq 0.9$ and $w(\forall x[c_1(x) \wedge c_2(x) \rightarrow d(x)]) \leq 0.8$ are contradictory. On the other hand, the pair of conditional constraints in P-$\mathcal{SROIQ}$ $(D|C_1)[0.9, 1]$ and $(D|C_1 \sqcap C_2)[0, 0.8]$ is perfectly satisfiable.[8]

The translation into universally quantified *conditional* formulas, i.e., formulas of the form $\forall x[l \leq w(d(x)|c(x)) \leq u]$ has more subtle issues. The idea of using them for capturing statistical assertions is originally due to Cheeseman [14]. It has been criticized by multiple authors (see esp. [4,15,3]) as it leads to intuitively unreasonable conflicts between statistics and beliefs. We will return to this point in the next section while here we can show that such translation is unfaithful in presence of named constants (i.e., nominals in $\mathcal{SROIQ}$) or classical ABoxes. For example, the PTBox $(\{a : \neg A\}, \{(A|\top)[1, 1]\})$ is satisfiable in P-$\mathcal{SROIQ}$ although the corresponding FOPL$_{II}$ theory $\{\neg A(a), \forall x(w(A(x)) = 1)\}$ is not. The problem is that this translation disregards the separation between classical and probabilistic individuals in P-$\mathcal{SROIQ}$.

In fact, the translation into quantified statements *does* work but requires a somewhat non-standard quantifier. It has to make bound variables act as *random designators*. This is precisely what we achieve by using the fresh constant r.

4.2   Representation of Statistics

The first question that has to be raised is whether P-$\mathcal{SROIQ}$ can be used to represent statistical knowledge given its subjective, interpretation based semantics. Here we prefer to distinguish between *practical* and *philosophical* difficulties. The former are the situations when some important statistical knowledge cannot be adequately represented, for example, all possible representations lead to statistically unsound conclusions. The latter are the situations which cause conceptual difficulties but do not lead to any erroneous entailments.

---

[8] In other words, conditional formulas do not constrain future beliefs after conditioning on new evidence. This is related to the lack of probabilistic inheritance from the cautious point of view, see the discussion of entailment strength in [13].

The main *philosophical* difficulty in P-$\mathcal{SROIQ}$ is that it enforces the separation between general statements, which are meant to capture statistics, and statements meant to represent beliefs about specific individuals. This is a well-known argument against representing both statistics and beliefs in FOPL$_{\mathrm{II}}$ (see [3,4]). Consider the following classical example:

$$PO = (\{Penguin \sqsubseteq Bird\},$$
$$\{(FlyingObject|Bird)[0.9, 1], (FlyingObject|Penguin)[0, 0.1]\},$$
$$(\{(Penguin|\top)[1, 1]\}_{Tweety}))$$

If all axioms above were combined in a single theory it would clearly be unsatisfiable. The TBox and PTBox axioms place restrictions on probability of *Penguin* (informally, penguins must be a "small" subclass of birds) which is violated by the PABox axiom. This means that an agent cannot *simultaneously* believe in the existence of a single flying penguin and the statistical knowledge that most penguins do not fly, which is unreasonable. Since there is no semantic separation (i.e., through different probability distributions as in FOPL$_{\mathrm{III}}$) between different kinds of axioms, they have to be separated syntactically. In addition, P-$\mathcal{SROIQ}$ has to include a special mechanism for combining these axioms for reasoning about individuals which has to be non-monotonic.

Interestingly, P-$\mathcal{SROIQ}$, as it stands, seems to avoid *practical* issues with handling statistics, but mostly because its language is quite limited rather than because its semantics is appropriate. The only probabilistic axioms provided by P-$\mathcal{SROIQ}$, conditional constraints of the form $(D|C)[l, u]$, express that "the probability that a random instance of C is an instance of D is in $[l, u]$". It does not allow specifying *how* that random instance was drawn as well as placing any other restrictions on probability distributions. It is easy to show that possible extensions in these directions could easily reveal the inadequacy of P-$\mathcal{SROIQ}$'s semantics for handling statistics.

Consider what happens if one wants to extend P-$\mathcal{SROIQ}$ to allow restricting probability functions to uniform distributions. This is useful if conditional constraints are to be interpreted as proportions (i.e., according to the frequentist interpretation of probability). Now consider the following PTBox where *marriedTo* is a functional role:

$$(\{Person \sqsubseteq Man \sqcup Woman, Man \sqcap Woman \sqsubseteq \bot,$$
$$Man \sqsubseteq \exists marriedTo.Woman\},$$
$$\{(Person|\top)[0.9, 0.9], (Man|Person)[0.5, 0.5]\})$$

This PTBox attempts to model a domain 90% of which consists of people. Every person is either a man or a woman. Furthermore, 50% of people are men and every man is *functionally* related to at least one woman, so the other half of people must be women. Due to the standard P-$\mathcal{SROIQ}$ semantics the PTBox will entail $[0, 1]$ as the tightest probability bounds for the query $(Woman|Person)[?, ?]$. This happens because the relationship between *extensions* of *Man* and *Woman* is ignored by the semantics of P-$\mathcal{SROIQ}$, i.e., it does not restrict the set of possible worlds in any way. Restricting probability functions to uniform distributions over possible worlds without changing the notion of possible world also does not achieve the goal (in this example there are fewer than 10 possible worlds, so constraints like $(Person|\top)[0.9, 0.9]$ would be unsatisfiable by themselves). Such considerations lead us to the conclusion that P-$\mathcal{SROIQ}$ is not well suited for representing *first-order* statistical statements.

4.3   Representation of Beliefs

Perhaps surprisingly the properties of P-$\mathcal{SROIQ}$, in particular, PII, lead to more practical difficulties with handling degrees of belief rather than representation of statistics. The prime issues are the separation between classical and probabilistic individuals and the lack of relational structures support.

The separation between different kinds of individuals precludes any combination between classical and probabilistic knowledge for the same individual. It is not possible, for example, to express that Mary is an instance of concept $Woman$ and has 90% chance of having BRCA1 gene mutation. Of course, it is possible to express that the *probability* that Mary is a woman is 1 but this is not always a satisfactory replacement for ABox statements. First, one may want to specify probabilistic facts about individuals already present in the ABox. Second, perhaps more importantly, specifying ABox axioms as PABox axioms does not lead to entailments which could be important. For example, if Mary is a woman and developed breast cancer, then her daughter, say, Jane, would be entailed as an instance of concept $WomanWithFamilyHistoryOfBRCA$. If Mary has to be a probabilistic individual then so does Jane, and the modeler will face the problem of representing their relationship. This is, in fact, the second issue with P-$\mathcal{SROIQ}$.

P-$\mathcal{SROIQ}$ does not support probabilistic relational structures in the sense that one cannot specify that one probabilistic individual has a certain probability of being related to another probabilistic individual. For example, if both Mary and Jane are probabilistic individuals one cannot specify that Mary is a mother of Jane with a probability of 1 (obviously, such a statement is most reasonably represented as a classical ABox axioms but this is also not possible due to the separation discussed above).[9] The reason, which is highlighted by PII, is that PABox statements do *not* correspond to ground probabilistic formulas in FOPL$_{\mathrm{II}}$. The information about individuals is present only on a meta-level, as labels of the corresponding PABoxes. As a consequence, knowledge about distinct probabilistic individuals has to be separated from each other, for example, by means of isolated PABoxes in P-$\mathcal{SROIQ}$ or disjoint signatures in FOPL$_{\mathrm{II}}$.

The second issue appears to be more difficult to overcome than the first. The separation between classical and probabilistic individuals can be eliminated by incorporating classical individuals in the description of possible worlds, for

---

[9] Note that the logic can represent probabilistic roles between a classical and a probabilistic individuals, e.g., as a PABox axiom $(\exists motherOf.\{Jane\}|\top)[1, 1]$ for $Mary$, but in this case no probabilistic facts can be specified for Jane.

example, by including nominal concept expressions in the set of basic concepts (the probabilistic signature). Relational structures, on the other hand, cannot be supported until PABox constraints are interpreted as PTBox constraints rather than ground statements bearing information about particular individuals on the *logical level*. However, that would require major semantic changes, at least a new direct inference mechanism to preserve interaction between PTBox and PABox knowledge (a discussion of such possibility can be found in [16]).

5   Summary

We presented a faithful translation of knowledge bases in P-$\mathcal{SROIQ}$ into theories in first-order probabilistic logic with Type II semantics. The translation places no restriction on expressiveness (e.g., use of nominals), uses only standard quantifiers and, most importantly, illuminates the probabilistic propositionality of P-$\mathcal{SROIQ}$ by using only a single probabilistic constant. That "propositionality" is the main culprit of the important limitations of P-$\mathcal{SROIQ}$, namely, the lack of support of probabilistic relational structures.

P-$\mathcal{SROIQ}$ can be seen as an *approximation* of FOPL$_{III}$ which trades separate probability distributions for statistical and belief statements for a practically implementable direct inference mechanism. There is nothing particularly wrong with such a design decision *per se* but it has to be understood by modelers. Our translation into FOPL$_{II}$ is an attempt to enhance that understanding analogously to how classical DLs are understood as fragments of classical FOL.

In conclusion, we present two examples that illustrate how viewing probabilistic DLs as fragments of FOPL helps their understanding and, on the contrary, how lack of such understanding can lead to errors. An example of the latter is the design of P-$\mathcal{SHOQ}$(D), a predecessor of P-$\mathcal{SROIQ}$ [17], which has domain-based (Type I) semantics. In that logic PABox axioms are represented using nominals, for example, $(C|\{a\})[0.5, 1]$ is supposed to model that a is an instance of C with probability at least 0.5. However, as proved by Halpern, closed first-order formulas can only have probability 0 or 1 in any Type I probabilistic model (see Lemma 2.3 in [4]) so the representation is unsatisfactory. It is easy to see that the probability of $(C|\{a\})$, which is equivalent to $\frac{Pr(C\sqcap\{a\})}{Pr(\{a\})}$, is 0 if $a^{\mathcal{I}} \notin C^{\mathcal{I}}$ or 1 if $a^{\mathcal{I}} \in C^{\mathcal{I}}$, if Pr is a probability distribution over $\Delta^{\mathcal{I}}$.

A positive example is the recent work of Lutz and Schröder [18]. They designed and presented a probabilistic DL, called Prob-$\mathcal{ALC}$, which supports probabilistic concepts of the form $P_{\geq\alpha}(C)$, which are interpreted as "the set of all domain objects which are instances of C with $\geq 90\%$ probability". The logic is designed as a fragment of FOPL$_{II}$ so a modeler can immediately realize that their ability to model statistical statements might be limited. And it is indeed the case, for example, if one tries to use axioms like $Bird \sqsubseteq P_{\geq 0.9}(FlyingObject)$ to capture the statistical statement that $\geq 90\%$ of birds fly, they will face the same difficulties as with using universally quantified probabilistic formulas in FOPL$_{II}$. In particular, such an axiom will be in conflict with statements asserting existence of a *specific* non-flying bird, i.e., $\{tweety : Bird, tweety : \neg FlyingObject\}$. On the other hand, and in contrast to P-$\mathcal{SROIQ}$, the logic fully supports relational structures in probabilistic ABoxes, something that could be expected from a FOPL$_{II}$ frament.

References

1. Lukasiewicz, T.: Expressive probabilistic description logics. Artificial Intelligence 172(6-7), 852–883 (2008)
2. Klinov, P., Parsia, B.: A Hybrid Method for Probabilistic Satisfiability. In: Bjørner, N., Sofronie-Stokkermans, V. (eds.) CADE 2011. LNCS, vol. 6803, pp. 354–368. Springer, Heidelberg (2011)
3. Bacchus, F.: Representing and reasoning with probabilistic knowledge. MIT Press (1990)
4. Halpern, J.Y.: An analysis of first-order logics of probability. Artificial Intelligence 46, 311–350 (1990)
5. Horrocks, I., Kutz, O., Sattler, U.: The even more irresistible $\mathcal{SROIQ}$. In: Knowledge Representation and Reasoning, pp. 57–67 (2006)
6. Lutz, C., Sattler, U., Tendera, L.: The complexity of finite model reasoning in description logics. Information and Computation 199(1-2), 132–171 (2005)
7. Klinov, P.: Practical Reasoning in Probabilistic Description Logic. PhD thesis, The University of Manchester (2011)
8. Abadi, M., Halpern, J.Y.: Decidability and expressiveness for first-order logics of probability. Information and Computation 112(1), 1–36 (1994)
9. Koller, D., Halpern, J.Y.: Irrelevance and conditioning in first-order probabilistic logic. In: Advances in Artificial Intelligence Conference, pp. 569–576 (1996)
10. Borgida, A.: On the relationship between description logic and predicate logic. In: International Conference on Information and Knowledge Management, pp. 219–225 (1994)
11. Klinov, P., Parsia, B., Sattler, U.: On correspondences between probabilistic first-order and description logics. In: International Workshop on Description Logic (2009)
12. Geffner, H., Pearl, J.: A framework for reasoning with defaults. In: Kyburg, H., Loui, R., Carlson, G. (eds.) Knowledge Representation and Defeasible Reasoning, pp. 69–87. Kluwer Academic Publishers (1990)
13. Lukasiewicz, T.: Nonmonotonic probabilistic logics under variable-strength inheritance with overriding: Complexity, algorithms, and implementation. International Journal of Approximate Reasoning 44(3), 301–321 (2007)
14. Cheeseman, P.: An inquiry into computer understanding. Computational Intelligence 4, 58–66 (1988)
15. Bacchus, F.: Lp, a logic for representing and reasoning with statistical knowledge. Computational Intelligence 6, 209–231 (1990)
16. Klinov, P., Parsia, B.: Relationships between probabilistic description and first-order logics. In: International Workshop on Uncertainty in Description Logics (2010)
17. Giugno, R., Lukasiewicz, T.: P-$\mathcal{SHOQ}$(**D**): A Probabilistic Extension of $\mathcal{SHOQ}$(**D**) for Probabilistic Ontologies in the Semantic Web. In: Flesca, S., Greco, S., Leone, N., Ianni, G. (eds.) JELIA 2002. LNCS (LNAI), vol. 2424, pp. 86–97. Springer, Heidelberg (2002)
18. Lutz, C., Schröder, L.: Probabilistic description logics for subjective uncertainty. In: International Conference on the Principles of Knowledge Representation and Reasoning (2010)

Pronto: A Practical Probabilistic Description Logic Reasoner*

Pavel Klinov[1] and Bijan Parsia[2]

[1] Institute of Artificial Intelligence
University of Ulm, Germany
pavel.klinov@uni-ulm.de
[2] School of Computer Science
University of Manchester, United Kingdom
bparsia@cs.man.ac.uk

Abstract. This paper presents Pronto—the first probabilistic Description Logic (DL) reasoner capable of processing knowledge bases containing about a thousand of probabilistic axioms. We describe in detail the novel probabilistic satisfiability (PSAT) algorithm which lies at the heart of Pronto's architecture. Its key difference from previously developed (propositional) PSAT algorithms is its interaction with the underlying DL reasoner which, first, enables applying well-known linear programming techniques to non-propositional PSAT and, second, is crucial to scaling with respect the amount of classical (non-probabilistic) knowledge. The latter is the key feature for dealing with probabilistic extensions of existing large ontologies. Finally we present the layered architecture of Pronto and demonstrate the experimental evaluation results on randomly generated instances of non-propositional PSAT.

1 Introduction

There are many proposed formalisms for combining Description Logics (DLs) with various sorts of uncertainty, although, to our knowledge, none have been used in a production environment. This is mostly due to two reasons: first, there is comparatively little knowledge about how to use these formalisms effectively (or even which are best suited for what purposes) and, second, there is a severe lack of tools, in particular, there have been no sufficiently effective reasoners.

This paper describes our work on the second problem. We present Pronto — the reasoner for the probabilistic extension of DL $\mathcal{SROIQ}$ (named P-$\mathcal{SROIQ}$) [1]. This logic can be viewed either as a generalization of the Nilsson's propositional probabilistic logic [2] or as a fragment of first-order probabilistic logic of Halpern and Bacchus [3] [4] (with certain non-monotonic extensions). One attractive feature of these probabilistic logics is that they allow modelers to declaratively describe their uncertain knowledge without fully specifying any probability distribution (in contrast to, for example, Bayesian networks). In other

---

* This work has been carried out when the first author was a doctoral student at the University of Manchester, UK.

F. Bobillo et al. (Eds.): URSW 2008-2010/UniDL 2010, LNAI 7123, pp. 59–79, 2013.
© Springer-Verlag Berlin Heidelberg 2013

words, they allow modelers to specify as little probabilistic knowledge as there is available thus avoiding the risk of overspecification (but at the expense of supporting weaker probabilistic entailments, the problem known as "inferential vacuity," see [5]). These logics are also proper generalizations of their classical counterparts which, in the case of P-$\mathcal{SROIQ}$, means that modelers can take an existing $\mathcal{SROIQ}$ ontology and add probabilistic axioms to capture uncertain, such as statistical, relationships.

In spite of their attractive features Nilsson-style logics have been criticized, partly for the intractability of probabilistic inference. Reasoning procedures are typically implemented via reduction to Linear Programming but it is well-known that corresponding linear programs are exponentially large so the scalability is very limited. Over the last two decades there have been several attempts to overcome that issue in the propositional case which led to some promising results, such as solving the probabilistic satisfiability problem (PSAT) for 800–1000 formulas [6]. It has been unclear whether the methods used to solve large propositional PSATs can be directly applied to PSATs in probabilistic DLs.

To the best of our knowledge, Pronto is the first reasoner for a Nilsson-style probabilistic DL whose PSAT algorithm is as scalable (and often faster) than those implemented by propositional PSAT solvers. In particular, it can solve propositional instances of PSAT at the same scale but also can effectively deal with KBs containing non-propositional classical knowledge, such as large OWL ontologies. We present experimental results which demonstrate the level of scalability when dealing with probabilistic extensions of real-life OWL ontologies. In addition, Pronto implements all the standard reasoning services for P-$\mathcal{SROIQ}$ as well as useful extra services, in particular, finding *all* minimal unsatisfiable fragments of a KB which is crucial for analyzing large bodies of conflicting probabilistic knowledge [7,8].

2   Background

P-$\mathcal{SROIQ}$ [1] is a probabilistic generalization of the DL $\mathcal{SROIQ}$ [9]. It provides means for expressing probabilistic relationships between arbitrary $\mathcal{SROIQ}$ concepts and a certain class of probabilistic relationships between classes and individuals. Any $\mathcal{SROIQ}$, and thus OWL 2 DL, ontology can be used as a basis for a P-$\mathcal{SROIQ}$ ontology, which facilitates transition from classical to probabilistic ontologies.

2.1   Syntax

The main additional syntactic feature in P-$\mathcal{SROIQ}$ is the conditional constraint.

Definition 1 (Conditional Constraint). *A conditional constraint is an expression of the form $(D|C)[l, u]$, where C and D are concept expressions in $\mathcal{SRIQ}$ (i.e., $\mathcal{SROIQ}$ without nominals) called **evidence** and **conclusion**, respectively, and $[l, u] \subseteq [0, 1]$ is a closed real-valued interval. In the case where C is $\top$ the constraint is called **unconditional**.*

Ontologies in P-$\mathcal{SROIQ}$ are separated into a classical and a probabilistic part. It is assumed that the set of individual names N_I is partitioned onto two sets: classical individuals N_{CI} and probabilistic individuals N_{PI}.

Definition 2 (PTBox, PABox, and Probabilistic Knowledge Base). *A* ***probabilistic TBox*** *(PTBox) is a pair $PT = (\mathcal{T}, \mathcal{P})$ where $\mathcal{T}$ is a classical $\mathcal{SROIQ}$ TBox and $\mathcal{P}$ is a finite set of conditional constraints. A* ***probabilistic ABox*** *(PABox) is a finite set of conditional constraints associated with a probabilistic individual $o_p \in N_{PI}$. A* ***probabilistic knowledge base*** *(or a probabilistic ontology) is a triple $PO = (\mathcal{T}, \mathcal{P}, \{\mathcal{P}_{o_p}\}_{o_p \in N_{PI}})$, where the first two components define a PTBox and the last is a set of PABoxes.*

Informally, a PTBox constraint $(D|C)[l, u]$ expresses a conditional statement of the form "if a *randomly* chosen individual is an instance of C, the probability of it being an instance of D is in $[l, u]$". A PABox constraint, which we write as $(D|C)_o[l, u]$ where o is a probabilistic individual, states that "if a *specific* individual (that is, o) is an instance of C, the probability of it being an instance of D is in $[l, u]$". That distinction is important for default reasoning in P-$\mathcal{SROIQ}$.

Definition 3 (Probabilistic Signature). *Given a probabilistic knowledge base PO let $\mathcal{CE}(PO)$ be the set of all concept expressions that appear either as evidence or conclusion in some conditional concept (in the PTBox or in a PABox). Then* ***probabilistic signature*** *of PO, denoted as $\Phi(PO)$, is the smallest set of concept expressions such that i) no expression is a union, intersection, or complement of other expressions and ii) its closure under union, intersection, and complementation is a superset of $\mathcal{CE}(PO)$.*

$\Phi(PO)$ (or simply Φ when the ontology is clear from context) is a finite set because $\mathcal{CE}(PO)$ is finite. It can be computed by starting from $\Phi = \mathcal{CE}(PO)$ and exhaustively applying the following rules, where C and D are concept expressions:

- If $C \sqcup D \in \Phi$, then $\Phi \leftarrow (\Phi \cup \{C, D\}) \setminus \{C \sqcup D\}$;
- If $C \sqcap D \in \Phi$, then $\Phi \leftarrow (\Phi \cup \{C, D\}) \setminus \{C \sqcap D\}$;
- If $\neg C \in \Phi$, then $\Phi \leftarrow (\Phi \cup \{C\}) \setminus \{\neg C\}$;

The process of applying the rules will terminate since every rule reduces the syntactic length of expressions in Φ.

2.2 Semantics

The semantics of P-$\mathcal{SROIQ}$ is based on the notion of a *world*.

Definition 4 (World). *Given the probabilistic signature Φ a* ***world*** *W is a subset of Φ. A concept $C \in \Phi$ occurs* ***positively***, *or is* ***satisfied***, *in a world W, written as $W \models C$, if $C \in W$, otherwise it is said to occur* ***negatively***. *The satisfaction relation is extended recursively to Boolean expressions over Φ in a standard way (e.g., $W \models A \sqcup B$ if $W \models A$ or $W \models B$).*

Finally, we extend the definition of satisfaction in a world to $\mathcal{SROIQ}$ TBoxes.

Definition 5 (Possible World, Index Set). *A world W is a **possible world** with respect to a $\mathcal{SROIQ}$ TBox $\mathcal{T}$, written as $W \models \mathcal{T}$, if $\mathcal{T} \cup \{\{o\} \sqsubseteq C | C \in W\} \cup \{\{o\} \sqsubseteq \neg C | C \notin W, C \in \Phi\}$ is satisfiable, where o is an individual name not occurring in $\mathcal{T}$. The set of all possible worlds over Φ with respect to $\mathcal{T}$, also called the **index set**, is denoted as $\mathcal{W}_\Phi(\mathcal{T})$.[1]*

Possible worlds correspond to what is commonly known as realizable concept types in the DL literature [10]. Each world W can be thought of as a conjunctive concept expression $X \equiv (\bigsqcap_{C \in W} C) \sqcap (\bigsqcap_{C \notin W, C \in \Phi} \neg C)$ so that the world is possible iff X is satisfiable (i.e., there is a realization of the concept type given a TBox).

In what follows we assume a *linear order* of basic concepts in Φ (the ordering will become important in Section 2.7). Since Φ is a finite set we can denote the i-th basic concept in Φ by C_i. For a given possible world W we also use the notation W_i to denote either C_i if C_i occurs positively in W or $\neg C_i$ if it occurs negatively. For a given PTBox the order of basic concepts is fixed across all possible worlds.

Definition 6 (Probabilistic Interpretation, Probability of a Concept). *A **probabilistic interpretation** Pr of a PTBox $(\mathcal{T}, \mathcal{P})$ is a function $Pr : \mathcal{W}_\Phi(\mathcal{T}) \to [0, 1]$ such that $\sum_{W \in \mathcal{W}_\Phi(\mathcal{T})} Pr(W) = 1$. The **probability of a concept** C, denoted as $Pr(C)$, is defined as $\sum_{W \models C} Pr(W)$. $Pr(D|C)$ is an abbreviation for $Pr(C \sqcap D)/Pr(C)$ if $Pr(C) > 0$ and undefined otherwise.*

In other words, a probabilistic interpretation is a probability distribution over possible worlds. It can be thought of as a function which maps each concept type Ct over Φ to the probability that a randomly chosen named individual is a realization of Ct.

Definition 7 (Satisfaction by Probabilistic Interpretation). *A probabilistic interpretation Pr satisfies (is a model of) a conditional constraint $(D|C)[l, u]$, written as $Pr \models (D|C)[l, u]$, if $Pr(C) = 0$ or $Pr(D|C) \in [l, u]$. Pr satisfies (is a model of) a set of conditional constraints $\mathcal{F}$ if it satisfies all constraints in $\mathcal{F}$. A PTBox $PT = (\mathcal{T}, \mathcal{P})$ is called **satisfiable** if there exists a probabilistic interpretation that satisfies $\mathcal{P}$.*

Observe that a conditional constraint is satisfied by all probabilistic interpretations that assign zero probability to the evidence. Given the definition of satisfaction we formulate logical consequence in a standard way.

Definition 8 (Logical Consequence). *A conditional constraint $(D|C)[l, u]$ is a **logical consequence** of a PTBox $(\mathcal{T}, \mathcal{P})$, written as $(\mathcal{T}, \mathcal{P}) \models (D|C)[l, u]$, if all models of $(\mathcal{T}, \mathcal{P})$ also satisfy $(D|C)[l, u]$. $(D|C)[l, u]$ is a **tight logical***

---

[1] We omit "w.r.t. $\mathcal{T}$" and simply write "possible world" (or $\mathcal{W}_\Phi$ instead of $\mathcal{W}_\Phi(\mathcal{T})$) when $\mathcal{T}$ is clear from context.

consequence of $(\mathcal{T}, \mathcal{P})$, written as $(\mathcal{T}, \mathcal{P}) \models_{tight} (D|C)[l, u]$ *if l (resp. u) is the minimum (resp. the maximum) of* $Pr(D|C)$ *over all models* Pr *of* $(\mathcal{T}, \mathcal{P})$ *such that* $Pr(C) > 0$.

2.3   Reasoning Problems

In this paper we consider only the monotonic problems of satisfiability (PSAT) and tight logical entailment (TLogEnt) formulated in the standard way. Ignored for space reasons are non-monotonic problems of default consistency and tight lexicographic entailment (TLexEnt) which are reducible, albeit non-trivially, to PSAT and TLogEnt. Details could be found in [1,7].

Probabilistic Satisfiability (PSAT): Given a PTBox PT decide if it is satisfiable.

Tight Logical Entailment (TLogEnt): Given a PTBox PT and two $\mathcal{SROIQ}$ concepts C and D, compute rational numbers $l, u \in [0, 1]$ such that $PT \models_{tight} (D|C)[l, u]$.

Both problems are reducible to classical reasoning in $\mathcal{SROIQ}$ and Linear Programming (LP), as will be shown in the next section.

2.4   The Probabilistic Satisfiability Algorithm

Pronto's level of scalability is due to the novel PSAT algorithm (see Section 2.9 for comparison with the previously developed methods) described in this section. For the sake of brevity we will consider a special case of PSAT where the PTBox is of the form $PT = (\mathcal{T}, \{(C_i|\top)[p_i, p_i]\})$ (i.e. all probabilistic statements are unconditional constraints with point-valued probabilities and all C_i are concept names). It is straightforward, but technically awkward, to generalize the procedure to handle conditional interval statements over arbitrary concept expressions (see [11]). Also, essentially the same algorithm can be applied to solve TLogEnt problem with the only difference that the linear program is optimized twice to get the lower and the upper probabilities.

A PTBox $PT = (\mathcal{T}, \{(C_i|\top)[p_i, p_i]\})$ is satisfiable iff the following system of linear inequalities is *feasible*, i.e. admits at least one solution (by generalization from propositional PSAT [6]):

$$\sum_{W \models C_i} x_W = p_i, \text{ for each } (C_i|\top)[p_i, p_i] \in \mathcal{P}, \tag{1}$$

$$\sum_{W \in \mathcal{W}_\Phi} x_W = 1 \text{ and all } x_W \geq 0$$

where $\mathcal{W}_\Phi$ is the set of all possible worlds for the set of concepts Φ in $\mathcal{T}$. Observe, that $\mathcal{W}_\Phi$ is finite and exponential in the size of Φ so it is not practical to try to explicitly generate this system in order to check whether it has a solution.

One successful approach to dealing with linear systems having an exponential number of variables is *column generation* [12]. It is based on the fundamental property of linear programming: any feasible program (i.e., a program that admits at least one solution) always has an optimal solution in which only a linear number of variables have non-zero values. Column generation exploits this property by trying to avoid an explicit representation of variables (columns) which will not have positive values in the finally discovered solution. The method is outlined in the next subsection.

2.5   Column Generation Basics

Consider the standard form of a linear program (2). Any linear program, in particular, a version of (1) with intervals can be reduced to it by adding slack variables.

$$\max z = cx \tag{2}$$
$$\text{s.t. } Ax = b \text{ and } x \geq 0$$

A denotes a $m \times n$ matrix of linear coefficients of (2). At every step of the simplex algorithm,[2] A is represented as a combination (B, N) where B and N are submatrices of the *basic* and *non-basic* variables, respectively. Values of non-basic variables are fixed to zero, and the solver proceeds by replacing one basic variable by a non-basic one until the optimal solution is found. Variables are represented as indexed columns of A. The index of a non-basic column which enters the basis is determined according to the following expression [6]:

$$j \in \{1, \ldots, |N|\} \text{ s.t. } c_j - u^T A^j \text{ is maximal} \tag{3}$$

where c_j is the objective coefficient for the new variable and u^T is the current dual solution of (2). The expression $c_j - u^T A^j$ is called *reduced cost*. At every iteration the column having the highest positive reduced cost is selected. If no such column exists the linear program is at an optimum and the simplex algorithm stops.

If the size of N is far too large, as is the case for the program (1), one should compute the index of the entering column according to (3) without examining all columns in N. This is done using the column generation technique in which (3) is treated as an optimization problem with the following objective function:

$$max \ (c_j - \sum_{i=1}^{m+1} u_i a_i^j), \ A^j = (a_i^j) \in \{0, 1\}^{m+1} \tag{4}$$

---

[2] Simplex method is a standard linear programming algorithm, see [13] for a detailed presentation. In spite of having EXPTIME worst-case complexity, it often performs better than worst-case optimal algorithms, e.g., interior-point methods.

where a_i^j are binary variables that represent linear coefficients of the entering column.

It is important to note that except for the way the entering column is obtained (i.e., generated vs selected) the simplex algorithm works along the same lines. Whether the column generation technique is successful or not is contingent upon the following criteria: i) there exists an efficient algorithm for the optimization problem (4), ii) an optimal solution of the program (2) can be found without the generation of an excessive number of columns (the number of generated columns characterizes *convergence* of the algorithm). In the next two subsections we present the PSAT algorithm and the optimized procedure it uses to generate improving columns.

2.6 Column Generation-Based PSAT Algorithm

In order to explain the PSAT algorithm we first rewrite the linear system (1) as the following linear program:

$$\max \sum_{W \in \mathcal{W}_\Phi} x_W \tag{5}$$

$$\text{s.t.} \sum_{W \models C_i} x_W = p_i \times \sum_{W \in \mathcal{W}_\Phi} x_W, \text{ for each } (C_i|\top)[p_i, p_i] \in \mathcal{P}$$

$$\sum_{W \in \mathcal{W}_\Phi} x_W \leq 1 \text{ and all } x_W \geq 0$$

This program has the optimal objective value of 1 if and only if the system (1) is feasible. The advantage of using this program is that it is feasible even if the PT-Box is not satisfiable which facilitates use of the column generation technique.[3] Algorithm 1 presents the PSAT algorithm based on column generation.

The algorithm follows the basic column generation procedure outlined in the previous subsection. It first constructs the so-called *restricted master problem* (RMP) which is a subprogram of (5) with a restricted set of variables (line 2). These initial variables are created by generating a subset of the index set $\mathcal{W}_\Phi$ (see Section 2.2). Next, the algorithm enters the main column generation loop (lines 3–10) during which it tries to generate an improving column (line 6). The column generation procedure *GenerateImprovingColumn*, which takes the PTBox and the current dual solution u^T, plays the central role and is explained in detail in the next subsection. If an improving column has been successfully generated, it is added to the linear program (line 9). The algorithm breaks out of the loop when no improving column can be generated. Finally, it checks the optimal value of the final RMP and returns *Yes* if it is equal to 1.

A number of implementation details have been omitted for the sake of presentation clarity. In particular, the algorithm forgets some previously generated

---

[3] In principle, it is possible to generate columns for an infeasible linear program but maintaining it feasible usually helps convergence.

Algorithm 1. PSAT algorithm based on column generation

    Input: PTBox $PT = (\mathcal{T}, \mathcal{P})$
    Output: Yes if PT is satisfiable, No otherwise
1 **if** $\mathcal{T}$ *is not satisfiable* **then return** No
2 $LP \leftarrow InitializeRMP(\mathcal{T}, \mathcal{P})$
3 **while** *true* **do**
4     $d \leftarrow \text{optimize}(LP)$
5     $u^T \leftarrow \text{dual\_solution}(LP)$
6     $A^j \leftarrow \text{GenerateImprovingColumn}((\mathcal{T}, \mathcal{P}), u^T)$
7   **if** $A^j = null$ **then break**
8   **else**
9 | Add A^j to LP as a new column
10   **end**
11 **end**
12 **if** $d = 1$ **then return** Yes **else return** No

---

columns which have not been in the basis for a large number of iterations in order to keep RMP tractable. This may potentially compromise the termination property (see the next subsection), so the algorithm implements special checks to detect the situation when previously forgotten columns are repeatedly regenerated.

2.7 Possible World Generation

This section explains the procedure for generating improving columns (possible worlds) for Algorithm 1. It describes in detail what each component of a PSAT column represents and how to set up and use the optimization problem (4) for generation of possible worlds.

Consider a_i^j, the i-th coefficient of some column A^j for the PSAT program (5). The column corresponds to some possible world $W^j = \{C_i\}_{i=1}^{|\Phi|}$, therefore $a_i^j = 1$ implies that C_i occurs positively in W^j while $a_i^j = 0$ implies that it occurs negatively (or equivalently, $W^j \models \neg C_i$). Thus it is possible to represent W^j as a *conjunctive* concept expression in $\mathcal{SROIQ}$ assuming a fixed linear ordering of concepts $\{C_i\}_{i=1}^{|\Phi|}$ in Φ (see Section 2.2). More formally, we define the following function η which maps columns, i.e. binary vectors, to conjunctions of basic concepts from Φ:

$$\eta(A^j) = \bigsqcap X_i, \text{ where } X_i = \begin{cases} C_i, & \text{if } a_i^j = 1 \\ \neg C_i, & \text{if } a_i^j = 0 \end{cases} \tag{6}$$

X_i are literals that denote either a basic concept or its negation.

Soundness of Algorithm 1 strongly depends on whether every solution of the optimization problem (4), which is added as a column to the main linear program (5), corresponds to a concept expression that is satisfiable w.r.t. $\mathcal{T}$, i.e., is a *possible* world. If this condition is true then soundness trivially follows because one may simply enumerate the set of all solutions (since the set of possible worlds is finite), so (5) will be equivalent to the original linear system (1). *Completeness* requires that every possible world for the given PTBox corresponds to some solution of (4). Therefore, for ensuring both soundness and completeness it is crucial to construct a set of constraints $\mathcal{H}$ for the problem (4) such that its set of solutions is in one-to-one correspondence with the set of all possible worlds $\mathcal{W}_\Phi$.

In what follows we will call columns which correspond to satisfiable expressions *valid* and the others *invalid*. More formally, given a $\mathcal{SROIQ}$ TBox $\mathcal{T}$, a column A^j is valid if $\mathcal{T} \not\models \eta(A^j) \sqsubseteq \bot$ and is invalid otherwise.

Validity can easily be ensured in the propositional case where each C_i is a clause. One possibility is to employ the well known formulation of SAT as a mixed-integer linear program (MILP) [14]. For example, if $C_i = c_1 \vee \neg c_2 \vee c_3$ then (4) will have the constraint $a_i = x_{c1} + (1 - x_{c2}) + x_{c3}$ where all variables x_{ck} are binary. In that case soundness and completeness follow from the reduction of SAT to MILP. Previously developed propositional PSAT algorithms take full advantage of that (see Section 2.9).

In the case of an expressive language, such as $\mathcal{SROIQ}$, there appears to be no easy way of determining the set of constraints $\mathcal{H}$. Furthermore, it is unclear whether such a set is polynomial in the size of $\mathcal{T}$. Informally, $\mathcal{H}$ must capture every entailment, such as $\mathcal{T} \models C_i \sqcap \cdots \sqcap C_j \sqsubseteq \bot$ in order to prevent generation of any column A^j such that $C_i \sqcap \cdots \sqcap C_j$ is a conjunctive subexpression of $\eta(A^j)$. All such entailments can be computed in a naive way by checking satisfiability of all conjunctions $C_i \sqcap \cdots \sqcap C_j$ over Φ but this is no better than trying to construct the full linear system (1).

Instead, Pronto implements a novel *hybrid*, *iterative* procedure to compute $\mathcal{H}$ (see Algorithm 2). The key lines are 7 and 10. On step 7 the algorithm invokes a $\mathcal{SROIQ}$ reasoner (in our case, Pellet [15]) to determine if the computed column corresponds to a *possible* world. This is critical for soundness. If yes, the column is valid and returned. If no, the current set of constraints $\mathcal{H}$ needs to be extended to exclude A^j from the set of solutions to (4). This step deserves a more detailed explanation which we present by first defining the notion of the minimal unsatisfiable core for an unsatisfiable conjunctive concept expression.

Definition 9 (Unsatisfiable Core). *Given a TBox $\mathcal{T}$ and unsatisfiable (w.r.t. $\mathcal{T}$) concept expression $\bigsqcap X_i$ represented as a set of conjuncts $X = \{X_i\}$, a minimal unsatisfiable subexpression (MUS) is a subset $X' = \{X_i'\} \subseteq \{X_i\}$ such that $\bigsqcap X_i$ is unsatisfiable w.r.t. $\mathcal{T}$ and any $X'' = \{X_i''\} \subset \{X_i'\}$ is satisfiable w.r.t. $\mathcal{T}$. The **unsatisfiable core** (UC) of $\bigsqcap X_i$ is the set of all its MUSes.*

Intuitively, our notion of UC for conjunctive $\mathcal{SROIQ}$ concepts corresponds to the standard notion of unsatisfiable core for propositional formulas in conjunctive normal form [16].

Each MUS can be regarded as a one "laconic justification" of the unsatisfiability of the original concept expression [17] (here "laconic" means that it contains no superfluous conjuncts). The UC is the set of all laconic justifications of the

---

Algorithm 2. Possible world generation algorithm

Input: PTBox $PT = (\mathcal{T}, \mathcal{P})$, current dual solution u^T of (5)
Output: New column A^j or *null*
1 $IP_{ColGen} \leftarrow$ initialize the integer program (4) using u^T and $\mathcal{P}$
2 $\mathcal{H} \leftarrow \emptyset$
3 **while** $A^j \neq null$ **do**
4 $\quad$ Solve IP_{ColGen} subject to $\mathcal{H}$ to optimality
5 $\quad$ $A^j \leftarrow$ some optimal solution of IP_{ColGen}
6 $\quad$ **if** $A^j \neq null$ **then**
7 $\quad\quad$ **if** $satisfiable(\eta(A^j), \mathcal{T})$ **then**
8 $\quad\quad\quad$ **return** A^j
9 $\quad\quad$ **end**
10 $\quad\quad$ $\mathcal{H} \leftarrow \mathcal{H} \cup$ inequalities that exclude A^j
11 $\quad$ **end**
12 **end**
13 **return** *null*

---

unsatisfiability. Clearly, to exclude the current, invalid column from the set of solutions to (4), it is sufficient to add to (4) a constraint that rules out *any* of the MUSes.

Next, we show how to translate MUSes into linear inequalities. A MUS is a set of conjuncts $\{X_i'\}$ each of which corresponds to a binary variable (observe that η, as defined in (6), is a bijective function). By a slight abuse of notation we write $a_i = \eta^{-1}(X_i')$ to denote the variable that corresponds to C_i. Then given a MUS $X' = \{X_i'\}_{i=1}^k$ we add the following linear constraint:

$$\sum_{i=1}^{k} a_i \leq k - 1, \text{ where } a_i = \begin{cases} \eta^{-1}(X_i'), & X_i' = C_i \\ 1 - \eta^{-1}(X_i'), & X_i = \neg C_i \end{cases} \tag{7}$$

If a conjunctive concept contains $\bigsqcap X_i$ as a subexpression then all binary expressions b_i, i.e. either a_i or $1 - a_i$ depending on whether X_i is a positive or a negative literal, are equal to 1. Therefore, $\sum_{i=1}^k a_i = k$ where k is the size of $\{X_i\}$. Constraining $\sum_{i=1}^k b_i$ to be less or equal to $k-1$ is equivalent to requiring at least one b_i to be equal to 0. According to the definition of η this is equivalent to removing of at least one conjunct from X' which makes it satisfiable (due to minimality of X', see Definition 9). Therefore, each of the constraints (7) is sufficient to exclude all columns, which correspond to concept expressions containing X', from the set of solutions to (4). Observe that the constraints do not exclude any columns which do *not* include X' since it is necessary to ensure completeness.

On line 10 the algorithm computes the unsatisfiable core for a concept expression that corresponds to the current solution of (4). Then it transforms each of the MUSes into a linear inequality according to (7) and adds them to the binary program (4).

We call our PSAT algorithm, which is composed of algorithms 1 and 2, "hybrid" because it combines invocations of an LP solver (to optimize (5)), MILP and $\mathcal{SROIQ}$ solvers (to optimize (4) and check satisfiability of concept expressions respectively). It is *iterative* because during the possible world generation phase it iteratively tightens the set of solutions to (4) until either a valid column is found or provably no such column exists.

Finally, we give a short example demonstrating our iterative technique for computing valid columns.

Example 1. Consider a PTBox where $\mathcal{T} = \{A \sqsubseteq \exists R.C, B \sqsubseteq \exists R.\neg C, \geq 2R.\top \sqsubseteq D\}$ and $\mathcal{P}$ contains some probabilistic constraints over the ordered set $\Phi = \{A, B, D\}$. Algorithm 2 starts out with an empty set of linear constraints for (4). The list of binary variables for (4) is (x_A, x_B, x_D). Assume that at some iteration the algorithm generates the following column: $A^j = (1, 1, 0, 1)$ (the last component of any column is always equal to 1 because of the normalization row in (5)). Then $\eta(A^j) = A \sqcap B \sqcap \neg D$.

It is not hard to see that $\mathcal{T} \models \eta(A^j) \sqsubseteq \bot$. The reason is that any instance o of $A \sqcap B$ must have two R-successors (domain elements which are connected to $o^{\mathcal{I}}$ by $R^{\mathcal{I}}$). Moreover, they are necessarily distinct because one is an instance of C and another is an instance of $\neg C$. Therefore, o is an instance of $\geq 2R.\top$ and consequently is an instance of D. This is a contradiction with $\neg D$ in $\eta(A^j)$.

The unsatisfiable core of $\eta(A^j)$ is $\{A, B, \neg D\}$. This MUS is converted into the following linear inequality $x_D \geq x_A + x_B - 1$ which is then added to the binary program (4). As a result, no invalid column containing this MUS will be computed on subsequent iterations.

2.8 Main Optimizations

We next describe several optimization techniques which play key roles for our implementation of the possible world generation algorithm.

Classical Modularity. It is possible that concepts appearing in conditional constraints, i.e., the probabilistic signature, are only a small fraction of those appearing in the classical part of the KB. This can happen especially when a large OWL ontology, such as the NCI Thesaurus, is only slightly augmented with probabilistic knowledge. Then it seems intuitively unreasonable to work with the full classical part when solving PSAT/TLOGENT because many OWL axioms do not interact with probabilistic knowledge but only slow down concept satisfiability checks during the column generation process.

Fortunately, we can employ *ontology modularity* techniques in order to extract a fragment of the classical part which is guaranteed to contain all relevant knowledge, i.e., a module [18,19]. More formally, given a signature Σ a $\Sigma-$module $\mathcal{O}'$ in an ontology $\mathcal{O}$, written as $\mathcal{M}(\mathcal{O}, \Sigma)$, is a subset of $\mathcal{O}$ such that any axiom α over Σ is entailed by $\mathcal{O}$ *iff* it is entailed by $\mathcal{O}'$ [18]. In the context of PSAT this means that any concept expression 6 that needs to be checked for satisfiability on step 10 of Algorithm 2 is unsatisfiable w.r.t. the whole classical part $\mathcal{T}$ *iff* it is unsatisfiable w.r.t. $\mathcal{M}(\mathcal{T}, \Phi)$, where Φ is the probabilistic signature.

Extracting modules enables us to substantially cut down the size of the classical part even though modules are not guaranteed to be minimal.[4] In fact, this optimization is not specific for our hybrid algorithm and even column generation. In particular, it can be used in propositional PSAT solvers as well. Note that modules are only guaranteed to contain all relevant knowledge w.r.t. a *fixed* signature so if the signature changes, for example, when a new conditional constraint is added or some new concept is used in an entailment query, the module needs to be recomputed.

Exploiting Concept Hierarchy. The first optimization stems from a natural observation that many inequalities for the binary program (4) can be added simply by examining the structure of the TBox. Virtually all modern DL reasoners can efficiently construct the so called *classification hierarchy* by finding all subsumptions between concept names that are logically entailed by the TBox. Such hierarchy can be used to construct an initial set of inequalities $\mathcal{H}_0$.

Consider the following TBox $\mathcal{T} = \{A \sqcup B \sqsubseteq C\}$. The classified version of $\mathcal{T}$ should include subsumptions $A \sqsubseteq C$ and $B \sqsubseteq C$. They can be directly translated to inequalities $x_A \leq x_C$ and $x_B \leq x_C$ to preclude computing an invalid column containing either $A \sqcap \neg C$ or $B \sqcap \neg C$ as subexpressions and converting these subexpressions into inequalities.

This idea helps to reduce the number of concept satisfiability tests. The effectiveness of this technique depends on the axiomatic richness of the TBox. For axiomatically weak TBoxes, where almost all subsumptions can be discovered by traversing the concept hierarchy, most of the set $\mathcal{H}$ is computed up front. More complex TBoxes may have non-trivial entailments involving concept expressions on both sides of subsumptions which can only be discovered when checking validity of some column candidate. One such examples is the subsumption $A \sqcap B \sqsubseteq D$ from Example 1.

A drawback of this optimization is that it might be *too eager* and generate more linear inequalities that can be fit in memory. One example of how this can happen is exploiting TBoxes which succinctly encode a quadratic number of disjointness axioms by using the `DisjointClasses` construct in OWL 2.[5] At the same time, as explained in the next subsection, there is a chance that enough valid columns can be generated *before* all the relationships entailed by the TBox are discovered and captured in linear constraints. Therefore a simple solution to this problem is to impose a limit on the total number of inequalities created up front. The value of the limit can be adapted depending on the amount of available memory and capabilities of the MILP solver. A more informative approach could be possible and is left for future investigation.

---

[4] Extracting strictly minimal models is undecidable for $\mathcal{SROIQ}$ so we rely on approximate solutions. See [18] for details.

[5] The axiom `DisjointClasses`$(C_1, \ldots, CE_n)$ asserts that the concepts are *pairwise* disjoint. It is equivalent to a quadratic number of binary disjointness axioms. Such axioms can be added by ontology editors transparently for the user. For more details refer to `http://www.w3.org/TR/owl2-syntax/#Disjoint_Classes`

Propositional Absorption. A large portion of knowledge in many real onto-logies is propositional. Thus it is natural to convert them into linear inequalities to avoid computing some invalid columns which violate propositional TBox ax-ioms. In the extreme case all propositional knowledge can be absorbed into the program (4). However, the algorithm tries to find a trade-off between eager ab-sorption (which can exhaust memory) and lazy generation of inequalities (which requires extra concept satisfiability checks). The balance depends on available memory and the number of propositional axioms.

This optimization can add extra variables to the column generation model (4). Normally, its variables correspond to concepts in the probabilistic signature Φ. Consider an example in which $\{C_1, C_2, C_3\} \subseteq \Phi$. If C_1 is itself defined as a Boolean combination over other concepts, e.g., $C_1 \equiv A \sqcup B$, it makes sense to add x_A and x_B to (4) (even if A and B do not appear in any conditional con-straints). The reason is that other concepts from Φ could appear in propositional expressions over $\{A, B, \dots\}$, for instance, $C_2 \equiv \neg A \sqcup B, B \sqsubseteq C_3$. In this case, adding extra variables and translating these axioms into linear inequalities will *automatically* enforce $C_1 \sqcap C_2 \sqsubseteq C_3$ for all future column candidates.

Optimistic Inequality Generation. One issue with a naive implementation of Algorithm 2 is that computing unsatisfiability cores may appear impractical for certain concept expressions and TBoxes. This may especially happen for long expressions which contain MUSes with little or no overlap. It is known from the model diagnosis theory [20] that finding all minimal unsatisfiable sets may require a number of SAT tests that is exponential in the total size of all sets.

To address this issue the algorithm imposes a time limit on the procedure that computes the UC. If at least one MUS has been found but finding others exceeds the timeout the procedure is interrupted. The found MUSes are then converted to linear inequalities and the algorithm moves on as if the full UC was computed.

This optimization does not cause a loss of either soundness or completeness. Completeness is trivially preserved because not adding some inequalities to the program (4) can only expand its set of solutions, so no possible world could be missed. Soundness is preserved because each computed column is still valid (SAT tests are never interrupted). The only possible negative impact of missing some MUSes is that they can appear in some future column candidates, so the algorithm might go through additional iterations. However, they do not *have to* appear because the optimal basis for the main program (5) can often be found before considering column candidates containing those MUSes. The algorithm behaves *optimistically* by hoping that additional iterations will not be required.

Unsurprisingly, timeouts typically occur when dealing with complex TBoxes with many non-trivial entailments. At the same time our experience shown that those TBoxes tend to improve convergence of column generation because their set of possible worlds is smaller. Therefore there is a higher chance that the column generation process will stop before all MUSes of some unsatisfiable concept ex-pression are discovered.

2.9   Comparison with Propositional PSAT Solvers

After Nilsson presented his basic propositional probabilistic logic [2] it became immediately clear that a straightforward approach to solving PSAT based on solving the system of linear inequalities (1) is intractable not just in the worst case but also in most practically relevant cases, i.e., when the number of probabilistic statements exceeds a few dozen. Most of approaches, including ours, have revolved around *column generation* to cope with this difficulty (see Section 2.5 and Chapter 26 in [13]). The early attempts to use that technique include works of Georgakopoulos, Kavvadias and Papadimitriou [21,22], Hooker [23], and Jaumard, Hansen and de Aragão [24,25,26]. They all use the standard simplex procedure to solve partially constructed instances of (5) (*master* problems) but differ in their methods of solving the auxiliary optimization problem (4) to generate new columns (variables).

More recent works on PSAT via global column generation not just use advanced heuristics to generate columns but also consider a range of extensions to the basic PSAT formulation. Hansen et al. [27] consider imprecise probabilities which include intervals and qualitative probabilistic constraints introduced by Coletti, i.e., formulas of the form $P(A) \leq$ (or $\geq)P(B)$ [28]. Ognjanovic et al. [29,30,31] deal with so called *weight formulas* which are probabilistic formulas of the form $a_1w(\alpha_1)+\cdots+a_nw(\alpha_n) \leq$ (or $\geq)b$, where $\{a_i\}$ and b are rational numbers, $\{\alpha_i\}$ are propositional formulas, and $w(\alpha_i)$ stands for "probability of α_i". They employ a highly efficient *variable neighborhood search* (VNS) technique [31], genetic algorithms [29], and their combination [30] for generating columns. These methods where the first to scale to 1000 probabilistic formulas [31]. Finally, we mention the recent work of de Souza Andrade et al. [32] who proposed yet another approach to producing columns by linearizing, but differently from [23], the column generation model (4).

The major differences between our PSAT algorithm and the previous ones manifest themselves in dealing with classical part of probabilistic knowledge bases. We consider two cases: when classical knowledge is propositional and when it is not.

To the best of our knowledge, our work is the first to describe a column generation based algorithm and its evaluation for *non-propositional* PSAT. The main difference between propositional and non-propositional PSAT problems in the context of column generation is that propositionality of the KB allows encoding of all its structure in a polynomial number of linear inequalities over a polynomial number of binary variables. This follows directly from the well known reduction of propositional SAT to integer programming [14,23]. Therefore, the column generation problem (4) is much easier to handle, either as a standard MILP instance [14,32] or as a non-linear pseudo-boolean program [25,6,30]. The full structure of a $\mathcal{SROIQ}$ TBox, on the other hand, cannot be captured as a system of linear inequalities.

The key property that enables our algorithm to deal with non-propositional KBs is its "hybridness". It interacts with a classical SAT solver for the target logic in order to ensure that all generated columns (possible worlds) are indeed possible, i.e., do not violate logical structure of the KB, which can be quite rich. This has several important advantages. First, it allows us to handle essentially *any* target logic for which a SAT solver is available, for example, we can use specialized reasoners for any profile of OWL 2. Second, this makes our algorithm more scalable with respect to the amount of classical knowledge. For example, modern DL reasoners can efficiently solve the concept satisfiability problem even for very large TBoxes containing thousands of axioms (a characteristic example is the NCI Thesaurus). The reasoners implement a range of optimizations and heavily exploit the logical structure of the KB, which is often lost if translated to linear inequalities (even when the translation is possible).

The behavior of our algorithm becomes closer to that of the previous methods when the classical knowledge is largely propositional. In that case, it can *absorb* the propositional part of the KB by converting it into a set of linear inequalities for the column generation problem (4), i.e., similarly to [14]. This helps to reduce the number of future calls to the classical SAT solver.

Pronto also implements a range of other algorithms, most notably, for finding minimal probabilistically unsatisfiable fragments in a PTBox and computing the tight lexicographic entailment, but they are omitted for space reasons. Please see [7] for their description, analysis, and evaluation.

3   Architecture

Pronto has a layered architecture, presented in Figure 1. Each layer has one or more components which invoke other components at the same level or at the next lower level. Lower level components never invoke upper level components but simply pass the requested information upwards.

3.1   Linear Program Layer

The main function of the components at the lowermost level is managing linear programs which are optimized in order to solve PSAT and TLogEnt problems. As mentioned earlier, these linear programs usually have exponentially many variables so it is futile to try to represent them explicitly. The linear program manager (LPM) and the column generator (CG) collectively implement the column generation algorithm described in Section 2.4 while other components, namely the various LP/MILP solvers and the DL reasoner provide the necessary optimization and $\mathcal{SROIQ}$ reasoning services.

The LPM is responsible for producing the initial version of the restricted master program (5), incorporating each new column into it, and checking the optimality (i.e. stopping) criteria. It interacts with a simplex solver, for example GLPK or CPLEX, which solves the current program (5) and returns its primal and dual solutions. The latter is supplied to the CG component in order to guide its search for a new, improving column.

The CG component implements Algorithm 2. It initializes and maintains the binary linear program (4), accepts the dual values u^T from the LPM, and generates new column candidates using a MILP solver. It then interacts with Pellet to

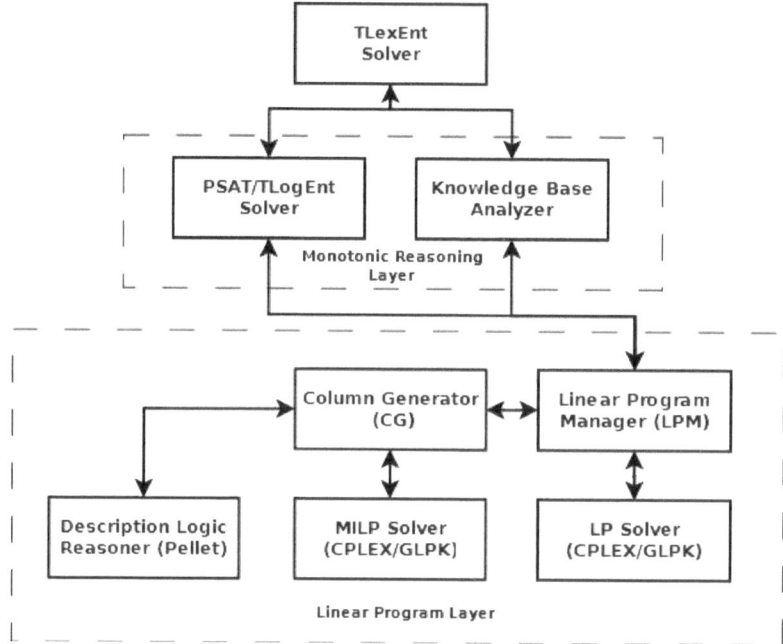

Fig. 1. The layered architecture of Pronto

check validity of each candidate and, if found valid, passes it back to the LPM. This component implements a number of optimizations described in Section 2.8 such as exploiting the concept hierarchy, multiple columns generation, optimistic generation and so on.

Currently, Pellet is the only $\mathcal{SROIQ}$ reasoner that Pronto can interact with. However, this is planned to be refactored to introduce an abstract interface, e.g., the OWL API, between CG and the reasoner. This would allow us to use other reasoners, such as FaCT++, HermiT, or RACER, as well as specialized reasoners for particular profiles of OWL 2 or other logics.

3.2 Monotonic Reasoning Layer

The components on the next layer use the underlying linear programs to solve PSAT and TLogEnt, and also analyze probabilistic KBs. The first two tasks are straightforward. They amount to checking if the generated linear program (5) has the optimal objective value less than 1 (PSAT) or optimizing it in both directions (TLogEnt).

The analyzer implements algorithms for finding one or all minimal unsatisfiable fragments of an unsatisfiable PTBox [7]. The analysis of an unsatisfiable probabilistic KB is a problem of finding all minimal unsatisfiable subsets of the KB where minimality is defined with respect to the set inclusion. This is essential for, first, computing all maximal satisfiable fragments of the KB during non-monotonic (lexicographic) reasoning, and, second, computing explanations for the results of probabilistic reasoning. The current implementation closely interacts with the LPM to perform a trial-and-error relaxation of the linear program (5) in order to discover all irreducible subprograms (IIS) which optimal value is less than 1 (they correspond to minimal conflicts).

3.3 Non-monotonic Reasoning Layer

The uppermost layer consists of a single component: the lexicographic reasoner. It implements the TLEXENT algorithm [7] which relies on the KB analyzer and the TLOGENT solver. TLEXENT is equivalent to solving TLOGENT for all lexicographically minimal subsets of the KB [1]. The latter require the auxiliary data structure called *z-partition*. The component implements two algorithms to compute z-partition: an optimized version of the original algorithm from [1] and the novel conflict-driven TLEXENT algorithm [7].

4   Evaluation of the Probabilistic Satisfiability Algorithm

Pronto is written in Java and compiled using Sun JDK 1.6. All evaluation tests have been performed on a PC with 2GHz CPU, 2GB of RAM, Sun JRE 1.6.0_07 running under Windows XP SP3. The only JVM option that was used for performance tuning was -Xmx to allow the reasoner use the maximal available amount of memory. All of the evaluation tests presented below use wall time as the main measure of performance ("Total Time"). We also also record the number of generated columns (to track convergence speed) and total column generation time ("CG Total").

Due to the lack of existing probabilistic knowledge bases, all of our experiments have to involve a generated probabilistic component. While there are hundreds to thousands of publicly available OWL ontologies of varying size, expressivity, and axiom sophistication, none of these have a probabilistic part. For our experiments, we selected six ontologies from the TONES repository: The NCI Anatomy Ontology (NCI), the Subcellular Anatomy Ontology (SAO), the Process Ontology, the Sequence Ontology with Composite Terms (SO-XP), the Teleost Anatomy Ontology (TAO), and the Cell Type ontology. None of these ontologies is propositional or small and simple enough to consider their propositionalization and are varied enough to give a reasonable feel for robustness. Of course, none of the previously developed PSAT algorithms is capable of dealing with thousands of classical axioms in addition to a comparable number of probabilistic formulas, much less non-propositional arguments.

The probabilistic parts of test PTBoxes are produced by a random generation process that takes a probabilistic signature as an argument. Common to all cases, we fix the number of unconditional statements to 10% of the size probabilistic part. We need to have some unconditional part for two reasons: First, it is necessary for realism; in our modelling experience, a small ratio of unconditional constraints likely common modeling pattern in P-$\mathcal{SROIQ}$, e.g., to represent probabilistic facts, or beliefs, about a specific individual [33]. Second, and relatedly, it is necessary to avoid *trivial* satisfiability. If all constraints are conditional, then we can obtain a vacuous model by assigning zero probability to all evidence concepts. This is undesirable from both a modeling perspective (that is, unrealistic) and does not engage the reasoning algorithm at all. As from modeling realism, we settle on 10% because in previous experiments wherein we varied the percentage of unconditionals from 10%-50% there was no significant performance differences (see [7] for details).

Ultimately it is the probability intervals attached to constraints that determine whether the resulting PTBox will be satisfiable or not. It has been reported that satisfiable KBs are typically harder for PSAT algorithms [32,6] so, we want to focus on (nontrivial) satisfiable problems. Unfortunately, random assignment of probabilities to generated constraints is likely to result in an unsatisfiable PTBox, provided that it contains unconditional statements [24,32,6]. Therefore, we use a standard technique based on generation of probabilistic interpretations which can then be used to assign probabilities to statements [24]. In that case satisfiability is guaranteed because satisfying interpretations (models) have been constructed explicitly. Its main advantage is that it works with any probabilistic KB, propositional or not, and does not impose any restrictions on its structure (such as cycle disallowance). The main disadvantage is that large cases become prohibitively more difficult to generate. For the current evaluation it has been implemented in the following steps: First, two sets of possible worlds $\mathcal{I}_\Phi^1, \mathcal{I}_\Phi^2$ of size $k \geq 2 \times |\mathcal{P}|$ are generated for a PTBox $(\mathcal{T}, \mathcal{P})$ with probabilistic signature Φ. Second, probabilistic interpretations Pr_1, Pr_2 are defined by generating two sequences of k random numbers summing to 1 which represent probabilities of possible worlds in $\mathcal{I}_\Phi^1$ and $\mathcal{I}_\Phi^2$. Third, the lower probability l (resp. the upper probability u) for each constraint $(D|C)[l, u]$ in $\mathcal{P}$ is determined as the smallest (resp. the largest) of values $Pr_i(D|C)$ ($i \in \{1, 2\}$).

The performance results are presented in Table 1. The total time is averaged over five runs on five different random PTBoxes. For each PTBox the size of the probabilistic signature was set to half the number of probabilistic statements.

The major conclusion that can be made from the evaluation results is that the algorithm is *robust*, i.e. it behaves quite predictively on satisfiable PTBoxes with varying parameters. No combination of the main parameters causes it to hit the worst case. It robustly scales to 1000 probabilistic statements defined over 500 concepts from expressive real ontologies (see more results in [7]). It is worth noting that the number of columns generated by Algorithm 2 does not seem to grow exponentially with either size of the PTBox or size of the probabilistic signature. This suggests that the PSAT algorithm may well scale beyond 1000 conditional constraints. We have not yet extended the experiments beyond 1000 since it is extremely time consuming to generate *satisfiable* probabilistic KBs of that size over complex ontologies because it requires computing a high number of possible worlds.

Table 1. PSAT performance on random PTBoxes

Ontology	Language	TBox size	PTBox size	Total time (s)	CG Total (s)	# columns
NCI	$\mathcal{ALE}+$	5423	250	100.21	32.79	83.4
			500	239.45	93.89	186.4
			750	429.13	157.37	301.4
			1000	745.15	231.62	418.4
SAO	$\mathcal{SHIN}$	2499	250	77.1	68.41	129.4
			500	178.16	149.29	276.4
			750	375.3	300.02	341.2
			1000	1360.21	1176.05	425.4
Process	$\mathcal{SHOF}$	2007	250	50.98	39.39	88.6
			500	119.92	87.02	176.4
			750	240.94	144.42	275.2
			1000	479.69	236.42	404.8
SO-XP	$\mathcal{SHI}$	1928	250	61.31	40.31	76
			500	197.05	144.17	189
			750	449.49	323.31	307.6
			1000	921.57	644.28	423.4
TAO	$\mathcal{EL}+$	3406	250	50.24	37.13	89.4
			500	125.76	89.38	179.8
			750	252.52	149.5	287.8
			1000	544.71	238.09	431.8
Cell Type	$\mathcal{EL}+$	1263	250	57.22	39.18	89.2
			500	137.89	91.66	182.6
			750	283.45	158.88	296.4
			1000	487.68	220.32	384.2

5 Summary

This paper presented Pronto: a computationally practical tool for modelers interested in probabilistic deductions over large OWL ontologies augmented with probabilistic statements. We have demonstrated that high worst-case complexity of P-$\mathcal{SROIQ}$ does not preclude reasonable performance in practice which puts P-$\mathcal{SROIQ}$ on the same ground as, for example, $\mathcal{SROIQ}$ (it must be noted, of course, that $\mathcal{SROIQ}$ reasoners have a longer history of optimization so their scalability is naturally better than Pronto's).

With the scalability limits pushed to around 1000 probabilistic statements, Pronto (as a tool) and P-$\mathcal{SROIQ}$ (as a formalism) become viable options for modeling real-life problems. One such example is our previous work on verifying consistency of the large medical decision support system named CADIAG-2 [8]. Having a large number of uncertain rules (around 20,000) verifying its consistency in an automated way has been a long-standing challenge. It has been solved by translating the classical (non-uncertain) part of CADIAG-2 into OWL and its uncertain rules into probabilistic statements, so that the resulting knowledge base was a P-$\mathcal{SROIQ}$ ontology. Not only Pronto has been able to prove its inconsistency but also extracted around 700 *distinct* minimal sets of unsatisfiable rules and proved that no other inconsistencies existed. We expect Pronto and P-$\mathcal{SROIQ}$ to be useful in other scenarios involving both classical and probabilistic knowledge, such as validation of ontology alignments [34,7].

References

1. Lukasiewicz, T.: Expressive probabilistic description logics. Artificial Intelligence 172(6-7), 852–883 (2008)
2. Nilsson, N.J.: Probabilistic logic. Artificial Intelligence 28(1), 71–87 (1986)
3. Halpern, J.Y.: An analysis of first-order logics of probability. Artificial Intelligence 46, 311–350 (1990)
4. Bacchus, F.: Representing and reasoning with probabilistic knowledge. MIT Press (1990)
5. Cozman, F.G., de Campos, C.P., da Rocha, J.C.F.: Probabilistic logic with independence. International Journal of Approximate Reasoning 49(1), 3–17 (2008)
6. Hansen, P., Perron, S.: Merging the local and global approaches to probabilistic satisfiability. International Journal of Approximate Reasoning 47(2), 125–140 (2008)
7. Klinov, P.: Practical Reasoning in Probabilistic Description Logic. PhD thesis, The University of Manchester (2011)
8. Klinov, P., Parsia, B., Picado-Muiño, D.: The consistency of the medical expert system CADIAG-2: A probabilistic approach. Journal of Information Technology Research 4(1), 1–20 (2011)
9. Horrocks, I., Kutz, O., Sattler, U.: The even more irresistible $\mathcal{SROIQ}$. In: Knowledge Representation and Reasoning, pp. 57–67 (2006)
10. Lutz, C., Sattler, U., Tendera, L.: The complexity of finite model reasoning in description logics. Information and Computation 199(1-2), 132–171 (2005)
11. Klinov, P., Parsia, B.: A Hybrid Method for Probabilistic Satisfiability. In: Bjørner, N., Sofronie-Stokkermans, V. (eds.) CADE 2011. LNCS, vol. 6803, pp. 354–368. Springer, Heidelberg (2011)
12. Dantzig, G.B., Wolfe, P.: Decomposition principle for linear programs. Operations Research 8, 101–111 (1960)
13. Chvátal, V.: Linear programming. A Series of Books in the Mathematical Sciences. W.H. Freeman and Company, New York (1983)
14. Hooker, J.N.: Quantitative approach to logical reasoning. Decision Support Systems 4, 45–69 (1988)
15. Sirin, E., Parsia, B., Grau, B.C., Kalyanpur, A., Katz, Y.: Pellet: A practical OWL-DL reasoner. Journal of Web Semantics 5(2), 51–53 (2007)
16. Lynce, I., Silva, J.P.M.: On computing minimum unsatisfiable cores. In: International Conference on Theory and Applications of Satisfiability Testing (2004)
17. Horridge, M., Parsia, B., Sattler, U.: Laconic and Precise Justifications in OWL. In: Sheth, A.P., Staab, S., Dean, M., Paolucci, M., Maynard, D., Finin, T., Thirunarayan, K. (eds.) ISWC 2008. LNCS, vol. 5318, pp. 323–338. Springer, Heidelberg (2008)
18. Grau, B.C., Horrocks, I., Kazakov, Y., Sattler, U.: Modular reuse of ontologies: Theory and practice. Journal of Artificial Intelligence Research 31, 273–318 (2008)
19. Sattler, U., Schneider, T., Zakharyaschev, M.: Which kind of module should I extract? In: International Workshop on Description Logic (2009)
20. Reiter, R.: A theory of diagnosis from first principles. Artificial Intelligence 32, 57–95 (1987)
21. Georgakopoulos, G.F., Kavvadias, D.J., Papadimitriou, C.H.: Probabilistic satisfiability. Journal of Complexity 4(1), 1–11 (1988)
22. Kavvadias, D.J., Papadimitriou, C.H.: A linear programming approach to reasoning about probabilities. Annals of Mathematics and Artificial Intelligence 1, 189–205 (1990)

23. Hooker, J.: A mathematical programming model for probabilistic logic. Working Paper 05-88-89, Graduate School of Industrial Administration, Carnegie Mellon University, Pittsburgh, PA 15213 (1988)
24. Jaumard, B., Hansen, P., de Aragão, M.P.: Column generation methods for probabilistic logic. In: Integer Programming and Combinatorial Optimization Conference, pp. 313–331 (1990)
25. Jaumard, B., Hansen, P., de Aragão, M.P.: Column generation methods for probabilistic logic. INFORMS Journal on Computing 3(2), 135–148 (1991)
26. Hansen, P., Jaumard, B., Nguetsé, G.B.D., de Aragão, M.P.: Models and algorithms for probabilistic and Bayesian logic. In: International Joint Conference on Artificial Intelligence, pp. 1862–1868 (1995)
27. Hansen, P., Jaumard, B., de Aragão, M.P., Chauny, F., Perron, S.: Probabilistic satisfiability with imprecise probabilities. International Journal of Approximate Reasoning 24(2-3), 171–189 (2000)
28. Coletti, G., Gilio, A., Scozzafava, R.: Comparative probability for conditional events: a new look through coherence. Theory and Decision 35, 237–258 (1993)
29. Ognjanović, Z., Midić, U., Kratica, J.: A Genetic Algorithm for Probabilistic SAT Problem. In: Rutkowski, L., Siekmann, J.H., Tadeusiewicz, R., Zadeh, L.A. (eds.) ICAISC 2004. LNCS (LNAI), vol. 3070, pp. 462–467. Springer, Heidelberg (2004)
30. Ognjanović, Z., Midić, U., Mladenović, N.: A Hybrid Genetic and Variable Neighborhood Descent for Probabilistic SAT Problem. In: Blesa, M.J., Blum, C., Roli, A., Sampels, M. (eds.) HM 2005. LNCS, vol. 3636, pp. 42–53. Springer, Heidelberg (2005)
31. Jovanovic, D., Mladenovic, N., Ognjanovic, Z.: Variable neighborhood search for the probabilistic satisfiability problem. In: International Conference on Metaheuristics, pp. 557–562 (2005)
32. de Souza Andrade, P.S., da Rocha, J.C.F., Couto, D.P., da Costa Teves, A., Cozman, F.G.: A toolset for propositional probabilistic logic. In: Encontro Nacional de Inteligencia Artificial, 1371–1380 (2007)
33. Klinov, P., Parsia, B.: Probabilistic modeling and OWL: A user oriented introduction into P-$\mathcal{SHIQ}$(D). In: OWL: Experiences and Directions (2008)
34. Castano, S., Ferrara, A., Lorusso, D., Näth, T.H., Möller, R.: Mapping Validation by Probabilistic Reasoning. In: Bechhofer, S., Hauswirth, M., Hoffmann, J., Koubarakis, M. (eds.) ESWC 2008. LNCS, vol. 5021, pp. 170–184. Springer, Heidelberg (2008)

Instance-Based Non-standard Inferences in $\mathcal{EL}$ with Subjective Probabilities

Rafael Peñaloza and Anni-Yasmin Turhan*

Institute for Theoretical Computer Science, TU Dresden, Germany
`last name@tcs.inf.tu-dresden.de`

Abstract. For practical ontology-based applications representing and reasoning with probabilities is an essential task. For Description Logics with subjective probabilities reasoning procedures for testing instance relations based on the completion method have been developed.

In this paper we extend this technique to devise algorithms for solving non-standard inferences for $\mathcal{EL}$ and its probabilistic extension Prob-$\mathcal{EL}_c^{01}$: computing the most specific concept of an individual and finding explanations for instance relations.

1 Introduction

The ontology language recommended for the semantic web OWL [11,25] is based on Description Logics (DLs) [4]. Description logics are knowledge representation formalisms with formal semantics. Based on these semantics, powerful reasoning services have been defined and reasoning algorithms have been investigated. In recent years, so-called lightweight DLs have been devised; these DLs have a limited expressiveness, which allows for efficient reasoning [6]. For the lightweight DL $\mathcal{EL}$, typical DL reasoning services such as classification of TBoxes, i.e., computation of all sub- / superconcept relations of named concepts, or the realization of ABoxes, i.e., computation of the named concepts each of the ABox individuals belongs to, can be done in polynomial time. The basis for ABox realization is *instance checking*, which tests whether a given individual from the ABox belongs to a given concept. In the so-called $\mathcal{EL}$-family of DLs, which are the tractable extensions of $\mathcal{EL}$, this inference can be computed using completion algorithms, which extend the ones for concept subsumption [2,3].

The DLs from the $\mathcal{EL}$-family are employed most prominently in the medical field, for instance in the well-known knowledge base SNOMED CT [23], as well as in context-aware applications. In both of these application areas, the need for characterizing uncertain observations, which are only known to hold with some probability, has been long recognized. While several probabilistic extensions of DLs have been proposed—see [14] for a survey—these are typically very expressive and thus no longer tractable and they cannot handle subjective probabilities. A simple probabilistic variant of $\mathcal{EL}$ that can express subjective probabilities is Prob-$\mathcal{EL}_c^{01}$, recently introduced in [15]. This logic allows only a fairly limited use of uncertainty. More precisely, it is only possible to

---

* Partially supported by the German Research Foundation (DFG) in the Collaborative Research Center 912 "Highly Adaptive Energy-Efficient Computing".

F. Bobillo et al. (Eds.): URSW 2008-2010/UniDL 2010, LNAI 7123, pp. 80–98, 2013.

express that a concept *may* hold ($P_{>0}C$), or that it holds *almost surely* ($P_{=1}C$). Despite its limited expressivity, this logic is interesting due to its nice algorithmic properties; as shown in [15], subsumption and instance checking can also be performed in polynomial time.

In this paper we employ the above mentioned completion algorithms to compute two non-standard inferences for DLs that allow to express subjective probability: the most specific concept and explanation of instance relations in Prob-$\mathcal{EL}_c^{01}$.

Many practical applications that need to represent observed information, such as medical applications or context-aware applications, need to characterize that these observations only hold with certain probability. Furthermore, these applications face the problem that information from different sources does not coincide, e.g., different diagnoses yield differing results. These applications need to "integrate" differing observations for the same state of affairs [24]. A way to determine what information the different information sources agree upon is to represent this information in the ABox by different individuals and then to find a common generalization of these individuals. A description of such a generalization of a group of ABox individuals can be obtained by applying the so-called *bottom-up approach* for constructing knowledge bases [5]. In this approach a set of individuals is generalized into a single concept description by first generating the most specific concept (msc) of each individual and then applying the least common subsumer (lcs) to the set of obtained concept descriptions to extract their commonalities.

The second step, i.e., a computation procedure for the approximate lcs has been investigated for $\mathcal{EL}$ and Prob-$\mathcal{EL}_c^{01}$ in [21]. In this paper we present a similar procedure for the msc. For the Description Logic $\mathcal{EL}$ the msc need not exist [1], if computed with respect to general $\mathcal{EL}$-TBoxes. However, it is still possible to find a concept description that is the msc up to a fixed role-depth. This so-called k-msc is still a generalization of the input, but not necessarily the least one—in this sense, it is only an approximation of the msc. We first describe a practical approach for computing the role-depth bounded msc, based on the polynomial-time completion algorithm for $\mathcal{EL}$, and then extend it to the probabilistic variant Prob-$\mathcal{EL}_c^{01}$. Our algorithms are based upon the completion algorithms for ABox realization in $\mathcal{EL}$ and in Prob-$\mathcal{EL}_c^{01}$ and thus can be easily implemented on top of reasoners of these DLs. All the proofs can be found in [18].

The second non-standard inference that we explore in this paper is the *explanation* of a given consequence. In case a large knowledge base is edited by hand, it is not trivial for the developer to see why a particular consequence holds [10,12]. In our case of instance checking, we want to identify those statements in the TBox and the ABox that cause an instance relationship to follow from the knowledge base. More precisely, we want to compute *minimal axiom sets* (MinAs) that entail the consequence. We compute these sets using a glass-box approach for axiom-pinpointing [22,7]. Even for ontology-based context-aware systems, which may operate on automatically generated ABoxes, the identification of MinAs that cause an unwanted consequence is crucial, since it is the first step to edit the knowledge base such that the consequence is resolved. More than in the crisp case, finding the axioms that entail a consequence for a knowledge base written in a DL with probabilities is a difficult task to do by hand.

A method to compute MinAs for subsumptions in $\mathcal{EL}$ was devised in [9] as an extension of the completion algorithm for TBox classification. In this paper we devise a method to compute MinAs for instance relationships as an extension of the completion algorithm for ABox realization for Prob-$\mathcal{EL}_c^{01}$.

This paper extends earlier work presented in [20,21] by algorithms for computing explanations for instance relationships in $\mathcal{EL}$ and Prob-$\mathcal{EL}_c^{01}$. To the best of our knowledge, explanation has not yet been investigated for DLs that allow to express probabilities. We start this undertaking by giving the basic notions in Section 2. In Section 3 we recall the completion algorithms for ABox realization. Section 4 discusses the computation algorithm for the role-depth bounded msc. In Section 5 we introduce the algorithm for computing explanations.

2 $\mathcal{EL}$ and Prob-$\mathcal{EL}$

In this section we introduce the DL $\mathcal{EL}$ and its probabilistic variant Prob-$\mathcal{EL}_c^{01}$. Let N_I, N_C and N_R be disjoint sets of *individual-*, *concept-* and *role names*, respectively. Prob-$\mathcal{EL}_c^{01}$*-concept descriptions* are built using the syntax rule

$$C, D ::= \top \mid A \mid C \sqcap D \mid \exists r.C \mid P_{>0}C \mid P_{=1}C,$$

where $A \in N_C$, and $r \in N_R$. $\mathcal{EL}$*-concept descriptions* are Prob-$\mathcal{EL}_c^{01}$-concept description that do not contain the constructors $P_{>0}$ or $P_{=1}$.

A *knowledge base* $\mathcal{K} = (\mathcal{T}, \mathcal{A})$ consists of a TBox $\mathcal{T}$ and an ABox $\mathcal{A}$. An $\mathcal{EL}$-(Prob-$\mathcal{EL}_c^{01}$-) *TBox* is a finite set of *general concept inclusions* (GCIs) of the form $C \sqsubseteq D$, where C, D are $\mathcal{EL}$- (Prob-$\mathcal{EL}_c^{01}$-) concept descriptions. An $\mathcal{EL}$-*ABox* is a set of assertions of the form $C(a)$ or $r(a, b)$, where C is an $\mathcal{EL}$-concept description, $r \in N_R$, and $a, b \in N_I$. A *Prob-$\mathcal{EL}_c^{01}$-ABox* is a set of assertions of the form $C(a)$, $r(a, b)$, $P_{>0}r(a, b)$, or $P_{=1}r(a, b)$, where C is a Prob-$\mathcal{EL}_c^{01}$-concept description, $r \in N_R$, and $a, b \in N_I$.

The semantics of $\mathcal{EL}$ is defined by means of interpretations $\mathcal{I} = (\Delta^{\mathcal{I}}, \cdot^{\mathcal{I}})$ consisting of a non-empty *domain* $\Delta^{\mathcal{I}}$ and an *interpretation function* $\cdot^{\mathcal{I}}$ that assigns binary relations on $\Delta^{\mathcal{I}}$ to role names, subsets of $\Delta^{\mathcal{I}}$ to concepts and elements of $\Delta^{\mathcal{I}}$ to individual names. For a more detailed description of this semantics, see [4].

We say that the interpretation $\mathcal{I}$ *satisfies* a general concept inclusion $C \sqsubseteq D$, denoted as $\mathcal{I} \models C \sqsubseteq D$, if $C^{\mathcal{I}} \subseteq D^{\mathcal{I}}$; it *satisfies* an assertion $C(a)$, denoted as $\mathcal{I} \models C(a)$ if $a^{\mathcal{I}} \in C^{\mathcal{I}}$ and it *satisfies* an assertion $r(a, b)$, denoted as $\mathcal{I} \models r(a, b)$ if $(a^{\mathcal{I}}, b^{\mathcal{I}}) \in r^{\mathcal{I}}$. It is a *model* of a knowledge base $\mathcal{K} = (\mathcal{T}, \mathcal{A})$ if it satisfies all GCIs in $\mathcal{T}$ and all assertions in $\mathcal{A}$.

The semantics of Prob-$\mathcal{EL}_c^{01}$ is a generalization of the semantics of $\mathcal{EL}$, that considers a set of possible worlds. A *probabilistic interpretation* is of the form

$$\mathcal{I} = (\Delta^{\mathcal{I}}, W, (\mathcal{I}_w)_{w \in W}, \mu),$$

where $\Delta^{\mathcal{I}}$ is the (non-empty) *domain*, W is a (non-empty) set of *worlds*, μ is a discrete probability distribution on W, and for each world $w \in W$, $\mathcal{I}_w$ is a classical $\mathcal{EL}$ interpretation with domain $\Delta^{\mathcal{I}}$, where $a^{\mathcal{I}_w} = a^{\mathcal{I}_{w'}}$ for all $a \in N_I$, $w, w' \in W$. The probability that a given element of the domain $d \in \Delta^{\mathcal{I}}$ belongs to the interpretation of a concept name A is

$$p_d^{\mathcal{I}}(A) := \mu(\{w \in W \mid d \in A^{\mathcal{I}_w}\}).$$

The functions $\mathcal{I}_w$ and $p_d^{\mathcal{I}}$ are extended to complex concepts in the usual way for the classical $\mathcal{EL}$-constructors, where the extension to the new constructors P_* is defined as

$$(P_{>0}C)^{\mathcal{I}_w} := \{d \in \Delta^{\mathcal{I}} \mid p_d^{\mathcal{I}}(C) > 0\},$$
$$(P_{=1}C)^{\mathcal{I}_w} := \{d \in \Delta^{\mathcal{I}} \mid p_d^{\mathcal{I}}(C) = 1\}.$$

The probabilistic interpretation $\mathcal{I}$ *satisfies* a general concept inclusion $C \sqsubseteq D$, denoted as $\mathcal{I} \models C \sqsubseteq D$, if for every $w \in W$ it holds that $C^{\mathcal{I}_w} \subseteq D^{\mathcal{I}_w}$. It is a *model* of a TBox $\mathcal{T}$ if it satisfies all general concept inclusions in $\mathcal{T}$. Let C, D be two Prob-$\mathcal{EL}_c^{01}$ concepts and $\mathcal{T}$ a TBox. We say that C is *subsumed* by D w.r.t. $\mathcal{T}$ ($C \sqsubseteq_{\mathcal{T}} D$) if for every model $\mathcal{I}$ of $\mathcal{T}$ it holds that $\mathcal{I} \models C \sqsubseteq D$. The concepts c and D are equivalent, if $C \sqsubseteq_{\mathcal{T}} D$ and $D \sqsubseteq_{\mathcal{T}} C$ holds. The probabilistic interpretation $\mathcal{I}$ *satisfies* the assertion $P_{>0}r(a,b)$ if $\mu(\{w \in W \mid \mathcal{I}_w \models r(a,b)\}) > 0$, and analogously for $P_{=1}r(a,b)$. $\mathcal{I}$ *satisfies* the ABox $\mathcal{A}$ if there is a $w \in W$ such that $\mathcal{I}_w \models \mathcal{A}$.

Finally, an individual $a \in N_I$ is an *instance* of a concept description C w.r.t. $\mathcal{K}$ ($\mathcal{K} \models C(a)$) if $\mathcal{I} \models C(a)$ for all models $\mathcal{I}$ of $\mathcal{K}$. The *ABox realization problem* is to compute for each individual a in $\mathcal{A}$ the set of named concepts from $\mathcal{K}$ that have a as an instance and that are least (w.r.t. $\sqsubseteq$). One of our main interests in this paper is to compute most specific concepts.

Definition 1 (most specific concept). *Let $\mathcal{L}$ be a DL, $\mathcal{K} = (\mathcal{T}, \mathcal{A})$ be a $\mathcal{L}$-knowledge base. The* most specific concept *(msc) of an individual a from $\mathcal{A}$ is the $\mathcal{L}$-concept description C s. t.*

1. $\mathcal{K} \models C(a)$, *and*
2. *for each $\mathcal{L}$-concept description D holds: $\mathcal{K} \models D(a)$ implies $C \sqsubseteq_{\mathcal{T}} D$.*

The msc depends on the DL in use. For the DLs with conjunction as concept constructor the msc is, if it exists, unique up to equivalence. Thus it is justified to speak of *the* msc.

3   Completion Algorithms for ABox Realization

In this section we briefly sketch the completion algorithms for instance checking in the DLs $\mathcal{EL}$ [2] and Prob-$\mathcal{EL}_c^{01}$ [15].

3.1   The Completion Algorithm for $\mathcal{EL}$

Assume we want to test for an $\mathcal{EL}$-knowledge base $\mathcal{K} = (\mathcal{T}, \mathcal{A})$ whether $\mathcal{K} \models D(a)$ holds. The completion algorithm first augments the knowledge base by introducing a concept name for the complex concept description D for the instance check; that is, it redefines the knowledge base to $\mathcal{K} = (\mathcal{T} \cup \{A_q \equiv D\}, \mathcal{A})$, where A_q is a new concept name not appearing in $\mathcal{K}$. The instance checking algorithm for $\mathcal{EL}$ works on knowledge

NF1 $C \sqcap \hat{D} \sqsubseteq E \longrightarrow \{ \hat{D} \sqsubseteq A, C \sqcap A \sqsubseteq E \}$

NF2 $\exists r.\hat{C} \sqsubseteq D \longrightarrow \{ \hat{C} \sqsubseteq A, \exists r.A \sqsubseteq D \}$

NF3 $\hat{C} \sqsubseteq \hat{D} \longrightarrow \{ \hat{C} \sqsubseteq A, A \sqsubseteq \hat{D} \}$

NF4 $B \sqsubseteq \exists r.\hat{C} \longrightarrow \{ B \sqsubseteq \exists r.A, A \sqsubseteq \hat{C} \}$

NF5 $B \sqsubseteq C \sqcap D \longrightarrow \{ B \sqsubseteq C, B \sqsubseteq D \}$

where $\hat{C}, \hat{D} \notin \mathsf{BC}_\mathcal{T}$ and A is a new concept name.

Fig. 1. $\mathcal{EL}$ normalization rules (from [2])

bases containing only axioms in a structured *normal form*. Every knowledge base can be transformed into a normalized one via a two-step procedure.

First the ABox is transformed into a simple ABox. An ABox $\mathcal{A}$ is a *simple ABox*, if for every concept assertion $C(a) \in \mathcal{A}$, C is a concept name. An arbitrary $\mathcal{EL}$-ABox $\mathcal{A}$ can be transformed into a simple ABox by first replacing each complex assertion $C(A)$ in $\mathcal{A}$ by $A(a)$ where A is a fresh concept name and, second, introducing $A \equiv C$ into the TBox.

After this step, the TBox is transformed into a normal form as well. For a concept description C let $\mathsf{CN}(C)$ denote the set of all concept names and $\mathsf{RN}(C)$ denote the set of all role names that appear in C. The *signature of a concept description C* (denoted $\mathsf{sig}(C)$) is given by $\mathsf{CN}(C) \cup \mathsf{RN}(C)$. Similarly, the set of concept (respectively role) names that appear in a TBox is denoted by $\mathsf{CN}(\mathcal{T})$ (respectively $\mathsf{RN}(\mathcal{T})$). The *signature of a TBox $\mathcal{T}$* (denoted $\mathsf{sig}(\mathcal{T})$) is $\mathsf{CN}(\mathcal{T}) \cup \mathsf{RN}(\mathcal{T})$. The *signature of an ABox $\mathcal{A}$* (denoted $\mathsf{sig}(\mathcal{A})$) is the set of concept (role / individual) names $\mathsf{CN}(\mathcal{A})$ ($\mathsf{RN}(\mathcal{A})/\mathsf{IN}(\mathcal{A})$ resp.) that appear in $\mathcal{A}$. The signature of a knowledge base $\mathcal{K} = (\mathcal{T}, \mathcal{A})$ (denoted $\mathsf{sig}(\mathcal{K})$) is $\mathsf{sig}(\mathcal{T}) \cup \mathsf{sig}(\mathcal{A})$.

An $\mathcal{EL}$-TBox $\mathcal{T}$ is in *normal form* if all concept axioms have one of the following forms, where $C_1, C_2 \in \mathsf{sig}(\mathcal{T})$ and $D \in \mathsf{sig}(\mathcal{T}) \cup \{\bot\}$:

$$C_1 \sqsubseteq D, \quad C_1 \sqcap C_2 \sqsubseteq D, \quad C_1 \sqsubseteq \exists r.C_2 \quad \text{or} \quad \exists r.C_1 \sqsubseteq D.$$

Any $\mathcal{EL}$-TBox can be transformed into normal form by introducing new concept names and by applying the normalization rules displayed in Figure 1 exhaustively, where $\mathsf{BC}_\mathcal{T}$ is the set containing all the concept names appearing in $\mathcal{T}$ and the concept $\top$. These rules replace the GCI on the left-hand side of the rules with the set of GCIs on the right-hand side. Clearly, for a knowledge base $\mathcal{K} = (\mathcal{T}, \mathcal{A})$ the signature of $\mathcal{A}$ may be changed only during the first of the two normalization steps and the signature of $\mathcal{T}$ may be extended during both of them. The normalization of the knowledge base can be done in linear time.

The completion algorithm for instance checking is based on the one for classifying $\mathcal{EL}$-TBoxes introduced in [2]. The completion algorithm constructs a representation of the minimal model of $\mathcal{K}$. Let $\mathcal{K} = (\mathcal{T}, \mathcal{A})$ be a normalized $\mathcal{EL}$-knowledge base, i.e., with a simple ABox $\mathcal{A}$ and a TBox $\mathcal{T}$ in normal form. The completion algorithm works on four kinds of *completion sets*: $S(a), S(a, r), S(C)$ and $S(C, r)$ for each $a \in \mathsf{IN}(\mathcal{A})$, $C \in \mathsf{CN}(\mathcal{K})$ and $r \in \mathsf{RN}(\mathcal{K})$. These completion sets contain concept names

CR1 If $C \in S(X), C \sqsubseteq D \in \mathcal{T}$, and $D \notin S(X)$
then $S(X) := S(X) \cup \{D\}$

CR2 If $C_1, C_2 \in S(X), C_1 \sqcap C_2 \sqsubseteq D \in \mathcal{T}$, and $D \notin S(X)$
then $S(X) := S(X) \cup \{D\}$

CR3 If $C \in S(X), C \sqsubseteq \exists r.D \in \mathcal{T}$, and $D \notin S(X,r)$
then $S(X,r) := S(X,r) \cup \{D\}$

CR4 If $Y \in S(X,r), C \in S(Y), \exists r.C \sqsubseteq D \in \mathcal{T}$, and
$D \notin S(X)$ then $S(X) := S(X) \cup \{D\}$

Fig. 2. $\mathcal{EL}$ completion rules

from $\mathsf{CN}(\mathcal{K})$. Intuitively, the completion rules make implicit subsumption and instance relationships explicit in the following sense:

- $D \in S(C)$ implies that $C \sqsubseteq_{\mathcal{T}} D$,
- $D \in S(C,r)$ implies that $C \sqsubseteq_{\mathcal{T}} \exists r.D$.
- $D \in S(a)$ implies that a is an instance of D w.r.t. $\mathcal{K}$,
- $D \in S(a,r)$ implies that a is an instance of $\exists r.D$ w.r.t. $\mathcal{K}$.

$S_{\mathcal{K}}$ denotes the set of all completion sets of a normalized $\mathcal{K}$. The completion sets are initialized for each $a \in \mathsf{IN}(\mathcal{A})$ and each $C \in \mathsf{CN}(\mathcal{K})$ as follows:

- $S(C) := \{C, \top\}$ for each $C \in \mathsf{CN}(\mathcal{K})$,
- $S(C,r) := \emptyset$ for each $r \in \mathsf{RN}(\mathcal{K})$,
- $S(a) := \{C \in \mathsf{CN}(\mathcal{A}) \mid C(a) \text{ appears in } \mathcal{A}\} \cup \{\top\}$, and
- $S(a,r) := \{b \in \mathsf{IN}(\mathcal{A}) \mid r(a,b) \text{ appears in } \mathcal{A}\}$ for each $r \in \mathsf{RN}(\mathcal{K})$.

Then these sets are extended by applying the completion rules shown in Figure 2 until no more rule applies. In these rules X and Y can refer to concept or individual names, while C, C_1, C_2 and D are concept names and r is a role name. After the completion has terminated, the following relations hold between an individual a, a role r and named concepts A and B:

- subsumption relation between A and B from $\mathcal{K}$ holds iff $B \in S(A)$
- instance relation between a and B from $\mathcal{K}$ holds iff $B \in S(a)$,

as shown in [2]. To decide the initial query: $\mathcal{K} \models D(a)$, one has to test whether A_q appears in $S(a)$. In fact, instance queries for all individuals and all named concepts from the knowledge base can be answered from the resulting completion sets; the completion algorithm does not only perform one instance check, but complete ABox realization. The completion algorithm runs in polynomial time in size of the knowledge base.

3.2 The Completion Algorithm for Prob-$\mathcal{EL}_c^{01}$

Before describing the completion algorithm for Prob-$\mathcal{EL}_c^{01}$, we modify the notion of basic concepts. The set $\mathsf{BC}_{\mathcal{T}}$ of Prob-$\mathcal{EL}_c^{01}$ *basic concepts* for a knowledge base $\mathcal{K}$ is the smallest set that contains

PR1	If $C' \in S_*(X,v)$ and $C' \sqsubseteq D \in \mathcal{T}$, then $S_*(X,v) := S_*(X,v) \cup \{D\}$
PR2	If $C_1, C_2 \in S_*(X,v)$ and $C_1 \sqcap C_2 \sqsubseteq D \in \mathcal{T}$, then $S_*(X,v) := S_*(X,v) \cup \{D\}$
PR3	If $C' \in S_*(X,v)$ and $C' \sqsubseteq \exists r.D \in \mathcal{T}$, then $S_*(X,r,v) := S_*(X,r,v) \cup \{D\}$
PR4	If $D \in S_*(X,r,v)$, $D' \in S_{\gamma(v)}(D, \gamma(v))$ and $\exists r.D' \sqsubseteq E \in \mathcal{T}$, then $S_*(X,v) := S_*(X,v) \cup \{E\}$
PR5	If $P_{>0}A \in S_*(X,v)$, then $S_*(X, P_{>0}A) := S_*(X, P_{>0}A) \cup \{A\}$
PR6	If $P_{=1}A \in S_*(X,v)$, $v \neq 0$, then $S_*(X,v) := S_*(X,v) \cup \{A\}$
PR7	If $A \in S_*(X,v)$ and $v \neq 0$, $P_{>0}A \in \mathcal{P}_0^{\mathcal{T}}$, then $S_*(X,v') := S_*(X,v') \cup \{P_{>0}A\}$
PR8	If $A \in S_*(X,1)$ and $P_{=1}A \in \mathcal{P}_1^{\mathcal{T}}$, then $S_*(X,v) := S_*(X,v) \cup \{P_{=1}A\}$
PR9	If $r(a,b) \in \mathcal{A}$, $C \in S(b,0), \exists r.C \sqsubseteq D \in \mathcal{T}$, then $S(a,0) := S(a,0) \cup \{D\}$
PR10	If $P_{>0}r(a,b) \in \mathcal{A}$, $C \in S(b, P_{>0}r(a,b))$ and $\exists r.C \sqsubseteq D \in \mathcal{T}$, then $S(a, P_{>0}r(a,b)) := S(a, P_{>0}r(a,b)) \cup \{D\}$
PR11	If $P_{=1}r(a,b) \in \mathcal{A}$, $C \in S(b,v)$ with $v \neq 0$ and $\exists r.C \sqsubseteq D \in \mathcal{T}$, then $S(a,v) := S(a,v) \cup \{D\}$

Fig. 3. Prob-$\mathcal{EL}_c^{01}$ completion rules

1. the concept $\top$,
2. all concept names used in $\mathcal{K}$, and
3. all concepts of the form $P_{>0}A$ or $P_{=1}A$,

where A is a concept name in $\mathcal{K}$. A Prob-$\mathcal{EL}_c^{01}$-TBox $\mathcal{T}$ is in *normal form* if all its axioms are of one of the following forms

$$C \sqsubseteq D, \quad C_1 \sqcap C_2 \sqsubseteq D, \quad C \sqsubseteq \exists r.A, \quad \exists r.A \sqsubseteq D,$$

where $C, C_1, C_2, D \in \mathsf{BC}_{\mathcal{T}}$ and A is a concept name. The normalization rules in Figure 1 can also be used to transform a Prob-$\mathcal{EL}_c^{01}$-TBox into this extended normal form. We still assume that the ABox $\mathcal{A}$ is a simple ABox; that is, for all assertions $C(a)$ in $\mathcal{A}$, C is a concept name. We denote as $\mathcal{P}_0^{\mathcal{T}}$ and $\mathcal{P}_1^{\mathcal{T}}$ the set of all concepts of the form $P_{>0}A$ and $P_{=1}A$ respectively, occurring in a normalized knowledge base $\mathcal{K}$. Analogously, $\mathcal{R}_0^{\mathcal{T}}$ denotes the set of all assertions of the form $P_{>0}r(a,b)$ appearing in $\mathcal{K}$.

The completion algorithm for Prob-$\mathcal{EL}_c^{01}$ follows the same idea as the algorithm for $\mathcal{EL}$, but uses several completion sets to deal with the information of what needs to be satisfied in the different worlds of a model. Intuitively, we will build a general description of all models, using the set of worlds $V := \{0, \varepsilon, 1\} \cup \mathcal{P}_0^{\mathcal{T}} \cup \mathcal{R}_0^{\mathcal{T}}$, where the probability distribution μ assigns a probability of 0 to the world 0, and the uniform probability $1/(|V|-1)$ to all other worlds. The main idea is that the world 1 will include all the entailments that hold with probability 1, and ε those that hold with probability greater than 0.

For each individual name a, concept name A, role name r and world v, we store the completion sets $S_0(A, v), S_\varepsilon(A, v), S_0(A, r, v), S_\varepsilon(A, r, v), S(a, v)$, and $S(a, r, v)$.

The algorithm initializes the sets as follows for every $A \in \mathsf{BC}_\mathcal{T}, r \in \mathsf{RN}(\mathcal{K})$, and $a \in \mathsf{IN}(\mathcal{A})$:

- $S_0(A, 0) = \{\top, A\}$ and $S_0(A, v) = \{\top\}$ for all $v \in V \setminus \{0\}$,
- $S_\varepsilon(A, \varepsilon) = \{\top, A\}$ and $S_\varepsilon(A, v) = \{\top\}$ for all $v \in V \setminus \{\varepsilon\}$,
- $S(a, 0) = \{\top\} \cup \{A \mid A(a) \in \mathcal{A}\}$, $S(a, v) = \{\top\}$ for all $v \neq 0$,
- $S_0(A, r, v) = S_\varepsilon(A, r, v) = \emptyset$ for all $v \in V$, $S(a, r, v) = \emptyset$ for $v \neq 0$,
- $S(a, r, 0) = \{b \in \mathsf{IN}(\mathcal{A}) \mid r(a, b) \in \mathcal{A}\}$.

These sets are then extended by exhaustively applying the rules shown in Figure 3, where X ranges over $\mathsf{BC}_\mathcal{T} \cup \mathsf{IN}(\mathcal{A})$, $S_*(X, v)$ stands for $S(X, v)$ if X is an individual and for $S_0(X, v), S_\varepsilon(X, v)$ if $X \in \mathsf{BC}_\mathcal{T}$, and $\gamma : V \to \{0, \varepsilon\}$ is defined by $\gamma(0) = 0$, and $\gamma(v) = \varepsilon$ for all $v \in V \setminus \{0\}$.

This algorithm terminates in polynomial time. After termination, the completion sets store all the information necessary to decide subsumption of concept names, as well as checking whether an individual is an instance of a given concept name [15]. For the former decision, it holds that for every pair A, B of concept names: $B \in S_0(A, 0)$ iff $A \sqsubseteq_\mathcal{K} B$. In the case of instance checking, we have that $\mathcal{K} \models A(a)$ iff $A \in S(a, 0)$.

4   Computing the k-MSC Using Completion

The msc was first investigated for $\mathcal{EL}$-concept descriptions and w.r.t. unfoldable TBoxes and possibly cyclic ABoxes in [13]. It was shown that the msc does not need to exists for cyclic ABoxes. Consider the ABox $\mathcal{A} = \{r(a, a), C(a)\}$. The msc of a is then

$$C \sqcap \exists r.(C \sqcap \exists r.(C \sqcap \exists r.(C \sqcap \cdots$$

and cannot be expressed by a finite concept description. For cyclic TBoxes it has been shown in [1] that the msc does not need to exists even if the ABox is acyclic.

To avoid infinite nestings in presence of cyclic ABoxes it was proposed in [13] to limit the role-depth of the concept description to be computed. This limitation yields an approximation of the msc, which is still a concept description with the input individual as an instance, but it does not need to be the least one (w.r.t. $\sqsubseteq$) with this property. We follow this idea to compute approximations of the msc also in presence of general TBoxes.

The *role-depth* of a concept description C (denoted $rd(C)$) is the maximal number of nested quantifiers of C. This allows us to define the msc with limited role-depth for $\mathcal{EL}$.

Definition 2 (role-depth bounded $\mathcal{EL}$-msc). *Let $\mathcal{K} = (\mathcal{T}, \mathcal{A})$ be an $\mathcal{EL}$-knowledge base and a an individual in $\mathcal{A}$ and $k \in \mathbb{N}$. Then the $\mathcal{EL}$-concept description C is the role-depth bounded $\mathcal{EL}$-most specific concept of a w.r.t. $\mathcal{K}$ and role-depth k (written $k\text{-}msc_\mathcal{K}(a)$) iff*

1. *$rd(C) \leq k$,*
2. *$\mathcal{K} \models C(a)$, and*
3. *for all $\mathcal{EL}$-concept descriptions E with $rd(E) \leq k$ holds: $\mathcal{K} \models E(a)$ implies $C \sqsubseteq_\mathcal{T} E$.*

Notice that in case the exact msc has a role-depth less or equal to k the role-depth bounded msc is the exact msc.

Example 3. As an example we consider the labeled knowledge base $\mathcal{K}_{ex} = (\mathcal{T}_{ex}, \mathcal{A}_{ex})$. In this labeled knowledge base each axiom and assertion is associated with a label (printed in the same line), which will be used later.

$$
\begin{aligned}
\mathcal{T}_{ex} = \{\, \exists r.\top \sqsubseteq A, && ax1 && \text{and} && \mathcal{A}_{ex} = \{\, B(a), && as1 \\
B \sqsubseteq \exists r.C, && ax2 &&&&& D(b), && as2 \\
D \sqsubseteq E\} && ax3 &&&&& r(a,b), && as3 \\
&&&&&&& s(a,c), && as4 \\
&&&&&&& r(c,a)\,\} && as5
\end{aligned}
$$

Obviously the ABox $\mathcal{A}_{ex}$ is cyclic due to the last two assertions. Note, that c is an instance of A due to $as5$ and $ax1$. Now, for $k = 3$ we obtain the following role-depth bounded msc for a:

$$
\begin{aligned}
3\text{-}msc_{\mathcal{K}_{ex}}(a) = \; & B \sqcap \\
& \exists r.D \sqcap \\
& \exists s.(A \sqcap \exists r.(B \sqcap \exists r.D \sqcap \exists s.A))).
\end{aligned}
$$

Next we describe how to obtain the k-msc in general.

4.1   Computing the k-msc in $\mathcal{EL}$ by Completion

The computation of the msc relies on a characterization of the instance relation. While in earlier works this was given by homomorphisms [13] or simulations [1] between graph representations of the knowledge base and the concept in question, we use the completion algorithm as such a characterization. Moreover, we construct the msc by traversing the completion sets to "collect" the msc. More precisely, the set of completion sets encodes a graph structure, where the sets $S(X)$ are the nodes and the sets $S(X, r)$ encode the edges. Traversing this graph structure, one can construct an $\mathcal{EL}$-concept. To obtain a finite concept in the presence of cyclic ABoxes or TBoxes one can limit the number of edges than can be traversed during this construction.

Definition 4 (traversal concept). *Let $\mathcal{K}$ be an $\mathcal{EL}$-knowledge base, $\mathcal{K}'$ be its normalized form, $S_{\mathcal{K}}$ the completion set obtained from $\mathcal{K}$ and $k \in \mathbb{N}$. Then the traversal concept of a named concept A (denoted k-$C_{S_{\mathcal{K}}}(A)$) with $\mathrm{sig}(A) \subseteq \mathrm{sig}(\mathcal{K}')$ is the concept obtained from executing the procedure call* traversal-concept-c(A, $S_{\mathcal{K}}$, k) *shown in Algorithm 1.*

The traversal concept of an individual a (denoted k-$C_{S_{\mathcal{K}}}(a)$) with $a \in \mathrm{sig}(\mathcal{K})$ is the concept description obtained from executing the procedure call traversal-concept-i(a, $S_{\mathcal{K}}$, k) *shown in Algorithm 1.*

The idea is that the traversal concept of an individual yields its msc. However, the traversal concept contains names from $\mathrm{sig}(\mathcal{K}') \setminus \mathrm{sig}(\mathcal{K})$, i.e., concept names that were introduced during normalization—we call this kind of concept names *normalization names* in the following. The returned msc should be formulated w.r.t. the signature of the original knowledge base, thus the normalization names need to be removed or replaced.

Algorithm 1. Computation of a role-depth bounded $\mathcal{EL}$-msc

Procedure k-msc $(a, \mathcal{K}, k)$
Input: a: individual from $\mathcal{K}$; $\mathcal{K} =(\mathcal{T}, \mathcal{A})$ an $\mathcal{EL}$-knowledge base; $k \in \mathbb{N}$
Output: role-depth bounded $\mathcal{EL}$-msc of a w.r.t. $\mathcal{K}$ and k.
1: $(\mathcal{T}', \mathcal{A}') := \mathsf{simplify\text{-}ABox}(\mathcal{T}, \mathcal{A})$
2: $\mathcal{K}' := (\mathsf{normalize}(\mathcal{T}'), \mathcal{A}')$
3: $\mathsf{S}_\mathcal{K} := \mathsf{apply\text{-}completion\text{-}rules}(\mathcal{K})$
4: **return** Remove-normalization-names (traversal-concept-i$(a, \mathsf{S}_\mathcal{K}, k)$)

Procedure traversal-concept-i (a, S, k)
Input: a: individual name from $\mathcal{K}$; S: set of completion sets; $k \in \mathbb{N}$
Output: role-depth traversal concept (w.r.t. $\mathcal{K}$) and k.
1: **if** $k = 0$ **then return** $\bigsqcap_{A \in \mathsf{S}(a)} A$
2: **else return** $\bigsqcap_{A \in \mathsf{S}(a)} A \sqcap$

$$\bigsqcap_{r \in \mathsf{RN}(\mathcal{K}')} \bigsqcap_{A \in \mathsf{CN}(\mathcal{K}') \cap \mathsf{S}(a,r)} \exists r. \text{ traversal-concept-c } (A, \mathsf{S}, k-1) \sqcap$$
$$\bigsqcap_{r \in \mathsf{RN}(\mathcal{K}')} \bigsqcap_{b \in \mathsf{IN}(\mathcal{K}') \cap \mathsf{S}(a,r)} \exists r. \text{ traversal-concept-i } (b, \mathsf{S}, k-1)$$

3: **end if**

Procedure traversal-concept-c (A, S, k)
Input: A: concept name from $\mathcal{K}'$; S: set of completion sets; $k \in \mathbb{N}$
Output: role-depth bounded traversal concept.
1: **if** $k = 0$ **then return** $\bigsqcap_{B \in \mathsf{S}(A)} B$
2: **else return** $\bigsqcap_{B \in \mathsf{S}(A)} B \sqcap \bigsqcap_{r \in \mathsf{RN}(\mathcal{K}')} \bigsqcap_{B \in \mathsf{S}(A,r)} \exists r.\text{traversal-concept-c } (B, \mathsf{S}, k-1)$
3: **end if**

Lemma 5. *Let $\mathcal{K}$ be an $\mathcal{EL}$-knowledge base, $\mathcal{K}'$ its normalized version, $\mathsf{S}_\mathcal{K}$ be the set of completion sets obtained for $\mathcal{K}$, $k \in \mathbb{N}$ a natural number and $a \in \mathsf{IN}(\mathcal{K})$. If $C = k\text{-}\mathsf{C}_{\mathsf{S}_\mathcal{K}}(a)$ and $\widehat{C}$ is obtained from C by removing the normalization names, then*

$$\mathcal{K}' \models C(a) \text{ iff } \mathcal{K} \models \widehat{C}(a).$$

This lemma guarantees that removing the normalization names from the traversal concept preserves the instance relationships. Intuitively, this lemma holds since the construction of the traversal concept conjoins exhaustively all named subsumers and all subsuming existential restrictions to a normalization name up to the role-depth bound. Thus removing the normalization name does not change the extension of the conjunction. The proof can be found in [18].

The procedure k-msc uses an individual a from a knowledge base $\mathcal{K}$, the knowledge base $\mathcal{K}$ itself and a number k for the role depth-bound as parameters. It first performs the two normalization steps on $\mathcal{K}$, then applies the completion rules from Figure 2 to the normalized knowledge base $\mathcal{K}'$, and then stores the set of completion sets in $\mathsf{S}_\mathcal{K}$. Afterwards it computes the traversal-concept of a from $\mathsf{S}_\mathcal{K}$ w.r.t. role-depth bound k. In a post-processing step it applies Remove-normalization-names to the traversal concept obtained in the previous step.

Example 6. We use the knowledge base from Example 3, to apply the algorithm k-msc to the individual a from $\mathcal{A}_{ex}$ again for $k = 3$. Since the TBox $\mathcal{T}_{ex}$ is in normal form and the ABox $\mathcal{A}_{ex}$ is simple, completion can be applied directly. After completion we have the following elements in the completion sets:

$$S(A) = \{\top, A\}$$
$$S(B) = \{\top, A, B\} \qquad S(a) = \{\top, A, B\} \qquad S(B, r) = \{\top, C\}$$
$$S(C) = \{\top, C\} \qquad S(b) = \{\top, D, E\} \qquad S(a, r) = \{\top, D, E\}$$
$$S(D) = \{\top, D, E\} \qquad S(c) = \{\top, A\} \qquad S(a, s) = \{\top, A\}$$
$$S(c, r) = \{\top, A\}$$

The here omitted completion sets do not change after initialization and are empty. We obtain:

$$\text{k-msc}(a, \mathcal{K}_{ex}, 3) = \top \sqcap A \sqcap B \sqcap$$
$$\exists r.(\top \sqcap D \sqcap E) \sqcap$$
$$\exists s.(\top \sqcap A \sqcap \exists r.(\top \sqcap A \sqcap B \sqcap \exists r.(\top \sqcap D \sqcap E) \sqcap \exists s.(\top \sqcap A))).$$

The resulting concept description is larger than the $k\text{-}msc$ derived in Example 3, since all the elements from the completion set are conjoined to the result concept description in traversal-concept-i and traversal-concept-c. However, it is easy to see that the result is a concept description equivalent to the $k\text{-}msc$ w.r.t. $\mathcal{K}_{ex}$.

Obviously, the concept description returned from the procedure k-msc has a role-depth less or equal to k. Thus the first condition of Definition 2 is fulfilled. As we prove next, the concept description obtained from the procedure k-msc fulfills also the second condition from Definition 2.

Lemma 7. *Let* $\mathcal{K} = (\mathcal{T}, \mathcal{A})$ *be an* $\mathcal{EL}$-*knowledge base and* a *an individual in* $\mathcal{A}$ *and* $k \in \mathbb{N}$. *If* $C = \text{k-msc}(a, \mathcal{K}, k)$, *then* $\mathcal{K} \models C(a)$.

The claim can be shown by induction on k. Each name in C is from a completion set of (1) an individual or (2) a concept, which is connected via existential restrictions to an individual. The full proof can be found in [18].

Lemma 8. *Let* $\mathcal{K} = (\mathcal{T}, \mathcal{A})$ *be an* $\mathcal{EL}$-*knowledge base,* a *an individual appearing in* $\mathcal{A}$, *and* $k \in \mathbb{N}$. *If* $C = \text{k-msc}(a, \mathcal{K}, k)$, *then for every* $\mathcal{EL}$-*concept description* E *with* $rd(E) \leq k$ *the following holds:* $\mathcal{K} \models E(a)$ *implies* $C \sqsubseteq_{\mathcal{T}} E$.

Again, the full proof can be found in [18]. Together, these two Lemmas yield the correctness of the overall procedure.

Theorem 9. *Let* $\mathcal{K} = (\mathcal{T}, \mathcal{A})$ *be an* $\mathcal{EL}$-*knowledge base and* a *an individual in* $\mathcal{A}$ *and* $k \in \mathbb{N}$.
Then $\text{k-msc}(a, \mathcal{K}, k) \equiv k\text{-}msc_{\mathcal{K}}(a)$.

It is important to notice that, while the completion sets can be computed in polynomial time, the $k\text{-}msc$ can grow exponential in the size of the knowledge base. In addition to that and as the example already indicated, the concept description obtained from k-msc contains a lot of redundant information and thus is quite larger. However for practical usability it is necessary to rewrite the concept to an equivalent, but smaller one. A heuristic for this has been proposed in [16]. The algorithm and the rewriting heuristic are implemented in the GEL system[1].

---

[1] See http://gen-el.sourceforge.net/

4.2 Computing the k-msc in Prob-$\mathcal{EL}_c^{01}$ by Completion

The role-bounded msc for a Prob-$\mathcal{EL}_c^{01}$-knowledge base can be computed in a similar fashion to the one described before for $\mathcal{EL}$. The knowledge base is first normalized and the completion procedure is executed to obtain all the completion sets.

In order to compute the msc, we simply accumulate all concepts to which the individual a belongs, given the information stored in the completion sets. This process needs to be done recursively in order to account for both, the successors of a explicitly encoded in the ABox, and the nesting of existential restrictions masked by normalization names. In the following we use the abbreviation $S^{>0}(a,r) := \bigcup_{v \in V \setminus \{0\}} S(a,r,v)$. We then define traversal-concept-i(a, S, k) as

$$\bigsqcap_{B \in S(a,0)} B \sqcap \bigsqcap_{r \in \mathsf{RN}(\mathcal{K}'')} \Big(\bigsqcap_{r(a,b) \in \mathcal{K}''} \exists r.\text{traversal-concept-i}(b, \mathsf{S}, k-1) \sqcap$$

$$\bigsqcap_{B \in \mathsf{CN}(\mathcal{K}'') \cap S(a,r,0)} \exists r.\text{traversal-concept-c}(B, \mathsf{S}, k-1) \sqcap$$

$$\bigsqcap_{B \in \mathsf{CN}(\mathcal{K}'') \cap S(a,r,1)} P_{=1}(\exists r.\text{traversal-concept-c}(B, \mathsf{S}, k-1)) \sqcap$$

$$\bigsqcap_{B \in \mathsf{CN}(\mathcal{K}'') \cap S^{>0}(a,r)} P_{>0}(\exists r.\text{traversal-concept-c}(B, \mathsf{S}, k-1))\Big),$$

where traversal-concept-c$(B, \mathsf{S}, k+1)$ is

$$\bigsqcap_{C \in S_0(B,0)} B \sqcap \bigsqcap_{r \in \mathsf{RN}} \Big(\bigsqcap_{C \in S_0(B,r,0)} \exists r.\text{traversal-concept-c}(C, \mathsf{S}, k) \sqcap$$

$$\bigsqcap_{C \in S_0(B,r,1)} P_{=1}(\exists r.\text{traversal-concept-c}(C, \mathsf{S}, k)) \sqcap$$

$$\bigsqcap_{C \in S_0^{>0}(B,r)} P_{>0}(\exists r.\text{traversal-concept-c}(C, \mathsf{S}, k))\Big)$$

and traversal-concept-c$(B, \mathsf{S}, 0) = \bigsqcap_{C \in S_0(B,0)} B$. Once the traversal concept has been computed, it is possible to remove all normalization names preserving the instance relation, which gives us the msc in the original signature of $\mathcal{K}$. As in the case for $\mathcal{EL}$, the proof of correctness of this method can be found in [18].

Theorem 10. *Let $\mathcal{K}$ a Prob-$\mathcal{EL}_c^{01}$-knowledge base, $a \in \mathsf{IN}(\mathcal{A})$, and $k \in \mathbb{N}$; then* Remove-normalization-names(traversal-concept-i(a, S, k)) $\equiv k$-$msc_{\mathcal{K}}(a)$.

5 Computing Explanations for Instance Relations in Prob-$\mathcal{EL}_c^{01}$

By definition, an individual a is always an instance of its (role-depth bounded) msc. However, it is not always obvious why this is the case. We thus provide a method for describing the axiomatic causes for a to be an instance of a concept name A.

Definition 11 (MinA). *Let $\mathcal{K} = (\mathcal{T}, \mathcal{A})$ be an Prob-$\mathcal{EL}_c^{01}$-knowledge base, a an individual in $\mathcal{A}$ and A a concept name such that $\mathcal{K} \models A(a)$. A minimal axiom set (MinA) for $\mathcal{K}$ w.r.t. $A(a)$ is a sub-knowledge base $\mathcal{K}' = (\mathcal{S}, \mathcal{B})$, with $\mathcal{S} \subseteq \mathcal{T}, \mathcal{B} \subseteq \mathcal{A}$ such that*

- $\mathcal{K}' \models A(a)$ and
- for all strict subsets $\mathcal{S}' \subset \mathcal{S}, \mathcal{B}' \subset \mathcal{B}$, it holds that (i) $(\mathcal{S}', \mathcal{B}) \not\models A(a)$ and (ii) $(\mathcal{S}, \mathcal{B}') \not\models A(a)$.

Intuitively, a MinA is a sub-ontology that still entails the instance relationship between a and A, and that is minimal (w.r.t. set inclusion) with this property. As the following example illustrates there may be several MinAs for one consequence.

Example 12. Continuing with our running example, we have that $\mathcal{K}_{ex} \models A(a)$, and there are two MinAs for $\mathcal{K}_{ex}$ w.r.t. this instance relationship, namely

$$\mathcal{K}_1 = (\{\exists r.\top \sqsubseteq A, B \sqsubseteq \exists r.C\}, \{B(a)\}), \text{ and}$$
$$\mathcal{K}_2 = (\{\exists r.\top \sqsubseteq A\}, \{r(a,b)\}).$$

It is a simple task to verify that indeed these two knowledge bases entail $A(a)$, and that they satisfy the minimality requirement w.r.t. set inclusion.

The process of computing MinAs is called *pinpointing*. As it has been done before for other kinds of reasoning problems, we show that the completion algorithm for Prob-$\mathcal{EL}_c^{01}$ can be modified into a *pinpointing algorithm*. Rather than directly computing the MinAs, we will construct a monotone Boolean formula—called the *pinpointing formula*—that encodes all these MinAs. To define this formula, we first assume that every axiom and every assertion α in $\mathcal{K}$ is labeled with a *unique* propositional variable $\mathsf{lab}(\alpha)$ and denote as $\mathsf{lab}(\mathcal{K})$ the set of all propositional variables labeling axioms and assertions in $\mathcal{K}$. A *monotone Boolean formula* over $\mathsf{lab}(\mathcal{K})$ is a Boolean formula that uses only variables from $\mathsf{lab}(\mathcal{K})$, the binary connectives conjunction ($\wedge$) and disjunction ($\vee$), and the constant t (for "truth"). As customary in propositional logic, we identify a *valuation* with the set of propositional variables that it makes true. Finally, given a valuation $\mathcal{V} \subseteq \mathsf{lab}(\mathcal{K})$, we define

$$\mathcal{K}_\mathcal{V} := (\{\alpha \in \mathcal{T} \mid \mathsf{lab}(\alpha) \in \mathcal{V}\}, \{\alpha \in \mathcal{A} \mid \mathsf{lab}(\alpha) \in \mathcal{V}\}).$$

Definition 13 (pinpointing formula). *Given a Prob-$\mathcal{EL}_c^{01}$-knowledge base $\mathcal{K} = (\mathcal{T}, \mathcal{A})$, an individual name a occurring in $\mathcal{A}$ and a concept name A, the monotone Boolean formula ϕ over $\mathsf{lab}(\mathcal{K})$ is a* pinpointing formula *for $\mathcal{K}$ w.r.t. $A(a)$ if for every valuation $\mathcal{V} \subseteq \mathsf{lab}(\mathcal{K})$ it holds that*

$$\mathcal{K}_\mathcal{V} \models A(a) \text{ iff } \mathcal{V} \text{ satisfies } \phi.$$

Example 14. Recall that we have given every axiom and assertion of $\mathcal{K}_{ex}$ a unique label, depicted in Example 3. Hence, for instance $\mathsf{lab}(\exists r.\top \sqsubseteq A) = ax1$. The following is a pinpointing formula for $\mathcal{K}_{ex}$ w.r.t. $A(a)$:

$$ax1 \wedge (as3 \vee (ax2 \wedge as1)).$$

The MinAs for an instance relation can be obtained from the pinpointing formula ϕ by computing the minimal valuations that satisfy ϕ.

Proposition 15. *If ϕ is a pinpointing formula for $\mathcal{K}$ w.r.t. $A(a)$, then the set*

$$\{\mathcal{K}_{\mathcal{V}} \mid \mathcal{V} \text{ is a minimal valuation satisfying } \phi\}$$

is the set of all MinAs for $\mathcal{K}$ w.r.t. $A(a)$.

We take advantage of this proposition and describe an algorithm that computes a pinpointing formula for a given instance relationship.[2] If one is interested in the specific MinAs, it is only necessary to find the minimal valuations that satisfy this formula. This can be done by e.g. bringing the pinpointing formula to disjunctive normal form first and then removing all the non-minimal disjuncts. In general, a pinpointing formula may yield a more compact representation of the set of all MinAs, and hence be of more practical use.

We will use a so-called glass-box approach for computing pinpointing formulas for all the instance relationships that follow from a knowledge base $\mathcal{K}$. The idea is to extend the completion algorithm for deciding instances in Prob-$\mathcal{EL}_c^{01}$ with a tracing mechanism that encodes all the axiomatic causes for a consequence—in this case, either a subsumption or an instance relation—to follow. Since $\mathcal{EL}$ is a sub-logic of Prob-$\mathcal{EL}_c^{01}$ and classification can be reduced to instance checking,[3] our approach can also find the pinpointing formulas for the different subsumption relations that follow from the knowledge base. Thus, we generalize previous results on axiom-pinpointing in $\mathcal{EL}$ [9] in two ways by developing explanations also for the entailed instance relationships and include the probabilistic concept constructors from Prob-$\mathcal{EL}_c^{01}$.

In order to describe the pinpointing algorithm, we assume first that the knowledge base $\mathcal{K}$ is already in normal form; recall that our example knowledge base $\mathcal{K}_{ex}$ is in normal form. The *pinpointing extension* of the completion algorithm for Prob-$\mathcal{EL}_c^{01}$ also stores completion sets $S(a, v), S(a, r, v), S_0(C, v), S_0(A, r, v), S_\varepsilon(A, v)$, and, $S_\varepsilon(A, r, v)$ for the different individual-, and role names a, r, respectively, and basic concept A appearing in the knowledge base. However, the elements of these sets are not only concept names from $\mathsf{CN}(\mathcal{K})$ as in Section 3, but rather pairs of the form (D, φ), where $D \in \mathsf{CN}(\mathcal{K})$ and φ is a monotone Boolean formula. Intuitively, $(D, \varphi) \in S(C)$ means that D is a subsumer of C w.r.t. $\mathcal{K}$, and φ stores information of the axioms responsible for this fact. For the other three kinds of completion sets the idea is analogous.

The pinpointing algorithm initializes these completion sets as follows: for every $A \in \mathsf{BC}_{\mathcal{T}}, r \in \mathsf{RN}(\mathcal{K})$, and $a \in \mathsf{IN}(\mathcal{A})$

- $S_0(A, 0) = \{(\top, \mathsf{t}), (A, \mathsf{t})\}$ and $S_0(A, v) = \{(\top, \mathsf{t})\}$ for all $v \in V \setminus \{0\}$,
- $S_\varepsilon(A, \varepsilon) = \{(\top, \mathsf{t}), (A, \mathsf{t})\}$ and $S_\varepsilon(A, v) = \{(\top, \mathsf{t})\}$ for all $v \in V \setminus \{\varepsilon\}$,
- $S(a, 0) = \{(\top, \mathsf{t})\} \cup \{(A, p) \mid A(a) \in \mathcal{A}, p = \mathsf{lab}(A(a))\}$,
- $S(a, v) = \{(\top, \mathsf{t})\}$ for all $v \neq 0$,
- $S_0(A, r, v) = S_\varepsilon(A, r, v) = \emptyset$ for all $v \in V$, $S(a, r, v) = \emptyset$ for $v \neq 0$,
- $S(a, r, 0) = \{(b, p) \in \mathsf{IN}(\mathcal{A}) \mid r(a, b) \in \mathcal{A}, p = \mathsf{lab}(A(a))\}$.

---

[2] In fact, our method produces pinpointing formulas for *all* instance relationships that follow from the knowledge base at once.

[3] Indeed, $A \sqsubseteq_{\mathcal{K}} B$ iff $\mathcal{K} \cup \{A(a)\} \models B(a)$, where a is an individual name not appearing in $\mathcal{K}$.

PpR1	If $(C', \varphi) \in S_*(X, v), \alpha = C' \sqsubseteq D \in \mathcal{T}$, and $\mathsf{lab}(\alpha) = p$ then $S_*(X, v) := S_*(X, v) \uplus (D, \varphi \wedge p)$
PpR2	If $(C_1, \varphi_1), (C_2, \varphi_2) \in S_*(X, v), \alpha = C_1 \sqcap C_2 \sqsubseteq D \in \mathcal{T}$, and $\mathsf{lab}(\alpha) = p$ then $S_*(X, v) := S_*(X, v) \uplus (D, \varphi_1 \wedge \varphi_2 \wedge p)$
PpR3	If $(C', \varphi) \in S_*(X, v), \alpha = C' \sqsubseteq \exists r.D \in \mathcal{T}$, and $\mathsf{lab}(\alpha) = p$ then $S_*(X, r, v) := S_*(X, r, v) \uplus (D, \varphi \wedge p)$
PpR4	If $(D, \varphi) \in S_*(X, r, v), (D', \varphi') \in S_{\gamma(v)}(D, \gamma(v)), \alpha = \exists r.D' \sqsubseteq E \in \mathcal{T}$, and $\mathsf{lab}(\alpha) = p$ then $S_*(X, v) := S_*(X, v) \uplus (E, \varphi \wedge \varphi' \wedge p)$
PpR5	If $(P_{>0}A, \varphi) \in S_*(X, v)$, then $S_*(X, P_{>0}A) := S_*(X, P_{>0}A) \uplus (A, \varphi)$
PpR6	If $(P_{=1}A, \varphi) \in S_*(X, v), v \neq 0$, then $S_*(X, v) := S_*(X, v) \uplus (A, \varphi)$
PpR7	If $(A, \varphi) \in S_*(X, v)$ and $v \neq 0, P_{>0}A \in \mathcal{P}_0^{\mathcal{T}}$ then $S_*(X, v') := S_*(X, v') \uplus (P_{>0}A, \varphi)$
PpR8	If $(A, \varphi) \in S_*(X, 1)$ and $P_{=1}A \in \mathcal{P}_1^{\mathcal{T}}$, then $S_*(X, v) := S_*(X, v) \uplus (P_{=1}A, \varphi)$
PpR9	If $\alpha_1 = r(a, b) \in \mathcal{A}, (C, \varphi) \in S(b, 0), \alpha_2 = \exists r.C \sqsubseteq D \in \mathcal{T}$, $\mathsf{lab}(\alpha_1) = p_1$, and $\mathsf{lab}(\alpha_2) = p_2$ then $S(a, 0) := S(a, 0) \uplus (D, \varphi \wedge p_1 \wedge p_2)$
PpR10	If $\alpha_1 = P_{>0}r(a, b) \in \mathcal{A}, (C, \varphi) \in S(b, P_{>0}r(a, b)), \alpha_2 = \exists r.C \sqsubseteq D \in \mathcal{T}$, $\mathsf{lab}(\alpha_1) = p_1$, and $\mathsf{lab}(\alpha_2) = p_2$ then $S(a, P_{>0}r(a, b)) := S(a, P_{>0}r(a, b)) \uplus (D, \varphi \wedge p_1 \wedge p_2)$
PpR11	If $\alpha_1 = P_{=1}r(a, b) \in \mathcal{A}, (C, \varphi) \in S(b, v)$ with $v \neq 0, \alpha_2 = \exists r.C \sqsubseteq D \in \mathcal{T}$, $\mathsf{lab}(\alpha_1) = p_1$, and $\mathsf{lab}(\alpha_2) = p_2$ then $S(a, v) := S(a, v) \uplus (D, \varphi \wedge p_1 \wedge p_2)$

Fig. 4. Prob-$\mathcal{EL}_c^{01}$ completion rules for axiom-pinpointing

For describing the extended completion rules, we need some more notation. For a set S and a pair (D, φ), the operation $S \uplus (D, \varphi)$ is defined as follows: if there exists a ψ such that $(D, \psi) \in S$, then $S \uplus (D, \varphi) := S \setminus \{(D, \psi)\} \cup \{(D, \psi \vee \varphi)\}$; otherwise, $S \uplus (D, \varphi) := S \cup \{(D, \varphi)\}$. In other words, if the concept name D already belongs to S with some associated formula ψ, we modify the formula by adding φ to it as a disjunct; otherwise, we simply add the pair (D, φ) to S.

The completion sets are then extended by exhaustively applying the rules shown in Figure 4, where X ranges over $\mathsf{BC}_{\mathcal{T}} \cup \mathsf{IN}(\mathcal{A})$, $S_*(X, v)$ stands for $S(X, v)$ if X is an individual and for $S_0(X, v), S_\varepsilon(X, v)$ if $X \in \mathsf{BC}_{\mathcal{T}}$, and $\gamma : V \to \{0, \varepsilon\}$ is defined by $\gamma(0) = 0$, and $\gamma(v) = \varepsilon$ for all $v \in V \setminus \{0\}$.

To ensure termination of this algorithm, the completion can only be applied if their application modifies at least one of the completion sets; that is, if either a new pair is added, or the second element of an existing pair is modified to a (strictly) more general Boolean formula. Under this applicability condition, this modified algorithm always terminates, although not necessarily in polynomial time. In fact, every completion set can contain at most as many pairs as there are concept names in $\mathcal{K}$, and hence polynomially many. Whenever the formula of a pair is changed, it is done so by generalizing it in the sense that it has more models than the previous one. As there are exponentially many models, such changes can only be done an exponential number of times. Thus, in total we can have at most exponentially many rule applications, which take each at most exponential time; that is, the pinpointing algorithm runs in exponential time in the size of $\mathcal{K}$.

As stated before, these completion sets make the subsumption and instance relationships explicit, together with a formula that describe which axioms are responsible for each of these relationships. It is easy to see that the concepts appearing in the completion sets are exactly the same that will be obtained by applying the *standard* completion rules from Section 3. We thus know that $A \sqsubseteq_{\mathcal{K}} B$ iff there is some ψ with $(B, \psi) \in S_0(A, 0)$ and $\mathcal{K} \models A(a)$ iff $(A, \psi) \in S(a, 0)$ for some monotone Boolean formula ψ. Moreover, the pinpointing algorithm maintains the following invariants:

- if $(B, \psi) \in S_0(A, 0)$, then for every valuation $\mathcal{V}$ satisfying ψ, $A \sqsubseteq_{\mathcal{K}_\mathcal{V}} B$,
- if $(A, \psi) \in S(a, 0)$, then for every valuation $\mathcal{V}$ satisfying ψ, $\mathcal{K}_\mathcal{V} \models A(a)$.

It can also be shown that when the algorithm has terminated, the converse implications also hold; this is a consequence of the results from [8].

Theorem 16. *Given a Prob-$\mathcal{EL}_c^{01}$-knowledge base in normal form, the pinpointing algorithm terminates in exponential time. After termination, the following holds for every concept name A and individual name a appearing in $\mathcal{K}$:*

$$\text{if } (A, \psi) \in S(a, 0), \text{ then } \psi \text{ is a pinpointing formula for } \mathcal{K} \text{ w.r.t. } A(a).$$

We have so far described how to find the MinAs of a normalized knowledge base w.r.t. instance and subsumption relations. We now show how to extend this method to deal also with non-normalized knowledge bases; that is, to obtain the MinAs referring to the original axioms of the knowledge base and not to their normalized versions. Before going into the details, it is worth noticing that the relationship between original axioms and normalized axioms is many-to-many: one axiom in the original knowledge base may produce several axioms in the normalized one, while one axiom in the normalized knowledge base can be due to the presence of several axioms from the original one. An example of the latter can be given by the two axioms $A \sqsubseteq B, A \sqsubseteq B \sqcap C$. The normalization rules change these axioms into $A \sqsubseteq B, A \sqsubseteq C$, but the first axiom has two sources; that is, it will appear in the normalized knowledge base whenever *any* of the two original axioms is present.

Let $\widehat{\mathcal{K}}$ be an arbitrary Prob-$\mathcal{EL}_c^{01}$-knowledge base and $\mathcal{K}$ its normalized version. If ϕ is a pinpointing formula for $\mathcal{K}$ w.r.t. an instance or subsumption relation, that uses only basic concepts appearing in $\widehat{\mathcal{K}}$, then we can modify ϕ into a pinpointing formula *for the original knowledge base* $\widehat{\mathcal{K}}$ as follows. As in the case of normalized knowledge bases, each axiom in $\widehat{\mathcal{K}}$ is associated with a unique propositional variable. Each normalized axiom in $\mathcal{K}$ has a finite number of original axioms that created it—at most as many as there were in the original knowledge base. We modify the pinpointing formula ϕ by replacing the propositional formula associated to each normalized axiom by the disjunction of the labels of all its sources. We thus obtain a new pinpointing formula that speaks of the original ontology $\widehat{\mathcal{K}}$. In the above example, let $\text{lab}(A \sqsubseteq B) = p_1$ and $\text{lab}(A \sqsubseteq B \sqcap C) = p_2$, and suppose that the labels of the normalized ontology are $\text{lab}(A \sqsubseteq B) = q_1, \text{lab}(A \sqsubseteq C) = q_2$, and that the knowledge base also contains an assertion $A(a)$ with label q_3. The pinpointing formula for the normalized ontology w.r.t. $B(a)$ is $q_1 \wedge q_3$. For the original ontology, this formula is changed to $(p_1 \vee p_2) \wedge q_3$.

It is worth commenting on the execution time of the pinpointing algorithm and the complexity of finding *all* MinAs. Recall that computing all MinAs is crucial when resolving an unwanted consequence of a knowledge base. As described before, the algorithm takes exponential time to compute all instance and subsumption relations between concept names and individual names, with their respective pinpointing formulas. These formulas may be exponential in the size of the knowledge base $\mathcal{K}$, however finding one or all the minimal valuations satisfying a formula is only exponential on the number of propositional variables appearing in that formula, hence, we can compute one or all MinAs from each of these pinpointing formulas in exponential time in the size of $\mathcal{K}$. Since classification of an $\mathcal{EL}$ TBox is a special case of our setting—where the ABox $\mathcal{A}$ is empty and no probabilistic concepts are used—our algorithm yields an optimal upper bound on the complexity of pinpointing for Prob-$\mathcal{EL}_c^{01}$. Indeed, it has been shown that finding all MinAs for one subsumption relation in $\mathcal{EL}$ requires already exponential time [17]. Additionally, other kinds of tasks like finding a MinA of least cardinality or the first MinA w.r.t. some underlying ordering, can be also solved by computing the related valuations over the pinpointing formula; this is in particular beneficial, as the various optimizations developed in the SAT community, and in particular the very efficient modern SAT/SMT-solvers, can be exploited.

6   Conclusions

In this paper we have presented a practical method for computing the role-depth bounded msc in $\mathcal{EL}$- and in Prob-$\mathcal{EL}_c^{01}$- w.r.t. a general TBox or cyclic ABoxes. Our approach is based on the completion sets that are computed during realization of a knowledge base. Thus, any of the available implementations of the $\mathcal{EL}$ completion algorithm, as for instance JCEL[4] [16] can be easily extended to an implementation of the (approximative) msc computation algorithm – as it is provided in the GEL system[5]. We also showed that the same idea can be adapted for the computation of the msc in the probabilistic DL Prob-$\mathcal{EL}_c^{01}$.

Together with the completion-based computation of role-depth bounded (least) common subsumers given in [19] these results complete the bottom-up approach for general $\mathcal{EL}$- and Prob-$\mathcal{EL}_c^{01}$-knowledge bases. This approach yields a practical method to compute commonalities for differing observations regarding individuals. To the best of our knowledge this has not been investigated for DLs that can express uncertainty.

We have also applied the ideas of axiom-pinpointing to compute explanations for instance relationships that follow from a Prob-$\mathcal{EL}_c^{01}$-knowledge base. To the best of our knowledge this is also the first time that axiom-pinpointing has been applied to instance relationships, even for crisp DLs. The glass-box approach proposed modifies the computation of the completion sets to include an encoding of the axiomatic causes for a concept to be added to each set. Understanding the causes for some unexpected instance relationships is an important first step towards correcting a knowledge base,

---

[4] http://jcel.sourceforge.net/
[5] http://gen-el.sourceforge.net/

specially in the case of automatically generated ones, as done through the bottom-up approach described before. In general, finding out the precise axioms responsible for an unwanted consequence is a very hard task, even for experts, due to the large number of axioms available. When dealing with uncertainty, the difficulty grows, as the probabilities may interact in unexpected ways. Thus, being able to explain the consequences of a Prob-$\mathcal{EL}_c^{01}$ ontology automatically is of special importance.

References

1. Baader, F.: Least common subsumers and most specific concepts in a description logic with existential restrictions and terminological cycles. In: Gottlob, G., Walsh, T. (eds.) Proc. of the 18th Int. Joint Conf. on Artificial Intelligence, IJCAI 2003, pp. 325–330. Morgan Kaufmann (2003)
2. Baader, F., Brandt, S., Lutz, C.: Pushing the $\mathcal{EL}$ envelope. In: Proc. of the 19th Int. Joint Conf. on Artificial Intelligence, IJCAI 2005, Edinburgh, UK. Morgan-Kaufmann Publishers (2005)
3. Baader, F., Brandt, S., Lutz, C.: Pushing the $\mathcal{EL}$ envelope further. In: Clark, K., Patel-Schneider, P.F. (eds.) Proc. of the OWLED Workshop (2008)
4. Baader, F., Calvanese, D., McGuinness, D., Nardi, D., Patel-Schneider, P. (eds.): The Description Logic Handbook: Theory, Implementation, and Applications. Cambridge University Press (2003)
5. Baader, F., Küsters, R., Molitor, R.: Computing least common subsumers in description logics with existential restrictions. In: Dean, T. (ed.) Proc. of the 16th Int. Joint Conf. on Artificial Intelligence, IJCAI 1999, Stockholm, Sweden, pp. 96–101. Morgan Kaufmann, Los Altos (1999)
6. Baader, F., Lutz, C., Turhan, A.-Y.: Small is again Beautiful in Description Logics. KI – Künstliche Intelligenz 24(1), 25–33 (2010)
7. Baader, F., Peñaloza, R.: Axiom pinpointing in general tableaux. Journal of Logic and Computation 20(1), 5–34 (2010); Special Issue: Tableaux and Analytic Proof Methods
8. Baader, F., Peñaloza, R.: Axiom pinpointing in general tableaux. Journal of Logic and Computation 20(1), 5–34 (2010); Special Issue: Tableaux and Analytic Proof Methods
9. Baader, F., Peñaloza, R., Suntisrivaraporn, B.: Pinpointing in the Description Logic $\mathcal{EL}^+$. In: Hertzberg, J., Beetz, M., Englert, R. (eds.) KI 2007. LNCS (LNAI), vol. 4667, pp. 52–67. Springer, Heidelberg (2007)
10. Baader, F., Suntisrivaraporn, B.: Debugging SNOMED CT using axiom pinpointing in the description logic $\mathcal{EL}^+$. In: Proceedings of the International Conference on Representing and Sharing Knowledge Using SNOMED, KR-MED 2008, Phoenix, Arizona (2008)
11. Bechhofer, S., van Harmelen, F., Hendler, J., Horrocks, I., McGuinness, D.L., Patel-Schneider, P.F., Stein, L.A.: OWL web ontology language reference. W3C Recommendation (February 2004), http://www.w3.org/TR/owl-ref/
12. Kalyanpur, A., Parsia, B., Horridge, M., Sirin, E.: Finding All Justifications of OWL DL Entailments. In: Aberer, K., Choi, K.-S., Noy, N., Allemang, D., Lee, K.-I., Nixon, L.J.B., Golbeck, J., Mika, P., Maynard, D., Mizoguchi, R., Schreiber, G., Cudré-Mauroux, P. (eds.) ISWC/ASWC 2007. LNCS, vol. 4825, pp. 267–280. Springer, Heidelberg (2007)
13. Küsters, R., Molitor, R.: Approximating most specific concepts in description logics with existential restrictions. AI Communications 15(1), 47–59 (2002)
14. Lukasiewicz, T., Straccia, U.: Managing uncertainty and vagueness in description logics for the semantic web. J. Web Sem. 6(4), 291–308 (2008)

15. Lutz, C., Schröder, L.: Probabilistic description logics for subjective probabilities. In: Lin, F., Sattler, U. (eds.) Proc. of the 12th Int. Conf. on the Principles of Knowledge Representation and Reasoning, KR 2010 (2010)

16. Mendez, J., Ecke, A., Turhan, A.-Y.: Implementing completion-based inferences for the $\mathcal{EL}$-family. In: Rosati, R., Rudolph, S., Zakharyaschev, M. (eds.) Proc. of the 2011 Description Logic Workshop, DL 2011, vol. 745. CEUR (2011)

17. Peñaloza, R., Sertkaya, B.: On the complexity of axiom pinpointing in the el family of description logics. In: Lin, F., Sattler, U., Truszczynski, M. (eds.) Proceedings of the Twelfth International Conference on Principles of Knowledge Representation and Reasoning, KR 2010. AAAI Press (2010)

18. Peñaloza, R., Turhan, A.-Y.: Completion-based computation of most specific concepts with limited role-depth for $\mathcal{EL}$ and prob-$\mathcal{EL}^{01}$. LTCS-Report LTCS-10-03, Chair f. Automata Theory, Inst. for Theoretical Computer Science, TU Dresden, Germany (2010)

19. Peñaloza, R., Turhan, A.-Y.: Role-depth bounded least common subsumers by completion for $\mathcal{EL}$- and Prob-$\mathcal{EL}$-TBoxes. In: Haarslev, V., Toman, D., Weddell, G. (eds.) Proc. of the 2010 Description Logic Workshop, DL 2010 (2010)

20. Peñaloza, R., Turhan, A.-Y.: Towards approximative most specific concepts by completion for $\mathcal{EL}^{01}$ with subjective probabilities. In: Lukasiewicz, T., Peñaloza, R., Turhan, A.-Y. (eds.) Proceedings of the First International Workshop on Uncertainty in Description Logics, UniDL 2010 (2010)

21. Peñaloza, R., Turhan, A.-Y.: A Practical Approach for Computing Generalization Inferences in $\mathcal{EL}$. In: Antoniou, G., Grobelnik, M., Simperl, E., Parsia, B., Plexousakis, D., De Leenheer, P., Pan, J. (eds.) ESWC 2011, Part I. LNCS, vol. 6643, pp. 410–423. Springer, Heidelberg (2011)

22. Schlobach, S., Cornet, R.: Non-standard reasoning services for the debugging of description logic terminologies. In: Gottlob, G., Walsh, T. (eds.) Proc. of the 18th Int. Joint Conf. on Artificial Intelligence, IJCAI 2003, Acapulco, Mexico, pp. 355–362. Morgan Kaufmann, Los Altos (2003)

23. Spackman, K.: Managing clinical terminology hierarchies using algorithmic calculation of subsumption: Experience with snomed-rt. Journal of the American Medical Informatics Assoc. (2000); Fall Symposium Special Issue

24. Springer, T., Turhan, A.-Y.: Employing description logics in ambient intelligence for modeling and reasoning about complex situations. Journal of Ambient Intelligence and Smart Environments 1(3), 235–259 (2009)

25. W3C OWL Working Group. OWL 2 web ontology language document overview. W3C Recommendation (October 27, 2009),
http://www.w3.org/TR/2009/REC-owl2-overview-20091027/

Finite Fuzzy Description Logics and Crisp Representations

Fernando Bobillo[1] and Umberto Straccia[2]

[1] Dpt. of Computer Science and Systems Engineering, University of Zaragoza, Spain
[2] Istituto di Scienza e Tecnologie dell'Informazione (ISTI - CNR), Pisa, Italy
fbobillo@unizar.es, straccia@isti.cnr.it

Abstract. Fuzzy Description Logics (DLs) are a formalism for the representation of structured knowledge that is imprecise or vague by nature. In fuzzy DLs, restricting to a finite set of degrees of truth has proved to be useful, both for theoretical and practical reasons. In this paper, we propose finite fuzzy DLs as a generalization of existing approaches. We assume a finite totally ordered set of linguistic terms or labels, which is very useful in practice since expert knowledge is usually expressed using linguistic terms. Then, we consider fuzzy DLs based on any smooth t-norm defined over this set. Initially we focus on the finite fuzzy DL $\mathcal{ALCH}$, studying some logical properties, and showing the decidability of the logic by presenting a reasoning preserving reduction to the classical case. Finally, we extend our logic in two directions: by considering non-smooth t-norms and by considering additional DL constructors.

1 Introduction

It has been widely pointed out that classical ontologies are not appropriate to deal with imprecise and vague knowledge, which is inherent to several real-world domains. Since fuzzy logic is a suitable formalism to handle these types of knowledge, there has been an important interest in generalizing ontologies to the fuzzy case. Description Logics (DLs) are a family of logics for representing structured knowledge [1], and many ontology languages are based on DLs [2]. Because of the need of managing imprecise and vagueness, several fuzzy DLs can be found in the literature. For a good survey, we refer the reader to [3]. Notice that the extension of ontologies and DLs with other formalisms to deal with imprecision and vagueness, such as rough set theory, has also been studied [4,5,6,7].

It is well known that different families of fuzzy operators (or fuzzy logics) lead to fuzzy DLs with different properties. For example, Gödel and Zadeh fuzzy logics have an idempotent conjunction, whereas Łukasiewicz and Product fuzzy logic do not. Clearly, different applications may need different fuzzy logics. For example, Łukasiewicz logic may not be suitable for combining information, as the conjunction easily collapses to zero [8]

Some recent results show that some fuzzy DLs with infinite model property are undecidable [9,11,10]. Also, in fuzzy DLs the infinite model property does not hold in relatively non expressive fuzzy DLs [12,13]. This makes the study of finite fuzzy DLs even more interesting.

F. Bobillo et al. (Eds.): URSW 2008-2010/UniDL 2010, LNAI 7123, pp. 99–118, 2013.
© Springer-Verlag Berlin Heidelberg 2013

In fuzzy DLs, assuming a finite set of degrees of truth is useful [14,15,16]. In Zadeh fuzzy logic it is interesting for computational reasons [14]. In Gödel logic, it is necessary to show that the logic satisfies the Witnessed Model Property [12]. In Łukasiewicz logic, it is necessary to obtain a classical representation of the fuzzy ontology [16]. The objective of our research is to study whether a finite set of degrees of truth we can assumed when fuzzy logics different to Zadeh, Gödel, and Łukasiewicz are considered. As we will see, the answer is positive.

There is a recent promising line of research that tries to fill the gap between mathematical fuzzy logic and fuzzy DLs [8,12,17,18,19]. Following this path, we build on the previous research on finite fuzzy logics [20,21,22,23] and propose a generalization of the existing approaches to fuzzy DLs under finite degrees of truth that have been proposed in the literature [14,15,16].

Instead of dealing with degrees of truth in $[0, 1]$, as usual in fuzzy DLs, we will assume a finite chain (a finite totally ordered set) of linguistic terms or labels. For instance, $\mathcal{N} = \{\mathtt{false}, \mathtt{closeToFalse}, \mathtt{neutral}, \mathtt{closeToTrue}, \mathtt{true}\}$. Then, we will start by considering any smooth t-norm defined over a chain of degrees of truth. Later on, we will also consider the non-smooth case.

In summary, the contributions of this paper are two-fold. On the one hand, we study the use of a finite chain of labels in fuzzy ontologies. This makes it possible to abstract from the numerical interpretations of these labels. This way, since experts' knowledge is usually expressed using a set of linguistic terms [20], the process of knowledge acquisition is easier. On other hand, we consider the general case of finite fuzzy DLs, starting from a finite smooth t-norm but also discussing the case of non-smooth t-norms. This makes it possible to use new fuzzy operators (e.g., QL-implications) in fuzzy DLs for the first time.

The use of linguistic labels as degrees in fuzzy DLs has already been proposed. [24] proposes to take the degrees from an uncertainty lattice. A recent extension of this work by other authors considers Zadeh $\mathcal{SHIN}$ [25]. Finite chains of degrees of truth have also been considered in the setting of fuzzy DLs. In [18,19] the authors use them as one of the building blocks of the first order t-norm based logic $L^*_\sim(\mathbf{S})\forall$, which can be used to define several related fuzzy DLs starting from a t-norm $*$. The difference with our work is that we directly consider fuzzy DLs and hence are to able to provide specific reasoning algorithms.

The remainder is organized as follows. Section 2 includes some preliminaries on finite fuzzy logics and classical DLs. Then, Section 3 defines a fuzzy extension of the DL $\mathcal{ALCH}$ based on finite fuzzy logics, discusses some logical properties, and shows the decidability of the logic by providing a reduction of fuzzy $\mathcal{ALCH}$ into crisp $\mathcal{ALCH}$. Section 4 discusses some extensions of this logic obtained by considering non-smooth t-norms or other DL constructors. Finally, Section 5 sets out some conclusions and ideas for future research.

2   Preliminaries

This section is split into two parts. Section 2.1 reviews some results about finite fuzzy logics, and Section 2.2 overviews the classical DL $\mathcal{ALCH}$.

2.1   Finite Fuzzy Logics

Fuzzy set theory and fuzzy logic were proposed by L. Zadeh [26] to manage imprecise and vague knowledge. Here, statements are not either true or false, but they are a matter of degree. Let X be a set of elements called the reference set, and let $\mathcal{S}$ be a totally ordered set with e as minimum element and u as maximum. A *fuzzy subset* A of X is defined by a membership function $A(x) : X \to \mathcal{S}$ which assigns any $x \in X$ to a value in $\mathcal{S}$. Similarly as in the classical case, e means no-membership and u full membership, but now a value between them represents to which extent x can be considered as an element of X.

In the following, we restrict to finite chains of degrees of truth. The rest of this section contains material from [20,21,22,23].

Definition 1. *A finite chain of degrees of truth is a totally ordered set* $\mathcal{N} = \{0 = \gamma_0 < \gamma_1 < \cdots < \gamma_p = 1\}$, *where* $p \geq 1$.

Example 1. {false, closeToFalse, weaklyFalse, weaklyTrue, closeToTrue, true} is a finite chain.

$\mathcal{N}$ can be understood as a set of linguistic terms or labels. For our purposes all finite chains with the same number of elements are equivalent.

In the rest of the paper, we will use the following notation: $\mathcal{N}^+ = \mathcal{N} \setminus \{\gamma_0\}$, $+\gamma_i = \gamma_{i+1}$, $-\gamma_i = \gamma_{i-1}$. Let us also denote by $[\gamma_i, \gamma_j]$ the finite chain given by the subinterval of all $\gamma_k \in \mathcal{N}$ such that $i \leq k \leq j$.

All crisp set operations are extended to fuzzy sets. The intersection, union, complement and implication are performed by a t-norm function, a t-conorm function, a negation function, and an implication function, respectively. These functions can be restricted to finite chains. Table 1 shows some popular examples: Zadeh, Gödel, and Łukasiewicz.

Definition 2. *A t-norm on* $\mathcal{N}$ *is a function* $\otimes : \mathcal{N}^2 \to \mathcal{N}$ *such that for all* $\gamma_i, \gamma_j, \gamma_k \in \mathcal{N}$ *the following conditions are satisfied:*

- $\gamma_i \otimes \gamma_j = \gamma_j \otimes \gamma_i$,
- $(\gamma_i \otimes \gamma_j) \otimes \gamma_k = \gamma_i \otimes (\gamma_j \otimes \gamma_k)$,
- $(\gamma_i \otimes \gamma_j) \leq (\gamma_i \otimes \gamma_k)$ *whenever* $\gamma_j \leq \gamma_k$,
- $\gamma_i \otimes \gamma_p = \gamma_i$.

Definition 3. *A function* $f : \mathcal{N} \to \mathcal{N}$ *is* smooth *iff it satisfies the following condition for all* $i \in \mathcal{N}^+$ $f(\gamma_i) = \gamma_j$ *implies that* $f(\gamma_{i-1}) = \gamma_k$ *with* $j - 1 \leq k \leq j + 1$. *A binary operator is smooth when it is smooth in each place.*

The *smoothness condition* is a discrete counterpart of continuity on $[0, 1]$. Smoothness for t-norms is equivalent to the divisibility condition in $[0, 1]$, i.e., $\gamma_i \leq \gamma_j$ if and only if there exists $\gamma_k \in \mathcal{N}$ such that $\gamma_j \otimes \gamma_k = \gamma_i$.

Definition 4. *A t-norm* $\otimes$ *is* Archimedean *iff* $\forall \gamma_1, \gamma_2 \in \mathcal{N} \setminus \{\gamma_0, \gamma_p\}$ *there is* $n \in \mathbb{N}$ *such that* $\gamma_1 \otimes \gamma_1 \cdots \otimes \gamma_1$ *(n times)* $< \gamma_2$.

Table 1. Popular fuzzy logics over a finite chain

Family	$\gamma_i \otimes \gamma_j$	$\gamma_i \oplus \gamma_j$	$\ominus \gamma_i$	$\gamma_i \Rightarrow \gamma_j$
Zadeh	$\min\{\gamma_i, \gamma_j\}$	$\max\{\gamma_i, \gamma_j\}$	γ_{p-i}	$\max\{\gamma_{p-i}, \gamma_j\}$
Gödel	$\min\{\gamma_i, \gamma_j\}$	$\max\{\gamma_i, \gamma_j\}$	$\begin{cases} \gamma_p, \gamma_i = 0 \\ \gamma_0, \gamma_i > 0 \end{cases}$	$\begin{cases} \gamma_p, \gamma_i \le \gamma_j \\ \gamma_j, \gamma_i > \gamma_j \end{cases}$
Lukasiewicz	$\gamma_{\max\{i+j-p,0\}}$	$\gamma_{\min\{i+j,p\}}$	γ_{p-i}	$\gamma_{\min\{p-i+j,p\}}$

Proposition 1. *There is one and only one Archimedean smooth t-norm on $\mathcal{N}$ given by $\gamma_i \otimes \gamma_j = \gamma_{\max\{0,i+j-p\}}$. Moreover, given any subset J of $\mathcal{N}$ containing γ_0, γ_p, there is one and only one smooth t-norm $\otimes^J$ on $\mathcal{N}$ that has J as the set of idempotent elements* [1]. *In fact, if J is the set $J = \{0 = \gamma_{i_0} < \gamma_{i_1} < \cdots < \gamma_{i_{m-1}} < \gamma_{i_m} = 1\}$ such a t-norm is given by:*

$$\gamma_i \otimes^J \gamma_j = \begin{cases} \gamma_{\max\{i_k,i+j-i_{k+1}\}} & \text{if } \gamma_i, \gamma_j \in [i_k, i_{k+1}] \text{ for some } 0 \le k \le m-1 \\ \gamma_{\min\{i,j\}} & \text{otherwise .} \end{cases}$$

Notice that the Archimedean smooth t-norm is obtained when $J = \{\gamma_0, \gamma_p\}$, and that the minimum is obtained when $J = \mathcal{N}$. It is also worth to note that, as a consequence of Proposition 1, a finite smooth product t-norm is not possible.

Example 2. Given the finite chain $\mathcal{N} = \{\gamma_0, \gamma_1, \gamma_2, \gamma_3, \gamma_4, \gamma_5\}$ and the set $J = \{\gamma_0, \gamma_3, \gamma_5\}$, $\otimes^J$ is defined as:

	γ_0	γ_1	γ_2	γ_3	γ_4	γ_5
γ_0	γ_0	γ_0	γ_0	γ_0	γ_0	γ_0
γ_1	γ_0	γ_0	γ_0	γ_1	γ_1	γ_1
γ_2	γ_0	γ_0	γ_1	γ_2	γ_2	γ_2
γ_3	γ_0	γ_1	γ_2	γ_3	γ_3	γ_3
γ_4	γ_0	γ_1	γ_2	γ_3	γ_3	γ_4
γ_5	γ_0	γ_1	γ_2	γ_3	γ_4	γ_5

Definition 5. *A strong negation on $\mathcal{N}$ is a function $\ominus : \mathcal{N} \to \mathcal{N}$ such that for all $\gamma_i, \gamma_j \in \mathcal{N}$ the following conditions are satisfied:*

- $\gamma_i < \gamma_j$ *implies* $\ominus \gamma_i > \ominus \gamma_j$,
- $\ominus \gamma_0 = \gamma_p, \ominus \gamma_p = \gamma_0$,
- $\ominus(\ominus \gamma_i) = \gamma_i$ *for all* $\gamma_i \in \mathcal{N}$.

There is only one strong negation on $\mathcal{N}$ and it is given by $\ominus \gamma_i = \gamma_{p-i}$

Definition 6. *A t-conorm on $\mathcal{N}$ is a function $\oplus : \mathcal{N}^2 \to \mathcal{N}$ such that for all $\gamma_i, \gamma_j, \gamma_k \in \mathcal{N}$ the following conditions are satisfied:*

- $\gamma_i \oplus \gamma_j = \gamma_j \oplus \gamma_i$,
- $(\gamma_i \oplus \gamma_j) \oplus \gamma_k = \gamma_i \oplus (\gamma_j \oplus \gamma_k)$,
- $(\gamma_i \oplus \gamma_j) \le (\gamma_i \oplus \gamma_k)$ *whenever* $\gamma_j \le \gamma_k$,
- $\gamma_i \oplus \gamma_0 = \gamma_i$.

---

[1] γ is idempotent iff $\gamma \otimes \gamma = \gamma$.

Proposition 2. *There is one and only one Archimedean smooth t-conorm on* $\mathcal{N}$ *given by* $\gamma_i \oplus \gamma_j = \gamma_{\min\{p, i+j\}}$. *Moreover, given any subset* J *of* $\mathcal{N}$ *containing* γ_0, γ_p, *there is one and only one smooth t-conorm* $\oplus^J$ *on* $\mathcal{N}$ *that has* J *as the set of idempotent elements. In fact, if* J *is the set* $J = \{0 = \gamma_{i_0} < \gamma_{i_1} < \cdots < \gamma_{i_{m-1}} < \gamma_{i_m} = 1\}$ *such a t-conorm is given by:*

$$\gamma_i \oplus^J \gamma_j = \begin{cases} \gamma_{\min\{i_{k+1}, i+j-i_k\}} & \text{if } \gamma_i, \gamma_j \in [i_k, i_{k+1}] \text{ for some } 0 \leq k \leq m-1 \\ \gamma_{\max\{i,j\}} & \text{otherwise .} \end{cases}$$

Note that the Archimedean smooth t-conorm is obtained when $J = \{\gamma_0, \gamma_p\}$, and that the maximum is obtained when $J = \mathcal{N}$.

Given a t-norm $\otimes$ and the strong negation $\ominus$, we can define the *dual* t-conorm $\oplus_\otimes$, as the function satisfying $\gamma_i \oplus \gamma_j = \ominus((\ominus\gamma_i) \otimes (\ominus\gamma_j))$.

Definition 7. *A binary operator* $\Rightarrow: \mathcal{N}^2 \to \mathcal{N}$ *is said to be an* implication, *if the following conditions are satisfied:*

- *if* $\gamma_i \leq \gamma_j$ *then* $(\gamma_i \Rightarrow \gamma_k) \geq (\gamma_j \Rightarrow \gamma_k)$ *for all* $\gamma_k \in \mathcal{N}$,
- *if* $\gamma_i \leq \gamma_j$ *then* $(\gamma_k \Rightarrow \gamma_i) \leq (\gamma_k \Rightarrow \gamma_j)$ *for all* $\gamma_k \in \mathcal{N}$,
- $\gamma_0 \Rightarrow \gamma_0 = \gamma_p \Rightarrow \gamma_p = \gamma_p$ *and* $\gamma_p \Rightarrow \gamma_0 = \gamma_0$.

Definition 8. *Given a t-norm* $\otimes$ *and the strong negation* $\ominus$, *an* S-implication $\Rightarrow_{s\otimes}$ *is the function satisfying* $\gamma_i \Rightarrow_{s\otimes} \gamma_j = \ominus(\gamma_i \otimes (\ominus\gamma_j))$.

Equivalently, an S-implication can also be defined as $\gamma_i \Rightarrow_{s\otimes} \gamma_j = (\ominus\gamma_i) \oplus \gamma_j$.

Proposition 3. *Let* $\otimes^J : \mathcal{N}^2 \to \mathcal{N}$ *be a smooth t-norm with* $J = \{0 = \gamma_{i_0} < \gamma_{i_1} < \cdots < \gamma_{i_{m-1}} < \gamma_{i_m} = 1\}$. *Then, the implication* $\Rightarrow_{s\otimes}$ *is given by:*

$$\gamma_i \Rightarrow_{s\otimes} \gamma_j = \begin{cases} \gamma_{\min\{p-i_k, i_{k+1}+j-i\}} & \text{if } \exists \gamma_{i_k} \in J \text{ such that } \gamma_{i_k} \leq \gamma_i, \gamma_{p-j} \leq \gamma_{i_{k+1}} \\ \gamma_{\max\{p-i, j\}} & \text{otherwise .} \end{cases}$$

The Łukasiewicz implication is obtained for the Archimedean t-norm. Similarly, the Kleene-Dienes implication $\gamma_i \Rightarrow \gamma_j = \max\{\gamma_{p-i}, \gamma_j\}$ is obtained for the minimum t-norm. This is the reason why we refer to the corresponding fuzzy logic that includes Kleene-Dienes implication as Zadeh fuzzy logic.

Definition 9. *Given a t-norm* $\otimes$, *an* R-implication $\Rightarrow_{r\otimes}$ *can be defined as* $\gamma_i \Rightarrow_{r\otimes} \gamma_j = \max\{\gamma_k \in \mathcal{N} \mid (\gamma_i \otimes \gamma_k) \leq \gamma_j\}$, *for all* $\gamma_i, \gamma_j \in \mathcal{N}$.

Proposition 4. *Let* $\otimes^J : \mathcal{N}^2 \to \mathcal{N}$ *be a smooth t-norm with* $J = \{0 = \gamma_{i_0} < \gamma_{i_1} < \cdots < \gamma_{i_{m-1}} < \gamma_{i_m} = 1\}$. *Then, the implication* $\Rightarrow_{r\otimes}$ *is given by:*

$$\gamma_i \Rightarrow_{r\otimes} \gamma_j = \begin{cases} \gamma_p & \text{if } \gamma_i \leq \gamma_j \\ \gamma_{i_{k+1}+j-i} & \text{if } \exists \gamma_{i_k} \in J \text{ such that } \gamma_{i_k} \leq \gamma_j < \gamma_i \leq \gamma_{i_{k+1}} \\ \gamma_j & \text{otherwise .} \end{cases}$$

Example 3. Given the t-norm in Example 2, $\Rightarrow_{r\otimes}$ is defined as follows, where the first column is the antecedent and the first row is the consequent:

	γ_0	γ_1	γ_2	γ_3	γ_4	γ_5
γ_0	γ_5	γ_5	γ_5	γ_5	γ_5	γ_5
γ_1	γ_2	γ_5	γ_5	γ_5	γ_5	γ_5
γ_2	γ_1	γ_2	γ_5	γ_5	γ_5	γ_5
γ_3	γ_0	γ_1	γ_2	γ_5	γ_5	γ_5
γ_4	γ_0	γ_1	γ_2	γ_4	γ_5	γ_5
γ_5	γ_0	γ_1	γ_2	γ_3	γ_4	γ_5

Gödel implication is obtained for the minimum t-norm, and the Łukasiewicz implication is obtained for the Archimedean t-norm.

Definition 10. *A* QL-implication *is an implication verifying* $\gamma_i \Rightarrow \gamma_j = (\ominus\gamma_i) \oplus (\gamma_i \otimes \gamma_j)$.

The following result shows that (in the smooth case) QL-implications only depend on a t-norm. In the non smooth case, this is not true.

Proposition 5. *Let* $\otimes : \mathcal{N}^2 \to \mathcal{N}$ *be a smooth t-norm. The operator* $\gamma_i \Rightarrow_{ql\otimes} \gamma_j = (\ominus\gamma_i) \oplus (\gamma_i \otimes \gamma_j)$ *is a* QL-implication *iff* $\oplus$ *is the Archimedean smooth t-conorm. Moreover, in this case,* $\gamma_i \Rightarrow_{ql\otimes} \gamma_j = \gamma_{p-i+z}$ *for all* $\gamma_i, \gamma_j \in \mathcal{N}$, *where* $\gamma_z = \gamma_i \otimes \gamma_j$.

Proposition 6. *Let* $\otimes^J : \mathcal{N} \times^J \mathcal{N} \to \mathcal{N}$ *be a smooth t-norm with* $J = \{0 = \gamma_{i_0} < \gamma_{i_1} < \cdots < \gamma_{i_{m-1}} < \gamma_{i_m} = 1\}$. *Then, the implication* $\Rightarrow_{ql\otimes}$ *is given by:*

$$\gamma_i \Rightarrow_{ql\otimes} \gamma_j = \begin{cases} \gamma_{\max\{p-i+i_k, p+j-i_{k+1}\}} & \text{if } \gamma_i, \gamma_j \in [i_k, i_{k+1}] \text{ for some } k \in [0, m-1] \\ \gamma_{p-i+j} & \text{if } \gamma_j \leq i_k \leq \gamma_i \text{ for some } i_k \in J \\ \gamma_p & \text{otherwise .} \end{cases}$$

The Łukasiewicz implication corresponds to the minimum t-norm, and the Kleene-Dienes implication corresponds to the Archimedean t-norm (note the difference with respect to S-implications).

Interestingly, $\Rightarrow_{s\otimes}$ and $\Rightarrow_{ql\otimes}$ are smooth if and only if so is $\otimes$, but the smoothness condition is not preserved in general for R-implications.

Another interesting operators are D-implications (also called NQL-implications), which generalize the Dishkant arrow in orthomodular lattices.

Definition 11. *A* D-implication *is an implication satisfying* $\gamma_i \Rightarrow \gamma_j = ((\ominus\gamma_i) \otimes (\ominus\gamma_j)) \oplus \gamma_j$ *for all* $\gamma_i, \gamma_j \in \mathcal{N}$.

However, if $\otimes$ is a smooth t-norm, then QL-implications and D-implications on $\mathcal{N}$ actually coincide. Given a set J and $J' = \{\gamma_{p-x} \mid \gamma_x \in J\}$, then $\Rightarrow_{ql\otimes^J}$ is equivalent to $\Rightarrow_{d\otimes^{J'}}$.

In the non-smooth case, a full characterization of the operators is still unknown, and only some partial results are available. However, this a interesting case as it includes popular operators such as the nilpotent minimum (Example 4).

Example 4. The nilpotent minimum is defined as $\gamma_x \otimes \gamma_y = \gamma_0$ if $x + y \leq p$, or $\min\{\gamma_x, \gamma_y\}$ otherwise. For $\mathcal{N} = \{\gamma_0, \gamma_1, \gamma_2, \gamma_3, \gamma_4, \gamma_5\}$ we have:

	γ_0	γ_1	γ_2	γ_3	γ_4	γ_5
γ_0	γ_0	γ_0	γ_0	γ_0	γ_0	γ_0
γ_1	γ_0	γ_0	γ_0	γ_0	γ_0	γ_1
γ_2	γ_0	γ_0	γ_0	γ_0	γ_2	γ_2
γ_3	γ_0	γ_0	γ_0	γ_3	γ_3	γ_3
γ_4	γ_0	γ_0	γ_2	γ_3	γ_4	γ_4
γ_5	γ_0	γ_1	γ_2	γ_3	γ_4	γ_5

The notions of fuzzy relation, inverse relation, composition of relations, reflexivity, symmetry and transitivity can trivially be restricted to $\mathcal{N}$.

2.2 The Description Logic $\mathcal{ALCH}$

Each DL is denoted by using a string of capital letters which identify the constructors of the logic and therefore its complexity. For instance, the language OWL 2 [2], the current W3C recommendation, is close equivalent to $\mathcal{SROIQ}(\mathbf{D})$ [27]. In this section we will quickly recap the main features of the DL $\mathcal{ALCH}$. For more details we refer the reader to [1].

Syntax. $\mathcal{ALCH}$ assumes three alphabets of symbols, for *concepts*, *roles* and *individuals*. In DLs, complex concepts and roles can be built using different concept and role constructors. A Knowledge Base (KB) comprises the intensional knowledge, i.e. axioms about the application domain (a Terminological Box or *TBox* $\mathcal{T}$ and a Role Box or *RBox* $\mathcal{R}$), and the extensional knowledge, i.e. particular knowledge about some specific situation (an Assertional Box or *ABox* $\mathcal{A}$ with axioms about individuals).

The syntax of concept, roles, and axioms of $\mathcal{ALCH}$ is shown in Table 2, where C, D are (possibly complex) concepts, A is an atomic concept, R is a role, and a, b are individuals.

Table 2. Syntax and semantics of the DL $\mathcal{ALCH}$

Element	Name	Syntax	Semantics
Concepts	Atomic concept	A	$A^{\mathcal{I}} \subseteq \Delta^{\mathcal{I}}$
	Top concept	$\top$	$\Delta^{\mathcal{I}}$
	Bottom concept	$\bot$	$\emptyset$
	Concept conjunction	$C \sqcap D$	$C^{\mathcal{I}} \cap D^{\mathcal{I}}$
	Concept disjunction	$C \sqcup D$	$C^{\mathcal{I}} \cup D^{\mathcal{I}}$
	Concept negation	$\neg C$	$\Delta^{\mathcal{I}} \setminus C^{\mathcal{I}}$
	Universal quantification	$\forall R.C$	$\{x \mid \forall y, (x,y) \notin R^{\mathcal{I}} \text{ or } y \in C^{\mathcal{I}}\}$
	Existential quantification	$\exists R.C$	$\{x \mid \exists y, (x,y) \subset R^{\mathcal{I}} \text{ and } y \in C^{\mathcal{I}}\}$
Roles	Atomic role	R	$R^{\mathcal{I}} \subseteq \Delta^{\mathcal{I}} \times \Delta^{\mathcal{I}}$
ABox axioms	Concept assertion	$a : C$	$a^{\mathcal{I}} \in C^{\mathcal{I}}$
	Role assertion	$(a,b) : R$	$(a^{\mathcal{I}}, b^{\mathcal{I}}) \in R^{\mathcal{I}}$
TBox axioms	GCI	$C \sqsubseteq D$	$C^{\mathcal{I}} \subseteq D^{\mathcal{I}}$
RBox axioms	RIA	$R_1 \sqsubseteq R_2$	$R_1^{\mathcal{I}} \subseteq R_2^{\mathcal{I}}$

In the KB, concept assertions represent that an individual a is an instance of a concept C, role assertions encode that (a, b) is an instance of R, a *general concept inclusion* (GCI) imposes that C is more specific than D, and a role inclusion axiom (RIA) says that R_1 is more specific than R_2.

Semantics. An interpretation $\mathcal{I}$ is a pair $(\Delta^{\mathcal{I}}, \cdot^{\mathcal{I}})$ consisting of a non empty set $\Delta^{\mathcal{I}}$ (the interpretation domain) and an interpretation function $\cdot^{\mathcal{I}}$ mapping:

- every individual a onto an element $a^{\mathcal{I}}$ of $\Delta^{\mathcal{I}}$,
- every concept C onto a set $C^{\mathcal{I}} \subseteq \Delta^{\mathcal{I}}$, and
- every role R onto a relation $R^{\mathcal{I}} \subseteq \Delta^{\mathcal{I}} \times \Delta^{\mathcal{I}}$.

The interpretation is defined as shown in Table 2. A Knowledge Base $K = \langle \mathcal{A}, \mathcal{T}, \mathcal{R} \rangle$ iff it satisfies each element in $\mathcal{A}$, $\mathcal{T}$ and $\mathcal{R}$.

3   Finite Smooth T-norm Based Fuzzy $\mathcal{ALCH}$

In this section we define fuzzy $\mathcal{ALCH}$, a fuzzy extension of $\mathcal{ALCH}$ where:

- concepts denote fuzzy sets of individuals;
- roles denote fuzzy binary relations;
- degrees of truth are taking from a finite chain $\mathcal{N}$;
- axioms have a degree of truth associated;
- the fuzzy connectives used are a smooth t-norm $\otimes$ on $\mathcal{N}$, the strong negation $\ominus$ on $\mathcal{N}$, the dual t-conorm $\oplus_{\otimes}$, and the implications $\Rightarrow_{s\otimes}, \Rightarrow_{r\otimes}, \Rightarrow_{ql\otimes}$.

3.1   Definition of the Logic

Notation. In the rest of this paper, C, D are (possibly complex) concepts, A is an atomic concept, R is a role, a, b are individuals, $\bowtie \in \{\geq, <, \leq, >\}$, $\vartriangleleft \in \{\geq, >\}$, $\vartriangleright \in \{\leq, <\}$, $\alpha \in \mathcal{N}^+, \beta \in \mathcal{N} \setminus \{\gamma_p\}$. We will also use $\equiv$ to denote semantical equivalence, and we will not write $\otimes$ in the subscripts of the implications.

Syntax. Finite fuzzy $\mathcal{ALCH}$ assumes three alphabets of symbols, for concepts, roles and individuals. A *Fuzzy Knowledge Base* (KB) contains a finite set of axioms organized in a fuzzy ABox $\mathcal{A}$ (axioms about individuals), a fuzzy TBox $\mathcal{T}$ (axioms about concepts), and a fuzzy RBox $\mathcal{T}$ (axioms about roles).
    The syntax of fuzzy concept, roles, and axioms is shown in Table 3. We will only allow axioms of the forms $\langle \tau \geq \alpha \rangle$, $\langle \tau > \beta \rangle$, $\langle \tau \leq \beta \rangle$, and $\langle \tau < \alpha \rangle$.

Example 5. The fact that it is likely true that Paul can be considered tall can be encoded using the axiom $\langle \texttt{paul: Tall} \geq \texttt{closeToTrue} \rangle$ without needing an explicit numerical degree.

Remark 1. As opposed to the crisp case, there are three types of universal restrictions, fuzzy GCIs, and fuzzy RIAs. In fact, the different subscripts $_s$, $_r$, and $_{ql}$ denote an S-implication, R-implication, and QL-implication, respectively.

Table 3. Syntax and semantics of finite fuzzy $\mathcal{ALCH}$

Element	Syntax	Semantics
Concepts	$\top$	γ_p
	$\bot$	γ_0
	A	$A^{\mathcal{I}}(x)$
	$C \sqcap D$	$C^{\mathcal{I}}(x) \otimes D^{\mathcal{I}}(x)$
	$C \sqcup D$	$C^{\mathcal{I}}(x) \oplus D^{\mathcal{I}}(x)$
	$\neg C$	$\ominus C^{\mathcal{I}}(x)$
	$\forall_s R.C$	$\inf_{y \in \Delta^{\mathcal{I}}} \{R^{\mathcal{I}}(x,y) \Rightarrow_s C^{\mathcal{I}}(y)\}$
	$\forall_r R.C$	$\inf_{y \in \Delta^{\mathcal{I}}} \{R^{\mathcal{I}}(x,y) \Rightarrow_r C^{\mathcal{I}}(y)\}$
	$\forall_{ql} R.C$	$\inf_{y \in \Delta^{\mathcal{I}}} \{R^{\mathcal{I}}(x,y) \Rightarrow_{ql} C^{\mathcal{I}}(y)\}$
	$\exists R.C$	$\sup_{y \in \Delta^{\mathcal{I}}} \{R^{\mathcal{I}}(x,y) \otimes C^{\mathcal{I}}(y)\}$
Roles	R	$R^{\mathcal{I}}(x,y)$
ABox axioms	$\langle a{:}C \bowtie \gamma \rangle$	$C^{\mathcal{I}}(a^{\mathcal{I}}) \bowtie \gamma$
	$\langle (a,b){:}R \bowtie \gamma \rangle$	$R^{\mathcal{I}}(a^{\mathcal{I}}, b^{\mathcal{I}}) \bowtie \gamma$
TBox axioms	$\langle C \sqsubseteq_s D \rhd \gamma \rangle$	$\inf_{x \in \Delta^{\mathcal{I}}} \{C^{\mathcal{I}}(x) \Rightarrow_s D^{\mathcal{I}}(x)\} \rhd \gamma$
	$\langle C \sqsubseteq_r D \rhd \gamma \rangle$	$\inf_{x \in \Delta^{\mathcal{I}}} \{C^{\mathcal{I}}(x) \Rightarrow_r D^{\mathcal{I}}(x)\} \rhd \gamma$
	$\langle C \sqsubseteq_{ql} D \rhd \gamma \rangle$	$\inf_{x \in \Delta^{\mathcal{I}}} \{C^{\mathcal{I}}(x) \Rightarrow_{ql} D^{\mathcal{I}}(x)\} \rhd \gamma$
RBox axioms	$\langle R_1 \sqsubseteq_s R_2 \rhd \gamma \rangle$	$\inf_{x,y \in \Delta^{\mathcal{I}}} \{R_1^{\mathcal{I}}(x) \Rightarrow_s R_2^{\mathcal{I}}(x)\} \rhd \gamma$
	$\langle R_1 \sqsubseteq_r R_2 \rhd \gamma \rangle$	$\inf_{x,y \in \Delta^{\mathcal{I}}} \{R_1^{\mathcal{I}}(x) \Rightarrow_r R_2^{\mathcal{I}}(x)\} \rhd \gamma$
	$\langle R_1 \sqsubseteq_{ql} R_2 \rhd \gamma \rangle$	$\inf_{x,y \in \Delta^{\mathcal{I}}} \{R_1^{\mathcal{I}}(x) \Rightarrow_{ql} R_2^{\mathcal{I}}(x)\} \rhd \gamma$

Semantics. A fuzzy interpretation $\mathcal{I}$ is a pair $(\Delta^{\mathcal{I}}, \cdot^{\mathcal{I}})$ where $\Delta^{\mathcal{I}}$ is a non empty set (the interpretation domain) and $\cdot^{\mathcal{I}}$ is a fuzzy interpretation function mapping

- every individual a to an element $a^{\mathcal{I}}$ of $\Delta^{\mathcal{I}}$,
- every concept C to a function $C^{\mathcal{I}} : \Delta^{\mathcal{I}} \to \mathcal{N}$, and
- every role R to a function $R^{\mathcal{I}} : \Delta^{\mathcal{I}} \times \Delta^{\mathcal{I}} \to \mathcal{N}$.

The fuzzy interpretation function is extended to fuzzy *complex concepts* and *axioms* as shown in Table 3. $C^{\mathcal{I}}$ denotes the membership function of the fuzzy concept C with respect to the fuzzy interpretation $\mathcal{I}$. $C^{\mathcal{I}}(x)$ gives us the degree of being x an element of the fuzzy concept C under $\mathcal{I}$. Similarly, $R^{\mathcal{I}}$ denotes the membership function of the fuzzy role R with respect to $\mathcal{I}$. $R^{\mathcal{I}}(x,y)$ gives us the degree of being (x,y) an element of the fuzzy role R.

Remark 2. Note an important difference with previous work in fuzzy DLs. Usually, $\cdot^{\mathcal{I}}$ maps every concept C onto a function $C^{\mathcal{I}} : \Delta^{\mathcal{I}} \to [0,1]$, and every role R onto $R^{\mathcal{I}} : \Delta^{\mathcal{I}} \times \Delta^{\mathcal{I}} \to [0,1]$. Consequently, a fuzzy KB $\{\langle a : C > 0.5\rangle, \langle a : C < 0.75\}$ is satisfiable, by taking $C^{\mathcal{I}}(a) \in (0.5, 0.75)$. But now, given $\mathcal{N} = \{\texttt{false}, \texttt{closeToFalse}, \texttt{neutral}, \texttt{closeToTrue}, \texttt{true}\}$, a fuzzy KB $\{\langle a : C > \texttt{closeToFalse}\rangle, \langle a : C < \texttt{neutral}\}$ is unsatisfiable, since $C^{\mathcal{I}}(a) \notin \mathcal{N}$.

Now we will briefly present an example where our logic has a behaviour which is different from both Zadeh, Gödel and Łukasiewicz fuzzy DLs. In some applications, this property could be more appropriate.

Example 6. Consider the finite chain in Example 1 and the pair of axioms $\langle a : C_1 \geq \gamma_2\rangle$ and $\langle a : C_2 \geq \gamma_2\rangle$. If we consider Gödel t-norm, $\langle a : C_1 \sqcap C_2 \geq \gamma_2\rangle$. If we consider Łukasiewicz finite t-norm, $\langle a : C_1 \sqcap C_2 \geq \gamma_0\rangle$. However, if we consider the t-norm in Example 2, we get an intermediate value: $\langle a : C_1 \sqcap C_2 \geq \gamma_1\rangle$.

3.2   Logical Properties

It can be easily shown that finite fuzzy $\mathcal{ALCH}$ is a sound extension [28] of crisp $\mathcal{ALCH}$, because fuzzy interpretations coincide with crisp interpretations if we restrict the membership degrees to $\{\gamma_0 = 0, \gamma_p = 1\}$.

Proposition 7. *Finite fuzzy $\mathcal{ALCH}$ interpretations coincide with crisp interpretations if we restrict the membership degrees to $\{\gamma_0 = 0, \gamma_p = 1\}$.*

The following properties are extensions to a finite chain $\mathcal{N}$ of properties for Zadeh fuzzy DLs [14] and Łukasiewicz fuzzy DLs [16].

1. *Concept simplification:* $C \sqcap \top \equiv C$, $C \sqcup \bot \equiv C$, $C \sqcap \bot \equiv \bot$, $C \sqcup \top \equiv \top$, $\exists R.\bot \equiv \bot$, $\forall_s R.\top \equiv \top$, $\forall_r R.\top \equiv \top$, $\forall_{ql} R.\top \equiv \top$.
2. *Involutive* negation: $\neg\neg C \equiv C$.
3. *Excluded middle and contradiction:* In general, $C \sqcup \neg C \not\equiv \top$, $C \sqcap \neg C \not\equiv \bot$.
4. *Idempotence* of conjunction/disjunction: In general, $C \sqcap C \not\equiv C$, $C \sqcup C \not\equiv C$.
5. *De Morgan* laws: $\neg(C \sqcup D) \equiv \neg C \sqcap \neg D$, $\neg(C \sqcap D) \equiv \neg C \sqcup \neg D$.
6. *Inter-definability of concepts:* $\bot \equiv \neg\top$, $\top \equiv \neg\bot$, $C \sqcap D \equiv \neg(\neg C \sqcap \neg D)$, $C \sqcup D \equiv \neg(\neg C \sqcup \neg D)$, $\forall_s R.C \equiv \neg\exists R.(\neg C)$, $\exists R.C \equiv \neg\forall_s R.(\neg C)$. However, in general, $C \sqcap D \not\equiv \neg(\neg C \sqcup \neg D)$, $C \sqcup D \not\equiv \neg(\neg C \sqcap \neg D)$, $\forall_r R.C \not\equiv \neg\exists R.(\neg C)$, $\exists R.C \not\equiv \neg\forall_r R.(\neg C)$, $\forall_{ql} R.C \not\equiv \neg\exists R.(\neg C)$, $\exists R.C \not\equiv \neg\forall_{ql} R.(\neg C)$.
7. *Inter-definability of axioms:* $\langle a : C \leq \alpha\rangle \equiv \langle a : \neg C \geq \ominus\alpha\rangle$, $\langle \tau > \beta\rangle \equiv \langle \tau > +\beta\rangle$, $\langle \tau < \alpha\rangle \equiv \langle \tau \leq -\alpha\rangle$.
8. *Contrapositive symmetry:* $C \sqsubseteq_s D \equiv \neg D \sqsubseteq_s \neg C$. However, in general, $C \sqsubseteq_r D \not\equiv \neg D \sqsubseteq_r \neg_s C$, $C \sqsubseteq_{ql} D \not\equiv \neg D \sqsubseteq_{ql} \neg_s C$.
9. *Modus ponens:* $\langle a : C \rhd \gamma_1\rangle$ and $\langle C \sqsubseteq_r D \rhd \gamma_2\rangle$ imply $\langle a : D \rhd \gamma_1 \otimes \gamma_2\rangle$, $\langle (a,b):R \rhd \gamma_1\rangle$ and $\langle R \sqsubseteq_r R' \rhd \gamma_2\rangle$ imply $\langle (a,b):R' \rhd \gamma_1 \otimes \gamma_2\rangle$.
10. *Self-subsumption:* $(C \sqsubseteq_r C)^{\mathcal{I}} = \gamma_p$, $(R \sqsubseteq_r R)^{\mathcal{I}} = \gamma_p$. However, in general, $(C \sqsubseteq_s C)^{\mathcal{I}} \neq \gamma_p$, $(R \sqsubseteq_s R)^{\mathcal{I}} \neq \gamma_p$, and $(C \sqsubseteq_{ql} C)^{\mathcal{I}} \neq \gamma_p$, $(R \sqsubseteq_{ql} R)^{\mathcal{I}} \neq \gamma_p$.

Remark 3. Property 7 makes it possible to restrict to fuzzy axioms $\langle \tau \geq \alpha\rangle$ and $\langle \tau \leq \beta\rangle$, as we will do in the rest of this paper.

A fuzzy interpretation $\mathcal{I}$ is *witnessed* iff, for every formula, the infimum corresponds to the minimum and the supremum corresponds to the maximum [12]. Finite fuzzy $\mathcal{ALCH}$ enjoys the Witnessed Model Property (WMP) (all interpretations are witnessed), because the number of degrees of truth in the fuzzy interpretations of the logic is finite [12].

3.3 Reasoning Tasks

Now we will define the most important reasoning tasks and show that all of them can be reduced to fuzzy KB satisfiability. We will use c to denote a new individual, which do not appear in a fuzzy KB $\mathcal{K}$.

- *Fuzzy KB satisfiability.* A fuzzy interpretation $\mathcal{I}$ *satisfies* (is a model of) a fuzzy KB $\mathcal{K} = \langle \mathcal{A}, \mathcal{T}, \mathcal{R} \rangle$ iff it satisfies each element in $\mathcal{A}$, $\mathcal{T}$ and $\mathcal{R}$.
- *Concept satisfiability.* C is α-satisfiable w.r.t. a fuzzy KB $\mathcal{K}$ iff $\mathcal{K} \cup \{\langle c{:}C \geq \alpha \rangle\}$ is satisfiable.
- *Entailment.* A fuzzy concept assertion $\langle a{:}C \geq \alpha \rangle$ is entailed by a fuzzy KB $\mathcal{K}$ (denoted $\mathcal{K} \models \langle a{:}C \geq \alpha \rangle$) iff $\mathcal{K} \cup \{\langle a{:}C < \alpha \rangle\}$ is unsatisfiable. Furthermore, $\mathcal{K} \models \langle (a,b){:}R \geq \alpha \rangle$ iff $\mathcal{K} \cup \{\langle b : B \geq \gamma_p \rangle\} \models \langle a : \exists R.B \geq \alpha \rangle$ where B is a new concept.
- *Greatest lower bound.* The greatest lower bound of a concept or role assertion τ is defined as the $\sup\{\alpha : \mathcal{K} \models \langle \tau \geq \alpha \rangle\}$. It can be computed performing at most $\log |\mathcal{N}|$ entailment tests [29].
- *Concept subsumption*: There are 3 cases depending on the fuzzy implication:
  - Under an S-implication, D α-subsumes C ($C \sqsubseteq_s D \geq \alpha$) w.r.t. a fuzzy KB $\mathcal{K}$ iff $\mathcal{K} \cup \{c{:}\neg C \sqcup D < \alpha\}$ is unsatisfiable.
  - Under an R-implication, D subsumes C ($C \sqsubseteq_r D \geq \alpha$) w.r.t. a fuzzy KB $\mathcal{K}$ iff $\mathcal{K} \cup \{c{:}C \geq \gamma_1\} \cup \{c{:}D < \gamma_2\}$ is unsatisfiable, for every $\gamma_1, \gamma_2 \in \mathcal{N}^+$ such that $\gamma_1 \otimes \alpha = \alpha_2$ [30].
  - Under a QL-implication, D α-subsumes C ($C \sqsubseteq_{ql} D \geq \alpha$) w.r.t. a fuzzy KB $\mathcal{K}$ iff $\mathcal{K} \cup \{c{:}\neg C \sqcup (C \sqcap D) < \alpha\}$ is unsatisfiable.

3.4 A Crisp Representation for Finite Fuzzy $\mathcal{ALCH}$

In this section we show how to reduce a fuzzy KB into a crisp KB. The procedure is satisfiability-preserving, so existing DL reasoners could be applied to the resulting KB. The basic idea is to create some new crisp concepts and roles, representing the α-cuts of the fuzzy concepts and relations. Next, every axiom in the ABox, the TBox and the RBox is represented, independently from other axioms, using these new crisp elements.

Before proceeding formally, we will illustrate this idea with an example.

Example 7. Consider the smooth t-norm on $\mathcal{N}$ used in Example 2, and let us compute some α-cuts of the fuzzy concept $A_1 \sqcap A_2$ (denoted $\rho(A_1 \sqcap A_2, \geq \alpha)$).

To begin with, let us consider $\alpha = \gamma_2$. By definition, this set includes the elements of the domain x satisfying $A_1^{\mathcal{I}}(x) \otimes A_2^{\mathcal{I}}(x) \geq \gamma_2$. There are two possibilities: *(i)* $A_1^{\mathcal{I}}(x) \geq \gamma_2$ and $A_2^{\mathcal{I}}(x) \geq \gamma_3$, or *(ii)* $A_1^{\mathcal{I}}(x) \geq \gamma_3$ and $A_2^{\mathcal{I}}(x) \geq \gamma_2$. Hence, $\rho(A_1 \sqcap A_2, \geq \gamma_2) = \big(\rho(A_1, \geq \gamma_2) \sqcap \rho(A_2, \geq \gamma_3)\big) \sqcup \big(\rho(A_1, \geq \gamma_3) \sqcap \rho(A_2, \geq \gamma_2)\big)$.

Now, let us consider $\alpha = \gamma_3$. Now, there is only one possibility: $A_1^{\mathcal{I}}(a^{\mathcal{I}}) \geq \gamma_3$ and $A_2^{\mathcal{I}}(a^{\mathcal{I}}) \geq \gamma_3$. Hence, $\rho(A_1 \sqcap A_2, \geq \gamma_3) = \rho(A_1, \geq \gamma_3) \sqcap \rho(A_2, \geq \gamma_3)$.

Observe that for idempotent degrees ($\alpha \in J$) the case is the same as in finite Zadeh and Gödel fuzzy logics [14,15], whereas for non-idempotent degrees it is similar to the case of finite Łukasiewicz fuzzy logic [16].

Adding New Elements. Let $\mathbf{A}$ be the set of atomic fuzzy concepts and $\mathbf{R}$ the set of atomic fuzzy roles in a fuzzy KB $\mathcal{K} = \langle \mathcal{A}, \mathcal{T}, \mathcal{R} \rangle$, respectively. For each $\alpha \in \mathcal{N}^+$, for each $A \in \mathbf{A}$, a new atomic concepts $A_{\geq \alpha}$ is introduced. $A_{\geq \alpha}$ represents the crisp set of individuals which are instance of A with degree higher or equal than α i.e the α-cut of A. Similarly, for each $R \in \mathbf{R}$, a new atomic role $R_{\geq \alpha}$ is created.

The semantics of these newly introduced atomic concepts and roles is preserved by some terminological and role axioms. For each $1 \leq i \leq p - 1$ and for each $A \in \mathbf{A}$, $T(\mathcal{N})$ is the smallest terminology containing the axioms $A_{\geq \gamma_{i+1}} \sqsubseteq A_{\geq \gamma_i}$. Similarly, for each $R_A \in \mathbf{R}$, $R(\mathcal{N})$ is the smallest terminology containing the axioms $R_{\geq \gamma_{i+1}} \sqsubseteq R_{\geq \gamma_i}$.

Remark 4. The atomic elements $A_{\geq \gamma_0}$ and $R_{\geq \gamma_0}$ are not considered because they are always equivalent to the $\top$ concept. Also, as opposite to previous works [14,15,16] we are not introducing elements of the forms $A_{>\beta}$ and $R_{>\beta}$ (for each $\beta \in \mathcal{N} \setminus \{\gamma_p\}$), since now $A_{>\gamma_i}$ is equivalent to $A_{\geq \gamma_{i+1}}$, and $R_{>\gamma_i}$ is equivalent to $R_{\geq \gamma_{i+1}}$. Hence, the number of new axioms needed here is smaller, since we do not need to deal with elements of the forms $A_{>\beta}$ and $R_{>\beta}$.

Mapping Fuzzy Concepts, Roles and Axioms. Fuzzy concept and roles are reduced using mapping ρ as shown in Table 4. Given a fuzzy concept C, $\rho(C, \geq \alpha)$ is a crisp set containing all the elements which belong to C with a degree greater or equal than α. The other cases $\rho(C, \bowtie \gamma)$ are similar. ρ is defined in a similar way for fuzzy roles. Furthermore, axioms are reduced as in the bottom part of Table 5, where $\kappa(\tau)$ maps a fuzzy axiom τ in finite fuzzy $\mathcal{ALCH}$ into a set of crisp axioms in $\mathcal{ALCH}$.

The reduction of the conjunction considers every pair $\gamma_x, \gamma_y \in (\gamma_{i_k}, \gamma_{i_{k+1}}]$ such that $\alpha \in (\gamma_{i_k}, \gamma_{i_{k+1}}]$, and $x + y = i_{k+1} + z$, with $\alpha = \gamma_z$. Note that the reduction does not consider a closed interval of the form $[\gamma_{i_k}, \gamma_{i_{k+1}}]$. The reason is that, if α is idempotent and we set $\gamma_{i_{k+1}} = \alpha$, the result is correct ($\gamma_x = \gamma_y = \alpha$). However, setting $\gamma_{i_k} = \alpha$ would yield an incorrect result.

Similarly, the reduction of the disjunction considers every pair $\gamma_1, \gamma_2 \in (\gamma_{i_k}, \gamma_{i_{k+1}}]$ such that $\alpha \in (\gamma_{i_k}, \gamma_{i_{k+1}}]$, and $\gamma_1 + \gamma_2 = \gamma_{i_k} + \alpha$, instead of a closed interval of the form $[\gamma_{i_k}, \gamma_{i_{k+1}}]$.

The crisp representations of R-implications and QL-implications only consider *optimal pairs* of elements, because we want to have efficient representations that avoid the inclusion of superfluous elements. To this end, we need to formally define the notions of optimality of pairs of degrees of truth.

Definition 12. *Let $\odot \in \{\otimes, \oplus\}$ be a fuzzy operator defined in $\mathcal{N}$, and $\gamma_x, \gamma_y \in \mathcal{N}$. The set of $(\odot_{\geq \alpha})$-optimal pairs is composed by every pair (γ_x, γ_y) such that:*

- $\gamma_x \odot \gamma_y \geq \alpha$, *and*
- $\nexists \gamma_x' \in \mathcal{N}$ *such that* $\gamma_x' \odot \gamma_y \geq \alpha$ *and* $\gamma_x' < \gamma_x$, *and*
- $\nexists \gamma_y' \in \mathcal{N}$ *such that* $\gamma_x \odot \gamma_y' \geq \alpha$ *and* $\gamma_y' < \gamma_y$.

Let us explain the intuition behind Definition 12 for a t-norm $\otimes$. Assume that there are $\gamma_x, \gamma_y \in \mathcal{N}$ such that $\gamma_x \otimes \gamma_y \geq \alpha$. Since t-norms are non-decreasing in both arguments, $\gamma'_x \otimes \gamma_y \geq \alpha$ trivially holds for every $\gamma'_x \in \mathcal{N}, \gamma'_x > \gamma_x$. Hence, we take the minimal degrees γ_x, γ_y that verify the condition.

The reduction of t-norms, t-conorms and S-implications is implicitly using optimal pairs as well. However, in these cases, we are able to give a general expression to compute these optimal pairs.

Table 4. Mapping of concepts and roles

$\rho(\top, \geq \alpha)$	$\top$
$\rho(\top, \leq \beta)$	$\bot$
$\rho(\bot, \geq \alpha)$	$\bot$
$\rho(\bot, \leq \beta)$	$\top$
$\rho(A, \geq \alpha)$	$A_{\geq \alpha}$
$\rho(A, \leq \beta)$	$\neg A_{\geq +\beta}$
$\rho(\neg C, \geq \alpha)$	$\rho(C, \leq \ominus \gamma)$
$\rho(\neg C, \leq \alpha)$	$\rho(C, \geq \ominus \gamma)$
$\rho(C \sqcap D, \geq \alpha)$	$\bigsqcup_{\gamma_x, \gamma_y} \{\rho(C, \geq \gamma_x) \sqcap \rho(D, \geq \gamma_y)\}$ for every γ_x, γ_y such that $\alpha, \gamma_x, \gamma_y \in (\gamma_{i_k}, \gamma_{i_{k+1}}]$, and $x + y = i_{k+1} + z$, with $\gamma_z = \alpha$
$\rho(C \sqcap D, \leq \beta)$	$\rho(\neg C \sqcup \neg D, \geq \ominus \beta)$
$\rho(C \sqcup D, \geq \alpha)$	$\rho(C, \geq \alpha) \sqcup \rho(D, \geq \alpha) \sqcup \bigsqcup_{\gamma_x, \gamma_y} \{\rho(C, \geq \gamma_x) \sqcap \rho(D, \geq \gamma_y)\}$ for every γ_x, γ_y such that $\alpha, \gamma_x, \gamma_y \in (\gamma_{i_k}, \gamma_{i_{k+1}}]$, and $x + y = i_k + z$, with $\gamma_z = \alpha$
$\rho(C \sqcup D, \leq \beta)$	$\rho(\neg C \sqcap \neg D, \geq \ominus \beta)$
$\rho(\exists R.C, \geq \alpha)$	$\bigsqcup_{\gamma_x, \gamma_y} \{\exists \rho(R, \geq \gamma_x).\rho(C, \geq \gamma_y)\}$ for every $\gamma_x, \gamma_y \in (\gamma_{i_k}, \gamma_{i_{k+1}}]$ such that $\gamma \in (\gamma_{i_k}, \gamma_{i_{k+1}}]$, and $x + y = i_{k+1} + z$, with $\gamma_z = \alpha$
$\rho(\exists R.C, \leq \beta)$	$\rho(\forall_s R.(\neg C), \geq \ominus \beta)$
$\rho(\forall_s R.C, \geq \alpha)$	$\bigsqcap_{\gamma_x, \gamma_y} \{\forall \rho(R, \geq \gamma_x).\rho(C, \geq \gamma_y)\}$ for every γ_x, γ_y such that $\gamma_x \in (\gamma_{i_k}, \gamma_{i_{k+1}}], \alpha, \gamma_y \in (\gamma_{p - i_{k+1}}, \gamma_{p - i_k}]$, and $y - i = z - i_{k+1}$, with $\gamma_z = \alpha$
$\rho(\forall_s R.C, \leq \beta)$	$\rho(\exists R.(\neg C), \geq \ominus \beta)$
$\rho(\forall_r R.C, \geq \alpha)$	$\bigsqcap_{\gamma_x, \gamma_y} \{\forall \rho(R, \geq \gamma_x).\rho(C, \geq \gamma_y)\}$ for every $\gamma_x, \gamma_y \in \mathcal{N}^+$ such that γ_x, γ_y are $(\Rightarrow_r \geq \alpha)$-optimal
$\rho(\forall_r R.C, \leq \beta)$	$\bigsqcup_{\gamma_x, \gamma_y} \{\exists \rho(R, \geq \gamma_x).\rho(C, \leq \gamma_y)\}$ for every $\gamma_x \in \mathcal{N}^+, \gamma_y \in \mathcal{N}$ such that γ_x, γ_y are $(\Rightarrow_r \leq \beta)$-optimal
$\rho(\forall_{ql} R.C, \geq \alpha)$	$\bigsqcap_{\gamma_x, \gamma_y} \{\forall \rho(R, \geq \gamma_x).\rho(C, \geq \gamma_y)\}$ for every $\gamma_x, \gamma_y \in \mathcal{N}^+$ such that γ_x, γ_y are $(\Rightarrow_{ql} \geq \alpha)$-optimal
$\rho(\forall_{ql} R.C, \leq \beta)$	$\bigsqcup_{\gamma_x, \gamma_y} \{\exists \rho(R, \geq \gamma_x).\rho(C, \leq \gamma_y)\}$ for every $\gamma_x \in \mathcal{N}^+, \gamma_y \in \mathcal{N}$ such that γ_x, γ_y are $(\Rightarrow_{ql} \leq \beta)$-optimal
$\rho(R, \geq \alpha)$	$R_{\geq \alpha}$
$\rho(R, \leq \beta)$	$\neg R_{\geq +\beta}$

Table 5. Mapping of axioms

$\kappa(\langle a : C \bowtie \gamma \rangle)$	$\{a : \rho(C, \bowtie \gamma)\}$
$\kappa(\langle (a,b) : R \bowtie \gamma \rangle)$	$\{(a,b) : \rho(R, \bowtie \gamma)\}$
$\kappa(\langle C \sqsubseteq_s D \geq \alpha \rangle)$	$\bigcup\{\rho(C, \geq \gamma_x) \sqsubseteq \rho(D, \geq \gamma_y)\}$ for every γ_x, γ_y such that $\gamma_x \in (\gamma_{i_k}, \gamma_{i_{k+1}}]$, $\alpha, \gamma_y \in (\gamma_{p-i_{k+1}}, \gamma_{p-i_k}]$, and $y - \gamma_i = z - \gamma_{i_{k+1}}$, with $\gamma_z = \alpha$
$\kappa(\langle C \sqsubseteq_r D \geq \alpha \rangle)$	$\bigcup\{\rho(C, \geq \gamma_x) \sqsubseteq \rho(D, \geq \gamma_y)\}$ for every $\gamma_x, \gamma_y \in \mathcal{N}^+$ such that γ_x, γ_y are $(\Rightarrow_r \geq \alpha)$-optimal
$\kappa(\langle C \sqsubseteq_{ql} D \geq \alpha \rangle)$	$\bigcup\{\forall \rho(C, \geq \gamma_x) \sqsubseteq \rho(D, \geq \gamma_y)\}$ for every $\gamma_x, \gamma_y \in \mathcal{N}^+$ such that γ_x, γ_y are $(\Rightarrow_{ql} \geq \alpha)$-optimal
$\kappa(\langle R_1 \sqsubseteq_s R_2 \geq \alpha \rangle)$	$\bigcup\{\rho(R_1, \geq \gamma_x) \sqsubseteq \rho(R_2, \geq \gamma_y)\}$ for every γ_x, γ_y such that $\gamma_x \in (\gamma_{i_k}, \gamma_{i_{k+1}}]$, $\alpha, \gamma_y \in (\gamma_{p-i_{k+1}}, \gamma_{p-i_k}]$, and $y - \gamma_i = z - \gamma_{i_{k+1}}$, with $\gamma_z = \alpha$
$\kappa(\langle R_1 \sqsubseteq_r R_2 \geq \alpha \rangle)$	$\bigcup\{\rho(R_1, \geq \gamma_x) \sqsubseteq \rho(R_2, \geq \gamma_y)\}$ for every $\gamma_x, \gamma_y \in \mathcal{N}^+$ such that γ_x, γ_y are $(\Rightarrow_r \geq \alpha)$-optimal
$\kappa(\langle R_1 \sqsubseteq_{ql} R_2 \geq \alpha \rangle)$	$\bigcup\{\rho(R_1, \geq \gamma_x) \sqsubseteq \rho(R_2, \geq \gamma_y)\}$ for every $\gamma_x, \gamma_y \in \mathcal{N}^+$ such that γ_x, γ_y are $(\Rightarrow_{ql} \geq \alpha)$-optimal

Definition 13. *Let $\Rightarrow$ be a fuzzy implication defined in $\mathcal{N}$, $\gamma_x, \gamma_y \in \mathcal{N}$. We define the set $S = \{(\gamma_x, \gamma_y) \mid \gamma_x \Rightarrow \gamma_y \geq \alpha\}$ to contain every pair of individuals whose implication is at least α. Let $X \subseteq \mathcal{N} \times \mathcal{N}$ be a set of pairs of degrees of truth. We define the mappings $R(X)$ and $L(X)$ as follows:*

- *$R(X) = \{(\gamma_x, \gamma_y) \in X \mid \nexists (\gamma_x, \gamma_y') \in X$ such that $\gamma_y' < \gamma_y\}$,*
- *$L(X) = \{(\gamma_x, \gamma_y) \in X \mid \nexists (\gamma_x', \gamma_y) \in X$ such that $\gamma_x' < \gamma_x\}$.*

The set of $(\Rightarrow_{\geq \alpha})$-optimal pairs is defined as: $L(R(S))$.

Similarly, we can define the notion of $(\odot_{\leq \beta})$-optimal pairs as follows.

Definition 14. *Let $\Rightarrow$ be a fuzzy implication defined in $\mathcal{N}$, $\gamma_x, \gamma_y \in \mathcal{N}$, $L(X)$ the mapping in Definition 13 and $S' = \{(\gamma_x, \gamma_y) \mid \gamma_x \Rightarrow \gamma_y \leq \beta\}$. Let $X \subseteq \mathcal{N} \times \mathcal{N}$ be a set of pairs of degrees of truth. We define the mapping $R'(X)$ as follows:*

- *$R'(X) = \{(\gamma_x, \gamma_y) \in X \mid \nexists (\gamma_x, \gamma_y') \in X$ such that $\gamma_y' > \gamma_y\}$,*

The set of $(\Rightarrow_{\leq \beta})$-optimal pairs is defined as: $L(R'(S'))$.

Example 8. Consider again the R-implication in Example 3. We can see that:

- The $(\Rightarrow_{\geq \gamma_3})$-optimal pairs in $\mathcal{N}^+$ are (γ_3, γ_3), (γ_2, γ_2), and (γ_1, γ_1).
- The $(\Rightarrow_{\leq \gamma_3})$-optimal pairs in $\mathcal{N}$ are (γ_5, γ_3), (γ_3, γ_2), (γ_2, γ_1), and (γ_1, γ_0).

Note that Definition 12 has an important difference with Definitions 13 and 14. In the two latter cases, not every $\gamma_x' < \gamma_x$ prevents γ_x from taking part of an optimal pair. For instance, in Example 8, (γ_3, γ_3) is an $(\Rightarrow_{\geq \gamma_3})$-optimal pair even if $\gamma_2 < \gamma_3$ and $\gamma_2 \otimes \gamma_3 \Rightarrow \gamma_3$. This definition is designed to use the fact that, for instance $\forall \rho(R, \geq \gamma_x).\rho(C, \geq \gamma_y)$ implies $\forall \rho(R, \geq \gamma'_x).\rho(C, \geq \gamma_y)$ for every $\gamma'_x \in \mathcal{N}, \gamma'_x > \gamma_x$.

Note also that R-implications are, in general, non smooth (see Example 3). Hence, a pair of elements γ_1, γ_y such that $\gamma_x \Rightarrow_r \gamma_y = \alpha$ might not exist, and thus we have to consider an inequality of the form $\gamma_x \Rightarrow_r \gamma_y \geq \alpha$. In smooth t-norms, t-conorms and QL-implications, $=$ and $\geq$ would yield the same result.

$\kappa(\mathcal{A})$ (resp. $\kappa(\mathcal{T})$, $\kappa(\mathcal{R})$) denotes the union of the reductions of every axiom in $\mathcal{A}$ (resp. $\mathcal{T}$, $\mathcal{R}$). $\mathsf{crisp}(\mathcal{K})$ denotes the reduction of a fuzzy KB $\mathcal{K}$. A fuzzy KB $\mathcal{K} = \langle \mathcal{A}, \mathcal{T}, \mathcal{R} \rangle$ is reduced into a KB $\mathsf{crisp}(\mathcal{K}) = \langle \kappa(\mathcal{A}), T(\mathcal{N}) \cup \kappa(\mathcal{T}), R(\mathcal{N}) \cup \kappa(\mathcal{R}) \rangle$.

Example 9. Let us show a full example of how the reduction works. To this end, consider the smooth t-norm used in Example 2 and the fuzzy KB $\mathcal{K} = \{ \langle a : A \sqcap B \geq \gamma_2 \rangle, \langle a : \neg B \geq \gamma_4 \rangle \}$.

The crisp representation $\mathsf{crisp}(\mathcal{K})$ is computed as follows. $T(\mathcal{N})$ is defined as the set containing the following axioms: $A_{\geq \gamma_2} \sqsubseteq A_{\geq \gamma_1}$, $A_{\geq \gamma_3} \sqsubseteq A_{\geq \gamma_2}$, $A_{\geq \gamma_4} \sqsubseteq A_{\geq \gamma_3}$, $A_{\geq \gamma_5} \sqsubseteq A_{\geq \gamma_4}$, $B_{\geq \gamma_2} \sqsubseteq B_{\geq \gamma_1}$, $B_{\geq \gamma_3} \sqsubseteq B_{\geq \gamma_2}$, $B_{\geq \gamma_4} \sqsubseteq B_{\geq \gamma_3}$, $B_{\geq \gamma_5} \sqsubseteq B_{\geq \gamma_4}$.

Now, let us compute $\kappa(\mathcal{A})$. To this end, we have to map every axiom in the fuzzy ontology. The first one is represented as $a : (A_{\geq \gamma_2} \sqcap B_{\geq \gamma_3}) \sqcup (A_{\geq \gamma_3} \sqcap B_{\geq \gamma_2})$, as shown in Example 7. The second axiom is represented as $a : \neg B_{\geq \gamma_2}$.

It is not difficult to check that both $\mathcal{K}$ and $\mathsf{crisp}(\mathcal{K})$ are unsatisfiable.

Properties of the Reduction

Correctness. The following theorem, showing the logic is decidable and that the reduction preserves reasoning, can be shown.

Theorem 1. *The satisfiability problem in finite fuzzy $\mathcal{ALCH}$ is decidable. Furthermore, a finite fuzzy $\mathcal{ALCH}$ fuzzy KB $\mathcal{K}$ is satisfiable iff $\mathsf{crisp}(\mathcal{K})$ is.*

Complexity. In general, the size of $\mathsf{crisp}(\mathcal{K})$ is $\mathcal{O}(|\mathcal{K}| \cdot |\mathcal{N}|^k)$, being k the maximal depth of the concepts appearing in $\mathcal{K}$. In the particular case of finite Zadeh fuzzy logic, the size of $\mathsf{crisp}(\mathcal{K})$ is $\mathcal{O}(|\mathcal{K}| \cdot |\mathcal{N}|)$ [14]. For other fuzzy operators the case is more complex because we cannot infer the exact values of the degrees of truth, so we need to build disjunctions or conjunctions over all possible degrees of truth.

Modularity. The reduction of an ontology can be reused when adding new axioms if they do not introduce new atomic concepts and roles. In this case, it remains to add the reduction of the new axioms. This allows to compute the reduction of the ontology off-line and update $\mathsf{crisp}(\mathcal{K})$ incrementally. The assumption that the basic vocabulary is fully expressed in the ontology is reasonable because ontologies do not usually change once that their development has finished.

4   Extending Finite Fuzzy $\mathcal{ALCH}$

In this section we will discuss how to extend the previous logic in two ways. Firstly, Section 4.1 will consider alternative fuzzy logic operators (in particular, the non smooth case). Then, Section 4.2 will consider alternative DL constructors, with the aim of obtaining more expressive logics than finite fuzzy $\mathcal{ALCH}$.

4.1 The Non-smooth Case

Up to know, we have defined all the fuzzy operators of the logic by starting from a finite chain $\mathcal{N}$ and from a finite smooth t-norm. In this section we will discuss what happens in the case of non-smooth t-norms.

The first observation is that a full characterization of the operators is unknown yet, and there are only some partial results [20,21,22,23].

Another important point is that in the non-smooth case QL-implications depend of both a t-norm and a t-conorm, so the t-conorm of the language should not be restricted to the dual of the t-norm.

Furthermore, D-implications cannot in general be defined by means of QL-implications, so they should be explicitly considered in the language. Hence, we must add the concept $\forall_d R.C$ and the axioms $C_1 \sqsubseteq_d C_2$ and $R_1 \sqsubseteq_d R_2$, with the obvious semantics.

Even if there is not a complete knowledge of these operators, we can provide a reasoning mechanism with them by using the same ideas as in the reduction of R-implications shown in Section 3. In fact, we can consider any pair of degrees that satisfy the semantics of the constructor, and them we can simplify the expression by only taking the optimal pairs.

Let us also denote by $t \in \{s, r, ql, d\}$ the type of the fuzzy implication function used. The reason is that we are going to provide a reduction of $\forall R.C$ concepts, TBox axioms and RBox axioms that can be used for every type of implication.

Now, we are ready to show the procedure to obtain a crisp representation. Table 6 shows the differences in mapping κ, whereas Table 7 shows the differences in mapping ρ. The concept/role constructors and axioms which are not included in these tables are reduced as shown in Section 3.

Table 6. Mapping of axioms in the case of non-smooth t-norms

$\kappa(\langle C \sqsubseteq_t D \geq \alpha \rangle)$	$\bigcup \{\rho(C, \geq \gamma_x) \sqsubseteq \rho(D, \geq \gamma_y)\}$,
	for every $\gamma_x, \gamma_y \in \mathcal{N}^+$ such that γ_x, γ_y are $(\Rightarrow_t {}_{\geq \alpha})$-optimal
$\kappa(\langle R_1 \sqsubseteq_t R_2 \geq \alpha \rangle)$	$\bigcup \{\rho(R_1, \geq \gamma_x) \sqsubseteq \rho(R_2, \geq \gamma_y)\}$,
	for every $\gamma_x, \gamma_y \in \mathcal{N}^+$ such that γ_x, γ_y are $(\Rightarrow_t {}_{\geq \alpha})$-optimal

Table 7. Mapping of concepts in the case of non-smooth t-norms

$\rho(C \sqcap D, \geq \alpha)$	$\bigsqcup_{\gamma_x, \gamma_y \in \mathcal{N}^+} \{\rho(C, \geq \gamma_x) \sqcap \rho(D, \geq \gamma_y)\}$, (γ_x, γ_y) $(\otimes_{\geq \alpha})$-optimal
$\rho(C \sqcup D, \geq \alpha)$	$\rho(C, \geq \alpha) \sqcup \rho(D, \geq \alpha) \sqcup_{\gamma_x, \gamma_y \in \mathcal{N}^+} \{\rho(C, \geq \gamma_x) \sqcap \rho(D, \geq \gamma_y)\}$,
	(γ_x, γ_y) $(\oplus_{\geq \alpha})$-optimal
$\rho(\exists R.C, \geq \alpha)$	$\bigsqcup_{\gamma_x, \gamma_y \in \mathcal{N}^+} \{\exists \rho(R, \geq \gamma_x).\rho(C, \geq \gamma_y)\}$, (γ_x, γ_y) $(\otimes_{\geq \alpha})$-optimal
$\rho(\forall_t R.C, \geq \alpha)$	$\bigsqcap_{\gamma_x, \gamma_y \in \mathcal{N}^+} \{\forall \rho(R, \geq \gamma_x).\rho(C, \geq \gamma_y)\}$, (γ_x, γ_y) $(\Rightarrow_{\geq \alpha})$-optimal
$\rho(\forall_t R.C, \leq \beta)$	$\bigsqcup_{\gamma_x \in \mathcal{N}^+, \gamma_y \in \mathcal{N}} \{\exists \rho(R, \geq \gamma_x).\rho(C, \leq \gamma_y)\}$, (γ_x, γ_y) $(\Rightarrow_{\leq \alpha})$-optimal

Example 10. Let us compare the reductions produced using the t-norms in Example 2 and Example 4 by looking at how they behave when reducing the axiom $\langle a : A \sqcap B \geq \gamma_2 \rangle$.

- On the one hand, using the t-norm in Example 2, we take $\gamma_x, \gamma_y \in (\gamma_0, \gamma_3]$ such that $x + y = 3 + 2 = 5$. Hence, we obtain:

$$a : (A_{\geq \gamma_2} \sqcap B_{\geq \gamma_3}) \sqcup (A_{\geq \gamma_3} \sqcap B_{\geq \gamma_2}) \ .$$

We recall the reader that this result was obtained from a more intuitive point of view in Example 7.
- On the other hand, using the t-norm in Example 4 the result is different. Now, have to consider the $(\otimes_{\geq \gamma_2})$-optimal pairs, and thus we have:

$$a : (A_{\geq \gamma_4} \sqcap B_{\geq \gamma_2}) \sqcup (A_{\geq \gamma_3} \sqcap B_{\geq \gamma_3}) \sqcup (A_{\geq \gamma_2} \sqcap B_{\geq \gamma_4}) \ .$$

Note that we have to consider $A_{\geq \gamma_3} \sqcap B_{\geq \gamma_3}$ even if $\gamma_3 \otimes \gamma_3 = \gamma_3 \neq \gamma_2$.

4.2 Other DL Constructors

Our reduction procedure is modular and it could be applied to more expressive DLs. In particular, adding some elements such that their semantics do not depend on any particular choice of fuzzy operators is straightforward because they can be dealt with in the same way as for the Zadeh family [14].

Let S denote a simple fuzzy role[2]. Firstly, we will consider two new concept constructors (fuzzy nominals and local reflexivity concepts) and one new role constructor (inverse roles). The syntax and semantics of these constructors are:

Syntax	Semantics
$\{\alpha/a\}$	α if $x = a^{\mathcal{I}}$, 0 otherwise
$\exists S.\mathbf{Self}$	$S^{\mathcal{I}}(x, x)$
R^-	$R^{\mathcal{I}}(y, x)$

Now, we will introduce some new axioms: disjoint, reflexive, irreflexive, symmetric, and asymmetric role axioms. The syntax and semantics of the axioms is defined as follows:

Syntax	Semantics
$\mathtt{dis}(S_1, \ldots, S_m)$	$\forall x, y \in \Delta^{\mathcal{I}}, \min\{S_i^{\mathcal{I}}(x, y), S_j^{\mathcal{I}}(x, y)\} = 0, \forall 1 \leq i < j \leq m$
$\mathtt{ref}(R)$	$\forall x \in \Delta^{\mathcal{I}}, R^{\mathcal{I}}(x, x) = 1$
$\mathtt{irr}(S)$	$\forall x \in \Delta^{\mathcal{I}}, S^{\mathcal{I}}(x, x) = 0$
$\mathtt{sym}(R)$	$\forall x, y \in \Delta^{\mathcal{I}}, R^{\mathcal{I}}(x, y) = R^{\mathcal{I}}(y, x)$
$\mathtt{asy}(S)$	$\forall x, y \in \Delta^{\mathcal{I}}, \text{if } S^{\mathcal{I}}(x, y) > 0 \text{ then } S^{\mathcal{I}}(y, x) = 0$

---

[2] Intuitively, simple roles are such that they do not take part in cyclic role inclusion axioms (see [14] for a formal definition in the context of fuzzy DLs). Simple roles are needed in some parts of a fuzzy KB to guarantee the decidability of the logic.

Mapping ρ can be extended in order to deal with these new concept and role constructors in the following way:

$\rho(\{\gamma/a\}, \geq \alpha)$	$\{a\}$ if $\gamma \geq \alpha$, $\perp$ otherwise
$\rho(\{\gamma/a\}, \leq \beta)$	$\neg\{a\}$ if $\gamma > \beta$, $\top$ otherwise
$\rho(\exists S.\text{Self}, \geq \alpha)$	$\exists\rho(S, \geq \alpha).\text{Self}$
$\rho(\exists S.\text{Self}, \leq \beta)$	$\neg\exists\rho(S, \neg \geq +\beta).\text{Self}$
$\rho(R^-, \geq \alpha)$	$\rho(R, \geq \alpha)^-$
$\rho(R^-, \leq \beta)$	$\rho(R, \leq \beta)^-$

Furthermore, mapping κ can be extended in order to deal with the new axioms in the following way:

$\kappa(\text{dis}(S_1, S_2))$	$\{\text{dis}(\rho(S_1, > \gamma_0), \rho(S_2, > \gamma_0))\}$
$\kappa(\text{ref}(R))$	$\{\text{ref}(\rho(R, \geq \gamma_p))\}$
$\kappa(\text{irr}(S))$	$\{\text{irr}(\rho(S, > \gamma_0))\}$
$\kappa(\text{sym}(R))$	$\bigcup_{\gamma \in \mathcal{N}^+}\{\text{sym}(\rho(R, \geq \gamma))\}$
$\kappa(\text{asy}(S))$	$\{\text{asy}(\rho(S, > \gamma_0))\}$

The logic obtained by extending $\mathcal{ALCH}$ with fuzzy nominals and inverse roles is called $\mathcal{ALCHOI}$. Clearly, finite fuzzy $\mathcal{ALCHOI}$ can be mapped into crisp $\mathcal{ALCHOI}$. After having added the other elements, it only remains to represent role inclusion axioms with role chains[3] and qualified cardinality restrictions, in order to cover a fuzzy extension of $\mathcal{SROIQ}$ (and hence OWL 2).

5   Conclusions and Future Work

This paper has set a general framework for fuzzy DLs with a finite chain of degrees of truth $\mathcal{N}$. $\mathcal{N}$ can be seen as a finite totally ordered set of linguistic terms or labels. This is very useful in practice, since expert knowledge is usually expressed using linguistic terms and avoiding their numerical interpretations.

Starting from a smooth finite t-norm on $\mathcal{N}$, we define the syntax and semantics of fuzzy $\mathcal{ALCH}$. The negation function and the t-conorm are imposed by the choice of the t-norm, but there are different options for the implication function. For this reason, whenever this is possible (i.e., in universal restriction concepts and in inclusion axioms), the language allows to use three different implications. We have studied some of the logical properties of the logic. This will help the ontology developers to use the implication that better suit their needs. Hence, our approach makes it possible to use in fuzzy DLs some fuzzy logical operators that have not been considered before.

The decidability of the logic has been shown by presenting a reasoning preserving reduction to the crisp case. Providing a crisp representation for a fuzzy ontology allows reusing current crisp ontology languages and reasoners, among

---

[3] Most of the works in the DL and fuzzy DL literature also consider transitive role axioms. We have not done so because transitive role axioms can be represented by using role inclusion axioms with role chains.

other related resources. The complexity of the crisp representation is higher than in finite Zadeh fuzzy DLs, because it is necessary to build disjunctions or conjunctions over all possible degrees of truth. However, Zadeh fuzzy DLs have some logical problems [14] which may not be acceptable in some applications, where alternative operators such as those introduced in this paper could be used.

We have also shown how to extend the logic in two directions, by considering non smooth operators, and by considering more expressive DL constructors, obtaining a closer logic to finite fuzzy $\mathcal{SROIQ}$ (and hence finite fuzzy OWL 2).

As future work we would like to study how to reduce qualified cardinality restrictions (see also [16]) and role inclusion axioms with role chains. This way, we will be able to provide the theoretical basis of a general fuzzy extension of OWL 2 under a finite chain of degrees of truth.

Acknowledgement. F. Bobillo acknowledges support from the Spanish Ministry of Science and Technology (project TIN2009-14538-C02-01) and Ministry of Education (program José Castillejo, grant JC2009-00337).

References

1. Baader, F., Calvanese, D., McGuinness, D., Nardi, D., Patel-Schneider, P.F.: The Description Logic Handbook: Theory, Implementation, and Applications. Cambridge University Press (2003)
2. Cuenca-Grau, B., Horrocks, I., Motik, B., Parsia, B., Patel-Schneider, P.F., Sattler, U.: OWL 2: The next step for OWL. Journal Web Semantics 6(4), 309–322 (2008)
3. Lukasiewicz, T., Straccia, U.: Managing uncertainty and vagueness in Description Logics for the Semantic Web. Journal of Web Semantics 6(4), 291–308 (2008)
4. Liau, C.J.: On rough terminological logics. In: Proceedings of the 4th International Workshop on Rough Sets, Fuzzy Sets and Machine Discovery, RSFD 1996, pp. 47–54 (1996)
5. Doherty, P., Grabowski, M., Łukaszewicz, W., Szałas, A.: Towards a framework for approximate ontologies. Fundamenta Informaticae 57(2-4), 147–165 (2003)
6. Schlobach, S., Klein, M.C.A., Peelen, L.: Description logics with approximate definitions: Precise modeling of vague concepts. In: Proceedings of the 20th International Joint Conference on Artificial Intelligence, IJCAI 2007, pp. 557–562 (2007)
7. Jiang, Y., Wang, J., Tang, S., Xiao, B.: Reasoning with rough description logics: An approximate concepts approach. Information Sciences 179(5), 600–612 (2009)
8. Cerami, M., Esteva, F., Bou, F.: Decidability of a Description Logic over infinite-valued product logic. In: Proceedings of the 12th International Conference on Principles of Knowledge Representation and Reasoning, KR 2010, pp. 203–213. AAAI Press (2010)
9. Baader, F., Peñaloza, R.: Are fuzzy Description Logics with General Concept Inclusion axioms decidable? In: Proceedings of the 20th IEEE International Conference on Fuzzy Systems, FUZZ-IEEE 2011, pp. 1735–1742. IEEE Press (2011)
10. Baader, F., Peñaloza, R.: GCIs make reasoning in fuzzy DL with the product t-norm undecidable. In: Proceedings of the 24th International Workshop on Description Logics, DL 2011. CEUR Workshop Proceedings, vol. 745 (2011)

11. Baader, F., Peñaloza, R.: On the Undecidability of Fuzzy Description Logics with GCIs and Product T-norm. In: Tinelli, C., Sofronie-Stokkermans, V. (eds.) FroCoS 2011. LNCS, vol. 6989, pp. 55–70. Springer, Heidelberg (2011)
12. Hájek, P.: Making fuzzy Description Logics more general. Fuzzy Sets and Systems 154(1), 1–15 (2005)
13. Bobillo, F., Bou, F., Straccia, U.: On the failure of the finite model property in some fuzzy Description Logics. Fuzzy Sets and Systems 172(1), 1–12 (2011)
14. Bobillo, F., Delgado, M., Gómez-Romero, J.: Crisp representations and reasoning for fuzzy ontologies. International Journal of Uncertainty, Fuzziness and Knowledge-Based Systems 17(4), 501–530 (2009)
15. Bobillo, F., Delgado, M., Gómez-Romero, J., Straccia, U.: Fuzzy Description Logics under Gödel semantics. International Journal of Approximate Reasoning 50(3), 494–514 (2009)
16. Bobillo, F., Straccia, U.: Reasoning with the finitely many-valued Łukasiewicz fuzzy Description Logic $\mathcal{SROIQ}$. Information Sciences 181(4), 758–778 (2011)
17. Hájek, P.: What does mathematical fuzzy logic offer to Description Logic? In: Capturing Intelligence: Fuzzy Logic and the Semantic Web, ch. 5, pp. 91–100. Elsevier (2006)
18. García-Cerdaña, A., Armengol, E., Esteva, F.: Fuzzy Description Logics and t-norm based fuzzy logics. International Journal of Approximate Reasoning 51(6), 632–655 (2010)
19. Cerami, M., García-Cerdaña, A., Esteva, F.: From classical Description Logic to n–graded fuzzy Description Logic. In: Proceedings of the 19th IEEE International Conference on Fuzzy Systems, FUZZ-IEEE 2010, pp. 1506–1513. IEEE Press (2010)
20. Mayor, G., Torrens, J.: On a class of operators for expert systems. International Journal of Intelligent Systems 8(7), 771–778 (1993)
21. Mayor, G., Torrens, J.: Triangular norms in discrete settings. In: Logical, Algebraic, Analytic, and Probabilistic Aspects of Triangular Norms, ch. 7, pp. 189–230. Elsevier (2005)
22. Mas, M., Monserrat, M., Torrens, J.: S-implications and R-implications on a finite chain. Kybernetika 40(1), 3–20 (2004)
23. Mas, M., Monserrat, M., Torrens, J.: On two types of discrete implications. International Journal of Approximate Reasoning 40(3), 262–279 (2005)
24. Straccia, U.: Description Logics over lattices. International Journal of Uncertainty, Fuzziness and Knowledge-Based Systems 14(1), 1–16 (2006)
25. Jiang, Y., Tang, Y., Wang, J., Deng, P., Tang, S.: Expressive fuzzy Description Logics over lattices. Knowledge-Based Systems 23(2), 150–161 (2010)
26. Zadeh, L.A.: Fuzzy sets. Information and Control 8, 338–353 (1965)
27. Horrocks, I., Kutz, O., Sattler, U.: The even more irresistible $\mathcal{SROIQ}$. In: Proceedings of the 10th International Conference of Knowledge Representation and Reasoning, KR 2006, pp. 452–457. IEEE Press (2006)
28. Stoilos, G., Stamou, G., Pan, J.Z.: Handling imprecise knowledge with fuzzy Description Logic. In: Proceedings of the 2006 International Workshop on Description Logics, DL 2006, pp. 119–126 (2006)
29. Straccia, U.: Reasoning within Fuzzy Description Logics. Journal of Artificial Intelligence Research 14, 137–166 (2001)
30. Stoilos, G., Stamou, G., Pan, J.Z.: Fuzzy extensions of OWL: Logical properties and reduction to fuzzy Description Logics. International Journal of Approximate Reasoning 51(6), 656–679 (2010)

Reasoning in Fuzzy OWL 2 with DeLorean

Fernando Bobillo[1], Miguel Delgado[2], and Juan Gómez-Romero[3]

[1] Dpt. of Computer Science and Systems Engineering, University of Zaragoza, Spain
[2] Dpt. of Computer Science and Artificial Intelligence, University of Granada, Spain
[3] Applied Artificial Intelligence Group, University Carlos III, Madrid, Spain
fbobillo@unizar.es, mdelgado@ugr.es, jgromero@inf.uc3m.es

Abstract. Classical ontologies are not suitable to represent imprecise or vague information, which has led to several extensions using non-classical logics. In particular, several fuzzy extensions have been proposed in the literature. In this paper, we present the fuzzy ontology reasoner DE-LOREAN, the first to support a fuzzy extension of OWL 2. We discuss how to use it for fuzzy ontology representation and reasoning, and describe some implementation details and optimization techniques. An empirical evaluation demonstrates that these optimizations considerably improve the performance of the reasoner.

1 Introduction

Ontologies have been successfully used as a formalism for knowledge representation in several applications. In particular, they are a core element in the layered architecture of the Semantic Web. In that regard, the language OWL 2 [1] has very recently become a W3C Recommendation for ontology representation.

Description Logics (DLs for short) [2] are a family of logics for representing structured knowledge. Each logic is denoted by using a string of capital letters which identify the constructors of the logic and therefore its complexity. They have proved to be very useful as ontology languages, in such a way that OWL 2 is closely equivalent to the DL $\mathcal{SROIQ}(\mathbf{D})$ [3].

Today, there is a growing interest in the development of knowledge representation formalisms able to deal with imprecise knowledge, a very common requirement in real world applications. Nevertheless, classical ontologies are not appropriate to deal with imprecision and vagueness in the knowledge, which is inherent to most real world application domains. Since fuzzy logic is a suitable formalism to handle these types of knowledge, several fuzzy extensions of DLs have been proposed [4].

The apparition of the new standard language OWL 2 has motivated a need to extend ontology editors, reasoners, and other supporting tools. The situation is similar in the fuzzy case, and having reasoners that are able to support fuzzy extensions of OWL 2 is of great importance. In this paper we report the implementation of DELOREAN (DEscription LOgic REasoner with vAgueNess)[1], the first reasoner that supports the fuzzy DL $\mathcal{SROIQ}(\mathbf{D})$, and hence fuzzy OWL 2.

---

[1] http://webdiis.unizar.es/~fbobillo/delorean

F. Bobillo et al. (Eds.): URSW 2008-2010/UniDL 2010, LNAI 7123, pp. 119–138, 2013.
© Springer-Verlag Berlin Heidelberg 2013

In a strict sense, DeLorean is not a reasoner but a *translator* from a fuzzy ontology language into a classical (i.e., non-fuzzy) ontology language –namely, the standard language OWL 2. The non-fuzzy ontology resulting from this translation, which preserves the semantics of the initial fuzzy representation, is afterwards processed by an integrated classical DL reasoner. According to this ability of combining the reduction procedure with the classical DL reasoning, we will simply refer to it as a reasoner.

This paper is organized as follows. Section 2 provides some background on fuzzy set theory and fuzzy logic. Then, Section 3 describes which fuzzy ontologies can be managed by the system and how they are represented. Next, Section 4 explains which reasoning tasks can be performed by using the reasoner and how they are accomplished. In Section 5, we give some implementation details. A use case and a preliminary evaluation of the implemented optimizations are discussed in Section 6. Finally, Section 7 sets out some conclusions and prospective directions for future work.

2 Fuzzy Logic

Fuzzy set theory and fuzzy logic were proposed by L. Zadeh [13] to manage imprecise and vague knowledge. While in classical set theory elements either belong to a set or not, in fuzzy set theory elements can belong to a set to some degree. More formally, let X be a set of elements called the reference set. A *fuzzy subset A* of X is defined by a membership function $\mu_A(x)$, or simply $A(x)$, which assigns any $x \in X$ to a value in the interval of real numbers between 0 and 1. As in the classical case, 0 means no membership and 1 full membership, but now a value between 0 and 1 represents the extent to which x can be considered an element of X.

Changing the usual true/false convention leads to a new type of propositions, called *fuzzy propositions*. Each fuzzy proposition may have a *degree of truth* in $[0, 1]$, denoting the compatibility of the fuzzy proposition with a given state of facts. For example, the truth of the proposition stating than a given tomato is a ripe tomato is clearly a matter of degree.

In this article we will consider *fuzzy formulae* (or fuzzy axioms) of the form $\phi \geq \alpha$ or $\phi \leq \beta$, where ϕ is a fuzzy proposition and $\alpha, \beta \in [0, 1]$ [14]. This imposes that the degree of truth of ϕ is *at least* α (resp. *at most* β). For example, x is a ripe tomato ≥ 0.9 says that we have a rather ripe tomato (the degree of truth of x being a ripe tomato is at least 0.9).

All classical set operations are extended to fuzzy sets. The intersection, union, complement and implication set operations are performed by corresponding functions: a t-norm, a t-conorm, a negation, and an implication, respectively. The combination of them is called a fuzzy logic.

There are three main fuzzy logics: Łukasiewicz, Gödel, and Product. The importance of these three fuzzy logics is due to the fact that any continuous t-norm can be obtained as a combination of Łukasiewicz, Gödel, and Product t-norms. It is also common to consider the fuzzy connectives originally considered by Zadeh (Gödel conjunction and disjunction, Łukasiewicz negation and Kleene-Dienes

Table 1. Popular fuzzy logics over [0,1]

Family	t-norm $\alpha \otimes \beta$	t-conorm $\alpha \oplus \beta$	negation $\ominus \alpha$	implication $\alpha \Rightarrow \beta$
Zadeh	$\min\{\alpha, \beta\}$	$\max\{\alpha, \beta\}$	$1 - \alpha$	$\max\{1 - \alpha, \beta\}$
Gödel	$\min\{\alpha, \beta\}$	$\max\{\alpha, \beta\}$	$\begin{cases} 1, & \alpha = 0 \\ 0, & \alpha > 0 \end{cases}$	$\begin{cases} 1 & \alpha \leq \beta \\ \beta, & \alpha > \beta \end{cases}$
Łukasiewicz	$\max\{\alpha + \beta - 1, 0\}$	$\min\{\alpha + \beta, 1\}$	$1 - \alpha$	$\min\{1 - \alpha + \beta, 1\}$
Product	$\alpha \cdot \beta$	$\alpha + \beta - \alpha \cdot \beta$	$\begin{cases} 1, & \alpha = 0 \\ 0, & \alpha > 0 \end{cases}$	$\begin{cases} 1 & \alpha \leq \beta \\ \beta/\alpha, & \alpha > \beta \end{cases}$

implication), which is known as Zadeh fuzzy logic. Table 1 shows these four fuzzy logics: Zadeh, Łukasiewicz, Gödel, and Product.

For every $\alpha \in [0, 1]$, the α-*cut* of a fuzzy set A is defined as the (crisp) set such that its elements belong to A with degree at least α, i.e. $\{x \mid \mu_A(x) \geq \alpha\}$.

Relations can also be extended to the fuzzy case. A (binary) *fuzzy relation* R over two countable sets X and Y is a function $R \colon X \times Y \to [0, 1]$. Several properties of the relations (such as reflexive, irreflexive, symmetric, asymmetric, transitive, or disjointness) and operations (inverse, composition) can be trivially extended to the fuzzy case.

3   Representing Fuzzy Ontologies

In this section we discuss the fuzzy ontologies that DeLorean is able to manage. Section 3.1 describes the elements of the supported fuzzy DL $\mathcal{SROIQ}(\mathbf{D})$. Section 3.2 discusses how to create these fuzzy ontologies.

3.1   Fuzzy $\mathcal{SROIQ}(\mathbf{D})$

Fuzzy $\mathcal{SROIQ}(\mathbf{D})$ [15,16], extends $\mathcal{SROIQ}(\mathbf{D})$ to the fuzzy case by letting concepts denote fuzzy sets of individuals and roles denote fuzzy binary relations.

Notation. In the following, we will use $\otimes$ for denoting a t-norm, $\oplus$ for a t-conorm, $\ominus$ for a negation, an $\Rightarrow$ for an implication. The subscript z denotes Zadeh fuzzy logic, and G denotes Gödel fuzzy logic.

We will assume that the *degrees of truth* are rational numbers of the form $\alpha \in (0, 1]$, $\beta \in [0, 1)$ and $\gamma \in [0, 1]$. Moreover, we will assume a set of *inequalities* $\bowtie \in \{\geq, >, \leq, <\}$, $\rhd \in \{\geq, >\}$, $\lhd \in \{\leq, <\}$.

Fuzzy $\mathcal{SROIQ}(\mathbf{D})$ has three *alphabets* of symbols for concepts, roles and individuals.

The *concepts* of the language are denoted as C, D (if they are complex concepts) and A (if atomic). Some complex concepts will use natural numbers n, m such that $n \geq 0, m > 0$.

The *roles* can be abstract (denoted R) or concrete (denoted T). R_A denotes an atomic abstract role, R the inverse role, S a simple role[2], and U the universal role (a relation which is true for every pair of individuals).

---

[2] Simple roles are needed to guarantee the decidability of the logic. Intuitively, simple roles cannot take part in cyclic role inclusion axioms (see [15] for a formal definition).

A *fuzzy concrete domain* (also called a fuzzy *datatype*) $\mathbf{D}$ is a pair $\langle \Delta_{\mathbf{D}}, \Phi_{\mathbf{D}} \rangle$, where $\Delta_{\mathbf{D}}$ is a concrete interpretation domain (for instance, the set of rational numbers in a given interval), and $\Phi_{\mathbf{D}}$ is a set of fuzzy concrete predicates $\mathbf{d}$ [18]. Typical examples of fuzzy concrete predicates are the *trapezoidal*, the *triangular*, the *left-shoulder*, and the *right-shoulder* membership functions. We will restrict to *trapezoidal* membership functions, which are more general than the other mentioned predicates.

The *individuals* are denoted as a, b (if abstract), and v (if concrete). Abstract individuals are elements of the interpretation domain, whereas concrete individuals are instances of the concrete interpretation domain.

Syntax. The syntax of fuzzy concepts, roles, and axioms is shown in Table 2. A *fuzzy Knowledge Base* (KB) is a finite set of fuzzy axioms, which can be grouped into a fuzzy ABox with axioms (A1)–(A7), a fuzzy TBox with axioms (A8)–(A9), and a fuzzy RBox with axioms (A10)–(A20). Note that the only syntactic differences with respect to crisp $\mathcal{SROIQ}(\mathbf{D})$ are (C11), (A1)–(A5), (A8), (A10)–(A11).

Most axioms are better known by a name. (A1) are named concept assertions, (A2)–(A5) are role assertions, (A6) are inequality assertions, (A7) are equality assertions, (A8) are General Concept Inclusions (GCIs, or subclass axioms), (A9) are concept equivalences, (A10)–(A11) are Role Inclusion Axioms (RIAs, or sub-role axioms), (A12)–(A13) are role equivalences, (A14) are transitive role axioms, (A15)–(16) are disjoint role axioms, (A17) are reflexive role axioms, (A18) are irreflexive role axioms, (A19) are symmetric role axioms, and (A20) are asymmetric role axioms.

As in the crisp case, GCIs can be used to express some interesting axioms, such as *disjointness* of concepts or *domain, range* and *functionality* of a role [17].

Example 1. The fuzzy concept assertion $\langle \mathsf{RoseDAnjou : RoseWine} \geq 0.75 \rangle$ states that it is almost true that the Rosé D'Anjou wine is a rose wine. RoseDAnjou is an abstract individual, and RoseWine is a fuzzy concept.   □

Semantics. The semantics of the logic is given using the notion of fuzzy interpretation. A *fuzzy interpretation* $\mathcal{I}$ with respect to a fuzzy concrete domain $\mathbf{D}$ is a pair $(\Delta^{\mathcal{I}}, \cdot^{\mathcal{I}})$ consisting of a non empty set $\Delta^{\mathcal{I}}$ (the interpretation domain) disjoint with $\Delta_{\mathbf{D}}$ and a fuzzy interpretation function $\cdot^{\mathcal{I}}$ mapping:

- An *abstract individual* a to an element $a^{\mathcal{I}} \subseteq \Delta^{\mathcal{I}}$.
- A *concrete individual* v to an element $v_{\mathbf{D}} \subseteq \Delta_{\mathbf{D}}$.
- A fuzzy *concept* C to a function $C^{\mathcal{I}} : \Delta^{\mathcal{I}} \to [0,1]$.
- A fuzzy *abstract role* R to a function $R^{\mathcal{I}} : \Delta^{\mathcal{I}} \times \Delta^{\mathcal{I}} \to [0,1]$.
- A fuzzy *concrete role* T to a function $T^{\mathcal{I}} : \Delta^{\mathcal{I}} \times \Delta_{\mathbf{D}} \to [0,1]$.
- An *n*-ary fuzzy *concrete predicate* $\mathbf{d}$ to a function $\mathbf{d}^{\mathcal{I}} : \Delta_{\mathbf{D}}^n \to [0,1]$.

$C^{\mathcal{I}}$ denotes the membership function of the fuzzy concept C w.r.t. $\mathcal{I}$. $C^{\mathcal{I}}(a^{\mathcal{I}})$ denotes to what extent the individual a can be considered an element of the fuzzy concept C. $R^{\mathcal{I}}$ denotes the membership function of the fuzzy role R w.r.t.

Table 2. Syntax and semantics of the fuzzy DL $\mathcal{SROIQ}(\mathbf{D})$

Concept	Syntax (C)	Semantics of $C^{\mathcal{I}}(x)$
(C1)	A	$A^{\mathcal{I}}(x)$
(C2)	$\top$	1
(C3)	$\bot$	0
(C4)	$C \sqcap D$	$C^{\mathcal{I}}(x) \otimes D^{\mathcal{I}}(x)$
(C5)	$C \sqcup D$	$C^{\mathcal{I}}(x) \oplus D^{\mathcal{I}}(x)$
(C6)	$\neg C$	$\ominus C^{\mathcal{I}}(x)$
(C7)	$\forall R.C$	$\inf_{y \in \Delta^{\mathcal{I}}} \{R^{\mathcal{I}}(x,y) \Rightarrow C^{\mathcal{I}}(y)\}$
(C8)	$\exists R.C$	$\sup_{y \in \Delta^{\mathcal{I}}} \{R^{\mathcal{I}}(x,y) \otimes C^{\mathcal{I}}(y)\}$
(C9)	$\forall T.\mathbf{d}$	$\inf_{v \in \Delta_{\mathbf{D}}} \{T^{\mathcal{I}}(x,v) \Rightarrow \mathbf{d}^{\mathcal{I}}(v)\}$
(C10)	$\exists T.\mathbf{d}$	$\sup_{v \in \Delta_{\mathbf{D}}} \{T^{\mathcal{I}}(x,v) \otimes \mathbf{d}^{\mathcal{I}}(v)\}$
(C11)	$\{\alpha/a\}$	α if $x = a^{\mathcal{I}}$, 0 otherwise
(C12)	$\geq m \, S.C$	$\sup_{y_1,\ldots,y_m \in \Delta^{\mathcal{I}}} \{\min_{i=1}^m \{S^{\mathcal{I}}(x,y_i) \otimes C^{\mathcal{I}}(y_i)\} \otimes$ $(\otimes_{1 \leq j < k \leq m} \{y_j \neq y_k\})\}$
(C13)	$\leq n \, S.C$	$\inf_{y_1,\ldots,y_{n+1} \in \Delta^{\mathcal{I}}} \{\min_{i=1}^{n+1} \{S^{\mathcal{I}}(x,y_i) \otimes C^{\mathcal{I}}(y_i)\} \Rightarrow$ $\oplus_{1 \leq j < k \leq n+1} \{y_j = y_k\}\}$
(C14)	$\geq m \, T.\mathbf{d}$	$\sup_{v_1,\ldots,v_m \in \Delta_{\mathbf{D}}} \{\min_{i=1}^m \{T^{\mathcal{I}}(x,v_i) \otimes \mathbf{d}^{\mathcal{I}}(v_i)\} \otimes$ $(\otimes_{j < k} \{v_j \neq v_k\})\{$
(C15)	$\leq n \, T.\mathbf{d}$	$\inf_{v_1,\ldots,v_{n+1} \in \Delta_{\mathbf{D}}} \{\min_{i=1}^{n+1} \{T^{\mathcal{I}}(x,v_i) \otimes \mathbf{d}^{\mathcal{I}}(v_i)\} \Rightarrow$ $(\oplus_{j < k} \{v_j = v_k\})\}$
(C16)	$\exists S.\mathtt{Self}$	$S^{\mathcal{I}}(x,x)$

Role	Syntax (R)	Semantics of $R^{\mathcal{I}}(x,y)$
(R1)	R_A	$R_A^{\mathcal{I}}(x,y)$
(R2)	R^-	$R^{\mathcal{I}}(y,x)$
(R3)	U	1
(R4)	T	$T^{\mathcal{I}}(x,y)$

Axiom	Syntax (τ)	Semantics ($\mathcal{I}$ satisfies τ if ...)
(A1)	$\langle a : C \bowtie \alpha \rangle$	$C^{\mathcal{I}}(a^{\mathcal{I}}) \bowtie \alpha$
(A2)	$\langle (a,b) : R \bowtie \alpha \rangle$	$R^{\mathcal{I}}(a^{\mathcal{I}}, b^{\mathcal{I}}) \bowtie \alpha$
(A3)	$\langle (a,b) : \neg R \bowtie \alpha \rangle$	$\ominus R^{\mathcal{I}}(a^{\mathcal{I}}, b^{\mathcal{I}}) \bowtie \alpha$
(A4)	$\langle (a,v) : T \bowtie \alpha \rangle$	$T^{\mathcal{I}}(a^{\mathcal{I}}, v_{\mathbf{D}}) \bowtie \alpha$
(A5)	$\langle (a,v) : \neg T \bowtie \alpha \rangle$	$\ominus T^{\mathcal{I}}(a^{\mathcal{I}}, v_{\mathbf{D}}) \bowtie \alpha$
(A6)	$\langle a \neq b \rangle$	$a^{\mathcal{I}} \neq b^{\mathcal{I}}$
(A7)	$\langle a = b \rangle$	$a^{\mathcal{I}} = b^{\mathcal{I}}$
(A8)	$\langle C \sqsubseteq D \rhd \alpha \rangle$	$\inf_{x \in \Delta^{\mathcal{I}}} \{C^{\mathcal{I}}(x) \Rightarrow D^{\mathcal{I}}(x)\} \rhd \alpha$
(A9)	$C_1 \equiv \cdots \equiv C_m$	$\forall_{x \in \Delta^{\mathcal{I}}} C_1^{\mathcal{I}}(x) = \cdots = C_m^{\mathcal{I}}(x)$
(A10)	$\langle R_1 \ldots R_m \sqsubseteq R \rhd \alpha \rangle$	$\inf_{x_1, x_{m+1} \in \Delta^{\mathcal{I}}} \{\sup_{x_2 \ldots x_m \in \Delta^{\mathcal{I}}} \{(R_1^{\mathcal{I}}(x_1, x_2) \otimes \cdots \otimes R_m^{\mathcal{I}}(x_m, x_{m+1})) \Rightarrow R^{\mathcal{I}}(x_1, x_{m+1})\}\} \rhd \alpha$
(A11)	$\langle T_1 \sqsubseteq T_2 \rhd \alpha \rangle$	$\inf_{x \in \Delta^{\mathcal{I}}, v \in \Delta_{\mathbf{D}}} \{T_1^{\mathcal{I}}(x,v) \Rightarrow T_2^{\mathcal{I}}(x,v)\} \rhd \alpha$
(A12)	$R_1 \equiv \cdots R_m$	$\forall_{x,y \in \Delta^{\mathcal{I}}} R_1^{\mathcal{I}}(x,y) = \cdots = R_m^{\mathcal{I}}(x,y)$
(A13)	$T_1 \equiv \cdots T_m$	$\forall_{x \in \Delta^{\mathcal{I}}, v \in \Delta_{\mathbf{D}}} R_1^{\mathcal{I}}(x,v) = \cdots = R_m^{\mathcal{I}}(x,v)$
(A14)	$\mathtt{trans}(R)$	$\forall x,y,z \in \Delta^{\mathcal{I}}, R^{\mathcal{I}}(x,z) \otimes R^{\mathcal{I}}(z,y) \leq R^{\mathcal{I}}(x,y)$
(A15)	$\mathtt{dis}(S_1, \ldots, S_m)$	$\forall x,y \in \Delta^{\mathcal{I}}, \min\{S_i^{\mathcal{I}}(x,y), S_j^{\mathcal{I}}(x,y)\} = 0, \forall 1 \leq i < j \leq m$
(A16)	$\mathtt{dis}(T_1, \ldots, T_m)$	$\forall x \in \Delta^{\mathcal{I}}, v \in \Delta_{\mathbf{D}}, \min\{T_i^{\mathcal{I}}(x,v), T_j^{\mathcal{I}}(x,v)\} = 0, \forall 1 \leq i < j \leq m$
(A17)	$\mathtt{ref}(R)$	$\forall x \in \Delta^{\mathcal{I}}, R^{\mathcal{I}}(x,x) = 1$
(A18)	$\mathtt{irr}(S)$	$\forall x \in \Delta^{\mathcal{I}}, S^{\mathcal{I}}(x,x) = 0$
(A19)	$\mathtt{sym}(R)$	$\forall x,y \in \Delta^{\mathcal{I}}, R^{\mathcal{I}}(x,y) = R^{\mathcal{I}}(y,x)$
(A20)	$\mathtt{asy}(S)$	$\forall x,y \in \Delta^{\mathcal{I}}, S^{\mathcal{I}}(x,y) > 0$ then $S^{\mathcal{I}}(y,x) = 0$

$\mathcal{I}$. $R^{\mathcal{I}}(a^{\mathcal{I}}, b^{\mathcal{I}})$ denotes to what extent (a, b) can be considered an element of the fuzzy role R.

Given the operators $\otimes, \oplus, \ominus, \Rightarrow$, the fuzzy interpretation function is defined for fuzzy concepts, roles, concrete domains and axioms as shown in Table 2. The fuzzy DL $Z \, \mathcal{SROIQ}(\mathbf{D})$ uses the operators in Zadeh fuzzy logic, whereas $G \, \mathcal{SROIQ}(\mathbf{D})$ uses the operators in Gödel fuzzy logic.

We say that a fuzzy interpretation $\mathcal{I}$ satisfies a fuzzy KB $\mathcal{K}$ iff $\mathcal{I}$ satisfies each element in $\mathcal{K}$.

3.2   A Fuzzy Ontology Editor

The input fuzzy ontologies supported by DeLorean can be created in 3 ways:

- Encoding the fuzzy ontology in the specific language of the reasoner (referred in this paper as "DeLorean syntax"). The details of DeLorean syntax can be found in the web page of the reasoner.
- Creating programmatically a new ontology by using the DeLorean API. The DeLorean API is a Java library that allows fuzzy ontology management (by loading existing ontologies or by creating and populating them) and solving reasoning tasks. The Javadoc documentation of the API can be found along with the distribution.
- Using a fuzzy ontology editor tool and then translating the ontology into DeLorean syntax.

In this section, we focus on the third option, which is the recommended one, since it allows us to edit fuzzy ontologies in a more abstract way.

A methodology for fuzzy ontology representation using OWL 2 has been recently proposed [26]. The key idea of this representation is to use an OWL 2 ontology and extend their elements with annotations representing the features of the fuzzy ontology that OWL 2 cannot directly encode. In order to separate the annotations including fuzzy information from other annotations, a new annotation property called `fuzzyLabel` is used, and every annotation is identified by the tag `fuzzyOwl2`.

Example 2. The fuzzy concept assertion of Example 1 is represented by annotating the axiom with the degree ≥ 0.75 as follows:

```
<ClassAssertion>
  <Class IRI='#RoseWine'/>
  <NamedIndividual IRI='#RoseDAnjou'/>
  <Annotation>
    <AnnotationProperty IRI='#fuzzyLabel'/>
    <Literal datatypeIRI='&rdf;PlainLiteral'>
      <fuzzyOwl2 fuzzyType="axiom">
        <Degree value="0.75" />
      </fuzzyOwl2>
    </Literal>
  </Annotation>
</ClassAssertion>
```

□

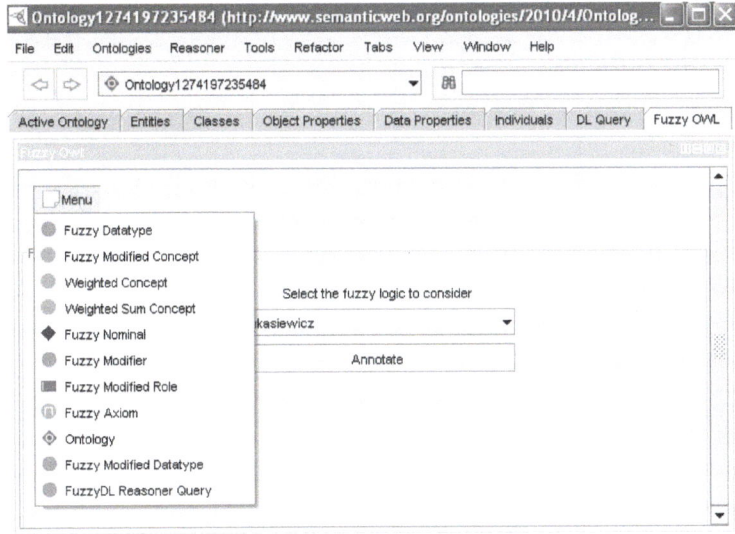

Fig. 1. Menu options of the plug-in

Since typing such annotations is a tedious and error-prone task, a Protégé plug-in has been implemented to make the syntax of the annotations transparent to the users. The plug-in is freely available[3]. Once it is installed, a new tab named Fuzzy OWL enables to use the plug-in. Figure 1 shows the available options.

It is important to remark that the plug-in is generic and not specific to our reasoner, so it offers the possibility of adding other elements that are not yet supported by DeLorean; e.g., weighted concepts, weighted sum concepts, fuzzy modified roles, fuzzy modified datatypes, Łukasiewicz fuzzy logic ...

Firstly, the user can create the non-fuzzy part of the ontology using the editor as usual. Then, the user can define the fuzzy elements of the ontology by using the plug-in; namely, fuzzy axioms, fuzzy datatypes, fuzzy modifiers, fuzzy modified concepts, and fuzzy nominals.

Figure 2 illustrates the plug-in use by showing how to create a new fuzzy datatype. The user specifies the name of the datatype, and the type of the membership function. Then, the plug-in asks for the necessary parameters according to the type. A picture is displayed to help the user recall the meaning of the parameters. Then, after some error checks, the new datatype is created and can be used in the ontology.

Once the fuzzy ontology has been created with Fuzzy OWL, it has to be translated into the language supported by a specific fuzzy DL reasoner –DeLorean, in this case– to allow reasoning with it. For instance, the datatype created in Figure 2 would be represented by using a trapezoidal function in DeLorean syntax.

---

[3] http://webdiis.unizar.es/~fbobillo/fuzzyOWL2

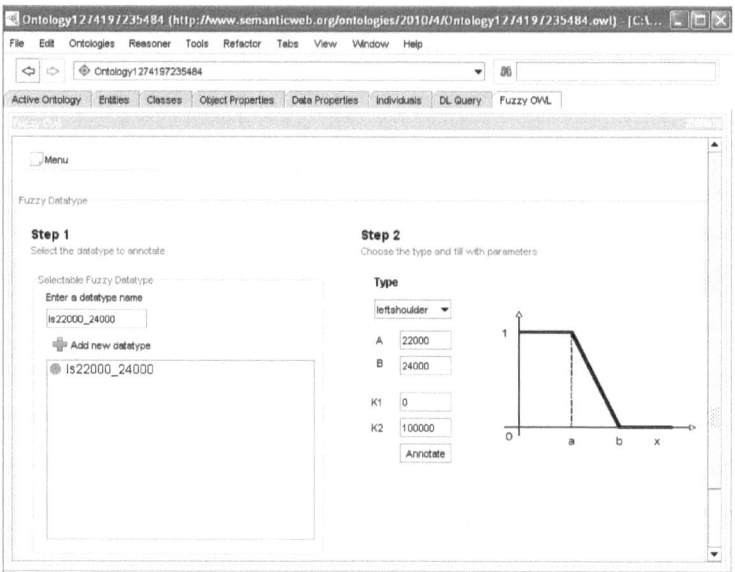

Fig. 2. Creation of a fuzzy datatype

For this purpose, the plug-in includes a general parser that can be customized to any reasoner by adapting a template code. The parser browses the contents of the ontology –with *OWL API 3*[4] [21], which allows iterating over the elements of the ontology in a transparent way– and prints an informative message. The output of the process is a fuzzy ontology that can be printed in the standard output or saved in a text file. If the user selects one of these elements, they will be discarded when translated into DELOREAN syntax, and an informative message will be displayed to the users.

The template code has been adapted to build two parsers, one for `fuzzyDL`, and one for `DeLorean`. Both the template and the parsers can be freely obtained from the plug-in web page. Furthermore, similar parsers for other fuzzy DL reasoners can be easily obtained. To do so, we can replace the default messages by well-formed axioms, according to the desired fuzzy ontology syntax.

4   Reasoning with DeLorean

In this section we discuss how to reason with the fuzzy ontologies by using DELOREAN. Firstly, Section 4.1 describes the supported reasoning tasks. Then, Section 4.2 explains how to interact with the application.

4.1   Reasoning Tasks

There are several reasoning tasks in fuzzy $\mathcal{SROIQ}(\mathbf{D})$ [19]. We will focus on the following ones:

---

[4] http://owlapi.sourceforge.net

- *Fuzzy KB satisfiability.* A fuzzy interpretation $\mathcal{I}$ *satisfies* (is a model of) a fuzzy KB $\mathcal{K}$ iff it satisfies each axiom in $\mathcal{K}$.
- *Concept satisfiability.* C is α-satisfiable w.r.t. a fuzzy KB $\mathcal{K}$ iff there exists a model $\mathcal{I}$ of $\mathcal{K}$ such that $C^{\mathcal{I}}(x) \geq \alpha$ for some $x \in \Delta^{\mathcal{I}}$.
- *Entailment*: An axiom τ of the forms (A1)–(A5) is entailed by a fuzzy KB $\mathcal{K}$ iff every model of $\mathcal{K}$ satisfies τ.
- *Concept subsumption*: D subsumes C (denoted $C \sqsubseteq D$) w.r.t. a fuzzy KB $\mathcal{K}$ iff every model $\mathcal{I}$ of $\mathcal{K}$ satisfies $\forall x \in \Delta^{\mathcal{I}}, C^{\mathcal{I}}(x) \leq D^{\mathcal{I}}(x)$.
- *Best degree bound* (BDB) of an axiom τ of the forms (A1)–(A5) is defined as the $\sup\{\alpha : \mathcal{K} \models \langle \tau \geq \alpha \rangle\}$.

DeLorean reasoning algorithms are based on the computation of a crisp ontology that preserves the semantics of the qoriginal fuzzy ontology, and therefore reasoning with the former is equivalent to reasoning with the latter. This kind of reduction has already been considered in the literature (see for instance [15] for Zadeh fuzzy DLs and [16] for Gödel fuzzy DLs).

The equivalent crisp ontology has a larger size than the original fuzzy ontology, because some axioms must be added to keep the same semantics. If we assume a fixed set of degrees of truth, the size of the equivalent crisp ontology depends linearly on the size of the fuzzy ontology.

An interesting property is that the computation of the equivalent crisp ontology can be reused when adding a new axiom. If the new axiom does not introduce new atomic concepts, atomic roles, nor a new degree of truth, we just need to add the reduction of the new axiom.

Example 3. Assume a set of degrees of truth $\mathcal{N} = \{0, 0.25, 0.5, 0.75, 1\}$. Consider the fuzzy KB $\mathcal{K} = \{\langle \mathsf{RoseDAnjou} : \neg \mathsf{RedWine} \geq 0.5 \rangle\}$, stating that it is almost true that Rosé D'Anjou is not a red wine. Let us show how to compute the crisp representation of $\mathcal{K}$ in Zadeh fuzzy logic.

To start with, we create 8 new crisp atomic concepts: $\mathsf{RoseDAnjou}_{\geq 0.25}$, $\mathsf{RedWine}_{\geq 0.25}$, $\mathsf{RoseDAnjou}_{\geq 0.5}$, $\mathsf{RedWine}_{\geq 0.5}$, $\mathsf{RoseDAnjou}_{\geq 0.75}$, $\mathsf{RedWine}_{\geq 0.75}$, $\mathsf{RoseDAnjou}_{\geq 1}$, $\mathsf{RedWine}_{\geq 1}$.

Next, we add some axioms keeping the semantics of these new concepts: $\mathsf{RoseDAnjou}_{\geq 0.5} \sqsubseteq \mathsf{RoseDAnjou}_{\geq 0.25}$, $\mathsf{RoseDAnjou}_{\geq 0.75} \sqsubseteq \mathsf{RoseDAnjou}_{\geq 0.5}$, $\mathsf{RoseDAnjou}_{\geq 1} \sqsubseteq \mathsf{RoseDAnjou}_{\geq 0.75}$, $\mathsf{RedWine}_{\geq 0.5} \sqsubseteq \mathsf{RedWine}_{\geq 0.25}$, $\mathsf{RedWine}_{\geq 0.75} \sqsubseteq \mathsf{RedWine}_{\geq 0.5}$, $\mathsf{RedWine}_{\geq 1} \sqsubseteq \mathsf{RedWine}_{\geq 0.75}$.

Finally, we represent the axiom in $\mathcal{K}$ as $\mathsf{RoseDAnjou} : \neg \mathsf{RedWine}_{\geq 0.75}$.

Now, we shall discuss the case of Gödel fuzzy logic. The procedure is very similar, but the representation of the axioms in the fuzzy ontology is different, as it must take into account the different semantics of the fuzzy logical operators. In particular, the axiom would be represented as $\mathsf{RoseDAnjou} : \neg \mathsf{RedWine}_{\geq 0.25}$.

□

4.2   Using DeLorean

DeLorean can be used as a stand-alone application. In addition, DeLorean reasoning services can also be used from other programs by means of the De-Lorean API. For details about the API, we refer the reader to the Javadoc

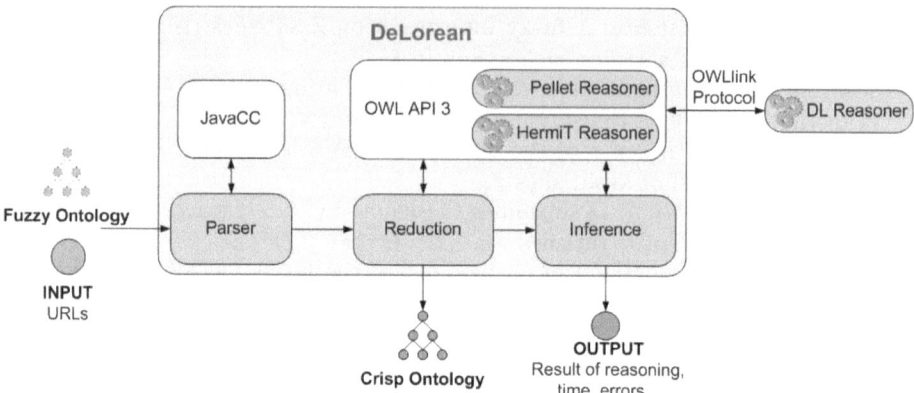

Fig. 3. Architecture of DeLorean reasoner

documentation of the package. In this section, we will focus on the user of the reasoner through its graphical interface.

Figure 3 illustrates the architecture of the system:

- The *Parser* reads an input file with a fuzzy ontology in DeLorean syntax and translates it into an internal representation[5]. It is important to remark that we can use any language as long as there is a parser that can obtain an internal representation. Also, we could have several parsers to support different input languages.
- The *Reduction* module implements the reduction procedures described in the previous section. It builds an OWL API model representing an equivalent crisp ontology that can be exported to an OWL 2 file. The implementation also takes into account all the optimizations already discussed along this document.
- The *Inference* module communicates with a non-fuzzy reasoner (either one of the integrated reasoners or a reasoner via the OWLlink [25][6] protocol) in order to perform the reasoning tasks.
- A simple *User interface* manages inputs and outputs (see details below).

Figure 4 shows a snapshot of the user interface, structured in 4 sections:

Input. Here, the user can specify the input fuzzy ontology and the DL reasoner that will be used in the reasoning. The possible choices are *HermiT* [24][7], *Pellet* [22][8], and an OWLlink-complaint reasoner. Once a fuzzy ontology is loaded, the reasoner will check that every degree of truth that appears in it belongs to the set specified in the section on the right.

---

[5] This parser should not be confused with the translator from the Protégé plug-in into DeLorean syntax discussed in Section 3.2.
[6] http://www.owllink.org
[7] http://hermit-reasoner.com
[8] http://clarkparsia.com/pellet

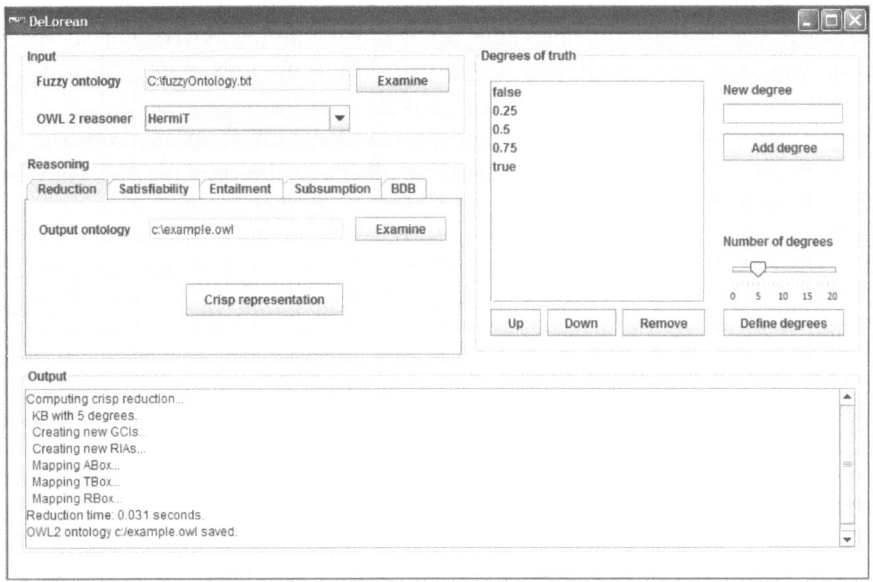

Fig. 4. User interface of DeLorean reasoner

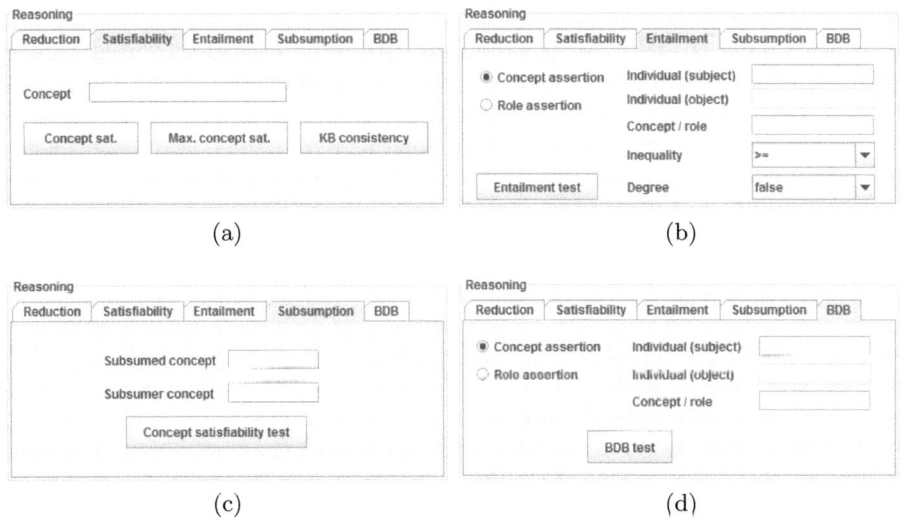

Fig. 5. (a) Concept satisfiability and KB consistency; (b) Concept and role entailment; (c) Concept subsumption; (d) BDB

Degrees of truth. The user can specify here the set of degrees of truth that will be considered. 0 (false) and 1 (true) are mandatory. Other degrees can be added, ordered (by moving them up or down), and removed. For the user's convenience, it is possible to directly specify a number of degrees of truth, and they will be automatically generated.

Output. Here, output messages are displayed. Some information about the reasoning is shown here, such as the time taken, or the result of the reasoning task process.

Reasoning. This part is used to perform the different reasoning tasks that DELOREAN supports. The panel is divided into five tabs, each of them dedicated to a specific reasoning task (we recall the reader that these reasoning tasks have been defined in Section 3.1).

> **Crisp representation.** The main reasoning task is the computation of the equivalent *crisp representation* of the fuzzy ontology, which is actually necessary for the other reasoning tasks. In this tab we can export the resulting non-fuzzy ontology into a new OWL 2 file. This tab can be seen in Figure 4.

> **Satisfiability.** In this tab (see Figure 5 (a)), the user can perform three tasks: *fuzzy KB consistency, fuzzy concept satisfiability* and the computation of the *maximum degree of satisfiability of a fuzzy concept*. In the two latter cases, the interface makes it possible to specify the name of the fuzzy concept for which the satisfiability test will be computed. Note that the interface expects the name of a fuzzy concept, and not a concept expression.

> **Entailment.** In this tab (see Figure 5 (b)), the user can compute, given the current fuzzy ontology, the *entailment* of a fuzzy concept assertion or a fuzzy role assertion. Firstly, the user has to specify the type of the assertion, and then the corresponding parameters. For fuzzy concept assertions, the parameters are: name of the individual, name of the fuzzy concept, inequality sign, and degree of truth. For fuzzy role assertions, the parameters are: name of the subject individual, name of the role, name of the object individual, inequality sign, and degree of truth for fuzzy role assertions.

> **Subsumption.** In this tab (see Figure 5 (c)), after specifying the names of the subsumed and the subsumer fuzzy concepts, it is possible to compute the *fuzzy concept subsumption*.

> **BDB.** Finally, in the fifth tab (see Figure 5 (d)), the user can compute the BDB of a fuzzy concept assertion or a fuzzy role assertion. As in the case of entailment, the user has to specify previously the type of the assertion and the corresponding parameters. For fuzzy concept assertions, the parameters are: name of the individual and name of the fuzzy concept. For fuzzy role assertions, the parameters are: name of the subject individual, name of the role and name of the object individual.

5   Implementation Details

In this section we briefly explain some implementation details of DeLorean. In Section 5.1 we run through the different versions of the reasoner and comment their main differences. Then, Section 5.2 summarizes some optimization techniques that are implemented in order to make the reasoning more efficient.

5.1   Some Historical Notes

The first version of the reasoner was based on Jena API[9]. It was developed in Java by relying on the parser generator JavaCC[10] and DIG 1.1 interface [20] (to communicate with crisp DL reasoners). The use of DIG limited the expressivity of the logic supported by DeLorean to $Z \, \mathcal{SHOIN}$ (OWL DL). From a historical point of view, this version was the first reasoner that supported a fuzzy extension of the OWL DL language. Only a few optimizations were implemented.

With the aim of augmenting the expressivity of the logic, we changed the subjacent API to OWL API 3 [21]. OWL API 3 is supported by several ontology reasoners, such as *Pellet* [22], *Fact++* [23] and *HermiT*. Now, DeLorean supports both $Z \, \mathcal{SROIQ}(\mathbf{D})$ and $G \, \mathcal{SROIQ}(\mathbf{D})$.

One of the most important differences of OWL API 3 is that it replaces DIG by OWLlink support. *OWLlink* is an extensible protocol for communication between OWL 2 systems that supersedes DIG 1.1 [25]. Since OWLlink is not widely supported yet, we have also integrated natively *Pellet* and *HermiT* reasoners with DeLorean. Hence, the user is free to choose either one of these reasoners or a generic one via OWLlink protocol.

5.2   Optimizations

We will summarize here the main optimizations that the reasoner implements. The interested reader is referred to [15] for details.

Optimizing the Number of New Elements and Axioms. As seen in Section 4.1, we must add some new concepts, roles, and axioms to compute an equivalent crisp ontology. DeLorean reduces the number of new concepts and roles introduced with respect to a direct translation. For instance, a crisp concept denoting individuals that belong to a fuzzy concept with a degree less than α is not needed, since we can use the negation of the α-cut of the fuzzy concept. As a consequence of the reduction of the number of concepts and roles, the number of new necessary axioms is reduced as well.

Optimizing GCI Reductions. In some particular cases, the crisp representation of fuzzy GCIs can be optimized. For example, domain role axioms, range role axioms, functional role axioms and disjoint concept axioms can be optimized

---

[9] http://jena.sourceforge.net
[10] https://javacc.dev.java.net

if we manage them as particular cases, instead of considering them as GCIs. Furthermore, some axioms in the resulting TBox may be unnecessary since they can be entailed by other axioms.

Allowing Crisp Concepts and Roles. In order to represent a fuzzy concept, we need to introduce several new concepts, and some new axioms keeping the semantics among them. In real applications, not all concepts and roles are fuzzy. If a concept is declared as crisp, we just need one concept to represent it and no new axioms. The case for fuzzy roles is exactly the same. DELOREAN makes it possible to define crisp concepts and roles. Of course, this optimization requires some manual intervention during the identification of the crisp elements.

Ignoring Superfluous Elements While Reasoning. The computation of the equivalent crisp ontology can be designed to promote reusing or efficiency. A direct translation into the crisp case makes ontology reuse easier when new axioms are added. The drawback is that reasoning is less efficient. Depending on the reasoning task, DELOREAN promotes reusing or avoiding superfluous elements and recomputing the crisp representation when necessary.

6   Use Case: A Fuzzy Wine Ontology

This section describes a concrete use case: a fuzzy extension of the well-known Wine ontology[11], a highly expressive $\mathcal{SHOIN}(\mathbf{D})$ ontology. Some metrics of the ontology are shown in the first column of Table 3.

There is a previous empirical evaluation of the reductions of fuzzy DLs to crisp DLs [27], but the only optimization thereby considered applies to the number of new elements and axioms. We will show that the additional optimizations hereby proposed, specially the (natural) assumption that there are some crisp elements, reduce significantly the number of axioms.

A Fuzzy Extension of the Ontology. We have defined a fuzzy version of the Wine ontology by adding a degree to the axioms. Given a variable set of degrees $N^{\mathcal{K}}$, the degrees of truth for fuzzy assertions is randomly chosen in $N^{\mathcal{K}}$. In the case of fuzzy GCIs and RIAs, the degree is always 1 in special GCIs (namely concept equivalences and disjointness, domain, range and functional role axioms) or if there is a crisp element in the left side; otherwise, the degree is 0.5.

Most fuzzy assertions are of the form $\langle \tau \rhd \beta \rangle$ with $\beta \neq 1$. This favors the use of elements of the forms $C_{\rhd\beta}$ and $R_{\rhd\beta}$, which reduces the number of superfluous concepts as seen in Section 5.2. Once again, this leads to the worst case from the point of view of the size of the resulting crisp ontology. Nonetheless, in practice we will be often able to say that an individual fully belongs to a fuzzy concept, or that two individuals are fully related by means of a fuzzy role, which is not the worst case.

---

[11] http://www.w3.org/TR/2003/CR-owl-guide-20030818/wine.rdf

Note that our objective is to build a test case to experiment with and not to build a fuzzy Wine ontology with interest in the real world. For a serious attempt to build a fuzzy Wine ontology, we refer the reader to [28].

Crisp Concepts and Roles. A careful analysis of the fuzzy $\mathcal{SROIQ}(\mathbf{D})$ KB brings about that most concepts and roles should be indeed interpreted as crisp –actually, the classification of concepts and roles into fuzzy or crisp in most cases may be very subjective and application-dependent.

For example, most subclasses of the class Wine refer to a well-defined geographical origin of the wines. For instance, Alsatian wine is a wine which has been produced in the French region of Alsace: AlsatianWine $\equiv$ Wine $\sqcap$ $\exists$locatedAt.{alsaceRegion}. In other applications there could be examples of fuzzy regions, but this is not our case.

Another important number of subclasses of Wine refer to the type of grape used, which is also a crisp concept. For instance, Riesling is a wine which has been produced from Riesling grapes: Riesling $\equiv$ Wine $\sqcap$ $\exists$madeFromGrape.{RieslingGrape} $\sqcap$ $\geq$ 1 madeFromGrape.$\top$.

The result of our study has identified 50 fuzzy concepts in the Wine ontology. The source of the vagueness is summarized in several categories[12]:

- Color of the wine: WineColor, RedWine, RoseWine, WhiteWine, RedBordeaux, RedBurgundy, RedTableWine, WhiteBordeaux, WhiteBurgundy, WhiteLoire, WhiteTableWine.
- Sweetness of the wine: WineSugar, SweetWine, SweetRiesling, WhiteNonSweetWine, DryWine, DryRedWine, DryRiesling, DryWhiteWine.
- Body of the wine: WineBody, FullBodiedWine.
- Flavor of the wine: WineFlavor, WineTaste.
- Age of the harvest: LateHarvest, EarlyHarvest.
- Spiciness of the food: NonSpicyRedMeat, NonSpicyRedMeatCourse, SpicyRedMeat, PastaWithSpicyRedSauce, PastaWithSpicyRedSauceCourse, PastaWithNonSpicyRedSauce, PastaWithNonSpicyRedSauceCourse, SpicyRedMeatCourse.
- Sweetness of the food: SweetFruit, SweetFruitCourse, SweetDessert, SweetDessertCourse, NonSweetFruit, NonSweetFruitCourse.
- Type of the meat: RedMeat, NonRedMeat, RedMeatCourse, NonRedMeatCourse. They are fuzzy because, according to the age of the animal, pork and lamb are classified as red (old animals) or white (young animals) meat.
- Heaviness of the cream: PastaWithHeavyCreamSauce, PastaWithLightCreamSauce. In this case the terms "heavy" and "light" depend on the fat percentage, and thus can be a matter of degree.
- Desserts: Dessert, DessertWine, CheeseNutsDessert, DessertCourse, CheeseNutsDessertCourse. We make these concepts fuzzy as the question whether something is a dessert or not does not have a clear answer.

---

[12] Clearly, these categories are not disjoint and some concepts may belong to more than one, meaning that they are fuzzy for several reasons. For example, DryRedWine is a fuzzy concept because both "dry" and "red" are vague terms.

As already discussed, the color, the sweetness, the body and the flavor of a wine are fuzzy. As a consequence, we can identify 5 fuzzy roles: hasColor, hasSugar, hasBody, hasFlavor, and hasWineDescriptor, where the role hasWineDescriptor is a super-role of the others.

Measuring the Importance of the Optimizations. Here, we will restrict to $Z \, \mathcal{SROIQ}$ (omitting the concrete role yearValue), but allowing the use of Kleene-Dienes and Gödel implications in the semantics of the axioms (A8), (A10), (A11).

Table 3 shows some metrics of the crisp ontologies obtained in the reduction of the fuzzy ontology after applying different optimizations.

1. Column "Original" shows some metrics of the original ontology.
2. "None" considers the reduction obtained after applying no optimizations.
3. "(NEW)" considers the reduction obtained after optimizing the number of new elements and axioms.
4. "(GCI)" considers the reduction obtained after optimizing GCI reductions.
5. "(C/S)" considers the reduction obtained after allowing crisp concepts and roles and ignoring superfluous elements.
6. Finally, "All" applies all the previous optimizations.

Table 3. Metrics of the Wine ontology and its fuzzy versions using 5 degrees

	Original	None	(NEW)	(GCI)	(C/S)	All
Individuals	206	206	206	206	206	206
Named concepts	136	2176	486	2176	800	191
Abstract roles	16	128	128	128	51	20
Concept assertions	194	194	194	194	194	194
Role assertions	246	246	246	246	246	246
Inequality assertions	3	3	3	3	3	3
Equality assertions	0	0	0	0	0	0
New GCIs	0	4352	952	4352	1686	324
Subclass axioms	275	1288	1288	931	390	390
Concept equivalences	87	696	696	696	318	318
Disjoint concepts	19	152	152	19	152	19
Domain role axioms	13	104	104	97	104	97
Range role axioms	10	80	80	10	80	10
Functional role axioms	6	48	48	6	48	6
New RIAs	0	136	119	136	34	34
Sub-role axioms	5	40	40	40	33	33
Role equivalences	0	0	0	0	0	0
Inverse role axioms	2	16	16	16	2	2
Transitive role axioms	1	8	8	8	1	1

Note that the size of the ABox is always the same, because every axiom in the fuzzy ABox generates exactly one axiom in the reduced ontology.

The number of new GCIs and RIAs added to preserve the semantics of the new elements is much smaller in the optimized versions. In particular, we reduce from 4352 to 324 GCIs (7.44%) and from 136 to 34 RIAs (25%). This shows the importance of reducing the number of new crisp elements and their corresponding axioms, defining crisp concepts and roles, and (to a lesser extent) handling superfluous concepts.

Optimizing GCI reductions turns out to be very useful in reducing the number of disjoint concepts, domain, range and functional role axioms: 152 to 19 (12.5 %), 104 to 97 (93.27 %), 80 to 10 (12.5 %), and 48 to 6 (12.5 %), respectively. In the case of domain role, axioms the reduction is not very high because we need an inverse role to be defined in order to apply the reduction. However, this happens for only one axiom.

Every fuzzy GCI or RIA generates several axioms in the reduced ontology. Combining the optimization of GCI reductions with the definition of crisp concepts and roles reduces the number of new axioms: from 1288 to 390 subclass axioms (30.28 %), from 696 to 318 concept equivalences (45.69 %), and from 40 to 33 sub-role axioms (82.5 %).

Finally, the number of inverse and transitive role axioms is reduced in the optimized version because fuzzy roles interpreted as crisp introduce one inverse or transitive axiom instead of several ones. This allows a reduction from 16 to 2 axioms, and from 8 to 1, respectively, which corresponds to the 12.5 %.

Table 4 shows the influence of the number of degrees on the size of the resulting crisp ontology, as well as on the reduction time (which is shown in seconds), when all the described optimizations are used. The reduction time is small enough to allow us to recompute the reduction of an ontology when necessary, thus allowing us to avoid superfluous concepts and roles.

Table 4. Influence of the number of degrees in the reduction

	Crisp	3	5	7	9	11	21
Number of axioms	811	1166	1674	2182	2690	3198	5738
Reduction time	-	0.343	0.453	0.64	0.782	0.859	1.75

7   Conclusions and Future Work

This paper has presented the main features of the fuzzy ontology reasoner DeLorean, the first one that supports a fuzzy extension of OWL 2. DeLorean integrates translation and reasoning tasks. Given a fuzzy ontology, DeLorean computes its equivalent non-fuzzy representation. Then, it uses a classical DL reasoner to perform the reasoning procedures.

We have also discussed how to create fuzzy ontologies by using a related Protégé plug-in, as well as the implemented optimizations. Optimizations allow us to define crisp concepts and roles and to remove superfluous concepts and roles before applying crisp reasoning. A preliminary evaluation shows that these optimizations help to reduce significantly the size of the resulting ontology.

Other fuzzy ontology reasoners can be found in the literature, e.g. FIRE [5][13], FUZZYDL [6][14], GURDL [7], GERDS [8][15], YADLR [9], DLMEDIA [11][16], FRESG [12] or ONTOSEARCH2 [10][17]. On the one hand, the advantages of DELOREAN are that it supports all the constructors of fuzzy $\mathcal{SROIQ}(\mathbf{D})$, and makes it possible to reuse existing ontology languages, editors, reasoners, and other existing resources. On the other hand, the equivalent crisp ontologies computed by DELOREAN are larger than the original fuzzy ontologies. Hence, other reasoners, such as ONTOSEARCH2 and DLMEDIA –which are specifically designed for scalable reasoning–, would very likely show a better performance for the fuzzy OWL 2 profiles. Furthermore, some of these reasoners implement some features that DELOREAN currently does not support. Some reasoners solve new alternative reasoning tasks, such as classification (FIRE) or defuzzification (FUZZYDL), or support different constructors, such as aggregation operators (FUZZYDL), extended fuzzy concrete domains (FRESG), or alternative uncertainty operators (GURDL).

In future work, we plan to develop a more detailed benchmark and, eventually, to compare DELOREAN against other fuzzy DL reasoners. This is a difficult task, since different reasoners support different features and expressivities. Moreover, as far as we know, nowadays there are no public real-world fuzzy ontologies to test with.

References

1. Cuenca-Grau, B., Horrocks, I., Motik, B., Parsia, B., Patel-Schneider, P.F., Sattler, U.: OWL 2: The next step for OWL. Journal of Web Semantics 6(4), 309–322 (2008)
2. Baader, F., Calvanese, D., McGuinness, D., Nardi, D., Patel-Schneider, P.F.: The Description Logic Handbook: Theory, Implementation, and Applications. Cambridge University Press (2003)
3. Horrocks, I., Kutz, O., Sattler, U.: The even more irresistible $\mathcal{SROIQ}$. In: Proceedings of the 10th International Conference of Knowledge Representation and Reasoning, KR 2006, pp. 452–457 (2006)
4. Lukasiewicz, T., Straccia, U.: Managing uncertainty and vagueness in Description Logics for the Semantic Web. Journal of Web Semantics 6(4), 291–308 (2008)
5. Stoilos, G., Simou, N., Stamou, G., Kollias, S.: Uncertainty and the Semantic Web. IEEE Intelligent Systems 21(5), 84–87 (2006)
6. Bobillo, F., Straccia, U.: fuzzyDL: An expressive fuzzy Description Logic reasoner. In: Proceedings of the 17th IEEE International Conference on Fuzzy Systems, FUZZ-IEEE 2008, pp. 923–930. IEEE Computer Society (2008)
7. Haarslev, V., Pai, H.I., Shiri, N.: A formal framework for Description Logics with uncertainty. Int. Journal of Approximate Reasoning 50(9), 1399–1415 (2009)

---

[13] http://www.image.ece.ntua.gr/~nsimou/FiRE

[14] http://straccia.info/software/fuzzyDL/fuzzyDL.html

[15] http://www1.osu.cz/home/habibal/page8.html

[16] http://straccia.info/software/DL-Media/DL-Media.html

[17] http://dipper.csd.abdn.ac.uk/OntoSearch

8. Habiballa, H.: Resolution strategies for fuzzy Description Logic. In: Proceedings of the 5th Conference of the European Society for Fuzzy Logic and Technology, EUSFLAT 2007, vol. 2, pp. 27–36 (2007)
9. Konstantopoulos, S., Apostolikas, G.: Fuzzy-DL Reasoning over Unknown Fuzzy Degrees. In: Meersman, R., Tari, Z., Herrero, P. (eds.) OTM 2007 Ws, Part II. LNCS, vol. 4806, pp. 1312–1318. Springer, Heidelberg (2007)
10. Pan, J.Z., Thomas, E., Sleeman, D.: ONTOSEARCH2: Searching and querying web ontologies. In: Proceeedings of the IADIS International Conference WWW/Internet 2006 (2006)
11. Straccia, U., Visco, G.: DL-Media: An ontology mediated multimedia information retrieval system. In: Proceedings of the 4th International Workshop on Uncertainty Reasoning for the Semantic Web, URSW 2008, vol. 423. CEUR Workshop Proceedings (2008)
12. Wang, H., Ma, Z.M., Yin, J.: FRESG: A Kind of Fuzzy Description Logic Reasoner. In: Bhowmick, S.S., Küng, J., Wagner, R. (eds.) DEXA 2009. LNCS, vol. 5690, pp. 443–450. Springer, Heidelberg (2009)
13. Zadeh, L.A.: Fuzzy sets. Information and Control 8, 338–353 (1965)
14. Hájek, P.: Metamathematics of Fuzzy Logic. Kluwer (1998)
15. Bobillo, F., Delgado, M., Gómez-Romero, J.: Crisp representations and reasoning for fuzzy ontologies. International Journal of Uncertainty, Fuzziness and Knowledge-Based Systems 17(4), 501–530 (2009)
16. Bobillo, F., Delgado, M., Gómez-Romero, J., Straccia, U.: Fuzzy Description Logics under Gödel semantics. Int. J. of Approximate Reasoning 50(3), 494–514 (2009)
17. Stoilos, G., Stamou, G., Pan, J.Z.: Fuzzy extensions of OWL: Logical properties and reduction to fuzzy description logics. International Journal of Approximate Reasoning 51(6), 656–679 (2010)
18. Straccia, U.: Description Logics with fuzzy concrete domains. In: Proc. of the 21st Conf. on Uncertainty in Artificial Intelligence, UAI 2005. AUAI Press (2005)
19. Straccia, U.: Reasoning within fuzzy Description Logics. Journal of Artificial Intelligence Research 14, 137–166 (2001)
20. Bechhofer, S., Möller, R., Crowther, P.: The DIG Description Logic interface: DIG/1.1. In: Proceedings of the 16th International Workshop on Description Logics, DL 2003 (2003)
21. Horridge, M., Bechhofer, S.: The OWL API: A Java API for OWL Ontologies. Semantic Web 2(1), 11–21 (2011)
22. Sirin, E., Parsia, B., Cuenca-Grau, B., Kalyanpur, A., Katz, Y.: Pellet: A practical OWL-DL reasoner. Journal of Web Semantics 5(2), 51–53 (2007)
23. Tsarkov, D., Horrocks, I.: FaCT++ Description Logic Reasoner: System Description. In: Furbach, U., Shankar, N. (eds.) IJCAR 2006. LNCS (LNAI), vol. 4130, pp. 292–297. Springer, Heidelberg (2006)
24. Shearer, R., Motik, B., Horrocks, I.: HermiT: A highly-efficient OWL reasoner. In: Proceedings of the 5th International Workshop on OWL: Experiences and Directions, OWLED 2008 (2008)
25. Liebig, T., Luther, M., Noppens, O., Wessel, M.: OWLlink. Semantic Web 2(1), 23–32 (2011)
26. Bobillo, F., Straccia, U.: Fuzzy ontology representation using OWL 2. International Journal of Approximate Reasoning 52(7), 1073–1094 (2011)

27. Cimiano, P., Haase, P., Ji, Q., Mailis, T., Stamou, G.B., Stoilos, G., Tran, T., Tzouvaras, V.: Reasoning with large A-Boxes in fuzzy Description Logics using DL reasoners: An experimental valuation. In: Proceedings of the 1st Workshop on Advancing Reasoning on the Web: Scalability and Commonsense, ARea 2008. CEUR Workshop Proceedings, vol. 350 (2008)
28. Carlsson, C., Brunelli, M., Mezei, J.: Fuzzy ontologies and knowledge mobilisation: Turning amateurs into wine connoisseurs. In: Proceedings of the 19th IEEE International Conference on Fuzzy Systems, FUZZ-IEEE 2010, pp. 1718–1723 (2010)

Dealing with Contradictory Evidence Using Fuzzy Trust in Semantic Web Data

Miklos Nagy[1] and Maria Vargas-Vera[2]

[1] Knowledge Media Institute (KMi)
The Open University
Walton Hall, Milton Keynes
MK7 6AA, United Kingdom
M.Nagy@open.ac.uk
[2] Universidad Adolfo Ibanez
Facultad de Ingenieria y Ciencias
Centro de Investigaciones en Informatica y Telecomunicaciones
Campus Vinia del Mar, Chile
maria.vargas-vera@uai.cl

Abstract. Term similarity assessment usually leads to situations where contradictory evidence support has different views concerning the meaning of a concept and how similar it is to other concepts. Human experts can resolve their differences through discussion, whereas ontology mapping systems need to be able to eliminate contradictions before similarity combination can achieve high quality results. In these situations, different similarities represent conflicting ideas about the interpreted meaning of the concepts. Such contradictions can contribute to unreliable mappings, which in turn worsen both the mapping precision and recall. In order to avoid including contradictory beliefs in similarities during the combination process, trust in the beliefs needs to be established and untrusted beliefs should be excluded from the combination. In this chapter, we propose a solution for establishing fuzzy trust to manage belief conflicts using a fuzzy voting model.

1 Introduction

Processing Semantic Web data inevitably leads to situations where interpretation can differ between different experts or algorithms that try to establish the meaning of such data. In these situations, correctly managing contradictory evidence could play an important role in providing higher quality information to the end users. Consider for example ontology mapping [1], which is an important aspect of the Semantic Web. One of the main challenges of Ontology Mapping [2] is how to handle conflicting information that stems from the interpretation of Semantic Web data. The source of conflict can range from missing or insufficient information to a contradictory description of the same or similar terms. During the mapping process each domain expert assesses the correctness of the sampled mappings and their assessments are then discussed before producing

F. Bobillo et al. (Eds.): URSW 2008-2010/UniDL 2010, LNAI 7123, pp. 139–157, 2013.
© Springer-Verlag Berlin Heidelberg 2013

a final assessment, which reflects a collective judgment. In case contradictions arise, experts discharge one of the opinions on the grounds of reliability or trust concerns.

Ontology mapping systems like DSSim (**D**empster-**S**hafer **Sim**ilarity) [3] try to mimic the aforementioned process, using different software agents as experts to evaluate and use the beliefs over similarities of different concepts in the source ontologies. In DSSim, mapping agents use WordNet as background knowledge to create a conceptual context for the words that are extracted from the ontologies and employ different syntactic and semantic similarities to create their subjective beliefs over the correctness of the mapping. DSSim addresses the uncertain nature of ontology mapping by considering different similarity measures as subjective probabilities for the correctness of the mapping. It employs the Dempster-Shafer theory of evidence in order to create and combine beliefs that have been produced by different similarity algorithms. For a detailed description of the DSSim algorithm one can refer to [4] [5].

Belief combination with Dempsters rule has its advantages compared to other belief combination approaches like Bayesian methods. However, the Dempster belief combination has received verifiable criticism from the research community. There is a problem with the belief combination if agents have conflicting beliefs over the solution. Since evidence theory was developed in the 1970s, a great deal of work has been done in the direction of changing the rule of combination [6]. This is because the rule of combination ignores contradictory evidence by the use of a normalisation factor. Perhaps the most important development on this direction is the work by Yager [7,8], which proposed to change the associative operator in the combination rule for a non-associative one. Recognising the problem of DST (Dempster-Shafer theory), in terms of dealing with highly conflicting evidence, the data fusion community has developed new approaches like the DSmT framework [9], which deals with the conflicting evidence efficiently in its context. Nevertheless, these approaches make assumptions that are only true in the context of data fusion, e.g., the possibility of reducing the size of the hyper power set due to the fact that most masses are zero. However, we argue that any solution that tries to redefine the combination rule or redistribute the masses in order to diminish the level of conflict only increases the complexity of the original algorithm. Our view is that, in the context of ontology mapping, it is better to eliminate conflicting evidence before it becomes part of the combined belief.

The main contribution of this chapter is a novel trust management approach for resolving conflict between beliefs in similarities, which is the core component of the DSSim ontology mapping system. The chapter is organised as follows. Section 2 gives an overview of the related work. Section 3 provides the description of the problem and its context, section 4 provides the source of conflict for ontology mapping and section 5 describes the voting model and how it is applied for determining trust during ontology mapping. In section 6 we present the experiments that have been carried out using the benchmarks of the Ontology Alignment Initiative. Finally, section 7 describes the conclusions found and future research that could be undertaken.

2   Related Work

To date, trust has not been thoroughly investigated in the context of ontology mapping. On-going research has mainly focused on how trust can be modelled in the Semantic Web context [10], where the trust of a users belief in statements supplied by any other user can be represented and combined.

Existing approaches for resolving belief conflicts are based on either negotiation or the definition of different combination rules that consider the possibility of belief conflict. Negotiation-based techniques are mainly proposed in the context of agent communication. In terms of conflicting ontology alignment, an argumentation based framework has been proposed [11], which can be applied for agent communication and web services where the agents are committed to an ontology and try to negotiate with other agents over the meaning of their concepts. Considering the existence of multi-agent systems on the Web, trust management approaches have successfully used fuzzy logic to represent trust between the agents from both individual [12] and community [13] perspectives. However, the main objective of these solutions is to create a reputation of an agent, which can be considered in future interactions.

Considering the different variants [14] [15] of combination rules that consider conflicting beliefs, a number of alternatives have been proposed. While these methods are based on a well-founded theoretical base they all modify the combination rule itself and, as such, these solutions do not consider the process in which these combinations take place. We believe that the conflict needs to be treated before the combination occurs. Further, our approach does not assume that any agent is committed to a particular ontology, instead that our agents are considered as "experts" in assessing similarities of terms in different ontologies and they need to reach conclusions over conflicting beliefs in similarities.

On the other side, different approaches to eliminate contradictions for ontology mapping have been proposed by the ontology mapping community. These approaches can be classified into two distinct categories.

The first category includes solutions that consider uncertainty and fuzziness as an inherent nature of the ontology mapping and tries to describe them accordingly. Ferrera et al. [16] models the whole ontology mapping problem as a fuzzy process where conflicts can occur. This kind of solution belongs to the family of probabilistic approaches for managing uncertainty and applies fuzzy description logic to address the problem of validating the mapping results. The main novelty of the inconsistency detection approach [16] is how mapping validation can be achieved by interpreting the mappings with certain fuzziness. Tang et al. [17] formalises the problem of ontology mapping as creating decision strategies, utilising the Bayesian theory. Also, Tang et al. have participated in several OAEI competitions with their proposed system, which is called "RiMOM". Their solution considers different kinds of conflicts in metadata, namely structure and name conflicts. However, they use a thesaurus and statistical techniques to eliminate them before combining the results.

The second category, however, differs conceptually because they mainly utilise data mining and logic reasoning techniques in pre- and post-processing stages of the mapping. Liu et al. [18] proposes an ontology mapping approach that has four phases. First, the system exploits the available data on labels and then verifies the matching of individuals. In the last two steps the system recalls its past experiences on similar mappings and combines them with structural comparison methods. The four stages are executed consecutively. Contradictions and mismatches are also eliminated before past experiences are applied using logic relation mining of the different attributes. A similar solution has been proposed by the ASMOV system [19], which automates the ontology alignment process using a weighted average of measurements. The proposed approach performs semantic validation of the resulting alignments using similarity along four different features of the ontologies. This system acknowledges that conflicting mappings are produced during the mapping process but they use an iterative post processing logic validation in order to filter them out.

3   Problem Description

This section introduces the notion of truth. If we assume that it is not possible to deduct an absolute truth from the available sources in the Semantic Web environment, then we need to evaluate content dependent trust levels by each application that processes the information on the Semantic Web, e.g., how particular information coming from one source compares the same or similar information that comes from other sources. As an example, consider two ontologies that describe conferences. Both contain concepts about the location of the event, where Ontology 1 contains the concept "Location" in the context of the "Event" and Ontology 2 contains "Place" in the context of the "Building", "Session room" or "Conference hall". Considering the extended contexts of these terms, e.g., Wordnet hypernyms, one can derive that both describe some kind of space or position. The issue is that this information cannot be explicitly derived from the ontologies, because the term "Place" is used to describe buildings and their parts, while "Location" describes a site, e.g., a city where the conference is organised. In order to resolve this contradiction, human experts can discuss their points of view and reach a consensus of whether a mapping can be made or not. Ontology mapping applications that operate without human intervention can use the aforementioned conflict resolution process, which can improve the quality of the mapping if the contradiction can effectively be eliminated.

Dominantly, the existing approaches that address the problem of the trustworthiness of the available data on the Semantic Web are reputation based, e.g., using digital signatures that would state who the publisher of the ontology is. However, another, probably most challenging, aspect of trust appears when we process the available information on the Semantic Web and we discover contradictory information from the evidence. Consider the following example in the context of ontology mapping. When we assess similarity between two terms, ontology mapping can use different linguistic and semantic [1] information to determine the similarity level, e.g., background knowledge or concept hierarchy. In practice, any similarity algorithm will produce good and bad mappings for the same domain, depending on the actual interpretation of the terms in the ontologies, e.g., using different background knowledge descriptions or class hierarchy. In order to overcome this shortcoming, a combination of different similarity measures is required. In recent years, a number of methods and strategies have been proposed [1] to combine these similarities. In practice, considering the overall results, these combination methods will perform well under different circumstances except when contradictory evidence occurs during the combination process.

In our ontology mapping framework, different agents assess similarities and their beliefs on these similarities need to be combined into a more coherent result. However, these individual beliefs in practice are often conflicting.

Conflicting beliefs in Dempster-Shafer theory can be defined in a qualitative manner when two sources support two different and completely disjunctive hypotheses. In these scenarios, where conflicting beliefs are present, using Dempsters combination rule can lead to an almost impossible choice, because the combination rule strongly relies on the agreement between multiple sources, i.e., it ignores all the conflicting evidence. In the context of the Semantic Web, trust can have different meanings; therefore, in the context of ontology mapping we define trust as follows:

Definition 1. *Trust: One mapping agents measurable belief in the competence of the other agents belief over the established similarities.*

Definition 2. *Content related trust: Dynamic trust measure that is dependent on the actual vocabulary of the mappings, which has been extracted from the ontologies and can change from mapping to mapping.*

Definition 3. *Belief: The state in which a software agent holds a proposition over a possible mapping of selected concept pair combinations to be true. Numerical representation of belief can be assigned to a value between [0..1].*

We argue that the problem of contradictions can only be handled from case to case by introducing trust for the similarity measures. This assumption means that trust assessment should only be applied for the selected mappings and can change from mapping to mapping during the process depending on the available evidences. We propose evaluating trust in the different beliefs that does not depend on the credentials of the ontology owner but it purely represents the trust in a proposed subjective belief. This belief is established by using different similarity algorithms.

4   Source of Conflict for Ontology Mapping

In our domain of interest, namely ontology mapping, several challenges were identified by Shvaiko and Euzenat [2] that are considered as major roadblocks for developing ontology mapping solutions that perform well in different domains. We have identified two problems that are the main sources of contradictions when algorithms need to "interpret" the meaning of the data represented by the different ontologies. These problems are "representation problems and uncertainty" and "quality of semantic web data".

4.1 Representation Problems and Uncertainty

The vision of the Semantic Web is to achieve machine-processable interoperability through annotation of the content. This implies that computer programs can achieve a certain degree of understanding of such data and use it to debate a user specific task like question answering or data integration.

Data on the Semantic Web is represented by ontologies, which typically consist of a number of classes, relations, instances and axioms. These elements are expressed using a logical language. The W3C has proposed RDF(S) [20] and OWL [21] as Web ontology languages. However, OWL has three sublanguages (OWL Lite, OWL DL, OWL Full) with different expressiveness and language constructs. In addition to the existing Web ontology languages, W3C has proposed other languages like SKOS [22], which is a standard to describe knowledge organisation systems (KOS). These systems typically contain domain thesauri, or taxonomies, in the context of the Semantic Web. SKOS have been constructed using the Resource Description Framework (RDF), which allows information exchange between different applications. Ontology designers can choose between these language variants depending on the intended purpose of the ontologies.

The problem of interpreting Semantic Web data, however, stems not only from the different language representations [23], but the fact that ontologies, especially OWL Full, have been designed as a general framework to represent domain knowledge, which in turn can differ from designer to designer. From the logical representation point of view these differences are valid separately and no person employing logical reason would find inconsistency in them individually. However, the problem occurs once we need to compare them in order to determine the similarities between classes and individuals. In these cases, systems that compare ontologies need to deal with a considerable amount of uncertainty and conflicting hypotheses related to the interpretation of the classes and their properties. As a result of these representation differences, ontology mapping systems will always need to consider the uncertain and conflicting aspects of how Semantic Web data can be interpreted and processed by different similarity algorithms.

4.2 Quality of Semantic Web Data

Data quality problems [24] [25] in the context of database integration [26] emerged long before the Semantic Web concept was proposed. One possible explanation is the increased number of databases that have been exposed to the Internet through web-based technologies in order to connect institutions, or companies with a large number of clients. For every organisation or individual, the context of the published data can be slightly different depending on how they want to use it. Therefore, from the exchange point of view, incompleteness of a particular data set is quite common. The problem is that fragmented data environments like the Semantic Web inevitably lead to data and information quality problems, causing the applications that process this data to deal with ill-defined, inaccurate or inconsistent information on the domain. Depending on the application context, incomplete data can pose numerous problems to both data owners and their clients. In traditional integration scenarios, resolving these data quality issues represents a vast amount of time and resources for human experts to evaluate before any integration can take place.

Data quality has two aspects, namely syntax and semantics. Data syntax describes the representation of the data, whereas data semantics deals with its meaning. Data syntax is not the main reason of concern, as it can be resolved independently from the context, because data consistency can be achieved by introducing changes in the data format for the different applications. For example consider defining a separation rule of compound terms like "MScThesis", "MSc_Thesis". The main problem that Semantic Web applications need to solve is how to resolve semantic data quality problems, i.e., what is useful and meaningful, because it would require more direct input from the users or creators of the ontologies. Clearly, considering any kind of designer support in the Semantic Web environment is unrealistic. Therefore, applications themselves need to have built in mechanisms to decide and reason whether the data is accurate, usable and useful in essence. Similarly, these applications need to determine whether their processes will deliver good information or function well for the required purpose. In any case, the semantic data quality can be considered to be low as the information is mostly dubious. Therefore, the Semantic Web application has to create its own hypotheses over the meaning of this data.

5 Fuzzy Trust Management for Conflicting Belief Combination

During the ontology mapping process, individual beliefs about similarities may vary. Nevertheless, the conflicting results of the different beliefs in similarity can only be resolved if the mapping algorithm can produce an agreed solution. We propose a solution for reaching this agreement by evaluating fuzzy trust between established beliefs through voting, which is a general method of reconciling differences. Voting is a mechanism where the opinions from a set of votes are evaluated in order to select the alternatives that best represent the collective preferences. Unfortunately, deriving binary trust in terms of being trustful or not trustful from the difference of belief functions is not so straightforward, since different voters express their opinions as having subjective probability over the similarities. For a particular mapping, this always involves a certain degree of vagueness; hence, the threshold between trust and distrust cannot be set definitely for all cases that can occur during the process. Additionally, there is no clear transition between characterising a particular belief highly or as being less trustful.

Fuzzy model is based on so-called linguistic variables, most commonly referred to as "fuzzy" variables. These variables correspond to linguistic terms, and not to concrete numbers, e.g., trust or belief conflict. The fuzzy variables are terms

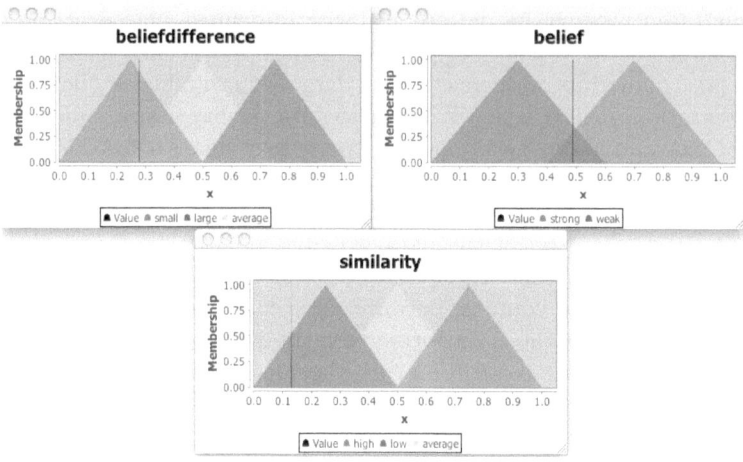

Fig. 1. Example fuzzy input membership functions

that define the meaning of slightly overlapping concepts, e.g., "high" trust, "low" trust. The proportion of each input compared to all possible values can be represented graphically by the membership functions. These membership functions can also be used to define the weight, overlap between inputs and describe the output that is interpreted as a response from the system.

The membership function can be defined differently and can take different shapes depending on the problem it has to represent. Typical membership functions are trapezoidal, triangle or exponential. The selection of our membership function is not arbitrary but can be derived directly from fact that our input, the belief difference, has to produce the trust level as an output. Each input has to produce an output that requires a Gaussian and overlapping membership function. Therefore, our argument is that the trust membership value, which is expressed by different voters, can be modelled properly by using fuzzy representation, as depicted in Figs. 1 and 2.

Consider the scenario in which before each agent evaluates the trust in an other agent's belief over the correctness of the mapping, the difference between its own and the other agents belief is calculated. The belief functions for each agent are derived from different similarity measures; therefore, the actual value might differ from agent to agent. We model these trust levels as fuzzy membership functions.

In fuzzy logic, the membership function $\mu(x)$ is defined on the universe of discourse U and represents a particular input value as a member of the fuzzy set, i.e., $\mu(x)$ is a curve that shows how each point in the U corresponds to a degree of membership. The values of the membership functions range between 0 and 1.

For representing trust (in beliefs over similarities), we have defined three overlapping Gaussian membership functions, which represent high, medium and low trust in the beliefs over concept and property similarities in our ontology mapping system.

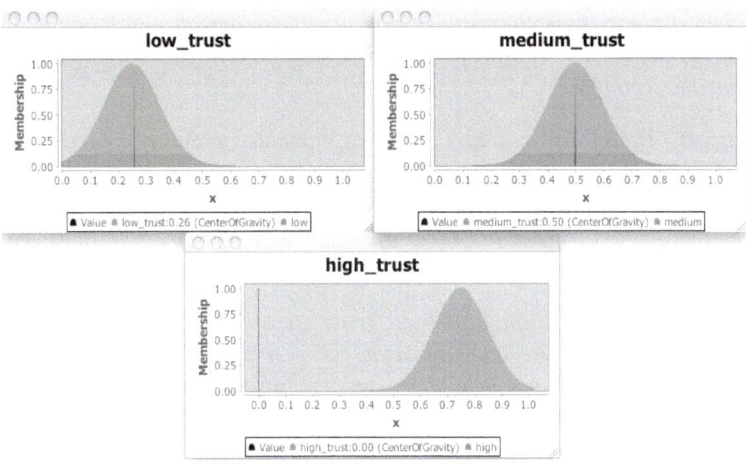

Fig. 2. Example fuzzy output membership functions

5.1   Fuzzy Voting Model

Fuzzification of Input and Output Variables. The fuzzification process of input and output variables transforms input and output variables into fuzzy sets. We have experimented with different types of curves, namely the triangular, trapezoidal and gaussian shaped membership functions.

Figures 1 and 2 show different fuzzy sets where the input variables are modelled triangularly and the output variables are seen with Gaussian membership functions. Membership in each fuzzy set can be determined using the graphical representation of the selected function. Our selected membership functions overlap in order to allow the selection of multiple linguistic values in the transitional areas. During the fuzzification process, rules are applied to the variables that are described with different linguistic terms. We have defined the following fuzzy variables:

Definition 4. *Similarity of entities in two different ontologies is defined as a numerical measure that is produced by certain metrics. These metrics can vary from simple syntactic string distances to more complex semantic sub-graph comparisons. We propose three values for the similarity fuzzy membership value:* $\xi(x) = \{low, average, high\}$

Definition 5. *Belief in the correctness of the mapping is an input variable, which describes the amount of justified support to A: that is, the lower probability function of Dempster, which accounts for all evidence E_k that supports the given proposition A*

$$belief_i(A) = \sum_{E_k \subseteq A} m_i(E_k) \tag{1}$$

where m is Dempsters belief mass function, which represents the strength of some evidence, i.e., $m(A)$ is our exact belief in a proposition represented by A.

The similarity algorithms themselves produce these assignments based on different similarity measures. We propose two values for the fuzzy membership value:
$\nu(x) = \{weak, strong\}$

Definition 6. *Belief difference is an input variable, which represents the agents own belief over the correctness of a mapping in order to establish mappings between concepts and properties in the ontology. During conflict resolution we need to be able to determine the level of difference. We propose three values for the fuzzy membership value:* $\mu(x) = \{small, average, large\}$

Definition 7. *Low, medium and high trust in other agents beliefs are output variables and represent the level of trust we can assign to the combination of our input variables. We propose three values for the fuzzy membership value:* $\tau(x) = \{low, medium, high\}$

Rule Set. Once fuzzy sets have been defined, the system uses a set of rules in order to deduce new information. Based on the input variables and the defined rules, fuzzy systems map the inputs to the outputs by a set of
conditions → *action* rules, i.e., rules that can be expressed in *If* − *Then* form.

For our conflict resolution problem we have defined four simple rules (Fig. 3) ensuring that each combination of the input variables produces output on more than one output, i.e., there is always more than one initial trust level assigned to any input variables.

RULE 1 : IF (beliefdifference IS large OR beliefdifference IS average) AND belief IS weak AND (similarity IS low OR similarity IS average) THEN low_trust IS low;

RULE 2 : IF (beliefdifference IS large OR beliefdifference IS average) AND belief IS weak AND (similarity IS low OR similarity IS average) THEN medium_trust IS medium;

RULE 3 : IF (beliefdifference IS small OR beliefdifference IS average) AND belief IS strong AND (similarity IS high OR similarity IS average) THEN high_trust IS high;

RULE 4 : IF (beliefdifference IS small OR beliefdifference IS average) AND belief IS strong AND (similarity IS high OR similarity IS average) THEN medium_trust IS medium;

Fig. 3. Fuzzy rules for trust assessment

Defuzzification Method. Defuzzification method: After the fuzzy reasoning has been finished, the output variables (linguistic in our case) need to be converted back into real numbers that are meaningful for further processing. The importance of this step is to come up with a single numeric value that corresponds to the linguistic variables that were represented as fuzzy values. In this way, the mapping between the fuzzy and real number domains is established in the fuzzy system. In our ontology mapping system, we have selected the centre of area defuzzification method. This method calculates the output variable by trimming, superimposing and balancing the fuzzy membership functions based on the linguistic variables. In our system, the trust levels are proportional to the area of the membership functions; therefore, other defuzzification methods like centre of maximum or mean of maximum methods do not correspond well to our requirements.

Our main objective is to be able to eliminate conflict between the two beliefs in Dempster-Shafer theory. Consider for example a situation where three agents have used WordNet as background knowledge and have built their beliefs considering a context of different concepts that was derived from said background knowledge, e.g., agent 1 used the direct hypernyms, agent 2 the sister terms and agent 3 the inherited hypernyms. Based on string similarity measures, a numerical belief value is calculated, which represents the strength of the confidence that the two terms are related to each other. The scenario is depicted in Table 1.

Table 1. Belief conflict detection

Conflict detection	Belief 1	Belief 2	Belief 3
Obvious	0.85	0.80	0.1
Difficult	0.85	0.65	0.45

The values given in Table 1 are demonstrative numbers just for the purpose of providing an example. In our ontology mapping framework, **Dempster-Shafer Sim**ilarity (DSSim), the similarities are considered as subjective beliefs, which is represented by belief mass functions that can be combined using Dempsters combination rule. This subjective belief is the outcome of a similarity algorithm, which is applied by a software agent in order to create mapping between two concepts in different ontologies. In our ontology mapping framework, different agents assess similarities and their beliefs in the similarities need to be combined into a more coherent result. However, these individual beliefs in practice are often conflicting. In this scenario, applying Dempsters combination rule to conflicting beliefs can lead to an almost impossible choice because the nature of the combination rule assumes agreement between the different sources. The impact of conflicting evidence is minimised due to the application of a normalisation factor. As mentioned in section 1, the counter-intuitive results that can occur with Dempsters rule of combination are well known and have generated a great deal of debate and suggested solutions within the uncertainty reasoning community. However, all of these solutions mainly focus on how the combination rule can be modified. Therefore, instead of proposing an additional combination rule, we have turned our attention to the root cause of the conflict itself, namely how and why conflicting and uncertain information was produced during the mapping process.

Fuzzy Voting. The fuzzy voting model was developed by Baldwin [27] and has been used in fuzzy logic applications. However, to our knowledge it has not been introduced in the context of trust management on the Semantic Web. In this section we will briefly introduce the fuzzy voting model theory using a simple example of 10 voters, voting against or in favour of the trustfulness of an another agents belief over the correctness of mapping. In our ontology mapping framework each mapping agent can request a number of voting agents to help assess how trustful the other mapping agents belief is.

According to Baldwin [27] linguistic variables are characterised by four elements. L represents the variable name, $T(L)$ describes the associated words for the linguistic variable that are the subset of all possible words U for the domain. Further, syntactic and semantic rules can be described by G and μ respectively. Based on this definition, we also adapt G as a syntactic rule and $T(L)$ as a finite set of words. A formalisation of the fuzzy voting model can be found [28].

Consider the set of words { Low_trust (L_t), Medium_trust (M_t) and High_trust (H_t) } as labels describing trust with possible values are defined as $U = [0, 1]$. Now assume that we have a set of voters represented by "m" and that these voters need to provide one or more terms from the possible variable set represented by $T(L)$. The voters can choose a term that can be described by the value u. Therefore, the calculated value of $\chi_{\mu_{(w)}(u)}$ is dependent on the number of responses that were selected from the possible labels.

We need to introduce more opinions into the system, i.e., we need to add the opinion of the other agents in order to vote for the best possible outcome. Therefore, we assume for the purpose of our example that we have 10 voters (agents) and a set of 3 linguistic words denoted $T(L)$. Formally, let us define:

$$V = \{A1, A2, A3, A4, A5, A6, A7, A8, A9, A10\} \qquad (2)$$
$$T(L) = \{L_t, M_t, H_t\}$$

The number of voters can differ. However, assuming 10 voters can ensure that:

1. The overlap between the membership functions can proportionally be distributed on the possible scale of the belief difference [0..1]
2. The workload of the voters does not slow the mapping process down

Let us start by illustrating the previous ideas with a small example. By definition, consider our linguistic variable L as TRUST and $T(L)$ the set of linguistic values as $T(L) = (Low\_trust, Medium\_trust, High\_trust)$. The universe of discourse is U, which is defined as $U = [0, 1]$. Then, we define the fuzzy sets $\{Low\_trust, Medium\_trust, High\_trust\}$ for the voters where each voter has different overlapping triangular membership functions. The difference in the membership functions represented by the different vertices of the triangle (one example of which is depicted in Fig. 1) ensures that voters can introduce different opinions as they pick the possible trust levels for the same difference in belief.

The possible set of trust levels $L = TRUST$ is defined by Table 2. Note that in the table we use a short notation: L_t means Low_trust, M_t means Medium_trust and H_t means High_trust. Once the fuzzy sets (membership functions) have been defined, the system is ready to assess the trust memberships for the input values. Based on the difference of beliefs in similarities the different voters will select the words they view as appropriate for the difference of belief. Assuming that the difference in beliefs (x) is 0.67 (one agents belief over similarities is 0.85 and an another agents belief is 0.18) the voters will select the labels representing the trust level as described in Table 2.

Table 2. Possible values for the voting

A1	A2	A3	A4	A5	A6	A7	A8	A9	A10
L_t	L_t	L_t	L_t	L_t	L_t	L_t	L_t	L_t	L_t
M_t	M_t	M_t	M_t	M_t	M_t				
H_t	H_t	H_t							

Note that each voter has their own membership function where the level of overlap is different for each voter. As an example, the belief difference of 0.67 can represent high, medium and low trust level for the first voter (A1) but it can only represent low trust for the last voter (A10).

Then we compute the membership value for each of the elements on set $T(L)$.

$$\chi_{\mu(Low\_trust)}(u) = 1 \tag{3}$$

$$\chi_{\mu(Medium\_trust)}(u) = 0.6 \tag{4}$$

$$\chi_{\mu(High\_trust)}(u) = 0.3 \tag{5}$$

and

$$L = \frac{Low\_trust}{1} + \frac{Medium\_trust}{0.6} + \frac{High\_trust}{0.3} \tag{6}$$

A value of x (actual belief difference between two agents) is presented and voters randomly pick exactly one word from a finite set to label x as depicted in Table 3. The number of voters will ensure that a realistic overall response will prevail during the process.

Table 3. Voting

A1	A2	A3	A4	A5	A6	A7	A8	A9	A10
H_t	M_t	L_t	L_t	M_t	M_t	L_t	L_t	L_t	L_t

Taken as a function of x in regard to these probabilities forming probability functions. They should, therefore, satisfy :

$$\sum_{w \in T(L)} Pr(L = w|x) = 1 \tag{7}$$

which gives a probability distribution on words:

$$\sum Pr(L = Low\_trust|x) = 0.6 \tag{8}$$

$$\sum Pr(L = Medium\_trust|x) = 0.3 \tag{9}$$

$$\sum Pr(L = High\_trust|x) = 0.1 \tag{10}$$

As a result of voting, we can conclude that, given the difference in belief, $x = 0.67$, the combination should not consider this belief in the similarity function since, based on its difference compared to another beliefs, it turns out to be a distrustful assessment. The aforementioned process is then repeated as many times as the many different beliefs we have for the similarity, i.e., as many as different similarity measures exist in the ontology mapping system.

5.2   Introducing Trust into Ontology Mapping

The problem of trustworthiness in the context of ontology mapping can be represented in different ways. In general, trust issues on the Semantic Web are associated with the source of the information, i.e., who said what and when and what credentials they had to say it. From this point of view the publisher of the ontology could greatly influence the outcome of the trust evaluation and the mapping process can prefer mappings that come from a more "trustful" source.

However, we believe that in order to evaluate trust it is better to look into our processes that map these ontologies, because from the similarity point of view it is more important to see how the information in the ontologies is conceived by our algorithms than who has created it, e.g., do our algorithms exploit all the available information in the ontologies or is that just part of it? The reason why we propose such trust evaluation is because ontologies of the Semantic Web usually represent a particular domain and support a specific need. Therefore, even if two ontologies describe the same concepts and properties, their relation to each other can differ depending on the conceptualisation of their creators, which is independent from the organisation that they belong to. In our ontology mapping method we propose that the trust in the provided similarity measures, which is assessed between the ontology entities that are associated to the actual understanding of the mapping entities differs from case to case, e.g., a similarity measure can be trusted in one case but not trusted in another case during the same process. Our mapping algorithm, which incorporates trust management into the process, is described by Algorithm 1.

```
     Input: Similarity belief matrices S_{n×m} = {S_1, .., S_k}
     Output: Mapping candidates
 1   for i=1 to n do
 2   │   BeliefVectors BeliefVectors ← GetBeliefVectors(S[i, 1 − m]) ;
 3   │   Concepts ← GetBestBeliefs(BeliefVectors BeliefVectors) ;
 4   │   Scenario ← CreateScenario(Concepts) ;
 5   │   for j=1 to size(Concepts) do
 6   │   │   Scenario ← AddEvidences (Concepts) ;
 7   │   end
 8   │   if Evidences are contradictory then
 9   │   │   for count=1 to numberOf(Experts) do
10   │   │   │   Voters ← CreateVoters(10) ;
11   │   │   │   TrustValues ← VoteTrustMembership(Evidences) ;
12   │   │   │   ProbabilityDistribution ← CalculateTrustProbability(TrustValues) ;
13   │   │   │   Evidences ← SelectTrustedEvidences(ProbabilityDistribution) ;
14   │   │   end
15   │   end
16   │   Scenario ← CombineBeliefs(Evidences) ;
17   │   MappingList ← GetMappings(Scenario) ;
18   end
```

Algorithm 1. Belief combination with trust

Our mapping algorithm receives the similarity matrices (both syntactic and se- mantic) as an input and produces the possible mappings as an output. The similarity matrices represent the assigned similarities between all concepts in ontologies 1 and 2. Our mapping algorithm iterates through all concepts in ontology 1 and selects the best possible candidate terms from ontology 2, which is represented as a vector of best beliefs (step 2). Once the best beliefs (highest belief values) have been selected we collect the terms that correspond to these selected beliefs in order to create the mapping scenario. This scenario contains all possible mapping pairs between the selected term in ontology 1 and the possible terms from ontology 2 (steps 3 and 4). Once we have built our mapping scenario we start adding evidence from the similarity matrices (step 6). This evidence might be contradictory because different similarity algorithms can assign different similarity measures for the same mapping candidates. In case this evidence is contradictory we need to evaluate which measure, i.e., mapping agents belief, we trust in this particular scenario (steps 8-15). The trust evaluation (see details in section 3.1) is invoked, which invalidates the evidence (agent beliefs), which cannot then be trusted in this scenario. Once the conflict resolution routine is finished, the valid beliefs can be combined and the possible mapping candidates can be selected from the scenario.

The advantage of our proposed solution is that the evaluated trust is independent from the source ontologies themselves and can change depending on the available information in the context.

6   Empirical Evaluation

The evaluation was measured using standard recall and precision measurements from the Information Retrieval community. Before we present our evaluation let us discuss what improvements one can expect considering the mapping precision or recall. Most people would expect that if the results can be doubled, i.e., increased by 100%, then this is a remarkable achievement. This might be the case for anything but ontology mapping. In reality, researchers are trying to push the limits of the existing matching algorithms and anything between 10% and 30% is considered to be a good improvement. The objective is always to make improvements in preferably both precision and recall.

We have carried out experiments with the benchmark ontologies of the Ontology Alignment Evaluation Initiative, (OAEI)[1], which is an international initiative that has been set up for evaluating ontology matching algorithms. The experiments were carried out to assess how trust management influences the results of our mapping algorithms. Our main objective was to evaluate the impact of establishing trust before combining beliefs in similarities between concepts and properties in the ontology. The OAEI benchmark contains tests that were created by modifying a reference ontology using different transformations. The main objective of this task was to measure how ontology mapping systems can cope with situations where certain information is missing or represented differently in the ontologies. The bibliographic reference ontology (different classifications of publications) contained 33 classes and 64 properties. Further, each generated ontology was aligned with the reference ontology.

The benchmark tests were created and grouped by the following criteria:

– Group 1xx: easy tests, comparing the ontologies that are identical or do not match because they are from different domains.
– Group 2xx: tests that were obtained by modifying the reference ontology. Several transformation rules were applied, e.g., hierarchy of the terms are removed, different naming conventions are used, and the terms with random strings are replaced. Additionally, structural transformations were also applied, e.g., expand the class hierarchy or remove the OWL restrictions.
– Group 3xx: four different bibliographic ontologies that were created and used by research institutions, e.g., MIT, UMBC.

As a basic comparison we have modified our algorithm (without trust), which does not evaluate trust before conflicting belief combination but just combines them using Dempsters combination rule. The recall and precision graphs for the algorithm with trust and without trust over the whole benchmarks are depicted in Fig. 4. Experiments have proved that by establishing trust one can reach a higher average precision and recall rate.

Fig. 4 shows the improvement in recall and precision that we have achieved by applying our trust model by combining contradictory evidence. From the precision point of view, the increased recall values have not impacted the results significantly, which is good because the objective is always the improvement of both recall and precision together. We have measured the average improvement for the whole benchmark test set, which contains 51 ontologies.

Based on the experiments, the average recall has increased by 12% and the precision by 16%. The relatively high increase in precision compared to recall is attributed to the fact that, in some cases, the precision has been increased by

---

[1] http://oaei.ontologymatching.org/

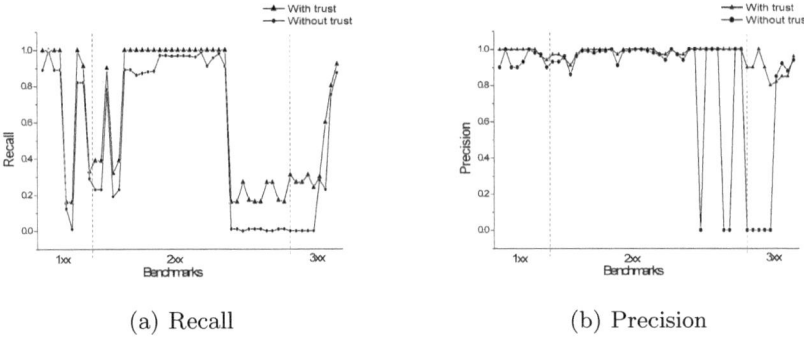

(a) Recall   (b) Precision

Fig. 4. Recall and Precision graphs

100% as a consequence of a small recall increase of 1%. This is perfectly normal because, if the recall increases from 0 to 1%, and the returned mappings are all correct (which is possible since the number of mappings are small), then the precision increases from 0 to 100%. Further, the increase in recall and precision greatly varies from test to test. Surprisingly, the precision decreased in some cases (5 out of 51). The maximum decrease in precision was 7% and maximum increase was 100%. The recalls never decreased in any of the tests and the minimum increase was 0.02% whereas the maximum increase was 37%.

As mentioned in our scenario, in our ontology mapping algorithm there are a number of mapping agents that carry out similarity assessments, hence creating belief mass assignments for the evidence. Before the belief mass function is combined, each mapping agent needs to dynamically calculate a trust value, which describes how confident the particular mapping agent is about the other mapping agents assessment. This dynamic trust assessment is based on the fuzzy voting model and depends on its own and other agents belief mass function. In our ontology mapping framework we assess trust between the mapping agents beliefs and determine which agents belief cannot be trusted, rejecting one, since the result of the trust assessment has become distrustful.

7   Conclusions and Future Research

In this chapter we have shown how the fuzzy voting model can be used to evaluate trust and we determine which belief is contradictory with other beliefs before combining them into a more coherent state. These aspects of ontology mapping are important because multi-agent systems that process data on the Semantic Web will end up in scenarios where the different agents have conflicting beliefs over a particular solution. In these situations, the agents need to resolve their conflicts in order to choose the best possible solution. Additionally, any application on the Semantic Web that operates with limited human intervention should be able to first eliminate conflicts independently from the users, which can improve the quality of the mappings. In order to resolve the aforementioned problems we have proposed new levels of trust in the context of ontology mapping, which is a prerequisite for any systems that make use of information available on the Semantic Web. Our system is flexible because the membership functions for the voters can be changed dynamically in order to influence the outputs according to the different similarity measures that can be used in the mapping system. We have described initial experimental results using the benchmarks of the Ontology Alignment Initiative, which demonstrates the effectiveness of our approach through improved recall and precision rates.

Concerning future work, there are many areas that can be further investigated. First of all, the proper selection of fuzzy membership functions relies heavily on experiments with a particular system. In our case, this means that a set of experiments could be carried out using different input and output membership functions and can then investigate how the selection impacts the reliability of the trust assessment and elimination process.

Secondly, the voting process involves the possibility of reaching an incorrect conclusion if the wrong opinions dominate the voters. Since the voting process does not get feedback on whether it works correctly or not, implementing any self-improvements within the voting process are not currently possible. In the future we intend to investigate how this feedback can be used to remove and add voters into the system in order to replace the ones that tend to produce wrong votes. Therefore, our primary focus concerning additional experimentation is to investigate different kinds of membership functions for the different voters and consider the effect of both the changing number of voters and the impact on mapping precision and recall.

References

1. Euzenat, J., Shvaiko, P.: Ontology matching. Springer, Heidelberg (2007)
2. Shvaiko, P., Euzenat, J.: Ten challenges for ontology matching. Technical Report DISI-08-042, University of Trento (2008)
3. Nagy, M., Vargas-Vera, M., Motta, E.: DSSim - managing uncertainty on the semantic web. In: Proceedings of the 2nd International Workshop on Ontology Matching (2007)
4. Nagy, M., Vargas-Vera, M., Motta, E.: Multi-agent ontology mapping with uncertainty on the semantic web. In: Proceedings of the 3rd IEEE International Conference on Intelligent Computer Communication and Processing (2007)
5. Nagy, M., Vargas-Vera, M., Stolarski, P.: DSSim results for OAEI 2009. In: Proceedings of the 4th International Workshop on Ontology Matching, OM 2009 (2009)
6. Sentz, K., Ferson, S.: Combination of evidence in dempster-shafer theory. Technical report, Systems Science and Industrial Engineering Department, Binghamton University (2002)
7. Yager, R.R.: On the dempster-shafer framework and new combination rules. Information Sciences: an International Journal 41, 93–137 (1987)
8. Yager, R.R.: Quasi-associative operations in the combination of evidence. Kybernetes 16, 37–41 (1987)
9. Smarandache, F., Dezert, J. (eds.): Advances and Applications of DSmT for Information Fusion (Collected Works), vol. 1. American Research Press (2004)

10. Richardson, M., Agrawal, R., Domingos, P.: Trust Management for the Semantic Web. In: Fensel, D., Sycara, K., Mylopoulos, J. (eds.) ISWC 2003. LNCS, vol. 2870, pp. 351–368. Springer, Heidelberg (2003)
11. Laera, L., Blacoe, I., Tamma, V., Payne, T., Euzenat, J., Bench-Capon, T.: Argumentation over ontology correspondences in MAS. In: AAMAS 2007: Proceedings of the 6th International Joint Conference on Autonomous Agents and Multiagent Systems, pp. 1–8. ACM, New York (2007)
12. Griffiths, N.: A Fuzzy Approach to Reasoning with Trust, Distrust and Insufficient Trust. In: Klusch, M., Rovatsos, M., Payne, T.R. (eds.) CIA 2006. LNCS (LNAI), vol. 4149, pp. 360–374. Springer, Heidelberg (2006)
13. Rehak, M., Pechoucek, M., Benda, P., Foltyn, L.: Trust in coalition environment: Fuzzy number approach. In: Proceedings of the 4th International Joint Conference on Autonomous Agents and Multi Agent Systems - Workshop Trust in Agent Societies, pp. 119–131 (2005)
14. Yamada, K.: A new combination of evidence based on compromise. Fuzzy Sets Syst. 159(13), 1689–1708 (2008)
15. Josang, A.: The consensus operator for combining beliefs. Artificial Intelligence 141(1), 157–170 (2002)
16. Ferrara, A., Lorusso, D., Stamou, G., Stoilos, G., Tzouvaras, V., Venetis, T.: Resolution of conflicts among ontology mappings: a fuzzy approach. In: Proceedings of the 3rd International Workshop on Ontology Matching (2008)
17. Tang, J., Li, J., Liang, B., Huang, X., Li, Y., Wang, K.: Using bayesian decision for ontology mapping. Web Semantics: Science, Services and Agents on the World Wide Web 4, 243–262 (2006)
18. Liu, X.J., Wang, Y.L., Wang, J.: Towards a semi-automatic ontology mapping - an approach using instance based learning and logic relation mining. In: Fifth Mexican International Conference (MICAI 2006) on Artificial Intelligence (2006)
19. Jean-Mary, Y.R., Kabuka, M.R.: ASMOV: Results for OAEI 2008. In: Proceedings of the 3rd International Workshop on Ontology Matching (2008)
20. Beckett, D.: RDF/XML syntax specification
21. McGuinness, D.L., van Harmelen, F.: OWL web ontology language
22. Miles, A., Bechhofer, S.: SKOS simple knowledge organization system
23. Lenzerini, M., Milano, D., Poggi, A.: Ontology representation & reasoning. Technical Report NoE InterOp (IST-508011), WP8, subtask 8.2, Universit di Roma La Sapienza, Roma, Italy (2004)
24. Wang, R.Y., Kon, H.B., Madnick, S.E.: Data quality requirements analysis and modeling. In: Proceedings of the Ninth International Conference on Data Engineering, pp. 670–677 (1993)
25. Wand, Y., Wang, R.Y.: Anchoring data quality dimensions in ontological foundations. Communications of the ACM, 86–95 (1996)
26. Batini, C., Lenzerini, M., Navathe, S.B.: A comparative analysis of methodologies for database schema integration. ACM Computing Surveys 18(4), 323–364 (1986)
27. Baldwin, J.F.: Mass Assignment Fundamentals for Computing with Words. In: L. Ralescu, A. (ed.) IJCAI-WS 1997. LNCS, vol. 1566, pp. 22–44. Springer, Heidelberg (1999)
28. Lawry, J.: A voting mechanism for fuzzy logic. International Journal of Approximate Reasoning 19, 315–333 (1998)

Storing and Querying Fuzzy Knowledge in the Semantic Web Using FiRE

Nikolaos Simou, Giorgos Stoilos, and Giorgos Stamou

Department of Electrical and Computer Engineering,
National Technical University of Athens,
Zographou 15780, Greece
{nsimou,gstoil,gstam}@image.ntua.gr

Abstract. An important problem for the success of ontology-based applications is how to provide persistent storage and querying. For that purpose, many RDF tools capable of storing and querying over a knowledge base, have been proposed. Recently, fuzzy extensions to ontology languages have gained considerable attention especially due to their ability to handle vague information. In this paper we investigate on the issue of using classical RDF storing systems in order to provide persistent storing and querying over large scale fuzzy information. To accomplish this we propose a novel way for serializing fuzzy information into RDF triples, thus classical storing systems can be used without any extensions. Additionally, we extend the existing query languages of RDF stores in order to support expressive fuzzy querying services over the stored data. All our extensions have been implemented in FiRE—an expressive fuzzy DL reasoner that supports the language fuzzy-$\mathcal{SHIN}$. Finally, the proposed architecture is evaluated using an industrial application scenario about casting for TV commercials and spots.

1 Introduction

Despite the great success of the World Wide Web during the last decade, it is still not uncommon that information is extremely difficult to find. Such cases include searching for popular names, multi-language words with different definitions in different languages and specific information that requires more sophisticated queries. The Semantic Web—an extension of the current Web—aims at alleviating these issues by adding semantics to the content that exists on the Web. Information in the (Semantic) Web would be linked forming a distributed knowledge base and documents would be semantically annotated.

Ontologies, through the OWL language [14], are expected to play a significant role in the Semantic Web. OWL is mainly based on Description Logics (DLs) [1], a popular family of knowledge representation languages. Their well-defined semantics, together with their decidable reasoning algorithms, have made them popular to a variety of applications [1]. However, despite their rich expressiveness, they are not capable of dealing with vague and uncertain information, which is commonly found in many real-world applications such as multimedia content,

F. Bobillo et al. (Eds.): URSW 2008-2010/UniDL 2010, LNAI 7123, pp. 158–176, 2013.
© Springer-Verlag Berlin Heidelberg 2013

medical informatics and more. For this purpose fuzzy extensions to Description Logics have been proposed [11,20,18,2].

Fuzzy ontologies have also gained considerable attention in many research applications. Similar to crisp ontologies, they can serve as basic semantic infrastructure, providing shared understanding of certain domains across different applications. Furthermore, the need for handling fuzzy and uncertain information is crucial to the Web. This is because information and data along the Web may often be uncertain or imperfect. Such cases are domains that may be described using concepts, like "near" for modeling distance or concept "tall" for modeling people and buildings height. The paradox that arises in the latter case is that there is no distinction between a person's and a building's height.

Clearly the information represented by those kinds of concepts is very important though imperfect. Therefore sophisticated uncertainty representation and reasoning are necessary for the alignment and integration of Web data from different sources. Recently, also fuzzy DL reasoners such as fuzzyDL [3] and FiRE [17] that can provide practical reasoning over imprecise information have been implemented. Despite these implementations little work has been done towards the persistent storage and querying of information. Closely related works are those on fuzzy DL-Lite [22,13], where the reasoning algorithms assume that the data have been stored previously in a relational database.

In this paper we follow a different paradigm and study the use of an RDF triple-store for the storage of fuzzy information. More precisely, we propose an architecture that can be used to store such information in an off-the-shelf triple-store and then show how the stored information can be queried using a fuzzy DL reasoner. The main contributions of this paper are the following:

1. It presents a novel framework for persistent storage and querying of expressive fuzzy knowledge bases,
2. It presents the first ever integration of fuzzy DL reasoners with RDF triple-stores, and
3. It provides experimental evaluation of the proposed architecture using a real-world industrial strength use-case scenario.

The rest of the paper is organized as follows. Firstly, in Section 2 the theoretical description of the fuzzy DL f-$\mathcal{SHIN}$ [18] is given. After that in Section 3 the fuzzy reasoning engine FiRE which supports the fuzzy DL f-$\mathcal{SHIN}$ and was used in our approach is presented. In the following Section (4) the proposed triples syntax accommodating the fuzzy element used for storing a fuzzy knowledge base in RDF-Stores, is presented. Additionally, the syntax and the semantics of expressive queries that have been proposed in the literature [13] to exploit fuzziness are briefly presented. In the last Section (5) the applicability of the proposed architecture is demonstrated, presenting a use case based on a database of human models. This database was used by a production company for the purposes of casting for TV commercials and spots. Some entries of the database were first fuzzified and then using an expressive knowledge base, abundant implicit knowledge was extracted. The extracted knowledge was stored to an RDF

Store, and various expressive queries were performed in order to benchmark the proposed architecture.

2   Preliminaries

2.1   The Fuzzy DL f$_{KD}$-$\mathcal{SHIN}$

In this section we briefly present the notation of DL f-$\mathcal{SHIN}$ which is a fuzzy extension of DL $\mathcal{SHIN}$ [8]. Similar to crisp description logic languages, a fuzzy description logic language consist of an alphabet of distinct concepts names ($\mathbf{C}$), role names ($\mathbf{R}$) and individual names ($\mathbf{I}$), together with a set of constructors to construct concept and role descriptions. If R is a role then R^- is also a role, namely the inverse of R. f-$\mathcal{SHIN}$-concepts are inductively defined as follows:

1. If $C \in \mathbf{C}$, then C is a f-$\mathcal{SHIN}$-concept,
2. If C and D are concepts, R is a role, S is a simple role and $n \in \mathbb{N}$, then $(\neg C)$, $(C \sqcup D)$, $(C \sqcap D)$, $(\forall R.C)$, $(\exists R.C)$, $(\geq nS)$ and $(\leq nS)$ are also f-$\mathcal{SHIN}$-concepts.

In contrast to crisp DLs, the semantics of fuzzy DLs are provided by a *fuzzy interpretation* [20]. A fuzzy interpretation is a pair $\mathcal{I} = \langle \Delta^{\mathcal{I}}, \cdot^{\mathcal{I}} \rangle$ where $\Delta^{\mathcal{I}}$ is a non-empty set of objects and $\cdot^{\mathcal{I}}$ is a fuzzy interpretation function, which maps an individual name $\mathbf{a}$ to elements of $\mathbf{a}^{\mathcal{I}} \in \Delta^{\mathcal{I}}$ and a concept name $\mathbf{A}$ (role name R) to a membership function $\mathbf{A}^{\mathcal{I}} : \Delta^{\mathcal{I}} \to [0,1]$ ($R^{\mathcal{I}} : \Delta^{\mathcal{I}} \times \Delta^{\mathcal{I}} \to [0,1]$).

By using fuzzy set theoretic operations the fuzzy interpretation function can be extended to give semantics to complex concepts, roles and axioms [9]. FiRE uses the standard fuzzy operators of $1 - x$, where x is a degree, for fuzzy negation (c), $\max(x, y)$, for fuzzy union (u), $\min(x, y)$ for fuzzy intersection (t) and $\max(1 - x, y)$ for fuzzy implication ($\mathcal{J}$). For example, if d_1 is the degree of membership of an object o to the set $A^{\mathcal{I}}$ and d_2 the degree of membership to $B^{\mathcal{I}}$, then the membership degree of o to $(A \sqcap B)^{\mathcal{I}}$ is given by $\min(d_1, d_2)$, while the membership degree to $(A \sqcup B)^{\mathcal{I}}$ is given by $\max(d_1, d_2)$. The complete set of semantics is depicted in Table 1.

A f-$\mathcal{SHIN}$ knowledge base Σ is a triple $\langle \mathcal{T}, \mathcal{R}, \mathcal{A} \rangle$, where $\mathcal{T}$ is a fuzzy *TBox*, $\mathcal{R}$ is a fuzzy *RBox* and $\mathcal{A}$ is a fuzzy *ABox*. *TBox* is a finite set of fuzzy concept axioms which are of the form $C \sqsubseteq D$ called fuzzy concept inclusion axioms and $C \equiv D$ called fuzzy concept equivalence axioms, where C, D are concepts, saying that C is a sub-concept or C is equivalent of D, respectively. In cases where C is allowed to be a complex concept we have a general concept inclusion axiom (GCI). At this point it is important to note that in our approach f$_{KD}$-$\mathcal{SHIN}$ without GCIs is considered. The interested reader is referred to [19,10] that present how GCIs are handled in fuzzy DLs and also to fuzzy DLs that have been proposed in the literature [7,18] and support GCIs. *RBox* is a finite set of fuzzy role axioms of the form $\mathsf{Trans}(R)$ called fuzzy transitive role axioms and $R \sqsubseteq S$ called fuzzy role inclusion axioms saying that R is transitive and R is a sub-role of S respectively. Finally, *ABox* is a finite set of fuzzy assertions of the

Table 1. Semantics of concepts and roles

Constructor	Syntax	Semantics
top	$\top$	$\top^{\mathcal{I}}(a) = 1$
bottom	$\bot$	$\bot^{\mathcal{I}}(a) = 0$
general negation	$\neg C$	$(\neg C)^{\mathcal{I}}(a) = c(C^{\mathcal{I}}(a))$
conjunction	$C \sqcap D$	$(C \sqcap D)^{\mathcal{I}}(a) = t(C^{\mathcal{I}}(a), D^{\mathcal{I}}(a))$
disjunction	$C \sqcup D$	$(C \sqcup D)^{\mathcal{I}}(a) = u(C^{\mathcal{I}}(a), D^{\mathcal{I}}(a))$
exists restriction	$\exists R.C$	$(\exists R.C)^{\mathcal{I}}(a) = \sup_{b \in \Delta^{\mathcal{I}}} \{t(R^{\mathcal{I}}(a,b), C^{\mathcal{I}}(b))\}$
value restriction	$\forall R.C$	$(\forall R.C)^{\mathcal{I}}(a) = \inf_{b \in \Delta^{\mathcal{I}}} \{\mathcal{J}(R^{\mathcal{I}}(a,b), C^{\mathcal{I}}(b))\}$
at-most	$\leq pR$	$\inf_{b_1,\ldots,b_{p+1} \in \Delta^{\mathcal{I}}} \mathcal{J}(t_{i=1}^{p+1} R^{\mathcal{I}}(a,b_i), u_{i<j}\{b_i = b_j\})$
at-least	$\geq pR$	$\sup_{b_1,\ldots,b_p \in \Delta^{\mathcal{I}}} t(t_{i=1}^{p} R^{\mathcal{I}}(a,b_i), t_{i<j}\{b_i \neq b_j\})$
inverse role	R^-	$(R^-)^{\mathcal{I}}(b,a) = R^{\mathcal{I}}(a,b)$
equivalence	$C \equiv D$	$\forall a \in \Delta^{\mathcal{I}}.C^{\mathcal{I}}(a) = D^{\mathcal{I}}(a)$
sub-concept	$C \sqsubseteq D$	$\forall a \in \Delta^{\mathcal{I}}.C^{\mathcal{I}}(a) \leq D^{\mathcal{I}}(a)$
transitive role	$\mathsf{Trans}(R)$	$\forall a,b \in \Delta^{\mathcal{I}}.R^{\mathcal{I}}(a,b) \geq \sup_{c \in \Delta^{\mathcal{I}}} \{t(R^{\mathcal{I}}(a,c), R^{\mathcal{I}}(c,b))\}$
sub-role	$R \sqsubseteq S$	$\forall a,b \in \Delta^{\mathcal{I}}.R^{\mathcal{I}}(a,b) \leq S^{\mathcal{I}}(a,b)$
concept assertions	$\langle a : C \bowtie n \rangle$	$C^{\mathcal{I}}(a^{\mathcal{I}}) \bowtie n$
role assertions	$\langle \langle a,b \rangle : R \bowtie n \rangle$	$R^{\mathcal{I}}(a^{\mathcal{I}}, b^{\mathcal{I}}) \bowtie n$

form $\langle a : C \bowtie n \rangle$, $\langle (a,b) : R \bowtie n \rangle$, where $\bowtie$ stands for $\geq, >, \leq$ or $<$, or $a \neq b$, for $a, b \in \mathbf{I}$. Intuitively, a fuzzy assertion of the form $\langle a : C \geq n \rangle$ means that the membership degree of a to the concept C is at least equal to n.

Example 1. An example of a fuzzy knowledge base Σ is shown below.
$\mathcal{T} = \{$MiddleAged $\equiv$ 40s $\sqcup$ 50s, TallChild $\equiv$ Child $\sqcap$ (Short $\sqcup$ Normal_Height),$\}$
$\mathcal{R} = \{$isFriendOf$^-$ = isFriendOf$\}$ and
$\mathcal{A} = \{\langle michalis1539 : $ Male$\rangle$, $\langle michalis1539 : $ (Tall $\sqcap$ GoodLooking) $\geq 0.8 \rangle$,
$\langle (michalis1539, maria1343) : $ isFriendOf $\geq 0.7 \rangle\}$

3   Fuzzy Reasoning Engine FiRE

In this section we present the graphical user interface, the syntax and the inference services of FiRE—an expressive fuzzy DL reasoner.

3.1   FiRE Interface

FiRE[1] is a Java based fuzzy reasoning engine. FiRE implements the tableau reasoning algorithm for fuzzy-$\mathcal{SHIN}$ presented in [18]. It can be used either as an API by another application or through its graphical user interface. The graphical user interface of FiRE consists of the editor panel, the inference services panel and the output panel (Figure 1). Hence the user has the ability to create or edit an existing fuzzy knowledge base using the editor panel, and to use the inference services panel to make different kinds of queries to the fuzzy knowledge

---

[1] http://www.image.ece.ntua.gr/~nsimou/FiRE/

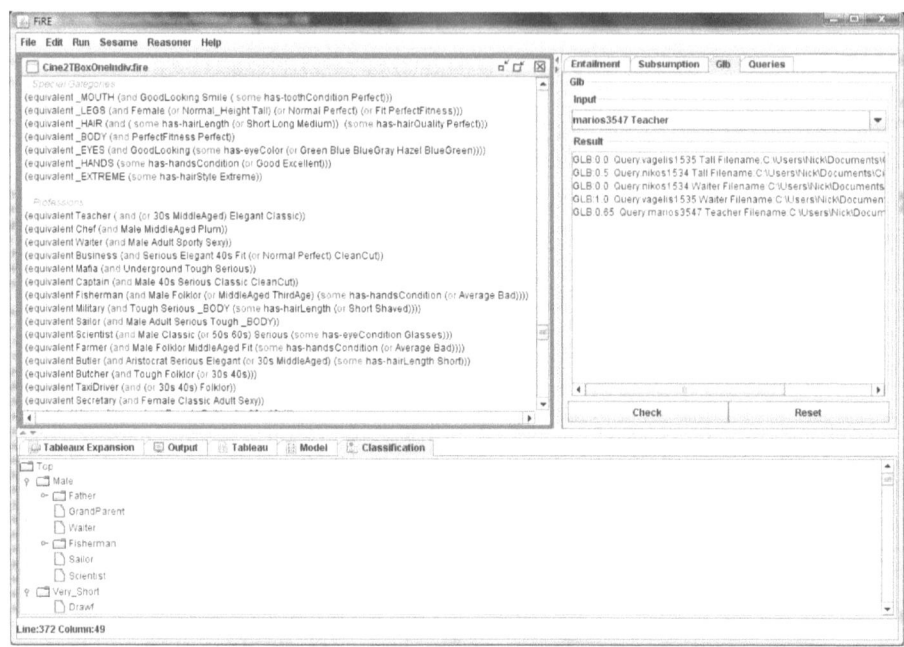

Fig. 1. The FiRE user interface consists of the editor panel (upper left), the inference services panel (upper right) and the output panel (bottom)

base. Finally, the output panel consists of four different tabs, each one displaying feedback depending on the user operation.

3.2   FiRE Syntax

As previously mentioned, a fuzzy knowledge base consists of three components *TBox*, *RBox* and *ABox*. The *TBox* and the *RBox* are defined using the Knowledge Representation System Specification [16] proposal since they do not include uncertainty. So, transitive roles or the sub-role of another role can be defined by using the keywords **transitive** and **parent** respectively and concept axioms by the keywords **implies** and **equivalent**. (Please refer to [16] for a full specification.)

On the contrary, since the assertions are extended in order to represent imperfect knowledge, the ABox specified differently. Instances in FiRE are defined using the keyword **instance** followed by the individual, the concept in which the individual participates, the inequality type (one of $<, <=, >, >=$) and the degree of confidence $degree \in [0, 1]$. Similarly, role assertions are defined by using the keyword **related** followed by subject and object individuals, the inequality type and the degree of confidence. In both cases the inequality type and the degree of confidence are required only for fuzzy assertions, if these are not mentioned then the assertions are assumed as crisp (i.e $>= 1$).

Example 2. The syntax of the assertions defined in example 1 are shown below in FiRE syntax.

```
(instance michalis1539 Male)
(instance michalis1539 (and Tall GoodLooking) >= 0.8)
(related michalis1539 maria1343 isFriendOf >= 0.7)
```

3.3   FiRE Reasoning Services

Description Logics offer a large range of inference services, which the user can query over a knowledge base. The main reasoning services offered are satisfiability checking, subsumption and entailment of concepts and axioms with respect to an ontology. For example, one is capable of asking queries like "Can the concept C have any instances in models of the ontology T?" (satisfiability of C), or "Is the concept D more general than the concept C in models of the ontology T?" (subsumption $C \sqsubseteq D$), or "Does axiom Ψ logically follow from the ontology?" (entailment of Ψ).

In addition to these reasoning services, fuzzy DLs also provide with *greatest lower bound queries* (GLB). In the case of fuzzy DL, satisfiability questions become of the form "Can the concept C have any instances with degree of participation $\bowtie n$ in models of the ontology T?". Furthermore, the incorporation of degrees in assertions makes the evaluation of the best lower and upper truth-value bounds of a fuzzy assertion vital. The term of *greatest lower bound* of a fuzzy assertion with respect to Σ was defined in [20]. Informally, greatest lower bound queries are queries like "What is the greatest degree n that our ontology entails an individual a to participate in a concept C?".

FiRE uses the tableau algorithm of f-$\mathcal{SHIN}$ presented in [18], in order to decide the key inference problems of a fuzzy ontology. Hence entailment queries that ask whether our knowledge base logically entails the membership of an individual to a specific concept and to a certain degree, are specified in the *Entailment* inference tab (see Figure 1). Their syntax is the same as the one used for the definition of a fuzzy instance. For example a statement of the form:

```
(instance michalis1539 (and Tall GoodLooking) > 0.8)
```

asks whether `michalis1539` is Tall and GoodLooking to a degree greater than or equal to 0.8. If there are assertions in the *ABox* of our Σ that satisfy this query (i.e. there is a model for our ontology) then FiRE will return true.

On the other hand subsumption queries that are specified in the *Subsumption* inference tab evaluate whether a concept is more general than another concept. Their syntax is of the following form:

```
(concept1) (concept2)
```

where `concept1` and `concept2` are f-$\mathcal{SHIN}$ concepts. Let us assume that the first concept is Father while the second concept is Male. Subsequently, assume that Father has been defined in the *TBox* using an equivalence axiom as follows: Father $\equiv$ Male $\sqcap$ MiddleAged. Then, the following subsumption query will always return true since Father will always be (in all models) a sub-concept of Male.

(Father) (Male)

Additionally, the user can perform a global concept classification procedure presenting the concept hierarchy tree in the *Classification* tab of the output panel.

Finally, FiRE supports GLB queries, which are evaluated by FiRE by performing entailment queries. During this procedure a set of entailment queries is constructed consisting of an entailment query for every degree contained in the ABox, using the individual and the concept of interest. These queries are performed using the binary search algorithm to reduce the degrees search space, resulting the GLB. The syntax of GLB queries is of the form:

individual (concept)

where concept can be any f-$\mathcal{SHIN}$-concept. In order to illustrate the operation of the GLB service we will present a trivial example using an atomic concept. Let the following *ABox* (in FiRE syntax)

```
(instance michalis1539 Tall > 0.8)
(instance maria231 GoodLooking > 0.6)
(instance nikos Male > 1)
(instance nikos Tall > 0.9)
```

We want to evaluate the GLB of individual michalis1539 in the concept Tall. In FiRE this is specified by issuing the query michalis1539 Tall. Then, the system proceeds as follows: Firstly, all the degrees that appear in *ABox* are sorted. FiRE then performs entailment queries for the individual michalis1539 with the concept Tall, using the binary search algorithm. This procedure is repeated until the entailment query is unsatisfiable. The greatest degree found before unsatisfiability is the greatest lower bound. In this example the following entailment queries are performed with the indicated results in order to evaluate that the greatest lower bound of michalis1539 participating in concept Tall is 0.8.

```
(instance michalis1539 Tall > 0.5) TRUE
(instance michalis1539 Tall > 0.8) TRUE
(instance michalis1539 Tall > 1)   FALSE
```

Finally, FiRE offers the possibility to perform a global GLB for the whole fuzzy knowledge base. Global GLB evaluates the greatest lower bound degree of all combinations of individuals and concepts in Σ.

4   Storing and Querying a Fuzzy Knowledge Base

In the current section we will present how FiRE stores and queries fuzzy knowledge. More precisely in the proposed architecture a triple-store is used as a back end for storing and querying RDF triples in a sufficient and convenient way, while the reasoner is the front end, which the user can use in order to store and query a fuzzy knowledge base. In that way, a user is able to access data from a repository, apply any of the available reasoning services on this data and then store back in the repository the implicit knowledge extracted from them.

Many triple-store systems have been developed in the context of the Semantic Web, like Sesame[2], Kowari[3], Jena[4] and more. These provide persistent storage of ontology data as well as querying services. In our approach FiRE was integrated with Sesame (Sesame 2 beta 6), an open source Java framework for storing and querying RDF/RDFS data. Sesame supports two ways of storing RDF data (called RDF Repositories). The first is the in-memory RDF Repository, which stores all the RDF triples in the main memory, while the second one is the native RDF Repository, which stores the RDF triples on the hard disk and uses B-Trees to index and access them.

4.1   Storage of a Fuzzy Knowledge Base

In order to use a triple-store for storing a fuzzy knowledge without enforcing any extensions on it, we needed to find a way to serialize fuzzy knowledge into RDF triples. For that purpose a fuzzy-OWL to RDF mapping was required, similar to the one provided in the OWL abstract syntax and semantics document [14]. In previous works Mazzieri and Dragoni [12] used RDF *reification* in order to store the membership degrees. However it is well-known that reification has weak, ill-defined model theoretic semantics and its support by RDF tools is limited. In another approach, Vaneková et al. [23] suggest the use of datatypes, but we argue that the use of a concrete feature like datatypes to represent abstract information such as fuzzy assertions is not appropriate.

Consequently, we were lead to propose a new way of mapping fuzzy-OWL into RDF triples. The technique makes use of RDF's blank nodes. First, we define three new entities, namely frdf:membership, frdf:degree and frdf:ineqType as types (i.e. rdf:type) of rdf:Property.

Using these new properties together with blank nodes we can represent fuzzy instances. Let's assume for example that we want to represent the assertion $\langle (michalis1539 : \mathsf{Tall}) \geq 0.8 \rangle$. The RDF triples representing this information are the following:

```
region1                     frdf:membership  _:michalis1539membTall .
_:michalis1539membTall      rdf:type         Tall .
_:michalis1539membTall      frdf:degree      "0.8^^xsd:float" .
_:michalis1539membTall      frdf:ineqType    "=" .
```

where _:michalis1539membTall is a blank node used to represent the fuzzy assertion of michalis1539 with the concept Tall.

Mapping fuzzy role assertions, however, is more tricky since RDF does not allow for blank nodes in the predicate position. To overcome this issue we de-

---

[2] http://www.openrdf.org/
[3] http://www.kowari.org/
[4] http://jena.sourceforge.net/

fine new properties for each assertion; for example, the fuzzy role assertion $\langle (michalis1539, maria1343) : \mathsf{isFriendOf} \geq 0.7 \rangle$ is mapped to

michalis1539	frdf:p1p2isFriendOf	maria1343 .
frdf:p1p2isFriendOf	rdf:type	isFriendOf .
frdf:p1p2isFriendOf	frdf:degree	"0.7^^xsd:float" .
frdf:p1p2isFriendOf	frdf:ineqType	"=" .

It is worth mentioning at that point that recently a fuzzy ontology representation has been proposed by Bobillo et. al [4] that it could be also used for the RDF serialization.

4.2   Fuzzy Queries

One of the main advantages of persistent storage systems, like relational databases and RDF storing systems, is their ability to support *conjunctive queries*. Conjunctive queries generalize the classical inference problem of *realization* of Description Logics [1], i.e. "get me all individuals of a given concept C", by allowing for the combination (conjunction) of concepts and roles. Formally, a conjunctive query is of the following form:

$$q(X) \leftarrow \exists Y.conj(X, Y) \tag{1}$$

or simply $q(X) \leftarrow conj(X, Y)$, where $q(X)$ is called the head, $conj(X, Y)$ is called the body, X are the *distinguished variables*, Y are the existentially quantified variables called *non-distinguished variables*, and $conj(X, Y)$ is a conjunction of atoms of the form $\mathsf{A}(v)$, $\mathsf{R}(v_1, v_2)$, where A, R are concept and role names, respectively, and v, v_1 and v_2 are *individual* variables in X and Y or individuals from the ontology.

Since in our case we extend classical assertions to fuzzy assertions, new methods of querying fuzzy information are possible. More precisely, in [13] the authors extend ordinary conjunctive queries to a family of significantly more expressive query languages, which are borrowed from the fields of fuzzy information retrieval [6]. These languages exploit the membership degrees of fuzzy assertions by introducing weights or thresholds in query atoms. In particular, the authors first define *conjunctive threshold queries* (CTQs) as:

$$q(X) \leftarrow \exists Y. \bigwedge_{i=1}^{n} (atom_i(X, Y) \geq k_i) \tag{2}$$

where $k_i \in [0, 1]$, $atom_i(X, Y)$ represents either a fuzzy-DL concept or role and all $k_i \in (0, 1]$ are thresholds. As it is obvious those answers of CTQs are a matter of true or false, in other words an evaluation either is or is not a solution to a query.

The authors also propose *General Fuzzy Conjunctive Queries* (GFCQs) that further exploit fuzziness and support degrees in query results. The syntax of a GFCQ is the following:

$$q(X) \leftarrow \exists Y. \bigwedge_{i=1}^{n} (atom_i(X,Y) : k_i) \tag{3}$$

where $atom_i(X,Y)$ is as above while k_i is the degree associated weight. As shown in [13], this syntax is general enough to allow various choices of semantics, which emerge by interpreting differently the degree of each fuzzy-DL atom $(atom_i(X,Y))$ with the associated weight (k_i). In what follows, we give some example of the semantic functions for conjunctions and degree-associated atoms.

1. **Fuzzy threshold queries:** As we mentioned earlier in a conjunctive threshold query an evaluation either satisfies the query entailment or not, thus providing only crisp answers. A straightforward extension will be instead of using crisp threshold we can use fuzzy ones with the aid of fuzzy R-implications. Hence, let a t-norm (t) as the semantic function for conjunctions and R-implications (ω_t) as the semantic function for degree-associated atoms, we get fuzzy threshold queries, in which the degree of truth of q_F under $\mathcal{I}$ is

$$d = \sup_{S' \in \Delta^{\mathcal{I}} \times \ldots \times \Delta^{\mathcal{I}}} \{ t_{i=1}^{n} \ \omega_t(k_i, atom_i^{\mathcal{I}}(\bar{v})_{[X \mapsto S, Y \mapsto S']}) \}.$$

   Given some S', if for all atoms we have $atom_i^{\mathcal{I}}(\bar{v})_{[X \mapsto S, Y \mapsto S']} \geq k_i$, since $\omega_t(x,y) = 1$ when $y \geq x$ [9], we have $d = 1$; this corresponds to threshold queries introduced earlier.
2. **Traditional conjunctive queries [21]:** These are the traditional conjunctive queries of the form 1, but instead of classical (boolean) conjunction between the atoms of the query we use the semantics of a t-norm. Then, the degree of truth of q_F under $\mathcal{I}$ is:

$$d = \sup_{S' \in \Delta^{\mathcal{I}} \times \ldots \times \Delta^{\mathcal{I}}} \{ t_{i=1}^{n} \ atom_i^{\mathcal{I}}(\bar{v})_{[X \mapsto S, Y \mapsto S']} \}.$$

   It is worth noting that this query language is a special case of the fuzzy threshold query language, where all $k_i = 1$ and since $\omega_t(1,y) = y$ [9].
3. **Fuzzy aggregation queries:** Another example of semantics for GFCTs would be to use fuzzy aggregation functions [9]. For example, let $G(x) = \sum_{i=1}^{n} x_i$, as a function for conjunctions and $a(k_i, y) = \frac{k_i}{\sum_{i=1}^{n} k_i} * y$ as the semantic function for degree-associated atoms. Then we get an instance of fuzzy aggregation queries, in which the degree of truth of q_F under $\mathcal{I}$ is

$$d = \sup_{S' \in \Delta^{\mathcal{I}} \times \ldots \times \Delta^{\mathcal{I}}} \frac{\sum_{i=1}^{n} k_i * atom_i^{\mathcal{I}}(\bar{v})_{[X \mapsto S, Y \mapsto S']}}{\sum_{i=1}^{n} k_i}.$$

4. **Fuzzy weighted queries:** If we use generalised weighted t-norms [5] as the semantic function for conjunction, we get fuzzy weighted queries, in which the degree of truth of q_F under $\mathcal{I}$ is

$$d = \sup_{S' \in \Delta^{\mathcal{I}} \times \ldots \times \Delta^{\mathcal{I}}} \{ \min_{i=1}^{n} \ u(\bar{k} - k_i, t(\bar{k}, atom_i^{\mathcal{I}}(\bar{v})_{[X \mapsto S, Y \mapsto S']})) \},$$

where $\bar{k} = \max_{i=1}^{n} k_i$ and u is a t-conorm (fuzzy union), such as $u(a, b) = \max(a, b)$. The main idea of this type of queries is that they provide an aggregation type of operation, on the other hand an entry with a low value for a low-weighted criterion should not be critically penalized. Moreover, lowering the weight of a criterion in the query should not lead to a decrease of the relevance score, which should mainly be determined by the high-weighted criteria. For more details see [5].

4.3   Fuzzy Queries Using FiRE

All of the above mentioned queries have been implemented in FiRE by using the SPARQL [15] query language for RDF. The queries are translated into SPARQL and are then issued on Sesame. A user can specify such queries using the *Queries* inference tab, and in the case of generalized fuzzy conjunctive queries a choice of the above mentioned semantics is possible.

Example 3. The following is a threshold query:

```
x,y <- Father(x) >= 0.8 ^ isFriendOf(x,y) >= 1.0
                          ^ Teacher(y) >= 0.7
```

Queries consist of two parts: the first one specifies the variables for which their bindings with individuals from the ABox will be returned as an answer, while the second one states the condition that has to be fulfilled for the individuals. The above query asks for all individuals that can be mapped to x and y, such that the individual mapped to x will be a Father to a degree at least 0.8, and then, also participate in the relation isFriendOf (to a degree 1.0) with another individual that y is mapped to, which also belongs to the concept Teacher to a degree at least 0.7.

The query is firstly converted from the FiRE conjunctive query syntax to the SPARQL query language. Based on the fuzzy OWL syntax in triples that we have defined previously, the query of Example 3 is as follows in SPARQL. (The query results are evaluated by the Sesame engine and visualized by FiRE.)

```
SELECT ?x WHERE {
        ?x frdf:membership ?Node1 .
        ?Node1 rdf:type ?Concept1 .
        ?Node1 frdf:ineqType ?IneqType1 .
        ?Node1 frdf:degree ?Degree1 .
        FILTER regex (?Concept1 , "CONCEPTS#Father")
        FILTER regex (?IneqType1 ,">")
        FILTER (?Degree1 >= "0.8^^xsd:float")

        ?BlankRole2 frdf:ineqType ?IneqType2 .
        ?BlankRole2 frdf:degree ?Degree2 .
        ?BlankRole2 rdf:type ?Role2 .
```

```
?x BlankRole2 ?y .
FILTER regex (?Role2 , "ROLES#isFriendOf")
FILTER regex (?IneqType1 ,">")
FILTER (?Degree2 >= "1^^xsd:float")

?y frdf:membership ?Node3,
?Node3 rdf:type ?Concept3 .
?Node3 frdf:ineqType ?IneqType3 .
?Node3 frdf:degree ?Degree3 .

FILTER regex (?Concept3 , "CONCEPTS#Short")
FILTER regex (?IneqType1 ,">")
FILTER (?Degree1 >= "0.7^^xsd:float")
}
```

Example 4. GFCQ are specified using the symbol ":" followed by the importance of participation for each condition statement. For example, we can ask for female persons having LongLegs and BeautifulEyes and rank higher those with larger degree for the latter:

```
x <- Female(x):1 ^ LongLegs(x) : 0.6  BeautifulEyes(x) : 0.8
```

In the case of GGCQs the operation is different. The SPARQL query is constructed in a way that retrieves the participation degrees of every role or concept used in the atoms criteria, for the results that satisfy all of the atoms. The participation degrees retrieved for each query atom together with its weight are then used by FiRE for the ranking procedure of the results based on the selected semantics. An excerpt of the SPARQL query for Example 4 follows.

```
SELECT ?x ?Degree1...
WHERE {
    ?x frdf:membership ?Node1 .
    ?Node1 rdf:type ?Concept1 .
    ?Node1 frdf:ineqType ?IneqType1 .
    ?Node1 frdf:degree ?Degree1 .
    FILTER regex (?Concept1 , "CONCEPTS#Female")
    FILTER regex (?IneqType1 ,">")
    FILTER (?Degree1 >= "0.0^^xsd:float")
    ...
}
```

Concluding this section, it should be noted that the proposed architecture (clearly) does not provide a complete query answering procedure for f-$\mathcal{SHIN}$, since queries are directly evaluated using the RDF triple-store and no f-$\mathcal{SHIN}$ reasoning is involved. However, prior to storing the information in the triple-store, FiRE applies a global GLB on the given ontology to materialize as much implied information as possible. This makes the proposed architecture complete for ground conjunctive queries, i.e., those queries that do not contain any non-distinguished variables, and that additionally do not allow for complex roles in the query body, i.e., roles that are transitive or inverse. On the one hand, this is a common practice for many proposed practical systems such as Jena,[5] OWLim,[6] and DLEJena[7], which first materialize implied knowledge and then store it into a triple-store, aiming at sacrificing some completeness in favour of performance. On the other hand, how to efficiently answer general conjunctive queries over expressive DLs is also still an open question and most DL reasoners, such as KAON2[8] and RacerPro[9] mainly only support ground conjunctive queries.

5   Evaluation

In the current section we present our use case scenario and the methods used to provide support through the previously presented architecture. First, we describe the use case and we show how we have fuzzified a certain number of database fields, in order to extract rich semantic information from numerical fields. Then we describe the models' knowledge base presenting the definitions of some new interesting concepts, which can be used to assist the process of casting.

5.1   Models Use Case

The data were taken from a production company database containing 2140 human models. The database contained information on each model regarding their height, age, body type, fitness type, tooth condition, eye-condition and color, hair quality, style, color and length, ending with the hands' condition. Apart from the above, there were some additional, special-appearance characteristics for certain models such as good-looking, sexy, smile, sporty, etc., introduced by the casting producer. Finally for a minority of models, a casting-video was stored in the database. The main objective of the production company was to pick a model, based on the above features, who would be suitable for a certain commercial spot. Furthermore, depending on the spot type, inquiries about models with some profession-like characteristics (like teacher, chef, mafia etc.) were also of interest.

Despite the fact that the database information on each model was relatively rich, there was great difficulty in querying models of appropriate characteristics and the usual casting process involved almost browsing over a large part of the database to find out for people matching the desired characteristics. One major issue is that information in the database is not semantically organised. Moreover, the organisation of the information in the various tables made searching for combined characteristics impossible. Additionally, the crisp approach of querying

---

[5] http://jena.sourceforge.net/
[6] http://www.ontotext.com/owlim
[7] http://lpis.csd.auth.gr/systems/DLEJena/
[8] http://kaon2.semanticweb.org/
[9] http://www.racer-systems.com/

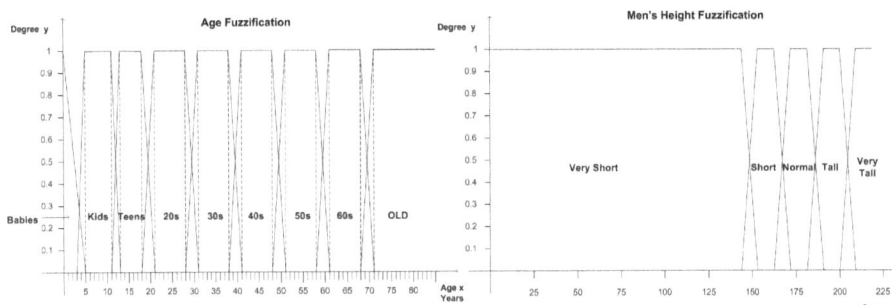

Fig. 2. Fuzzification graphs

over databases makes it difficult to perform queries over fields such as age, height and weight. If a person does not match exactly the specified criteria it will not be returned by the database. Hence, casting people usually set those criteria much lower than the wanted in order to avoid having a low recall and then browsed through the returned set to exclude false positive answers. Our goal in the current use case was to semantically enrich the database in order to make this process more easy. Moreover, we aimed at using the fuzzy technologies to alleviate the problem of precise matching of query criteria.

The process of constructing the fuzzy knowledge base was split into two parts. The first part involved the generation of the fuzzy *ABox*. The characteristics given by numerical values in the database, like height and age, were fuzzified giving rise to new (fuzzy) concepts, while remaining characteristics were used as crisp assertions. For example, the fuzzification process of the field age was performed by setting fuzzy partitions depending on age and by defining the concepts Kid, Teen, Baby, 20s, 30s, 40s, 50s, 60s and Old. As can be observed from the age fuzzification graph, a model who is 29 years old participates in both concepts 20s and 30s with degrees 0.35 and 0.65 respectively. Similarly for the fuzzification of field height, the concepts Very_Short, Short, Normal_Height, Tall and Very_Tall were defined. In the case of the height, the fuzzy partition used for female models was different from the one used for males, since the average height of females is lower than that of males. The fuzzification graphs of age and men's height are shown in Figure 2. An example of the produced assertions is shown in Example 5.

Example 5. An excerpt of the ABox for the model *michalis*1539:

$\langle michalis1539 : 20s \geq 0.66 \rangle$, $\langle michalis1539 : 30s \geq 0.33 \rangle$,

$\langle michalis1539 : Normal\_Height \geq 0.5 \rangle$,

$\langle michalis1539 : Tall > 0.5 \rangle$, $\langle michalis1539 : GoodLooking \geq 1 \rangle$,

$\langle (michalis1539, good) : has - toothCondition \geq 1 \rangle$,

$\langle good : Good \geq 1 \rangle$

Table 2. Concepts and Roles defined for the various characteristics (fields in the database)

Concepts	
Gender:	Male, Female
Height:	Very_Short, Short, Normal_Height, Tall, Very_Tall
Age:	Baby, Kid, Teen, 20s,30s, 40s, 50s, 60s, Old
Body Type:	Fat, Normal, Perfect, Plum, Slim
Fitness:	Fit, Unfit, PerfectFitness
Hair Length:	Long, Medium, Shaved, Tonsure, Bold
Hair Color:	Black, BrownLight, Brown, BlondLight, Grey, BlondDark, BlondRed, BrownRed, Blond, Red, White, Brown − Grey, 2 − Colour, Platinum, Black − Grey
Hair Style:	Wavy, Straight, Curly, Extreme, Rasta, Frizy
Hair Quality:	Natural, Dyed, Highlights
Eyes Color:	Green, Blue, BlueGray, Hazel, BlueGreen
Eyes Condition:	Healthy, NotHealthy, Glasses, ContactLenses
Tooth Condition:	MissingTooth, Brace, Good, Bad, MissingTeeth
Hands Condition:	Excellent, Average
Special Characteristics:	Sexy, GoodLooking, Smile, Sporty, Ugly, Serious, Funny, Tough, Aristocrat, Artistic, Folklor, Romantic, Elegant, Classic, CleanCut, Underground
Roles	
has − hairLength, has − hairColour, has − eyeColor, has − eyeCondition, has − handsCondition	

5.2   The Fuzzy Knowledge Base

In order to permit knowledge-based retrieval of human models we have implemented an expressive terminology for a fuzzy knowledge base. The alphabet of concepts used for the fuzzy knowledge base consists of the features described above while some characteristics like hair length, hair condition etc. were represented by the use of roles. The most important of roles and concepts are shown below. In the concept set the concept features are highlighted in bold.

The set of individuals consist of the models name along with an ID.

The effective extraction of implicit knowledge from the explicit one requires an expressive terminology capable of defining higher concepts. In our case the higher domain concepts defined for human models lie into five categories: age, height, family, some special categories and the professions. Hence, the profession Scientist has been defined as male, between their 50s or 60s, with classic appearance who also wears glasses. In a similar way we have defined 33 domain concepts; an excerpt of the *TBox* can be found in Table 3.

5.3   Results

All the experiments were conducted under Windows XP on a Pentium 2.40 GHz computer with 2. GB of RAM.

The described fuzzy knowledge base was used in the evaluation of our approach. Implicit knowledge was extracted using the greatest lower bound service of FiRE, asking for the degree of participation of all individuals, in all the defined

Table 3. An excerpt of the Knowledge Base (*TBox*)

$\mathcal{T} = \{$MiddleAged $\equiv$ 40s $\sqcup$ 50s,
TallChild $\equiv$ Child $\sqcap$ (Short $\sqcup$ Normal_Height),
Father $\equiv$ Male $\sqcap$ (30s $\sqcup$ MiddleAged),
Legs $\equiv$ Female $\sqcap$ (Normal_Height $\sqcup$ Tall)
$\sqcap$(Normal $\sqcup$ Perfect) $\sqcap$ (Fit $\sqcup$ PerfectFitness),
Teacher $\equiv$ (30s $\sqcup$ MiddleAged) $\sqcap$ Elegant $\sqcap$ Classic,
Fisherman $\equiv$ Male $\sqcap$ Folklor $\sqcap$ (MiddleAged $\sqcup$ ThirdAge)
$\sqcap\exists$has $-$ handsCondition.(Average $\sqcup$ Bad),
Military $\equiv$ Male $\sqcap$ Tough $\sqcap$ Serious $\sqcap$ _BODY
$\sqcap\exists$has $-$ hairLength.(Short $\sqcap$ Shaved),
Scientist $\equiv$ Male $\sqcap$ Classic $\sqcap$ (50s $\sqcup$ 60s)
$\sqcap$Serious $\sqcap \exists$has $-$ eyeCondition.Glasses $\}$
Butler $\equiv$ Male $\sqcap$ Aristocrat $\sqcap$ Serious $\sqcap$ Elegant
$\sqcap$(30s $\sqcup$ MiddleAged) $\sqcap \exists$has $-$ hairLength.Short ,
Butcher $\equiv$ Male $\sqcap$ Tough $\sqcap$ Folklor $\sqcap$ (30s $\sqcup$ 40s) ,
TaxiDriver $\equiv$ Folklor $\sqcap$ (30s $\sqcup$ 40s) ,
Secretary $\equiv$ Female $\sqcap$ Classic $\sqcap$ Adult $\sqcap$ Sexy ,
HouseCleaner $\equiv$ Female $\sqcap$ Folklor $\sqcap$ (30s $\sqcup$ 40s) $\}$

domain concepts. The average number of explicit assertions per individual was 13 while the defined concepts were 33, that together with the 2140 individuals (i.e entries of the database) resulted to 29460 explicit assertions and the extraction of 2430 implicit. These results, together with concept and role axioms, were stored to a Sesame repository using the proposed fuzzy OWL triples syntax to form a repository of 529.926 triples.

The average time for the GLB reasoning process and the conversion of explicit and implicit knowledge to fuzzy OWL syntax in triples was 1112 milliseconds. The time required for uploading the knowledge to a Sesame repository depends on the type of repository (Memory or Native) and also on repository's size. Based on our experiments, we have observed that the upload time is polynomial to the size of the repository but without significant differences. Therefore, the average minimum upload time to an almost empty repository (0-10.000 triples) is 213 milliseconds while the average maximum upload time to a full repository (over 500.000 triples) is 700 milliseconds.

FiRE and Sesame were also tested on expressive fuzzy queries. We have used the following set of test queries:

$$q_1 : \ x \leftarrow \text{Scientist}(x)$$
$$q_2 : \ x \leftarrow \text{Father}(x) \geq 1 \wedge \text{Teacher}(x) \geq 0.8$$
$$q_3 : \ x \leftarrow \text{Legs}(x) \geq 1 \wedge \text{Eyes}(x) \geq 0.8 \wedge 20s(x) \geq 0.5$$
$$q_4 : \ x \leftarrow \text{Scientist}(x) : 0.8$$
$$q_5 : \ x \leftarrow \text{Father}(x) : 0.6 \wedge \text{Teacher}(x) : 0.7$$
$$q_6 : \ x \leftarrow \text{Legs}(x) : 0.8 \wedge \text{Eyes}(x) : 1 \wedge 20s(x) : 0.6$$

Table 4. Evaluation of the fuzzy queries (time in ms)

Query	Native			Memory		
	100.000	250.000	500.000	100.000	250.000	500.000
q_1	1042	2461	3335	894	2364	3332
q_2	1068	2694	3935	994	2524	3732
q_3	3667	8752	21348	4267	7348	9893
q_4	2562	4173	5235	3042	4543	6027
q_5	4318	6694	8935	4341	7896	9306
q_6	9906	29831	66251	12164	16421	20489

The performance in this case mainly depended on the complexity of the query but also on the type and size of the repository. Queries using role names in combination with large repositories can dramatically slow down the response. Table 4 illustrates the response times in milliseconds using both types of repositories and different repository sizes. Repository sizes was set by adjusting the number of assertions. As it can be observed, very expressive queries seeking for young female models with beautiful legs and eyes as well as long hair, a popular demand in commercial spots, can be easily performed. It is worth mentioning that these queries consist of higher domain concepts defined in our fuzzy knowledge base.

Since our system is not a sound and complete query answering system for f-$\mathcal{SHIN}$, the GLB service performed before uploading the triples is employed in order to use as much of the expressivity of the language as possible producing new implied assertions.

Furthermore, the results regarding query answering time are also very encouraging, at least for the specific application. Although, compared to crisp querying, over crisp knowledge bases, our method might require several more seconds to be answered (mainly due to post processing steps for GFCQs or due to very lengthy SPARQL queries for CTQs) this time is significantly less, compared to the time spent by producers on casting (usually counted in days), since they usually have to browse through a very large number of videos and images before they decide.

6   Conclusions

Due to the fact that imperfect information is inherent in many web-based applications, the effective management of imperfect knowledge is very important for the realisation of the Semantic Web. In this paper, we have proposed an architecture that can be used for storing and querying fuzzy knowledge bases for the Semantic Web. Our proposal which is based on the DL f-$\mathcal{SHIN}$, integrates the Sesame RDF triple-store (through a proposed serialisation of f-OWL to RDF triples), the fuzzy reasoning engine FiRE and implements very expressive fuzzy queries on top of them.

The proposed architecture was evaluated using an industrial application scenario about casting for TV commercials and spots. The obtained results are very promising from the querying perspective. From the initial 29460 explicit assertions made by database instances for models, 2430 new implicit assertions where extracted and both uploaded in the Sesame repository. In this way expressive semantic queries like "Find me young female models with beautiful legs and eyes as well as long hair", that might have proved very difficult or even impossible using the producing company's database, are applicable through FiRE. This reveals both the strength of knowledge-based applications, and technologies for managing fuzzy knowledge, since a wealth of the information of the databases, like age, height, as well as many high level concepts of the specific application, like "beautiful eyes", "perfect fitness" and "scientist look" are inherently fuzzy.

As far as future directions are concerned, we intend to further investigate on different ways of performing queries using expressive fuzzy description logics. Finally, it would be of great interest to attempt a comparison between the proposed architecture and approaches using fuzzy DL-lite ontologies and approximation.

References

1. Baader, F., McGuinness, D., Nardi, D., Patel-Schneider, P.F.: The Description Logic Handbook: Theory, implementation and applications. Cambridge University Press (2002)
2. Bobillo, F., Delgado, M., Gómez-Romero, J.: Crisp representations and reasoning for fuzzy ontologies. International Journal of Uncertainty, Fuzziness and Knowledge-Based Systems 17(4), 501–530 (2009)
3. Bobillo, F., Straccia, U.: fuzzydl: An expressive fuzzy description logic reasoner. In: Proceedings of the 17th IEEE International Conference on Fuzzy Systems, FUZZ-IEEE 2008, pp. 923–930 (2008)
4. Bobillo, F., Straccia, U.: Fuzzy ontology representation using owl 2. International Journal of Approximate Reasoning 52(7), 1073–1094 (2011)
5. Chortaras, A., Stamou, G., Stafylopatis, A.: Adaptation of Weighted Fuzzy Programs. In: Kollias, S.D., Stafylopatis, A., Duch, W., Oja, E. (eds.) ICANN 2006. LNCS, vol. 4132, pp. 45–54. Springer, Heidelberg (2006)
6. Cross, V.: Fuzzy information retrieval. Journal of Intelligent Information Systems 3, 29–56 (1994)
7. Bobillo, F., Delgado, M., Gómez-Romero, J.: A Crisp Representation for Fuzzy $\mathcal{SHOIN}$ with Fuzzy Nominals and General Concept Inclusions. In: da Costa, P.C.G., d'Amato, C., Fanizzi, N., Laskey, K.B., Laskey, K.J., Lukasiewicz, T., Nickles, M., Pool, M. (eds.) URSW 2005-2007. LNCS (LNAI), vol. 5327, pp. 174–188. Springer, Heidelberg (2008)
8. Horrocks, I., Sattler, U., Tobies, S.: Reasoning with Individuals for the Description Logic $\mathcal{SHIQ}$. In: McAllester, D. (ed.) CADE 2000. LNCS (LNAI), vol. 1831, pp. 482–496. Springer, Heidelberg (2000)
9. Klir, G.J., Yuan, B.: Fuzzy Sets and Fuzzy Logic: Theory and Applications. Prentice-Hall (1995)
10. Li, Y., Xu, B., Lu, J., Kang, D.: Discrete tableau algorithms for $\mathcal{FSHI}$. In: Proceedings of the International Workshop on Description Logics, DL 2006, Lake District, UK (2006)
11. Lukasiewicz, T., Straccia, U.: Managing uncertainty and vagueness in description logics for the semantic web. Web Semant. 6(4), 291–308 (2008)

12. Mazzieri, M., Dragoni, A.F.: A fuzzy semantics for semantic web languages. In: ISWC-URSW, pp. 12–22 (2005)

13. Pan, J.Z., Stamou, G., Stoilos, G., Thomas, E.: Expressive querying over fuzzy DL-Lite ontologies. In: Proceedings of the International Workshop on Description Logics, DL 2007 (2007)

14. Patel-Schneider, P.F., Hayes, P., Horrocks, I.: OWL Web Ontology Language Semantics and Abstract Syntax. Technical report, W3C. W3C Recommendation (February 2004), http://www.w3.org/TR/2004/REC-owl-semantics-20040210/

15. Prud'hommeaux, E., Seaborne, A.: SPARQL query language for RDF. W3C Working Draft (2006), http://www.w3.org/TR/rdf-sparql-query/

16. Schneider, P.P., Swartout, B.: Description-logic knowledge representation system specification from the KRSS group of the ARPA knowledge sharing effort (1993), http://www.bell-labs.com/user/pfps/papers/krss-spec.ps

17. Simou, N., Kollias, S.: Fire: A fuzzy reasoning engine for impecise knowledge. K-Space PhD Students Workshop, Berlin, Germany (September 14, 2007)

18. Stoilos, G., Stamou, G., Tzouvaras, V., Pan, J.Z., Horrocks, I.: Reasoning with very expressive fuzzy description logics. Journal of Artificial Intelligence Research 30(5), 273–320 (2007)

19. Stoilos, G., Straccia, U., Stamou, G., Pan, J.Z.: General concept inclusions in fuzzy description logics. In: European Conference on Artificial Intelligence, ECAI 2006, Riva del Garda, Italy (2006)

20. Straccia, U.: Reasoning within fuzzy description logics. Journal of Artificial Intelligence Research 14, 137–166 (2001)

21. Straccia, U.: Answering vague queries in fuzzy DL-Lite. In: Proceedings of the 11th International Conference on Information Processing and Management of Uncertainty in Knowledge-Based Systems, IPMU 2006, pp. 2238–2245 (2006)

22. Straccia, U., Visco, G.: DLMedia: an ontology mediated multimedia information retrieval system. In: Proceeedings of the International Workshop on Description Logics, DL 2007, Insbruck, Austria, vol. 250. CEUR (2007)

23. Vaneková, V., Bella, J., Gurský, P., Horváth, T.: Fuzzy RDF in the semantic web: Deduction and induction. In: Proceedings of Workshop on Data Analysis, WDA 2005, pp. 16–29 (2005)

Transforming Fuzzy Description Logic $\mathcal{ALC}_{\mathcal{FL}}$ into Classical Description Logic $\mathcal{ALCH}$

Yining Wu

University of Luxembourg, Luxembourg

Abstract. In this paper, we present a satisfiability preserving transformation of the fuzzy Description Logic $\mathcal{ALC}_{\mathcal{FL}}$ into the classical Description Logic $\mathcal{ALCH}$. We can use the already existing DL systems to do the reasoning of $\mathcal{ALC}_{\mathcal{FL}}$ by applying the result of this paper. This work is inspired by Straccia, who has transformed the fuzzy Description Logic f$\mathcal{ALCH}$ into the classical Description Logic $\mathcal{ALCH}$.

1 Introduction

The Semantic Web is a vision for the future of the Web in which information is given explicit meaning, making it easier for machines to automatically process and integrate information available on the Web. While as a basic component of the Semantic Web, an ontology is a collection of information and is a document or file that formally defines the relations among terms. OWL[1] is a *Web Ontology Language* and is intended to provide a language that can be used to describe the classes and relations between them that are inherent in Web documents and applications. The OWL language provides three increasingly expressive sublanguages: OWL Lite, OWL DL, OWL Full. OWL DL is so named due to its correspondence with description logics. OWL DL was designed to support the existing Description Logic business segment and has desirable computational properties for reasoning systems. According to the corresponding relation between axioms of OWL ontology and terms of Description Logic, we can represent the knowledge base contained in the ontology in syntax of DLs.

Description Logics (DLs) [1] have been studied and applied successfully in a lot of fields. The concepts in classical DLs are usually interpreted as crisp sets, i.e., an individual either belongs to the set or not. In the real world, the answers to some questions are often not only yes or no, rather we may say that an individual is an instance of a concept only to some certain degree. We often say linguistic terms such as "Very", "More or Less" etc. to distinguish, e.g. between a young person and a very young person. In 1970s, the theory of approximate reasoning based on the notions of linguistic variable and fuzzy logic was introduced and developed by Zadeh [21–23]. Adverbs as "Very", "More or Less" and "Possibly"

---

[1] Please visit http://www.w3.org/TR/owl-guide/ for more details.

F. Bobillo et al. (Eds.): URSW 2008-2010/UniDL 2010, LNAI 7123, pp. 177–196, 2013.

are called hedges in fuzzy DLs. Some approaches to handling uncertainty and vagueness in DL for the Semantic Web are described in [12].

A well known feature of DLs is the emphasis on reasoning as a central service. Some reasoning procedures for fuzzy DLs have been proposed in [18]. A transformation of $f\mathcal{ALCH}$ into $\mathcal{ALCH}$ has been presented by Straccia [19]. T

In this paper we consider the fuzzy linguistic description logic $\mathcal{ALC}_{\mathcal{FL}}$ [9] which is an instance of the description logic framework $\mathcal{L} - \mathcal{ALC}$ with the certainty lattice characterized by a hedge algebra (HA) and allows the modification by hedges. Because the certainty lattice is characterized by a HA, the modification by hedges becomes more natural than that in $\mathcal{ALC}_{\mathcal{FH}}$ [10] and $\mathcal{ALC}_{\mathcal{FLH}}$ [16] which extend fuzzy $\mathcal{ALC}$ by allowing the modification by hedges of HAs. We will present a satisfiability preserving transformation of $\mathcal{ALC}_{\mathcal{FL}}$ into $\mathcal{ALCH}$ which makes the reuse of the technical results of classical Dls for $\mathcal{ALC}_{\mathcal{FL}}$ feasible.

The remaining part of this paper is organized in the following way. First we state some preliminaries on $\mathcal{ALCH}$, hedge algebra and $\mathcal{ALC}_{\mathcal{FL}}$. Then we present the transformation of $\mathcal{ALC}_{\mathcal{FL}}$ into $\mathcal{ALCH}$. Finally we discuss the main result of the paper and identify some possibilities for further work.

2   Preliminaries

2.1 $\mathcal{ALCH}$

We consider the language $\mathcal{ALCH}$ (Attributive Language with Complement and role Hierarchy). In abstract notation, we use the letters A and B for concept names, the letter R for role names, and the letters C and D for concept terms.

Definition 1. *Let N_R and N_C be disjoint sets of role names and concept names. Let $A \in N_C$ and $R \in N_R$. Concept terms in $\mathcal{ALCH}$ are formed according to the following syntax rule:*

$$A|\top|\bot|C \sqcap D|C \sqcup D|\neg C|\forall R.C|\exists R.C$$

The semantics of concept terms are defined formally by interpretations.

Definition 2. *An interpretation $\mathcal{I}$ is a pair $(\Delta^{\mathcal{I}}, \cdot^{\mathcal{I}})$, where $\Delta^{\mathcal{I}}$ is a nonempty set (interpretation domain) and $\cdot^{\mathcal{I}}$ is an interpretation function which assigns to each concept name A a set $A^{\mathcal{I}} \subseteq \Delta^{\mathcal{I}}$ and to each role name R a binary relation $R^{\mathcal{I}} \subseteq \Delta^{\mathcal{I}} \times \Delta^{\mathcal{I}}$. The interpretation of complex concept terms is extended by the following inductive definitions:*

$$\top^{\mathcal{I}} = \Delta^{\mathcal{I}}$$
$$\bot^{\mathcal{I}} = \emptyset$$
$$(C \sqcap D)^{\mathcal{I}} = C^{\mathcal{I}} \cap D^{\mathcal{I}}$$
$$(C \sqcup D)^{\mathcal{I}} = C^{\mathcal{I}} \cup D^{\mathcal{I}}$$
$$(\neg C)^{\mathcal{I}} = \Delta^{\mathcal{I}} \setminus C^{\mathcal{I}}$$
$$(\forall R.C)^{\mathcal{I}} = \{d \in \Delta^{\mathcal{I}} \mid \forall d'.(d, d') \notin R^{\mathcal{I}} \text{ or } d' \in C^{\mathcal{I}}\}$$
$$(\exists R.C)^{\mathcal{I}} = \{d \in \Delta^{\mathcal{I}} \mid \exists d'.(d, d') \in R^{\mathcal{I}} \text{ and } d' \in C^{\mathcal{I}}\}$$

A concept term C is *satisfiable* iff there exists an interpretation $\mathcal{I}$ such that $C^{\mathcal{I}} \neq \emptyset$, denoted by $\mathcal{I} \models C$. Two concept terms C and D are *equivalent* (denoted by $C \equiv D$) iff $C^{\mathcal{I}} = D^{\mathcal{I}}$ for all interpretation $\mathcal{I}$.

We have seen how we can form complex descriptions of concepts to describe classes of objects. Now, we introduce *terminological axioms*, which make statements about how concept terms and roles are related to each other respectively.

In the most general case, *terminological axiom* have the form $C \sqsubseteq D$ or $R \sqsubseteq S$, where C, D are concept terms, R, S are role names. This kind of terminological axioms are also called *inclusions*. A set of axioms of the form $R \sqsubseteq S$ is called *role hierarchy*. An interpretation $\mathcal{I}$ *satisfies* an inclusion $C \sqsubseteq D$ ($R \sqsubseteq S$) iff $C^{\mathcal{I}} \subseteq D^{\mathcal{I}}$ ($R^{\mathcal{I}} \subseteq S^{\mathcal{I}}$), denoted by $\mathcal{I} \models C \sqsubseteq D$ ($\mathcal{I} \models R \sqsubseteq S$).

A *terminology*, i.e., *TBox*, is a finite set of terminological axioms. An interpretation $\mathcal{I}$ *satisfies* (is a *model* of) a terminology $\mathcal{T}$ iff $\mathcal{I}$ *satisfies* each element in $\mathcal{T}$, denoted by $\mathcal{I} \models \mathcal{T}$.

Assertions define how individuals relate with each other and how individuals relate with concept terms. Let N_I be a set of individual names which is disjoint to N_R and N_C. An *assertion* α is an expression of the form $a : C$ or $(a, b) : R$, where $a, b \in N_I$, $R \in N_R$ and $C \in N_C$. A finite set of *assertions* is called *ABox*. An interpretation $\mathcal{I}$ *satisfies* a concept assertion $a : C$ iff $a^{\mathcal{I}} \in C^{\mathcal{I}}$, denoted by $\mathcal{I} \models a : C$. $\mathcal{I}$ *satisfies* a role assertion $(a, b) : R$ iff $(a^{\mathcal{I}}, b^{\mathcal{I}}) \in R^{\mathcal{I}}$, denoted by $\mathcal{I} \models (a, b) : R$. An interpretation $\mathcal{I}$ *satisfies* (is a *model* of) an ABox $\mathcal{A}$ iff $\mathcal{I}$ *satisfies* each assertion in $\mathcal{A}$, denoted by $\mathcal{I} \models \mathcal{A}$.

A *knowledge base* is of the form $\langle \mathcal{T}, \mathcal{A} \rangle$ where $\mathcal{T}$ is a TBox and $\mathcal{A}$ is an ABox. An interpretation $\mathcal{I}$ *satisfies* (is a *model* of, denoted by $\mathcal{I} \models \mathcal{K}$) a knowledge base $\mathcal{K} = \langle \mathcal{T}, \mathcal{A} \rangle$ iff $\mathcal{I}$ *satisfies* both $\mathcal{T}$ and $\mathcal{A}$. We say that a knowledge base $\mathcal{K}$ *entails* an assertion α, denoted $\mathcal{K} \models \alpha$ iff each model of $\mathcal{K}$ satisfies α. Furthermore, let $\mathcal{T}$ be a TBox and let C, D be two concept terms. We say that D *subsumes* C with respect to $\mathcal{T}$ (denoted by $C \sqsubseteq_{\mathcal{T}} D$) iff for each model of $\mathcal{T}$, $\mathcal{I} \models C^{\mathcal{I}} \subseteq D^{\mathcal{I}}$.

The problem of determining whether $\mathcal{K} \models \alpha$ is called *entailment problem*; the problem of determining whether $C \sqsubseteq_{\mathcal{T}} D$ is called *subsumption problem*; and the problem of determining whether $\mathcal{K}$ is satisfiable is called *satisfiability problem*. Entailment problem and subsumption problem can be reduced to satisfiability problem.

2.2 Linear Symmetric Hedge Algebra

In this section, we introduce linear symmetric Hedge Algebras (HAs). For general HAs, please refer to [13–15].

Let us consider a linguistic variable $TRUTH$ with the domain $dom(TRUTH) =$ {*True, False, VeryTrue, VeryFalse,MoreTrue, MoreFalse, PossiblyTrue*, ...}. This domain is an infinite partially ordered set, with a natural ordering $a < b$ meaning that b describes a larger degree of truth if we consider $True > False$. This set is generated from the basic elements (*generators*) $G = \{ True, False \}$ by using *hedges*, i.e., unary operations from a finite set $H = \{ Very, Possibly, More \}$. The $dom(TRUTH)$ which is a set of linguistic values can be represented as $X = \{ \delta c \mid c \in G, \delta \in H^* \}$ where H^* is the Kleene star of H, From the algebraic point of view, the truth domain can be described as an abstract algebra $AX = (X, G, H, >)$.

To define relations between hedges, we introduce some notations first. We define that $H(x) = \{\sigma x \mid \sigma \in H^*\}$ for all $x \in X$. Let I be the identity hedge, i.e., $\forall x \in X.Ix = x$. The identity I is the least element. Each element of H is an *ordering operation*, i.e., $\forall h \in H, \forall x \in X$, either $hx > x$ or $hx < x$.

Definition 3. *[14] Let $h, k \in H$ be two hedges, for all $x \in X$ we define:*

- h, k are converse *if $hx < x$ iff $kx > x$;*
- h, k are compatible *if $hx < x$ iff $kx < x$;*
- h modifies terms stronger or equal than k, *denoted by $h \geq k$ if $hx \leq kx \leq x$ or $hx \geq kx \geq x$;*
- $h > k$ *if $h \geq k$ and $h \neq k$;*
- h is positive wrt k *if $hkx < kx < x$ or $hkx > kx > x$;*
- h is negative wrt k *if $kx < hkx < x$ or $kx > hkx > x$.*

$\mathcal{ALC}_{\mathcal{FL}}$ only considers symmetric HAs, i.e., there are exactly two generators as in the example $G = \{\mathit{True}, \mathit{False}\}$. Let $G = \{c^+, c^-\}$ where $c^+ > c^-$. c^+ and c^- are called *positive* and *negative generators* respectively. Because there are only two generators, the relations presented in Definition 3 divides the set H into two subsets $H^+ = \{h \in H \mid hc^+ > c^+\}$ and $H^- = \{h \in H \mid hc^+ < c^+\}$, i.e., every operation in H^+ is converse w.r.t. any operation in H^- and vice-versa, and the operations in the same subset are compatible with each other.

Definition 4. *[9] An abstract algebra $AX = (X, G, H, >)$, where $H \neq \emptyset, G = \{c^+, c^-\}$ and $X = \{\sigma c \mid c \in G, \sigma \in H^*\}$ is called a* linear symmetric hedge algebra *if it satisfies the properties* (A1)-(A5).

(A1) Every hedge in H^+ is a converse operation of all operations in H^-.
(A2) Each hedge operation is either positive or negative w.r.t. the others, including itself.
(A3) The sets $H^+ \cup \{I\}$ and $H^- \cup \{I\}$ are linearly ordered with the I.
(A4) If $h \neq k$ and $hx < kx$ then $h'hx < k'kx$, for all $h, k, h', k' \in H$ and $x \in X$.
(A5) If $u \notin H(v)$ and $u \leq v$ ($u \geq v$) then $u \leq hv$ ($u \geq hv$), for any hedge h and $u, v \in X$.

Let $AX = (X, G, H, >)$ be a linear symmetric hedge algebra and $c \in G$. We define that, $\bar{c} = c^+$ if $c = c^-$ and $\bar{c} = c^-$ if $c = c^+$. Let $x \in X$ and $x = \sigma c$, where $\sigma \in H^*$. The *contradictory element* to x is $y = \sigma \bar{c}$ written $y = -x$.

[14] gave us the following proposition to compare elements in X.

Proposition 5. *Let $AX = (X, G, H, >)$ be a linear symmetric HA, $x = h_n \cdots h_1 u$ and $y = k_m \cdots k_1 u$ are two elements of X where $u \in X$. Then there exists an index $j \leq \min\{n, m\} + 1$ such that $h_i = k_i$ for all $i < j$, and*

(i) *$x < y$ iff $h_j x_j < k_j x_j$, where $x_j = h_{j-1} \cdots h_1 u$;*
(ii) *$x = y$ iff $n = m = j$ and $h_j x_j = k_j x_j$.*

In order to define the semantics of the hedge modification, we only consider monotonic HAs defined in [9] which also extended the order relation on $H^+ \cup \{I\}$ and $H^- \cup \{I\}$ to one on $H \cup \{I\}$. We will use "hedge algebra" instead of "linear symmetric hedge algebra" in the rest of this paper.

2.3   Inverse Mapping of Hedges

Fuzzy description logics represent the assessment "It is true that Tom is very old" by

$$(VeryOld)^{\mathcal{I}}(Tom)^{\mathcal{I}} = True. \tag{1}$$

In a fuzzy linguistic logic [21–23], the assessment "It is true that Tom is very old" and the assessment "It is very true that Tom is old" are equivalent, which means

$$(Old)^{\mathcal{I}}(Tom)^{\mathcal{I}} = VeryTrue, \tag{2}$$

and (1) has the same meaning. (In other word, a fuzzy interpretation $\mathcal{I}$ (Definition 8) satisfies an assertion $Tom : VeryOld \geq True$ if and only if $\mathcal{I}$ satisfies the assertion $Tom : Old \geq VeryTrue$.) This signifies that the modifier can be moved from concept term to truth value and vice versa. For any $h \in H$ and for any $\sigma \in H^*$, the rules of moving hedges [13] are as follows,

$$RT1 : (hC)^{\mathcal{I}}(d) = \sigma c \rightarrow (C)^{\mathcal{I}}(d) = \sigma h c$$
$$RT2 : (C)^{\mathcal{I}}(d) = \sigma h c \rightarrow (hC)^{\mathcal{I}}(d) = \sigma c.$$

where C is a concept term and $d \in \Delta^{\mathcal{I}}$.

Definition 6. *[9] Consider a monotonic HA $AX = (X, \{c^+, c^-\}, H, >)$ and a $h \in H$. A mapping $h^- : X \rightarrow X$ is called an* inverse mapping *of h iff it satisfies the following two properties,*

1. $h^-(\sigma h c) = \sigma c$.
2. $\sigma_1 c_1 > \sigma_2 c_2 \Leftrightarrow h^-(\sigma_1 c_1) > h^-(\sigma_2 c_2)$.

where $c, c_1, c_2 \in G$, $h \in H$ and $\sigma_1, \sigma_2 \in H^*$.

2.4 $\mathcal{ALC}_{\mathcal{FL}}$

$\mathcal{ALC}_{\mathcal{FL}}$ is a Description Logic in which the truth domain of interpretations is represented by a hedge algebra. The syntax of $\mathcal{ALC}_{\mathcal{FL}}$ is similar to that of $\mathcal{ALCH}$ except that $\mathcal{ALC}_{\mathcal{FL}}$ allows concept modifiers and does not include role hierarchy.

Definition 7. *Let H be a set of hedges. Let A be a concept name and R a role, complex concept terms denoted by C, D in $\mathcal{ALC}_{\mathcal{FL}}$ are formed according to the following syntax rule:*

$$A \mid \top \mid \bot \mid C \sqcap D \mid C \sqcup D \mid \neg C \mid \delta C \mid \forall R.C \mid \exists R.C$$

where $\delta \in H^*$.

In [15], HAs are extended by adding two artificial hedges inf and sup defined as $\inf(x) = \text{infimum}(H(x))$, $\sup(x) = \text{supremum}(H(x))$. If $H = \emptyset$, $H(c^+)$ and $H(c^-)$ are infinite, according to [15] $\inf(c^+) = \sup(c^-)$. Let $W = \inf(\textit{True}) = \sup(\textit{False})$ and let $\sup(\textit{True})$ and $\inf(\textit{False})$ be the greatest and the least elements of X respectively.

The semantics is based on the notion of interpretations.

Definition 8. *Let AX be a monotonic HA such that $AX = (X, \{\textit{True}, \textit{False}\}, H, >)$. A fuzzy interpretation (f-interpretation) $\mathcal{I}$ for $\mathcal{ALC}_{\mathcal{FL}}$ is a pair $(\Delta^{\mathcal{I}}, \cdot^{\mathcal{I}})$, where $\Delta^{\mathcal{I}}$ is a nonempty set and $\cdot^{\mathcal{I}}$ is an interpretation function mapping:*

- *individuals to elements in $\Delta^{\mathcal{I}}$;*
- *a concept C into a function $C^{\mathcal{I}} : \Delta^{\mathcal{I}} \to X$;*
- *a role R into a function $R^{\mathcal{I}} : \Delta^{\mathcal{I}} \times \Delta^{\mathcal{I}} \to X$.*

For all $d \in \Delta^{\mathcal{I}}$ the interpretation function satisfies the following equations

$$\top^{\mathcal{I}}(d) = \sup(\textit{True}),$$
$$\bot^{\mathcal{I}}(d) = \inf(\textit{False}),$$
$$(\neg C)^{\mathcal{I}}(d) = -C^{\mathcal{I}}(d),$$
$$(C \sqcap D)^{\mathcal{I}}(d) = \min(C^{\mathcal{I}}(d), D^{\mathcal{I}}(d)),$$
$$(C \sqcup D)^{\mathcal{I}}(d) = \max(C^{\mathcal{I}}(d), D^{\mathcal{I}}(d)),$$
$$(\delta C)^{\mathcal{I}}(d) = \delta^-(C^{\mathcal{I}}(d)),$$
$$(\forall R.C)^{\mathcal{I}}(d) = \inf_{d' \in \Delta^{\mathcal{I}}}\{\max(-R^{\mathcal{I}}(d, d'), C^{\mathcal{I}}(d'))\},$$
$$(\exists R.C)^{\mathcal{I}}(d) = \sup_{d' \in \Delta^{\mathcal{I}}}\{\min(R^{\mathcal{I}}(d, d'), C^{\mathcal{I}}(d'))\},$$

where $-x$ is the contradictory element of x, and δ^- is the inverse of the hedge chain δ.

Definition 9. *A* fuzzy assertion (fassertion) *is an expression of the form $\langle \alpha \bowtie \sigma c \rangle$ where α is of the form $a : C$ or $(a, b) : R$, $\bowtie \in \{\geq, >, \leq, <\}$ and $\sigma c \in X$.*

Formally, an f-interpretation $\mathcal{I}$ satisfies a fuzzy assertion $\langle a : C \geq \sigma c \rangle$ (respectively $\langle (a, b) : R \geq \sigma c \rangle$) iff $C^{\mathcal{I}}(a^{\mathcal{I}}) \geq \sigma c$ (respectively $R^{\mathcal{I}}(a^{\mathcal{I}}, b^{\mathcal{I}}) \geq \sigma c$). An f-interpretation $\mathcal{I}$ satisfies a fuzzy assertion $\langle a : C \leq \sigma c \rangle$ (respectively $\langle (a, b) : R \leq \sigma c \rangle$) iff $C^{\mathcal{I}}(a^{\mathcal{I}}) \leq \sigma c$ (respectively $R^{\mathcal{I}}(a^{\mathcal{I}}, b^{\mathcal{I}}) \leq \sigma c$). Similarly for $>$ and $<$.

Concerning terminological axioms, an $\mathcal{ALC}_{\mathcal{FL}}$ terminology axiom is of the form $C \sqsubseteq D$, where C and D are $\mathcal{ALC}_{\mathcal{FL}}$ concept terms. From a semantics point of view, a f-interpretation $\mathcal{I}$ satisfies a fuzzy concept inclusion $C \sqsubseteq D$ iff $\forall d \in \Delta^{\mathcal{I}}.C^{\mathcal{I}}(d) \leq D^{\mathcal{I}}(d)$. Two concept terms C, D are said to be *equivalent*, denoted by $C \equiv D$ iff $C^{\mathcal{I}} = D^{\mathcal{I}}$ for all f-interpretations $\mathcal{I}$. Some properties concerning the hedge modification are showed in the following proposition [9].

Proposition 10. *We have the following semantical equivalence:*

$$\delta(C \sqcap D) \equiv \delta(C) \sqcap \delta(D)$$
$$\delta(C \sqcup D) \equiv \delta(C) \sqcup \delta(D)$$
$$\delta_1(\delta_2 C) \equiv (\delta_1 \delta_2)C.$$

A *fuzzy knowledge base* (f*KB*) is $\langle \mathcal{T}, \mathcal{A} \rangle$, where $\mathcal{T}$ and $\mathcal{A}$ are finite sets of terminological axioms and fassertions respectively.

Example 11. *A* f*KB* f$\mathcal{K} = \langle \{A \sqsubseteq \forall R.\neg B\}, \{a : \forall R.C \geq \textit{VeryTrue}\} \rangle$.

An f-interpretation $\mathcal{I}$ *satisfies* (is a *model* of) a TBox $\mathcal{T}$ iff $\mathcal{I}$ satisfies each element in $\mathcal{T}$. $\mathcal{I}$ *satisfies* (is a *model* of) an ABox $\mathcal{A}$ iff $\mathcal{I}$ satisfies each element in $\mathcal{A}$. $\mathcal{I}$ *satisfies* (is a *model* of) a fKB f$\mathcal{K} = \langle \mathcal{T}, \mathcal{A} \rangle$ iff $\mathcal{I}$ satisfies both $\mathcal{A}$ and $\mathcal{T}$. Given a fKB f$\mathcal{K}$ and a fassertion fα. We say that f$\mathcal{K}$ *entails* fα (denoted f$\mathcal{K} \models$ fα) iff each model of f$\mathcal{K}$ satisfies fα.

3   Transforming $\mathcal{ALC}_{\mathcal{FL}}$ into $\mathcal{ALCH}$

We will introduce a satisfiability preserving transformation from $\mathcal{ALC}_{\mathcal{FL}}$ into $\mathcal{ALCH}$ in this section. First, we illustrate the basic idea which is similar to the one in [19] which is the first efforts in this direction. There is also other more efficient representation in [3].

Consider a monotonic HA $AX = (X, \{\textit{True}, \textit{False}\}, H, >)$. In the following, we assume that $c \in \{c^+, c^-\}$ where $c^+ = \textit{True}, c^- = \textit{False}, \sigma \in H^*, \sigma c \in X$ and $\bowtie \in \{\geq, >, \leq, <\}$. Assume we have an $\mathcal{ALC}_{\mathcal{FL}}$ knowledge base, f$\mathcal{K} = \langle \mathcal{T}, \mathcal{A} \rangle$, where $\mathcal{A} = \{$f$\alpha_1,$ f$\alpha_2,$ f$\alpha_3,$ f$\alpha_4\}$ and f$\alpha_1 = \langle a : A \geq \textit{True} \rangle$, f$\alpha_2 = \langle b : A \geq \textit{VeryTrue} \rangle$, f$\alpha_3 = \langle a : B \leq \textit{False} \rangle$, and f$\alpha_4 = \langle b : B \leq \textit{VeryFalse} \rangle$ where A, B are concept names. We introduce four new concept names: $A_{\geq \textit{True}}$, $A_{\geq \textit{VeryTrue}}$, $B_{\leq \textit{False}}$ and $B_{\leq \textit{VeryFalse}}$. The concept name $A_{\geq \textit{True}}$ represents the set of individuals that are instances of A with degree greater and equal to *True*. The concept name $B_{\leq \textit{VeryFalse}}$ represents the set of individuals that are instances of B with degree less and equal to *VeryFalse*. We can map the fuzzy assertions into classical assertions:

$$\langle a : A \geq \textit{True} \rangle \rightarrow \langle a : A_{\geq \textit{True}} \rangle,$$
$$\langle b : A \geq \textit{VeryTrue} \rangle \rightarrow \langle b : A_{\geq \textit{VeryTrue}} \rangle,$$
$$\langle a : B \leq \textit{False} \rangle \rightarrow \langle a : B_{\leq \textit{False}} \rangle,$$
$$\langle b : B \leq \textit{VeryFalse} \rangle \rightarrow \langle b : B_{\leq \textit{VeryFalse}} \rangle.$$

We also need to consider the relationships among the newly introduced concept names. Because *VeryTrue* > *True*, it is easy to get if a truth value $\sigma c \geq \textit{VeryTrue}$ then $\sigma c \geq \textit{True}$. Thus, we obtain a new inclusion $A_{\geq \textit{VeryTrue}} \sqsubseteq A_{\geq \textit{True}}$. Similarly for B, because *VeryFalse* < *False*, a truth value $\sigma c \leq \textit{VeryFalse}$ implies $\sigma c \leq \textit{False}$ too. Then the inclusion $B_{\leq \textit{VeryFalse}} \sqsubseteq B_{\leq \textit{False}}$ is obtained.

Now, let us proceed with the mappings. Let f$\mathcal{K} = \langle \mathcal{T}, \mathcal{A} \rangle$ be an $\mathcal{ALC}_{\mathcal{FL}}$ knowledge base. We are going to transform f$\mathcal{K}$ into an $\mathcal{ALCH}$ knowledge base K. We assume $\sigma c \in [\inf(\textit{False}), \sup(\textit{True})]$ and $\bowtie \in \{>, >, <, <\}$.

3.1   The Transformation of ABox

In order to transform $\mathcal{A}$, we define two mappings θ and ρ to map all the assertions in $\mathcal{A}$ into classical assertions. Notice that we do not allow assertions of the forms

$(a, b) : R < \sigma c$ and $(a, b) : R \leq \sigma c$ although they are legal forms of assertions in $\mathcal{ALC}_{\mathcal{FL}}$ because they related to 'negated role' which is not part of classical $\mathcal{ALCH}$.

We use the mapping ρ to encode the basic idea we present at the beginning of this section. The mapping ρ combines the $\mathcal{ALC}_{\mathcal{FL}}$ concept term, the $\bowtie$ and the fuzzy value σc together into one $\mathcal{ALCH}$ concept term.

Let A be a concept name, C, D be concept terms and R be a role name. For roles we have simply

$$\rho(R, \bowtie \sigma c) = R_{\bowtie \sigma c}.$$

For concept terms, the mapping ρ is inductively defined on the structures of concept terms:
For $\top$,

$$\rho(\top, \bowtie \sigma c) = \begin{cases} \top \text{ if } \bowtie \sigma c = \geq \sigma c \\ \top \text{ if } \bowtie \sigma c = > \sigma c, \sigma c < \sup(c^+) \\ \bot \text{ if } \bowtie \sigma c = > \sup(c^+) \\ \top \text{ if } \bowtie \sigma c = \leq \sup(c^+) \\ \bot \text{ if } \bowtie \sigma c = \leq \sigma c, \sigma c < \sup(c^+) \\ \bot \text{ if } \bowtie \sigma c = < \sigma c. \end{cases}$$

For $\bot$,

$$\rho(\bot, \bowtie \sigma c) = \begin{cases} \top \text{ if } \bowtie \sigma c = \geq \inf(c^-) \\ \bot \text{ if } \bowtie \sigma c = \geq \sigma c, \sigma c > \inf(c^-) \\ \bot \text{ if } \bowtie \sigma c = > \sigma c \\ \top \text{ if } \bowtie \sigma c = \leq \sigma c \\ \top \text{ if } \bowtie \sigma c = < \sigma c, \sigma c > \inf(c^-) \\ \bot \text{ if } \bowtie \sigma c = < \inf(c^-). \end{cases}$$

For concept name A,

$$\rho(A, \bowtie \sigma c) = A_{\bowtie \sigma c}.$$

For concept conjunction $C \sqcap D$,

$$\rho(C \sqcap D, \bowtie \sigma c) = \begin{cases} \rho(C, \bowtie \sigma c) \sqcap \rho(D, \bowtie \sigma c) \text{ if } \bowtie \in \{\geq, >\} \\ \rho(C, \bowtie \sigma c) \sqcup \rho(D, \bowtie \sigma c) \text{ if } \bowtie \in \{\leq, <\}. \end{cases}$$

For concept disjunction $C \sqcup D$,

$$\rho(C \sqcup D, \bowtie \sigma c) = \begin{cases} \rho(C, \bowtie \sigma c) \sqcup \rho(D, \bowtie \sigma c) \text{ if } \bowtie \in \{\geq, >\} \\ \rho(C, \bowtie \sigma c) \sqcap \rho(D, \bowtie \sigma c) \text{ if } \bowtie \in \{\leq, <\}. \end{cases}$$

For concept negation $\neg C$,

$$\rho(\neg C, \bowtie \sigma c) = \rho(C, \neg \bowtie \sigma \bar{c}),$$

where $\neg \geq \, = \, \leq, \neg > \, = \, <, \neg \leq \, = \, \geq, \neg < \, = \, >$.

For modifier concept δC,

$$\rho(\delta C, \bowtie \sigma c) = \rho(C, \bowtie \sigma \delta c).$$

For existential quantification $\exists R.C$,

$$\rho(\exists R.C, \bowtie \sigma c) = \begin{cases} \exists \rho(R, \bowtie \sigma c).\rho(C, \bowtie \sigma c) & \text{if } \bowtie \in \{\geq, >\} \\ \forall \rho(R, - \bowtie \sigma c).\rho(C, \bowtie \sigma c) & \text{if } \bowtie \in \{\leq, <\}, \end{cases}$$

where $- \leq = >$ and $- < = \geq$.

For universal quantification $\forall R.C$,

$$\rho(\forall R.C, \bowtie \sigma c) = \begin{cases} \forall \rho(R, + \bowtie \sigma \bar{c}).\rho(C, \bowtie \sigma c) & \text{if } \bowtie \in \{\geq, >\} \\ \exists \rho(R, \neg \bowtie \sigma \bar{c}).\rho(C, \bowtie \sigma c) & \text{if } \bowtie \in \{\leq, <\}, \end{cases}$$

where $+ \geq = >$ and $+ > = \geq$.

θ maps fuzzy assertions into classical assertions using ρ. Let $\mathfrak{f}\alpha$ be a fassertion in $\mathcal{A}$, we define it as follows.

$$\theta(\mathfrak{f}\alpha) = \begin{cases} a : \rho(C, \bowtie \sigma c) & \text{if } \mathfrak{f}\alpha = \langle a : C \bowtie \sigma c \rangle \\ (a, b) : \rho(R, \bowtie \sigma c) & \text{if } \mathfrak{f}\alpha = \langle (a, b) : R \bowtie \sigma c \rangle. \end{cases}$$

Example 12. *Let* $\mathfrak{f}\alpha = \langle a : Very(A \sqcap B) \leq LessFalse \rangle$, *then*

$$\begin{aligned} \theta(\mathfrak{f}\alpha) &= a : \rho(Very(A \sqcap B), \leq LessFalse) \\ &= a : \rho((A \sqcap B), \leq LessVeryFalse) \\ &= a : \rho(A, \leq LessVeryFalse) \sqcup \rho(B, \leq LessVeryFalse) \\ &= a : A_{\leq LessVeryFalse} \sqcup B_{\leq LessVeryFalse}. \end{aligned}$$

We extend θ to a set of fassertions $\mathcal{A}$ point-wise,

$$\theta(\mathcal{A}) = \{\theta(\mathfrak{f}\alpha) \mid \mathfrak{f}\alpha \in \mathcal{A}\}.$$

According to the rules above, we can see that $|\theta(\mathcal{A})|$ is linearly bounded by $|\mathcal{A}|$.

4   The Transformation of TBox

The new TBox is a union of two terminologies. One is the newly introduced TBox (denoted by $\mathcal{T}(N^{\mathfrak{f}\mathcal{K}})$) which is the terminology relating to the newly introduced concept names and role names. The other one is $\kappa(\mathfrak{f}\mathcal{K}, \mathcal{T})$ which is reduced by a mapping κ from the TBox of an $\mathcal{ALC}_{\mathcal{FL}}$ knowledge base.

4.1   The Newly Introduced TBox

Many new concept names and new role names are introduced when we transform an ABox. We need a set of terminological axioms to define the relationships among those new names.

We need to collect all the linguist terms σc that might be the subscript of a concept name or a role name. It means that not only the set of linguistic terms that appears in the original ABox but also the set of new linguist terms which are produced by applying the ρ for modifier concepts should be included. Let A be a concept name, R be a role name.

$$X^{\mathfrak{fK}} = \{\sigma c \mid \langle \alpha \bowtie \sigma c \rangle \in A\} \cup \{\sigma\delta c \mid \rho(\delta C, \bowtie \sigma c) = \rho(C, \bowtie \sigma\delta c)\}.$$

such that δC occurs in $\mathfrak{fK}$.

We define a sorted set of linguistic terms,

$$\begin{aligned}
N^{\mathfrak{fK}} &= \{\inf(False), W, \sup(True)\} \cup X^{\mathfrak{fK}} \cup \{\sigma\bar{c} \mid \sigma c \in X^{\mathfrak{fK}}\} \\
&= \{n_1, \ldots, n_{|N^{\mathfrak{fK}}|}\}
\end{aligned}$$

where $n_i < n_{i+1}$ for $1 \leq i \leq |N^{\mathfrak{fK}}| - 1$ and $n_1 = \inf(False), n_{|N^{\mathfrak{fK}}|} = \sup(True)$.

Example 13. *Consider Example 11, the sorted set is,*

$$N^{\mathfrak{fK}} = \{\inf(False), VeryFalse, W, VeryTrue, \sup(True)\}.$$

Let $\mathcal{T}(N^{\mathfrak{fK}})$ be the set of terminological axioms relating to the newly introduced concept names and role names.

Definition 14. *Let $\mathcal{A}^{\mathfrak{fK}}$ and $\mathcal{R}^{\mathfrak{fK}}$ be the sets of concept names and role names occurring in $\mathfrak{fK}$ respectively. For each $A \in \mathcal{A}^{\mathfrak{fK}}$, for each $R \in \mathcal{R}^{\mathfrak{fK}}$, for each $1 \leq i \leq |N^{\mathfrak{fK}}| - 1$ and for each $2 \leq j \leq |N^{\mathfrak{fK}}| - 1$, $\mathcal{T}(N^{\mathfrak{fK}})$ contains*

$$\begin{aligned}
A_{\geq n_{i+1}} &\sqsubseteq A_{>n_i}, \ A_{>n_j} \sqsubseteq A_{\geq n_j}, \\
R_{\geq n_{i+1}} &\sqsubseteq R_{>n_i}, \ R_{>n_j} \sqsubseteq R_{\geq n_j}.
\end{aligned}$$

where $n \in N^{\mathfrak{fK}}$.

$n_{i+1} > n_i$ because $N^{\mathfrak{fK}}$ is a sorted set. Then if an individual is an instance of a concept name with degree $\geq n_{i+1}$ then the degree is also $> n_i$. The first terminological axiom shows that if an individual is an instance of $A_{\geq n_{i+1}}$ then it is an instance of $A_{>n_i}$ as well. Similarly, if an individual is an instance of a concept name with degree $> n_i$ then the degree is also $\geq n_i$. The second terminological axiom shows that if an individual is an instance of $A_{>n_i}$ then it is also an instance of $A_{\geq n_i}$.

$\mathcal{T}(N^{\mathfrak{fK}})$ contains $2|\mathcal{A}^{\mathfrak{fK}}|(|N^{\mathfrak{fK}}| - 1)$ plus $2|\mathcal{R}^{\mathfrak{fK}}|(|N^{\mathfrak{fK}}| - 1)$ terminological axioms.

Example 15. *Consider the $\mathcal{ALC}_{\mathcal{FL}}$ knowledge base in Example 11, the following is an excerpt of the $\mathcal{T}(N^{\mathfrak{fK}})$,*

$$\begin{aligned}
\mathcal{T}(N^{\mathfrak{fK}}) = \{&A_{\geq \sup(True)} \sqsubseteq A_{>VeryTrue}, \ A_{\geq VeryTrue} \sqsubseteq A_{>W}, \\
&A_{\geq W} \sqsubseteq A_{>VeryFalse}, \ A_{\geq VeryFalse} \sqsubseteq A_{>\inf(False)}\} \\
\cup \{&A_{>VeryTrue} \sqsubseteq A_{\geq VeryTrue}, A_{>W} \sqsubseteq A_{\geq W}, \\
&A_{>VeryFalse} \sqsubseteq A_{\geq VeryFalse}\} \\
\cup \{&\ldots, R_{\geq \sup(True)} \sqsubseteq R_{>VeryTrue}, \ldots\}.
\end{aligned}$$

4.2   The Mapping κ

κ maps the fuzzy TBox into the classical TBox.

Definition 16. *Let C, D be two concept terms and $C \sqsubseteq D \in \mathcal{T}$. For all $n \in N^{\mathfrak{f}\mathcal{K}}$*

$$\kappa(\mathfrak{f}\mathcal{K}, C \sqsubseteq D) = \bigcup_{n \in N^{\mathfrak{f}\mathcal{K}}, \bowtie \in \{\geq, >\}} \{\rho(C, \bowtie n) \sqsubseteq \rho(D, \bowtie n)\} \\ \bigcup_{n \in N^{\mathfrak{f}\mathcal{K}}, \bowtie \in \{\leq, <\}} \{\rho(D, \bowtie n) \sqsubseteq \rho(C, \bowtie n)\} \qquad (3)$$

We extend κ to a terminology $\mathcal{T}$ point-wise. For all $\tau \in \mathcal{T}$

$$\kappa(\mathfrak{f}\mathcal{K}, \mathcal{T}) = \bigcup_{\tau \in \mathcal{T}} \kappa(\mathfrak{f}\mathcal{K}, \tau).$$

κ reduces a terminological axiom in $\mathcal{ALC}_{\mathcal{FL}}$ into a set of $\mathcal{ALCH}$ terminology axioms.

4.3   The Satisfiability Preserving Theorem

Now we can define the *reduction* of $\mathfrak{f}\mathcal{K}$ into an $\mathcal{ALCH}$ knowledge base, denoted $\mathcal{K}(\mathfrak{f}\mathcal{K})$,

$$\mathcal{K}(\mathfrak{f}\mathcal{K}) = \langle \mathcal{T}(N^{\mathfrak{f}\mathcal{K}}) \cup \kappa(\mathfrak{f}\mathcal{K}, \mathcal{T}), \theta(\mathcal{A}) \rangle.$$

The transformation can be done in polynomial time. The soundness and completeness of the algorithm is guaranteed by the following satisfiability preserving theorem.

Theorem 17. *Let $\mathfrak{f}\mathcal{K}$ be an $\mathcal{ALC}_{\mathcal{FL}}$ knowledge base. Then $\mathfrak{f}\mathcal{K}$ is satisfiable iff the $\mathcal{ALCH}$ knowledge base $\mathcal{K}(\mathfrak{f}\mathcal{K})$ is satisfiable.*

Proof. Let $\mathfrak{f}\mathcal{K} = \langle \mathcal{T}, \mathcal{A} \rangle$ be an $\mathcal{ALC}_{\mathcal{FL}}$ knowledge base , $\mathcal{K}(\mathfrak{f}\mathcal{K}) = \langle \mathcal{T}', \mathcal{A}' \rangle$ be the transformed $\mathcal{ALCH}$ knowledge base, where $\mathcal{T}' = \mathcal{T}(N^{\mathfrak{f}\mathcal{K}}) \cup \kappa(\mathfrak{f}\mathcal{K}, \mathcal{T})$ and $\mathcal{A}' = \theta(\mathcal{A})$. We define that $\rhd \in \{\geq, >\}$ and $\lhd \in \{\leq, <\}$.

Our goal is to prove that there exists an interpretation $\mathcal{I}$ such that $\mathcal{I} \models \mathfrak{f}\mathcal{K}$ if and only if there exists an interpretation $\mathcal{I}'$ such that $\mathcal{I}' \models \mathcal{K}(\mathfrak{f}\mathcal{K})$, where $\mathcal{I}$ is a fuzzy interpretation and $\mathcal{I}'$ is an $\mathcal{ALCH}$ interpretation.

$\Rightarrow$.) Assume $\mathcal{I}$ is an interpretation such that $\mathcal{I} \models \mathfrak{f}\mathcal{K}$. So $\mathcal{I} \models \mathcal{A}$ and $\mathcal{I} \models \mathcal{T}$. We construct an $\mathcal{ALCH}$ interpretation $\mathcal{I}'$:

- $\Delta^{\mathcal{I}'} := \Delta^{\mathcal{I}}$,
- $a^{\mathcal{I}'} := a^{\mathcal{I}}$ for all individual a,
- $A_{\bowtie \sigma c}^{\mathcal{I}'} := \{d \in \Delta^{\mathcal{I}'} \mid A^{\mathcal{I}}(d) \bowtie \sigma c\}$, for all concept name $A_{\bowtie \sigma c}$,
- $R_{\bowtie \sigma c}^{\mathcal{I}'} := \{(d, d') \in \Delta^{\mathcal{I}'} \times \Delta^{\mathcal{I}'} \mid R^{\mathcal{I}}(d, d') \bowtie \sigma c\}$, for all role name $R_{\bowtie \sigma c}$.

In order to show $\mathcal{I}' \models \mathcal{K}(\mathfrak{f}\mathcal{K})$, we have to show that $\mathcal{I}' \models \theta(\mathcal{A})$ and $\mathcal{I}' \models \mathcal{T}(N^{\mathfrak{f}\mathcal{K}}) \cup \kappa(\mathfrak{f}\mathcal{K}, \mathcal{T})$. Then it is sufficient to prove that:

1. for each $\alpha \bowtie \sigma c \in \mathcal{A}$, $\mathcal{I}' \models \theta(\alpha \bowtie \sigma c)$, and
2. $\mathcal{I}' \models \mathcal{T}(N^{\mathfrak{f}\mathcal{K}})$ and for each $C \sqsubseteq D \in \mathcal{T}$, $\mathcal{I}' \models \kappa(\mathfrak{f}\mathcal{K}, C \sqsubseteq D)$.

First, we need the following Lemma.

Lemma 18. *Let C be a concept term in $\mathcal{ALC}_{\mathcal{FL}}$. $C \neq \top$ and $C \neq \bot$. It follows that $(\rho(C, \bowtie \sigma c))^{\mathcal{I}'} = \{d \in \Delta^{\mathcal{I}'} \mid C^{\mathcal{I}}(d) \bowtie \sigma c\}$.*

Proof. We use proof by induction.

Basic step:

Let R be a role name. Then
$(\rho(R, \bowtie \sigma c))^{\mathcal{I}'} = R_{\bowtie \sigma c}^{\mathcal{I}'} = \{(d, d') \in \Delta^{\mathcal{I}'} \times \Delta^{\mathcal{I}'} \mid R^{\mathcal{I}}(d, d') \bowtie \sigma c\}$.

Let A be a concept name. Then
$(\rho(A, \bowtie \sigma c))^{\mathcal{I}'} = A_{\bowtie \sigma c}^{\mathcal{I}'} = \{d \in \Delta^{\mathcal{I}'} \mid A^{\mathcal{I}}(d) \bowtie \sigma c\}$.

Inductive step:

Let C, D be concept terms. Assume
$(\rho(C, \bowtie \sigma c))^{\mathcal{I}'} = \{d \in \Delta^{\mathcal{I}'} \mid C^{\mathcal{I}}(d) \bowtie \sigma c\}$ and
$(\rho(D, \bowtie \sigma c))^{\mathcal{I}'} = \{d \in \Delta^{\mathcal{I}'} \mid D^{\mathcal{I}}(d) \bowtie \sigma c\}$.

we prove inductively on the structures of concept terms.

Case $\neg C$.

$$\begin{aligned}
(\rho(\neg C, \bowtie \sigma c))^{\mathcal{I}'} &= (\rho(C, \neg \bowtie \sigma \bar{c}))^{\mathcal{I}'} \\
&= \{d \in \Delta^{\mathcal{I}'} \mid C^{\mathcal{I}}(d) \neg \bowtie \sigma \bar{c}\} \\
&= \{d \in \Delta^{\mathcal{I}'} \mid (\neg C)^{\mathcal{I}}(d) \bowtie \sigma c\}.
\end{aligned}$$

Case δC.

$$\begin{aligned}
(\rho(\delta C, \bowtie \sigma c))^{\mathcal{I}'} &= (\rho(C, \bowtie \sigma \delta c))^{\mathcal{I}'} \\
&= \{d \in \Delta^{\mathcal{I}'} \mid C^{\mathcal{I}}(d) \bowtie \sigma \delta c\} \\
&= \{d \in \Delta^{\mathcal{I}'} \mid (\delta C)^{\mathcal{I}}(d) \bowtie \sigma c\}.
\end{aligned}$$

Case $C \sqcap D$.

$$\begin{aligned}
(\rho(C \sqcap D, \triangleright \sigma c))^{\mathcal{I}'} &= (\rho(C, \triangleright \sigma c) \sqcap \rho(D, \triangleright \sigma c))^{\mathcal{I}'} \\
&= \{d \in \Delta^{\mathcal{I}'} \mid C^{\mathcal{I}}(d) \triangleright \sigma c\} \cap \{d \in \Delta^{\mathcal{I}'} \mid D^{\mathcal{I}}(d) \triangleright \sigma c\}. \\
&= \{d \in \Delta^{\mathcal{I}'} \mid C^{\mathcal{I}}(d) \triangleright \sigma c \wedge D^{\mathcal{I}}(d) \triangleright \sigma c\}. \\
&= \{d \in \Delta^{\mathcal{I}'} \mid \min(C^{\mathcal{I}}(d), D^{\mathcal{I}}(d)) \triangleright \sigma c\}. \\
&= \{d \in \Delta^{\mathcal{I}'} \mid (C \sqcap D)^{\mathcal{I}}(d) \triangleright \sigma c\}.
\end{aligned}$$

$$\begin{aligned}
(\rho(C \sqcap D, \triangleleft \sigma c))^{\mathcal{I}'} &= (\rho(C, \triangleleft \sigma c) \sqcup \rho(D, \triangleleft \sigma c))^{\mathcal{I}'} \\
&= \{d \in \Delta^{\mathcal{I}'} \mid C^{\mathcal{I}}(d) \triangleleft \sigma c\} \cup \{d \in \Delta^{\mathcal{I}'} \mid D^{\mathcal{I}}(d) \triangleleft \sigma c\} \\
&= \{d \in \Delta^{\mathcal{I}'} \mid C^{\mathcal{I}}(d) \triangleleft \sigma c \vee D^{\mathcal{I}}(d) \triangleleft \sigma c\} \\
&= \{d \in \Delta^{\mathcal{I}'} \mid \min(C^{\mathcal{I}}(d), D^{\mathcal{I}}(d)) \triangleleft \sigma c\} \\
&= \{d \in \Delta^{\mathcal{I}'} \mid (C \sqcap D)^{\mathcal{I}}(d) \triangleleft \sigma c\}.
\end{aligned}$$

Case $C \sqcup D$.

$$\begin{aligned}
(\rho(C \sqcup D, \triangleright \sigma c))^{\mathcal{I}'} &= (\rho(C, \triangleright \sigma c) \sqcup \rho(D, \triangleright \sigma c))^{\mathcal{I}'} \\
&= \{d \in \Delta^{\mathcal{I}'} \mid C^{\mathcal{I}}(d) \triangleright \sigma c\} \cup \{d \in \Delta^{\mathcal{I}'} \mid D^{\mathcal{I}}(d) \triangleright \sigma c\} \\
&= \{d \in \Delta^{\mathcal{I}'} \mid C^{\mathcal{I}}(d) \triangleright \sigma c \vee D^{\mathcal{I}}(d) \triangleright \sigma c\} \\
&= \{d \in \Delta^{\mathcal{I}'} \mid \max(C^{\mathcal{I}}(d), D^{\mathcal{I}}(d)) \triangleright \sigma c\} \\
&= \{d \in \Delta^{\mathcal{I}'} \mid (C \sqcup D)^{\mathcal{I}}(d) \triangleright \sigma c\}.
\end{aligned}$$

$$\begin{aligned}
(\rho(C \sqcup D, \lhd \sigma c))^{\mathcal{I}'} &= (\rho(C, \lhd \sigma c) \sqcap \rho(D, \lhd \sigma c))^{\mathcal{I}'} \\
&= \{d \in \Delta^{\mathcal{I}'} \mid C^{\mathcal{I}}(d) \lhd \sigma c\} \cap \{d \in \Delta^{\mathcal{I}'} \mid D^{\mathcal{I}}(d) \lhd \sigma c\} \\
&= \{d \in \Delta^{\mathcal{I}'} \mid C^{\mathcal{I}}(d) \lhd \sigma c \wedge D^{\mathcal{I}}(d) \lhd \sigma c\} \\
&= \{d \in \Delta^{\mathcal{I}'} \mid \max(C^{\mathcal{I}}(d), D^{\mathcal{I}}(d)) \lhd \sigma c\} \\
&= \{d \in \Delta^{\mathcal{I}'} \mid (C \sqcup D)^{\mathcal{I}}(d) \lhd \sigma c\}.
\end{aligned}$$

Case $\forall R.C$.

$$\begin{aligned}
(\rho(\forall R.C, \rhd \sigma c))^{\mathcal{I}'} &= (\forall \rho(R, + \rhd \sigma\bar{c}).\rho(C, \rhd \sigma c))^{\mathcal{I}'} \\
&= \{d \in \Delta^{\mathcal{I}'} \mid \forall d' \in \Delta^{\mathcal{I}'}.(d,d') \notin R^{\mathcal{I}'}_{+\rhd\sigma\bar{c}} \vee C^{\mathcal{I}}(d') \rhd \sigma c\} \\
&= \{d \in \Delta^{\mathcal{I}'} \mid \forall d' \in \Delta^{\mathcal{I}'}.(d,d') \in R^{\mathcal{I}'}_{\neg\rhd\sigma\bar{c}} \vee C^{\mathcal{I}}(d') \rhd \sigma c\} \\
&= \{d \in \Delta^{\mathcal{I}'} \mid \bigwedge_{d' \in \Delta^{\mathcal{I}'}}(R^{\mathcal{I}}(d,d')\neg \rhd \sigma\bar{c} \vee C^{\mathcal{I}}(d') \rhd \sigma c)\} \\
&= \{d \in \Delta^{\mathcal{I}'} \mid \bigwedge_{d' \in \Delta^{\mathcal{I}'}}(-R^{\mathcal{I}}(d,d') \rhd \sigma c \vee C^{\mathcal{I}}(d') \rhd \sigma c)\} \\
&= \{d \in \Delta^{\mathcal{I}'} \mid \bigwedge_{d' \in \Delta^{\mathcal{I}'}}(\max(-R^{\mathcal{I}}(d,d'), C^{\mathcal{I}}(d')) \rhd \sigma c)\} \\
&= \{d \in \Delta^{\mathcal{I}'} \mid \inf_{d' \in \Delta^{\mathcal{I}'}}(\max(-R^{\mathcal{I}}(d,d'), C^{\mathcal{I}}(d')\} \rhd \sigma c)\} \\
&= \{d \in \Delta^{\mathcal{I}'} \mid (\forall R.C)^{\mathcal{I}}(d) \rhd \sigma c)\}.
\end{aligned}$$

$$\begin{aligned}
(\rho(\forall R.C, \lhd \sigma c))^{\mathcal{I}'} &= (\exists \rho(R, \neg \lhd \sigma\bar{c}).\rho(C, \lhd \sigma c))^{\mathcal{I}'} \\
&= \{d \in \Delta^{\mathcal{I}'} \mid \exists d' \in \Delta^{\mathcal{I}'}.(d,d') \in R^{\mathcal{I}'}_{\neg\lhd\sigma\bar{c}} \wedge C^{\mathcal{I}}(d') \lhd \sigma c\} \\
&= \{d \in \Delta^{\mathcal{I}'} \mid \bigvee_{d' \in \Delta^{\mathcal{I}'}}(R^{\mathcal{I}}(d,d')\neg \lhd \sigma\bar{c} \wedge C^{\mathcal{I}}(d') \lhd \sigma c)\} \\
&= \{d \in \Delta^{\mathcal{I}'} \mid \bigvee_{d' \in \Delta^{\mathcal{I}'}}(-R^{\mathcal{I}}(d,d') \lhd \sigma c \wedge C^{\mathcal{I}}(d') \lhd \sigma c)\} \\
&= \{d \in \Delta^{\mathcal{I}'} \mid \bigvee_{d' \in \Delta^{\mathcal{I}'}}(\max(-R^{\mathcal{I}}(d,d'), C^{\mathcal{I}}(d')) \lhd \sigma c)\} \\
&= \{d \in \Delta^{\mathcal{I}'} \mid \inf_{d' \in \Delta^{\mathcal{I}'}}(\max(-R^{\mathcal{I}}(d,d'), C^{\mathcal{I}}(d')\} \lhd \sigma c)\} \\
&= \{d \in \Delta^{\mathcal{I}'} \mid (\forall R.C)^{\mathcal{I}}(d) \lhd \sigma c)\}.
\end{aligned}$$

Case $\exists R.C$.

$$\begin{aligned}
(\rho(\exists R.C, \rhd \sigma c))^{\mathcal{I}'} &= (\exists \rho(R, \rhd \sigma c).\rho(C, \rhd \sigma c))^{\mathcal{I}'} \\
&= \{d \in \Delta^{\mathcal{I}'} \mid \exists d' \in \Delta^{\mathcal{I}'}.(d,d') \in R^{\mathcal{I}'}_{\rhd\sigma c} \wedge C^{\mathcal{I}}(d') \rhd \sigma c\} \\
&= \{d \in \Delta^{\mathcal{I}'} \mid \bigvee_{d' \in \Delta^{\mathcal{I}'}}(R^{\mathcal{I}'}(d,d') \rhd \sigma c \wedge C^{\mathcal{I}}(d') \rhd \sigma c)\} \\
&= \{d \in \Delta^{\mathcal{I}'} \mid \bigvee_{d' \in \Delta^{\mathcal{I}'}}(\min(R^{\mathcal{I}}(d,d'), C^{\mathcal{I}}(d')) \rhd \sigma c)\} \\
&= \{d \in \Delta^{\mathcal{I}'} \mid \sup_{d' \in \Delta^{\mathcal{I}'}}\{\min(R^{\mathcal{I}}(d,d'), C^{\mathcal{I}}(d')\} \rhd \sigma c)\} \\
&= \{d \in \Delta^{\mathcal{I}'} \mid (\exists R.C)^{\mathcal{I}}(d) \rhd \sigma c)\}.
\end{aligned}$$

$$\begin{aligned}
(\rho(\exists R.C, \lhd \sigma c))^{\mathcal{I}'} &= (\forall \rho(R, - \lhd \sigma c).\rho(C, \lhd \sigma c))^{\mathcal{I}'} \\
&= \{d \in \Delta^{\mathcal{I}'} \mid \forall d' \in \Delta^{\mathcal{I}'}.(d,d') \notin R^{\mathcal{I}'}_{-\lhd\sigma c} \vee C^{\mathcal{I}}(d') \lhd \sigma c\} \\
&= \{d \in \Delta^{\mathcal{I}'} \mid \forall d' \in \Delta^{\mathcal{I}'}.(d,d') \in R^{\mathcal{I}'}_{\lhd\sigma c} \vee C^{\mathcal{I}}(d') \lhd \sigma c\} \\
&= \{d \in \Delta^{\mathcal{I}'} \mid \forall d' \in \Delta^{\mathcal{I}'}.(R^{\mathcal{I}}(d,d') \lhd \sigma c \vee C^{\mathcal{I}}(d') \lhd \sigma c)\} \\
&= \{d \in \Delta^{\mathcal{I}'} \mid \bigwedge_{d' \in \Delta^{\mathcal{I}'}}(\min(R^{\mathcal{I}}(d,d'), C^{\mathcal{I}}(d')) \lhd \sigma c)\} \\
&= \{d \in \Delta^{\mathcal{I}'} \mid \sup_{d' \in \Delta^{\mathcal{I}'}}(\min(R^{\mathcal{I}}(d,d'), C^{\mathcal{I}}(d')\} \lhd \sigma c)\} \\
&= \{d \in \Delta^{\mathcal{I}'} \mid (\exists R.C)^{\mathcal{I}}(d) \lhd \sigma c\}.
\end{aligned}$$

In the following, we use $C_{\bowtie\sigma c}$ to represent $\rho(C, \bowtie \sigma c)$.

(1) Now we prove that $\mathcal{I}' \models \theta(\mathcal{A})$. Let $\alpha \bowtie \sigma c \in \mathcal{A}$. Then $\mathcal{I} \models \alpha \bowtie \sigma c$ because $\mathcal{I} \models \mathcal{A}$.

If σ is a role assertion of the form $(a, b) : R$, then

$$
\begin{aligned}
\mathcal{I} \models (a, b) : R \bowtie \sigma c &\Rightarrow R^{\mathcal{I}}(a^{\mathcal{I}}, b^{\mathcal{I}}) \bowtie \sigma c \\
&\Rightarrow (a^{\mathcal{I}'}, b^{\mathcal{I}'}) \in R^{\mathcal{I}'}_{\bowtie \sigma c} \\
&\Rightarrow \mathcal{I}' \models (a, b) : R_{\bowtie \sigma c}.
\end{aligned}
$$

For concept assertions, we inductively prove on the structure of concept term:

Case $\top$. For all interpretation $\mathcal{I}$ and for all $d \in \Delta^{\mathcal{I}}$, $\top^{\mathcal{I}}(d) = \sup(True)$, so $a : \top \geq \sigma c, a : \top > \sigma c$ if $\sigma c < \sup(True)$ and $a : \top \leq \sup(True)$ are valid, $a : \top$ is valid too. While $a : \top > \sup(True), a : \top \leq \sigma c$ if $\sigma c < \sup(True)$ and $a : \top < \sigma c$ are unsatisfiable, $a : \bot$ is unsatisfiable as well.

Case $\bot$. For all interpretation $\mathcal{I}$ and for all $d \in \Delta^{\mathcal{I}}$, $\bot^{\mathcal{I}}(d) = \inf(False)$, so $a : \bot \geq \inf(False), a : \bot < \sigma c$ if $\sigma c > \inf(False)$ and $a : \bot \leq \sigma c$ are valid, so is $a : \top$. While $a : \bot < \inf(False), a : \bot \geq \sigma c$ if $\sigma c > \inf(False)$ and $a : \bot > \sigma c$ are unsatisfiable. $a : \bot$ is also unsatisfiable.

Case concept name A. $\mathcal{I} \models a : A \bowtie \sigma c \Rightarrow A^{\mathcal{I}}(a^{\mathcal{I}}) \bowtie \sigma c \Rightarrow a^{\mathcal{I}'} \in A^{\mathcal{I}'}_{\bowtie \sigma c} \Rightarrow \mathcal{I}' \models a : A_{\bowtie \sigma c}$.

Case concept negation $\neg C$.

$$
\begin{aligned}
\mathcal{I} \models a : \neg C \bowtie \sigma c &\Rightarrow (\neg C)^{\mathcal{I}}(a^{\mathcal{I}}) \bowtie \sigma c \\
&\Rightarrow -C^{\mathcal{I}}(a^{\mathcal{I}}) \bowtie \sigma c \\
&\Rightarrow C^{\mathcal{I}}(a^{\mathcal{I}})\neg \bowtie \sigma \bar{c} \\
&\Rightarrow a^{\mathcal{I}} \in C^{\mathcal{I}'}_{\neg \bowtie \sigma \bar{c}} \\
&\Rightarrow a^{\mathcal{I}'} \in C^{\mathcal{I}'}_{\neg \bowtie \sigma \bar{c}} \\
&\Rightarrow \mathcal{I}' \models a : C_{\neg \bowtie \sigma \bar{c}}.
\end{aligned}
$$

Case modifier concept δC.

$$
\begin{aligned}
\mathcal{I} \models a : \delta C \bowtie \sigma c &\Rightarrow \mathcal{I} \models a : C \bowtie \sigma \delta c \\
&\Rightarrow C^{\mathcal{I}}(a^{\mathcal{I}}) \bowtie \sigma \delta c \\
&\Rightarrow a^{\mathcal{I}} \in C^{\mathcal{I}'}_{\bowtie \sigma \delta c} \\
&\Rightarrow a^{\mathcal{I}'} \in C^{\mathcal{I}'}_{\bowtie \sigma \delta c} \\
&\Rightarrow \mathcal{I}' \models a : C_{\bowtie \sigma \delta c}.
\end{aligned}
$$

Case concept conjunction $C \sqcap D$.

$$
\begin{aligned}
\mathcal{I} \models a : C \sqcap D \triangleright \sigma c &\Rightarrow (C \sqcap D)^{\mathcal{I}}(a^{\mathcal{I}}) \triangleright \sigma c \\
&\Rightarrow \min(C^{\mathcal{I}}(a^{\mathcal{I}}), D^{\mathcal{I}}(a^{\mathcal{I}})) \triangleright \sigma c \\
&\Rightarrow (C^{\mathcal{I}}(a^{\mathcal{I}}) \triangleright \sigma c) \wedge (D^{\mathcal{I}}(a^{\mathcal{I}}) \triangleright \sigma c) \\
&\Rightarrow a^{\mathcal{I}} \in C^{\mathcal{I}'}_{\triangleright \sigma c} \wedge a^{\mathcal{I}} \in D^{\mathcal{I}'}_{\triangleright \sigma c} \\
&\Rightarrow a^{\mathcal{I}'} \in C^{\mathcal{I}'}_{\triangleright \sigma c} \wedge a^{\mathcal{I}'} \in D^{\mathcal{I}'}_{\triangleright \sigma c} \\
&\Rightarrow a^{\mathcal{I}'} \in C^{\mathcal{I}'}_{\triangleright \sigma c} \sqcap D^{\mathcal{I}'}_{\triangleright \sigma c} \\
&\Rightarrow \mathcal{I}' \models a : C_{\triangleright \sigma c} \sqcap D_{\triangleright \sigma c}.
\end{aligned}
$$

$$\mathcal{I} \models a : C \sqcap D \lhd \sigma c \Rightarrow (C \sqcap D)^{\mathcal{I}}(a^{\mathcal{I}}) \lhd \sigma c$$
$$\Rightarrow \min(C^{\mathcal{I}}(a^{\mathcal{I}}), D^{\mathcal{I}}(a^{\mathcal{I}})) \lhd \sigma c$$
$$\Rightarrow (C^{\mathcal{I}}(a^{\mathcal{I}}) \lhd \sigma c) \vee (D^{\mathcal{I}}(a^{\mathcal{I}}) \lhd \sigma c)$$
$$\Rightarrow a^{\mathcal{I}} \in C^{\mathcal{I}'}_{\lhd \sigma c} \vee a^{\mathcal{I}} \in D^{\mathcal{I}'}_{\lhd \sigma c}$$
$$\Rightarrow a^{\mathcal{I}'} \in C^{\mathcal{I}'}_{\lhd \sigma c} \vee a^{\mathcal{I}'} \in D^{\mathcal{I}'}_{\lhd \sigma c}$$
$$\Rightarrow a^{\mathcal{I}'} \in C^{\mathcal{I}'}_{\lhd \sigma c} \cup D^{\mathcal{I}'}_{\lhd \sigma c}$$
$$\Rightarrow \mathcal{I}' \models a : C_{\lhd \sigma c} \sqcup D_{\lhd \sigma c}.$$

Case concept disjunction $C \sqcup D$.

$$\mathcal{I} \models a : C \sqcup D \rhd \sigma c \Rightarrow (C \sqcup D)^{\mathcal{I}}(a^{\mathcal{I}}) \rhd \sigma c$$
$$\Rightarrow \max(C^{\mathcal{I}}(a^{\mathcal{I}}), D^{\mathcal{I}}(a^{\mathcal{I}})) \rhd \sigma c$$
$$\Rightarrow (C^{\mathcal{I}}(a^{\mathcal{I}}) \rhd \sigma c) \vee (D^{\mathcal{I}}(a^{\mathcal{I}}) \rhd \sigma c)$$
$$\Rightarrow a^{\mathcal{I}} \in C^{\mathcal{I}'}_{\rhd \sigma c} \vee a^{\mathcal{I}} \in D^{\mathcal{I}'}_{\rhd \sigma c}$$
$$\Rightarrow a^{\mathcal{I}'} \in C^{\mathcal{I}'}_{\rhd \sigma c} \vee a^{\mathcal{I}'} \in D^{\mathcal{I}'}_{\rhd \sigma c}$$
$$\Rightarrow a^{\mathcal{I}'} \in C^{\mathcal{I}'}_{\rhd \sigma c} \cup D^{\mathcal{I}'}_{\rhd \sigma c}$$
$$\Rightarrow \mathcal{I}' \models a : C_{\rhd \sigma c} \sqcup D_{\rhd \sigma c}.$$

$$\mathcal{I} \models a : C \sqcup D \lhd \sigma c \Rightarrow (C \sqcup D)^{\mathcal{I}}(a^{\mathcal{I}}) \lhd \sigma c$$
$$\Rightarrow \max(C^{\mathcal{I}}(a^{\mathcal{I}}), D^{\mathcal{I}}(a^{\mathcal{I}})) \lhd \sigma c$$
$$\Rightarrow (C^{\mathcal{I}}(a^{\mathcal{I}}) \lhd \sigma c) \wedge (D^{\mathcal{I}}(a^{\mathcal{I}}) \lhd \sigma c)$$
$$\Rightarrow a^{\mathcal{I}} \in C^{\mathcal{I}'}_{\lhd \sigma c} \wedge a^{\mathcal{I}} \in D^{\mathcal{I}'}_{\lhd \sigma c}$$
$$\Rightarrow a^{\mathcal{I}'} \in C^{\mathcal{I}'}_{\lhd \sigma c} \wedge a^{\mathcal{I}'} \in D^{\mathcal{I}'}_{\lhd \sigma c}$$
$$\Rightarrow a^{\mathcal{I}'} \in C^{\mathcal{I}'}_{\lhd \sigma c} \cap D^{\mathcal{I}'}_{\lhd \sigma c}$$
$$\Rightarrow \mathcal{I}' \models a : C_{\lhd \sigma c} \sqcap D_{\lhd \sigma c}.$$

Case universal quantification $\forall R.C$.

$$\mathcal{I} \models a : \forall R.C \rhd \sigma c$$
$$\Rightarrow (\forall R.C)^{\mathcal{I}}(a^{\mathcal{I}}) \rhd \sigma c$$
$$\Rightarrow \inf_{d' \in \Delta^{\mathcal{I}}} \{\max(-R^{\mathcal{I}}(a^{\mathcal{I}}, d'), C^{\mathcal{I}}(d'))\} \rhd \sigma c$$
$$\Rightarrow \bigwedge_{d' \in \Delta^{\mathcal{I}}} (\max(-R^{\mathcal{I}}(a^{\mathcal{I}}, d'), C^{\mathcal{I}}(d')) \rhd \sigma c)$$
$$\Rightarrow \bigwedge_{d' \in \Delta^{\mathcal{I}}} ((-R^{\mathcal{I}}(a^{\mathcal{I}}, d') \rhd \sigma c) \vee (C^{\mathcal{I}}(d') \rhd \sigma c))$$
$$\Rightarrow \bigwedge_{d' \in \Delta^{\mathcal{I}}} ((R^{\mathcal{I}}(a^{\mathcal{I}}, d') \neg \rhd \sigma \bar{c}) \vee (C^{\mathcal{I}}(d') \rhd \sigma c))$$
$$\Rightarrow \forall d' \in \Delta^{\mathcal{I}}.(((a^{\mathcal{I}}, d') \in R^{\mathcal{I}'}_{\neg \rhd \sigma \bar{c}}) \vee (d' \in C^{\mathcal{I}'}_{\rhd \sigma c}))$$
$$\Rightarrow \forall d' \in \Delta^{\mathcal{I}'}.(((a^{\mathcal{I}}, d') \notin R^{\mathcal{I}'}_{+ \rhd \sigma \bar{c}}) \vee (d' \in C^{\mathcal{I}'}_{\rhd \sigma c}))$$
$$\Rightarrow a^{\mathcal{I}} - \{d \in \Delta^{\mathcal{I}} \mid \forall d' \in \Delta^{\mathcal{I}} : (d, d') \notin R^{\mathcal{I}'}_{+ \rhd \sigma \bar{c}} \vee d' \in C^{\mathcal{I}'}_{\rhd \sigma c}\}$$
$$\Rightarrow a^{\mathcal{I}'} = \{d \in \Delta^{\mathcal{I}'} \mid \forall d' \in \Delta^{\mathcal{I}'} : (d, d') \notin R^{\mathcal{I}'}_{+ \rhd \sigma \bar{c}} \vee d' \in C^{\mathcal{I}'}_{\rhd \sigma c}\}$$
$$\Rightarrow a^{\mathcal{I}'} \in (\forall R_{+ \rhd \sigma \bar{c}}.C_{\rhd \sigma c})^{\mathcal{I}'}$$
$$\Rightarrow \mathcal{I}' \models a : \forall R_{+ \rhd \sigma \bar{c}} C_{\rhd \sigma c}$$

$$\mathcal{I} \models a : \forall R.C \lhd \sigma c$$
$$\Rightarrow (\forall R.C)^{\mathcal{I}}(a^{\mathcal{I}}) \lhd \sigma c$$
$$\Rightarrow \inf_{d' \in \Delta^{\mathcal{I}}} \{\max(-R^{\mathcal{I}}(a^{\mathcal{I}}, d'), C^{\mathcal{I}}(d'))\} \lhd \sigma c$$
$$\Rightarrow \bigvee_{d' \in \Delta^{\mathcal{I}}} (\max(-R^{\mathcal{I}}(a^{\mathcal{I}}, d'), C^{\mathcal{I}}(d')) \lhd \sigma c)$$
$$\Rightarrow \bigvee_{d' \in \Delta^{\mathcal{I}}} ((-R^{\mathcal{I}}(a^{\mathcal{I}}, d') \lhd \sigma c) \wedge (C^{\mathcal{I}}(d') \lhd \sigma c))$$
$$\Rightarrow \bigvee_{d' \in \Delta^{\mathcal{I}}} ((R^{\mathcal{I}}(a^{\mathcal{I}}, d')\neg \lhd \sigma \bar{c}) \wedge (C^{\mathcal{I}}(d') \lhd \sigma c))$$
$$\Rightarrow \exists d' \in \Delta^{\mathcal{I}}.(((a^{\mathcal{I}}, d') \in R^{\mathcal{I}'}_{\neg \lhd \sigma \bar{c}}) \wedge (d' \in C^{\mathcal{I}'}_{\lhd \sigma c}))$$
$$\Rightarrow a^{\mathcal{I}} = \{d \in \Delta^{\mathcal{I}} \mid \exists d' \in \Delta^{\mathcal{I}} : (d, d') \in R^{\mathcal{I}'}_{\neg \lhd \sigma \bar{c}} \wedge d' \in C^{\mathcal{I}'}_{\lhd \sigma c}\}$$
$$\Rightarrow a^{\mathcal{I}'} = \{d \in \Delta^{\mathcal{I}'} \mid \exists d' \in \Delta^{\mathcal{I}'} : (d, d') \in R^{\mathcal{I}'}_{\neg \lhd \sigma \bar{c}} \wedge d' \in C^{\mathcal{I}'}_{\lhd \sigma c}\}$$
$$\Rightarrow a^{\mathcal{I}'} \in (\exists R_{\neg \lhd \sigma \bar{c}}.C_{\lhd \sigma c})^{\mathcal{I}'}$$
$$\Rightarrow \mathcal{I}' \models a : \exists R_{\neg \lhd \sigma \bar{c}}.C_{\lhd \sigma c}.$$

Case existential quantification $\exists R.C$.

$$\mathcal{I} \models a : \exists R.C \rhd \sigma c$$
$$\Rightarrow (\exists R.C)^{\mathcal{I}}(a^{\mathcal{I}}) \rhd \sigma c$$
$$\Rightarrow \sup_{d' \in \Delta^{\mathcal{I}}} \{\min(R^{\mathcal{I}}(a^{\mathcal{I}}, d'), C^{\mathcal{I}}(d'))\} \rhd \sigma c$$
$$\Rightarrow \bigvee_{d' \in \Delta^{\mathcal{I}}} (\min(R^{\mathcal{I}}(a^{\mathcal{I}}, d'), C^{\mathcal{I}}(d')) \rhd \sigma c)$$
$$\Rightarrow \bigvee_{d' \in \Delta^{\mathcal{I}}} ((R^{\mathcal{I}}(a^{\mathcal{I}}, d') \rhd \sigma c) \wedge (C^{\mathcal{I}}(d') \rhd \sigma c))$$
$$\Rightarrow \exists d' \in \Delta^{\mathcal{I}}.(((a^{\mathcal{I}}, d') \in R^{\mathcal{I}'}_{\rhd \sigma c}) \wedge (d' \in C^{\mathcal{I}'}_{\rhd \sigma c}))$$
$$\Rightarrow a^{\mathcal{I}} = \{d \in \Delta^{\mathcal{I}} \mid \exists d' \in \Delta^{\mathcal{I}} : (d, d') \in R^{\mathcal{I}'}_{\rhd \sigma c} \wedge d' \in C^{\mathcal{I}'}_{\rhd \sigma c}\}$$
$$\Rightarrow a^{\mathcal{I}'} = \{d \in \Delta^{\mathcal{I}'} \mid \exists d' \in \Delta^{\mathcal{I}'} : (d, d') \in R^{\mathcal{I}'}_{\rhd \sigma c} \wedge d' \in C^{\mathcal{I}'}_{\rhd \sigma c}\}$$
$$\Rightarrow a^{\mathcal{I}'} \in (\exists R_{\rhd \sigma c}.C_{\rhd \sigma c})^{\mathcal{I}'}$$
$$\Rightarrow \mathcal{I}' \models a : \exists R_{\rhd \sigma c}.C_{\rhd \sigma c}.$$

$$\mathcal{I} \models a : \exists R.C \lhd \sigma c$$
$$\Rightarrow (\exists R.C)^{\mathcal{I}}(a^{\mathcal{I}}) \lhd \sigma c$$
$$\Rightarrow \sup_{d' \in \Delta^{\mathcal{I}}} \{\min(R^{\mathcal{I}}(a^{\mathcal{I}}, d'), C^{\mathcal{I}}(d'))\} \lhd \sigma c$$
$$\Rightarrow \bigwedge_{d' \in \Delta^{\mathcal{I}}} (\min(R^{\mathcal{I}}(a^{\mathcal{I}}, d'), C^{\mathcal{I}}(d')) \lhd \sigma c)$$
$$\Rightarrow \bigwedge_{d' \in \Delta^{\mathcal{I}}} ((R^{\mathcal{I}}(a^{\mathcal{I}}, d') \lhd \sigma c) \vee (C^{\mathcal{I}}(d') \lhd \sigma c))$$
$$\Rightarrow \forall d' \in \Delta^{\mathcal{I}}.(((a^{\mathcal{I}}, d') \in R^{\mathcal{I}'}_{\lhd \sigma c}) \vee (d' \in C^{\mathcal{I}'}_{\lhd \sigma c}))$$
$$\Rightarrow \forall d' \in \Delta^{\mathcal{I}}.(((a^{\mathcal{I}}, d') \notin R^{\mathcal{I}'}_{-\lhd \sigma c}) \vee (d' \in C^{\mathcal{I}'}_{\rhd \sigma c}))$$
$$\Rightarrow a^{\mathcal{I}} = \{d \in \Delta^{\mathcal{I}} \mid \forall d' \in \Delta^{\mathcal{I}} : (d, d') \notin R^{\mathcal{I}'}_{-\lhd \sigma c} \vee d' \in C^{\mathcal{I}'}_{\lhd \sigma c}\}$$
$$\Rightarrow a^{\mathcal{I}'} = \{d \in \Delta^{\mathcal{I}'} \mid \forall d' \in \Delta^{\mathcal{I}'} : (d, d') \notin R^{\mathcal{I}'}_{-\lhd \sigma c} \vee d' \in C^{\mathcal{I}'}_{\lhd \sigma c}\}$$
$$\Rightarrow a^{\mathcal{I}'} \in (\forall R_{-\lhd \sigma c}.C_{\lhd \sigma c})^{\mathcal{I}'}$$
$$\Rightarrow \mathcal{I}' \models a : \forall R_{-\lhd \sigma c}.C_{\lhd \sigma c}.$$

The proof shows that for each $\alpha \bowtie \sigma c \in \mathcal{A}$ if $\mathcal{I} \models \alpha \bowtie \sigma c$ then $\mathcal{I}' \models \theta(\alpha \bowtie \sigma c)$ which implies that $\mathcal{I} \models \mathcal{A} \Rightarrow \mathcal{I}' \models \theta(\mathcal{A})$.

(2) Now we prove that $\mathcal{I}' \models \mathcal{T}(N^{\mathfrak{f}\mathcal{K}}) \cup \kappa(\mathfrak{f}\mathcal{K}, \mathcal{T})$.

It is trivial that $\mathcal{I}' \models \mathcal{T}(N^{\mathfrak{f}\mathcal{K}})$ according to our basic idea.

Let $C \sqsubseteq D \in \mathcal{T}$, then for all $\sigma c \in \mathcal{N}^{\mathfrak{f}\mathcal{K}}$, $C_{\rhd \sigma c} \sqsubseteq D_{\rhd \sigma c} \in \kappa(\mathfrak{f}\mathcal{K}, C \sqsubseteq D)$ and $D_{\lhd \sigma c} \sqsubseteq C_{\lhd \sigma c} \in \kappa(\mathfrak{f}\mathcal{K}, C \sqsubseteq D)$.

$$\mathcal{I} \models C \sqsubseteq D \Rightarrow \forall d \in \Delta^{\mathcal{I}}.C^{\mathcal{I}}(d) \le D^{\mathcal{I}}(d)$$
$$\Rightarrow \text{if } C^{\mathcal{I}}(d) \rhd \sigma c \text{ then } D^{\mathcal{I}}(d) \rhd \sigma c$$
$$\Rightarrow \text{if } d \in C^{\mathcal{I}'}_{\rhd \sigma c} \text{ then } d \in D^{\mathcal{I}'}_{\rhd \sigma c}$$
$$\Rightarrow C^{\mathcal{I}'}_{\rhd \sigma c} \subseteq D^{\mathcal{I}'}_{\rhd \sigma c}$$
$$\Rightarrow \mathcal{I}' \models C_{\rhd \sigma c} \sqsubseteq D_{\rhd \sigma c}.$$

$$\mathcal{I} \models C \sqsubseteq D \Rightarrow \forall d \in \Delta^{\mathcal{I}}.C^{\mathcal{I}}(d) \le D^{\mathcal{I}}(d)$$
$$\Rightarrow \text{if } D^{\mathcal{I}}(d) \lhd \sigma c \text{ then } C^{\mathcal{I}}(d) \lhd \sigma c$$
$$\Rightarrow \text{if } d \in D^{\mathcal{I}'}_{\lhd \sigma c} \text{ then } d \in C^{\mathcal{I}'}_{\lhd \sigma c}$$
$$\Rightarrow D^{\mathcal{I}'}_{\lhd \sigma c} \subseteq C^{\mathcal{I}'}_{\lhd \sigma c}$$
$$\Rightarrow \mathcal{I}' \models D_{\lhd \sigma c} \sqsubseteq C_{\lhd \sigma c}.$$

So for each $C \sqsubseteq D \in \mathcal{T}$, if $\mathcal{I} \models C \sqsubseteq D$ then $\mathcal{I}' \models \{C_{\rhd \sigma c} \sqsubseteq D_{\rhd \sigma c}, D_{\lhd \sigma c} \sqsubseteq C_{\lhd \sigma c}\}$. It follows that $\mathcal{I}' \models \kappa(\mathfrak{f}\mathcal{K}, C \sqsubseteq D)$. So $\mathcal{I}' \models \mathcal{T}(N^{\mathfrak{f}\mathcal{K}}) \cup \kappa(\mathfrak{f}\mathcal{K}, C \sqsubseteq D)$.

$\Leftarrow$.) Let $\mathcal{I}'$ be a finite model of $\mathcal{K}(\mathfrak{f}\mathcal{K})$ whose domain $\Delta^{\mathcal{I}'}$ is finite. We build an $\mathcal{ALC}_{\mathcal{FL}}$ interpretation $\mathcal{I}$ such that

- $\Delta^{\mathcal{I}} := \Delta^{\mathcal{I}'}$,
- $a^{\mathcal{I}} := a^{\mathcal{I}'}$ for all individual a,
- $\forall d \in \Delta^{\mathcal{I}}.A^{\mathcal{I}}(d) := \sigma' c'$ for all concept name A, where
  Let $\sigma_1 c_1 = \sup\{\sigma c \mid d \in A^{\mathcal{I}'}_{\rhd \sigma c}\}$, $\sigma_2 c_2 = \inf\{\sigma c \mid d \in A^{\mathcal{I}'}_{\lhd \sigma c}\}$ and $\delta \in H^*$ such that for all $\delta' \in H^*$ and $\delta' \ne \delta$, $\delta' \sigma c > \delta \sigma c > \sigma c$.
  1. Since $\mathcal{K}(\mathfrak{f}\mathcal{K})$ is satisfiable, if $\sigma_1 c_1 = \sigma_2 c_2$ then $\sigma' c' = \sigma_1 c_1 = \sigma_2 c_2$,
  2. otherwise if $\sigma_1 c_1 < \sigma_2 c_2$, $\sigma' c' = \delta \sigma_1 c_1$.
  If $\forall \sigma c.d \notin A^{\mathcal{I}'}_{\bowtie \sigma c}$, $\sigma' c' = \inf(False)$.
- $\forall d, d' \in \Delta^{\mathcal{I}}.R^{\mathcal{I}}(d, d') := \sigma' c'$ for all role name R, where
  Let $\sigma_1 c_1 = \sup\{\sigma c \mid (d, d') \in R^{\mathcal{I}'}_{\rhd \sigma c}\}$, $\sigma_2 c_2 = \inf\{\sigma c \mid (d, d') \in R^{\mathcal{I}'}_{\lhd \sigma c}\}$ and $\delta \in H^*$ such that for all $\delta' \in H^*$ and $\delta' \ne \delta$, $\delta' \sigma c > \delta \sigma c > \sigma c$.
  1. Since $\mathcal{K}(\mathfrak{f}\mathcal{K})$ is satisfiable, if $\sigma_1 c_1 = \sigma_2 c_2$ then $\sigma' c' = \sigma_1 c_1 = \sigma_2 c_2$,
  2. otherwise if $\sigma_1 c_1 < \sigma_2 c_2$, $\sigma' c' = \delta \sigma_1 c_1$.
  If $\forall \sigma c.(d, d') \notin R^{\mathcal{I}'}_{\bowtie \sigma c}$, $\sigma' c' = \inf(False)$.

We have the following Lemma from our basic idea and the definition of the interpretation $\mathcal{I}$.

Lemma 19. *For all σc and for all $d, d' \in \Delta^{\mathcal{I}'}$, $d \in U^{\mathcal{I}'}_{\bowtie \sigma c} \Rightarrow U^{\mathcal{I}}(d) \bowtie \sigma c$ and $(d, d') \in R^{\mathcal{I}'}_{\bowtie \sigma c} \Rightarrow R^{\mathcal{I}}(d, d') \bowtie \sigma c$.*

Proof. Please refer to [20].

(1) For ABox, the proof is exactly the reverse processes of that of the $\Rightarrow$.) from which we can prove that if $\mathcal{I}' \models \theta(\mathcal{A}')$ then $\mathcal{I} \models \mathcal{A}$.

(2) For all $\sigma c \in N^{\mathfrak{f}\mathcal{K}}$, $C_{\rhd \sigma c} \sqsubseteq D_{\rhd \sigma c} \in \kappa(\mathfrak{f}\mathcal{K}, \mathcal{T})$, then for all $d \in C^{\mathcal{I}'}_{\rhd \sigma c}$, $d \in D^{\mathcal{I}'}_{\rhd \sigma c}$. Therefore, if $C^{\mathcal{I}}(d) \ge \sigma c$ then $D^{\mathcal{I}}(d) \ge \sigma c$.

Assume $\mathcal{I}' \models \mathcal{T}'$ and $\mathcal{I} \nvDash C \sqsubseteq D$ where $C \sqsubseteq D \in \mathcal{T}$. So there exists a $d' \in \Delta^{\mathcal{I}}$ such that $C^{\mathcal{I}}(d') > D^{\mathcal{I}}(d')$. Consider $C^{\mathcal{I}}(d') = \sigma' c'$. Of course $C^{\mathcal{I}}(d') \ge \sigma' c'$. Therefore, $D^{\mathcal{I}}(d') \ge \sigma' c'$. From the hypothesis it follows that $\sigma' c' = C^{\mathcal{I}}(d') > D^{\mathcal{I}}(d') \ge \sigma' c'$, which contradicts the hypothesis. So $\mathcal{I} \models \mathcal{T}$.

5 Discussion

In this paper, we have presented a satisfiability preserving transformation of $\mathcal{ALC}_{\mathcal{FL}}$ into $\mathcal{ALCH}$ which is with general TBox and role hierarchy. Since all other reasoning tasks such as entailment problem and subsumption problem can be reduced to satisfiability problem, this result allows for algorithms and complexity results that were found for $\mathcal{ALCH}$ to be applied to $\mathcal{ALC}_{\mathcal{FL}}$.

As for the complexity of the transformation, we know that,

1. $|\theta(\mathcal{A})|$ is linearly bounded by $|\mathcal{A}|$;
2. $|\mathcal{T}(N^{\mathfrak{f}\mathcal{K}})| = 2|\mathcal{A}^{\mathfrak{f}\mathcal{K}}|(|\mathcal{N}^{\mathfrak{f}\mathcal{K}}| - 1) + 2|\mathcal{R}^{\mathfrak{f}\mathcal{K}}|(|\mathcal{N}^{\mathfrak{f}\mathcal{K}}| - 1)$;
3. $\kappa(\mathfrak{f}\mathcal{K}, \mathcal{T})$ contains at most $4|\mathcal{T}||N^{\mathfrak{f}\mathcal{K}}|$.

Therefore, the resulted classical knowledge base (at most polynomial size) can be constructed in polynomial time.

The work of Straccia [19] transforms fuzzy $\mathcal{ALCH}$ into classical $\mathcal{ALCH}$. The truth domains of fuzzy $\mathcal{ALCH}$ is different from that of $\mathcal{ALC}_{\mathcal{FL}}$. $\mathcal{ALC}_{\mathcal{FL}}$ uses hedges as the fuzzy extension and the truth domain of interpretations is represented by a hedge algebra. Moreover, the hedges occur not only in the fuzzy values but also in concept terms. Thus there is one more rule for dealing with modifier concept terms in our current work.

Many approaches to transformation various fuzzy DLs into classical DLs have been proposed. Boillo et al. [3] proposed a reasoning preserving reduction for the fuzzy DL $\mathcal{SROIQ}$ under Gödel semantics to the crisp case. In the reduction, concept and role modifiers are allowed. While the truth domains of fuzzy DL $\mathcal{SROIQ}$ is not represented by a hedge algebra either. Bobillo and Straccia [5] have proposed a general framework for fuzzy DLs with a finite chain of degrees of truth N which can be seen as a finite totally ordered set of linguistic terms or labels. They also provided a a reasoning preserving reduction to the crisp case. Bobillo and Straccia [6] have shown that a fuzzy extension of SROIQ is decidable over a finite set of truth values by presenting a reasoning preserving procedure to obtain a non-fuzzy representation for the logic. This fuzzy extension of the logic SROIQ is the logic behind the language OWL 2. This reduction makes it possible to reuse current representation languages as well as currently available reasoners for ontologies.

There exist some reasoners for fuzzy DLs, e.g. *FiRE* [17], *GURDL* [7], *DeLorean* [2], *GERDS* [8], *YADLR* [11] and *fuzzyDL* [4]. Among them, *fuzzyDL* allows modifiers defined in terms of linear hedges and triangular functions and *DeLorean* supports triangularly-modified concept. So the approaches to transformation variety of fuzzy DLs into classical DLs make it possible to use the already existing resources for classical DL systems to do the reasoning of fuzzy DLs without adapting fuzzy DLs to some other fuzzy language.

References

1. Baader, F., Calvanese, D., McGuinness, D.L., Nardi, D., Patel-Schneider, P.F. (eds.): The Description Logic Handbook: Theory, Implementation, and Applications. Cambridge University Press (2003)

2. Bobillo, F., Delgado, M., Gómez-Romero, J.: Optimizing the Crisp Representation of the Fuzzy Description Logic $\mathcal{SROIQ}$. In: da Costa, P.C.G., d'Amato, C., Fanizzi, N., Laskey, K.B., Laskey, K.J., Lukasiewicz, T., Nickles, M., Pool, M. (eds.) URSW 2005-2007. LNCS (LNAI), vol. 5327, pp. 189–206. Springer, Heidelberg (2008)

3. Bobillo, F., Delgado, M., Gómez-Romero, J., Straccia, U.: Fuzzy description logics under gödel semantics. Int. J. Approx. Reasoning 50(3), 494–514 (2009)

4. Bobillo, F., Straccia, U.: fuzzyDL: An expressive fuzzy description logic reasoner. In: 2008 International Conference on Fuzzy Systems, FUZZ 2008. IEEE Computer Society (2008)

5. Bobillo, F., Straccia, U.: Finite fuzzy description logics: A crisp representation for finite fuzzy ALCH. In: URSW, pp. 61–72 (2010)

6. Bobillo, F., Straccia, U.: Reasoning with the finitely many-valued lukasiewicz fuzzy description logic SROIQ. Inf. Sci. 181(4), 758–778 (2011)

7. Haarslev, V., Pai, H.-I., Shiri, N.: Optimizing tableau reasoning in alc extended with uncertainty. In: Description Logics (2007)

8. Habiballa, H.: Resolution strategies for fuzzy description logic. In: EUSFLAT Conf. (2), pp. 27–36 (2007)

9. Hölldobler, S., Dzung, D.-K., Dinh-Khang, T.: The fuzzy linguistic description logic $\mathcal{ALC_{FL}}$. In: Proceedings of the Eleventh International Conference on Information Processing and Management of Uncertainty in Knowledge-Based Systems, IPMU, pp. 2096–2103 (2006)

10. Hölldobler, S., Störr, H.-P., Tran, D.K.: The fuzzy description logic $\mathcal{ALC_{FH}}$ with hedge algebras as concept modifiers. Journal of Advanced Computational Intelligence and Intelligent Informatics, JACIII 7(3), 294–305 (2003)

11. Konstantopoulos, S., Apostolikas, G.: Fuzzy-DL Reasoning over Unknown Fuzzy Degrees. In: Meersman, R., Tari, Z., Herrero, P. (eds.) OTM 2007 Ws, Part II. LNCS, vol. 4806, pp. 1312–1318. Springer, Heidelberg (2007)

12. Lukasiewicz, T., Straccia, U.: Managing uncertainty and vagueness in description logics for the semantic web. Journal of Web Semantics 6, 291–308 (2008)

13. Nguyen, C.-H., Tran, D.-K., Huynh, V.-N., Nguyen, H.-C.: Hedge algebras, linguistic-valued logic and their application to fuzzy reasoning. International Journal of Uncertainty, Fuzziness and Knowledge-Based Systems 7(4), 347–361 (1999)

14. Nguyen, C.-H., Wechler, W.: Hedge algebras: an algebraic approach to structure of sets of linguistic truth values. Fuzzy Sets and Systems 35(3), 281–293 (1990)

15. Nguyen, C.-H., Wechler, W.: Extended hegde algebras and their application to fuzzy logic. International Journal of Uncertainty, Fuzziness and Knowledge-Based Systems 52, 259–281 (1992)

16. Nguyen, H.-N., Hölldobler, S., Tran, D.-K.: The fuzzy description logic $\mathcal{ALC_{FCH}}$. In: Proc. 9th IASTED International Conference on Artificial Intelligence and Soft Computing, pp. 99–104 (2005)

17. Simou, N., Kollias, S.: Fire: A fuzzy reasoning engine for impecise knowledge. K-Space PhD Students Workshop, Berlin, Germany (September 14, 2007)

18. Straccia, U.: Reasoning within fuzzy description logics. JAIR 14, 137–166 (2001)

19. Straccia, U.: Transforming Fuzzy Description Logics into Classical Description Logics. In: Alferes, J.J., Leite, J. (eds.) JELIA 2004. LNCS (LNAI), vol. 3229, pp. 385–399. Springer, Heidelberg (2004)

20. Wu, Y.: Transforming fuzzy description logic $\mathcal{ALC_{FL}}$ into classical description logic $\mathcal{ALCH}$. Master Thesis (2007)

21. Zadeh, L.A.: The concept of a linguistic variable and its application to approximate reasoning - I. Information Sciences 8(3), 199–249 (1975)
22. Zadeh, L.A.: The concept of a linguistic variable and its application to approximate reasoning - II. Information Sciences 8(4), 301–357 (1975)
23. Zadeh, L.A.: The concept of a linguistic variable and its application to approximate reasoning - III. Information Sciences 9(1), 43–80 (1975)

A Fuzzy Logic-Based Approach to Uncertainty Treatment in the Rule Interchange Format: From Encoding to Extension

Jidi Zhao[1,3], Harold Boley[2], and Jing Dong[3]

[1] Shanghai Jiao Tong University, Shanghai 200052, China,
JudyZhao33@gmail.com, JingDong@sjtu.edu.cn
[2] Institute for Information Technology, National Research Council of Canada,
Fredericton NB E3B 9W4, Canada
Harold.Boley@nrc.gc.ca
[3] East China Normal University, Shanghai 200062, China

Abstract. The Rule Interchange Format (RIF) is a W3C recommendation that allows rules to be exchanged between rule systems. Uncertainty is an intrinsic feature of real world knowledge, hence it is important to take it into account when building logic rule formalisms. However, the set of truth values in the RIF Basic Logic Dialect (RIF-BLD) currently consists of only two values (t and f), although the RIF Framework for Logic Dialects (RIF-FLD) allows for more. In this paper, we first present two techniques of encoding uncertain knowledge and its fuzzy semantics in RIF-BLD presentation syntax. We then propose an extension leading to an Uncertainty Rule Dialect (RIF-URD) to support a direct representation of uncertain knowledge. In addition, rules in Logic Programs (LP) are often used in combination with the other widely-used knowledge representation formalism of the Semantic Web, namely Description Logics (DL), in many application scenarios of the Semantic Web. To prepare DL as well as LP extensions, we present a fuzzy extension to Description Logic Programs (DLP), called Fuzzy DLP, and discuss its mapping to RIF. Such a formalism not only combines DL with LP, as in DLP, but also supports uncertain knowledge representation.

Keywords: Rule Interchange Format, Uncertainty, Fuzzy Logic.

1   Introduction

Description Logics (DL) and Logic Programs (LP)[1] are the two main paradigms of knowledge representation formalisms for the Semantic Web, both of which are based on subsets of first-order logic [19]. DL and LP cover different but overlapping areas of knowledge representation. They are complementary to some degree, for example, typical DL cannot directly express LP's n-ary function applications (complex terms) while classic LP cannot express DL's disjunctions (in the head).

---

[1] In this paper, we only consider the Horn Logic subset of LP, without negation-as-failure.

F. Bobillo et al. (Eds.): URSW 2008-2010/UniDL 2010, LNAI 7123, pp. 197–216, 2013.

Combining DL with LP in order to "build rules on top of ontologies" or, "build ontologies on top of rules" has become an emerging topic for various applications of the Semantic Web. It is therefore important to research the combination of DL and LP with different strategies. There have been various achievements in this area, including several proposed combination frameworks [9, 11, 17, 22, 23]. As a minimal approach in this area, the Description Logic Program (DLP) 'intersection' of DL and LP has been studied, along with mappings from DL to LP [11]. Both [9] and [22] studied the combination of standard Datalog inference procedures with intermediate DL satisfiability checking.

On the other hand, as evidenced by Fuzzy RuleML [6] and W3C's Uncertainty Reasoning for the World Wide Web (URW3) Incubator Group [20], handling uncertain knowledge is becoming a critical research direction for the (Semantic) Web. For example, many concepts needed in business ontology modeling lack well-defined boundaries or, precisely defined criteria of relationships with other concepts. To take care of these knowledge representation needs, different approaches for integrating uncertain knowledge into traditional rule languages and DL languages have been studied [7, 18, 19, 21, 26–29, 31, 34].

The Rule Interchange Format (RIF) has been developed by W3C's Rule Interchange Format (RIF) Working Group to support the exchange of rules between rule systems [4]. In particular, the Basic Logic Dialect (RIF-BLD) corresponds to the language of definite Horn rules with equality and a first-order semantics. While RIF's Framework for Logic-based Dialects (RIF-FLD) [5] permits multi-valued logics, RIF-BLD instantiates RIF-FLD with the set of truth values consisting of only two values, t and f, hence is not designed for expressing uncertain knowledge.

According to the final report from the URW3 Incubator group, uncertainty is a term intended to include different types of uncertain knowledge, including incompleteness, vagueness, ambiguity, randomness, and inconsistency [20]. Mathematical theories for representing uncertain knowledge include, but are not limited to, Probability, Fuzzy Sets, Belief Functions, Random Sets, Rough Sets, and combinations of several models (Hybrid). The uncertain knowledge representations and interpretations discussed in this paper are limited to Fuzzy Sets and Fuzzy Logic (a multi-valued logic based on Fuzzy set theory); other approaches should be studied in future work.

The main contributions of this paper are: (1) two techniques of encoding uncertain information in RIF as well as an uncertainty extension to RIF; (2) an extension of DLP to Fuzzy DLP and the mapping of Fuzzy DLP to RIF.

Two earlier uncertainty extensions to the combination of DL and LP that we can expand on are [30] and [32]. While our approach emphasizes the interoperation in the intersection of fuzzy DL and fuzzy LP allows DL atoms in the head of hybrid rules and DL subsumption axioms in hybrid rules, the approach of [30] does not allow the expressiveness. Our approach deals with fuzzy subsumption of fuzzy concepts of the form $C \sqsubseteq D = c$ whereas [32] deals with crisp subsumption of fuzzy concepts of the form $C \sqsubseteq D$. Also, we do not limit hybrid

knowledge bases to the intersection of (fuzzy) DL and (fuzzy) LP. We extend
[32] and study the decidable union of DL and LP.

The rest of this paper is organized as follows. Section 2 reviews earlier work
on the interoperation between DL and LP in the intersection of these two formalisms (known as DLP) and represents the DL-LP mappings in RIF. Section 3
addresses the syntax and semantics of fuzzy Logic Programs, and then presents
two techniques of bringing uncertainty into the RIF presentation syntax (and
then into its semantics and XML syntax), using encodings as RIF functions
and RIF predicates. Section 4 adapts the definition of the set of truth values
in RIF-FLD for the purpose of representing uncertain knowledge directly, and
proposes the new Uncertainty Rule Dialect (RIF-URD), extending RIF-BLD.
Section 5 extends DLP to Fuzzy DLP, supporting mappings between fuzzy DL
and fuzzy LP, and gives representations of Fuzzy DLP in RIF and RIF-URD.
Finally, Section 6 summarizes our main results and gives an outlook on future
research.

2   Description Logic Programs and Their Representation in RIF

In this section, we summarize the work on Description Logic Programs (DLP)
[11] and then show how to represent the mappings between DL and LP in RIF
presentation syntax.

The paper [11] studied the intersection between the leading Semantic Web
approaches to rules in LP and ontologies in DL, and showed how to interoperate
between DL and LP in the intersection known as DLP. A DLP knowledge base
permits:

1. stating that a class C is a *Subclass* of a class D, $C \sqsubseteq D$;
2. stating that the *Domain* of a property R is a class C, $\top \sqsubseteq \forall R^-.C$;
3. stating that the *Range* of a property R is a class C, $\top \sqsubseteq \forall R.C$;
4. stating that a property R is a *Subproperty* of a property P, $R \sqsubseteq P$;
5. stating that an individual a is an Instance of a class C, $C(a)$;
6. stating that a pair of individuals (a, b) is an Instance of a property R, $R(a, b)$;
7. using the *Intersection* connective (conjunction) within class descriptions, $C_1 \sqcap C_2$;
8. using the *Union* connective (disjunction) within subclass descriptions, $C_1 \sqcup C_2 \sqsubseteq D$;
9. using *Universal quantification* within superclass descriptions, $C \sqsubseteq \forall R.D$;
10. using *Existential quantification* within subclass descriptions $\exists R.C \sqsubseteq D$;
11. stating that a property R is *Transitive*, $R^+ \sqsubseteq R$;
12. stating that a property R is the Inverse of a property P.

Here C, D, C_1, C_2 are concepts, $\top$ is the universal concept, R, P are roles, R^-
and R^+ are the inverse role and the transitive role of R, respectively, and a, b
are individuals.

In RIF presentation syntax, the quantifiers Exists and Forall are made explicit, rules are written with a ":-" infix, variables start with a "?" prefix, and whitespace is used as a separator. Table 1 summarizes the mappings in [11] between DL and LP in the DLP intersection, and shows its representation in RIF. Note that in DLP, a complex concept expression which is a disjunction (e.g. $C_1 \sqcup C_2$) or an existential (e.g. $\exists R.C$) is not allowed in the right side of a concept subsumption axiom (superclass).

Table 1. Mapping between LP and DL

LP Syntax	DL Syntax	RIF
$D(x) \leftarrow C(x)$	$C \sqsubseteq D$	Forall ?x (D(?x) :- C(?x))
$C(y) \leftarrow R(x,y)$	$\top \sqsubseteq \forall R.C$	Forall ?x ?y (C(?y) :- R(?x ?y))
$C(x) \leftarrow R(x,y)$	$\top \sqsubseteq \forall R^-.C$	Forall ?x ?y (C(?x) :- R(?x ?y))
$P(x,y) \leftarrow R(x,y)$	$R \sqsubseteq P$	Forall ?x ?y (P(?x ?y) :- R(?x ?y))
$C(a)$	$C(a)$	C(a)
$R(a,b)$	$R(a,b)$	R(a,b)
$D(x) \leftarrow C_1(x) \wedge C_2(x)$	$C_1 \sqcap C_2 \sqsubseteq D$	Forall ?x (D(?x) :- And(C_1(?x) C_2(?x)))
$D_1(x) \leftarrow C(x),$ $D_2(x) \leftarrow C(x),$	$C \sqsubseteq D_1 \sqcap D_2$	Forall ?x (D_1(?x) :- C(?x)) Forall ?x (D_2(?x) :- C(?x))
$D(x) \leftarrow C_1(x),$ $D(x) \leftarrow C_2(x)$	$C_1 \sqcup C_2 \sqsubseteq D$	Forall ?x (D(?x) :- C_1(?x)) Forall ?x (D(?x) :- C_2(?x))
$D(y) \leftarrow C(x), R(x,y)$	$C \sqsubseteq \forall R.D$	Forall ?x ?y (D(?y) :- And(C(?x) R(?x ?y)))
$D(x) \leftarrow C(y), R(x,y)$	$\exists R.C \sqsubseteq D$	Forall ?x ?y (D(?x) :- And(C(?y) R(?x ?y)))
$R(x,z) \leftarrow R(x,y), R(y,z)$	$R^+ \sqsubseteq R$	Forall ?x ?y ?z(R(?x ?z) :- And(R(?x ?y) R(?y ?z)))
$R(x,y) \leftarrow P(y,x),$ $P(y,x) \leftarrow R(x,y)$	$P \equiv R^-$	Forall ?x ?y(R(?x ?y) :- P(?y ?x)) Forall ?x ?y(P(?y ?x) :- R(?x ?y))

3   Encoding Uncertainty in RIF

Fuzzy set theory was introduced in [37] as an extension of the classical notion of sets to capture the inherent vagueness (the lack of crisp boundaries) of real-world sets. Formally, a fuzzy set A with respect to a set of elements X (also called a universe) is characterized by a membership function $\mu_A(x)$ which assigns a value in the real unit interval [0,1] to each element $x \in X$. $\mu_A(x)$ gives the degree to which an element x belongs to the set A. Fuzzy logic is a form of multi-valued logic derived from fuzzy set theory to deal with reasoning that is approximate rather than precise. In Fuzzy Logic the degree of truth of a statement can range between 0 and 1 and is not constrained to the two truth values, t and f, as in classic predicate logic [24]. Such degrees can be computed based on various specific membership functions, for example, a trapezoidal function.

Fuzzy Logic extends the Boolean operations defined on crisp sets and relations for fuzzy sets and fuzzy relations. Basic operations in Fuzzy Logic apply to fuzzy sets include negation, intersection, union, and implication. Today, in the broader sense, Fuzzy Logic is actually a family of fuzzy operations [35] [13] divided into different classes, among which, the most widely known include Zadeh Logic [37], Lukasiewicz Logic [16], Product Logic [13], Gödel Logic [1, 10], and Yager Logic [36]. For example, in Zadeh Logic, the membership function of the union of two fuzzy sets is defined as the *maximum* of the two membership functions for the two fuzzy sets (the *maximum* criterion); the membership function of the intersection of two fuzzy sets is defined as the *minimum* of the two membership functions (the *minimum* criterion); while the membership function of the complement of a fuzzy set is defined as the negation of the specified membership function (the *negation* criterion).

In this section, we first present the syntax and semantics for fuzzy Logic Programs based on Fuzzy Sets and Fuzzy Logic [37] and on previous work on fuzzy LP [31, 33, 34], and then propose two techniques of encoding the semantics of uncertain knowledge based on Fuzzy Logic in the presentation syntax of RIF-BLD using BLD functions and BLD predicates respectively.

3.1 Fuzzy Logic Programs

Rules in van Emden's formalism for fuzzy LP have the syntactic form

$$H \leftarrow_c B_1, \cdots, B_n \tag{1}$$

where H, B_i are atoms, $n \geq 0$, and the factor c is a real number in the interval [0,1] [31]. For $n = 0$, such fuzzy rules degenerate to fuzzy facts.

The fuzzy LP language proposed by [33, 34] is a generalization of van Emden's work [31]. Rules are constructed from an implication ($\leftarrow$) with a corresponding t-norm adjunction operator (f_1), and another t-norm operator denoted by f_2. A t-norm is a generalization to the many-valued setting of the conjunction connective. In their setting, a rule is of the form $H \leftarrow_{f_1} f_2(B_1, \cdots, B_n)$ *withCF c*, where the confidence factor c is a real number in the unit interval [0,1] and H, B_i are atoms with truth values in (0, 1]. If we take the operator f_1 as the product following Goguen implication and the operator f_2 as the Gödel t-norm (minimum), this is exactly of the form by van Emden [31].

In [40], we presented norm-parameterized fuzzy Description Logics. In this paper, we follow this norm-parameterized approach when considering the DL counterpart of the DLP and propose a corresponding norm-parameterized fuzzy extension to Logic Programs, more precisely, to the Horn Logic subset of Logic Programs. We call it norm-parameterized as we integrate different norms from the Fuzzy Logic family into the fuzzy extension. A fuzzy LP knowledge base consists of these norm parameters and a finite set of fuzzy rules. The norm parameters, $\overline{F}_{IN}$, $\overline{F}_{U}$, and F_{IM}, define the intersection, union, and implication operators respectively. Since only Horn Logic is considered, we can ignore the negation operation for now. A fuzzy rule has the following form:

$$H(\overrightarrow{x}) \leftarrow B_1(\overrightarrow{x_1}), \cdots, B_n(\overrightarrow{x_n}) \quad /c \tag{2}$$

Here $H(\vec{x})$, $B_i(\vec{x_i})$ are atoms, $\vec{x}$, $\vec{x_i}$ are vectors of variables or constants, $n \geq 0$ and the confidence factor c (also called certainty degree) is a real number in the interval $[0,1]$. For the special case of fuzzy facts this becomes $H \quad /c$. These forms with a "/" symbol have the advantages of avoiding possible confusion with the equality symbol usually used for functions in logics with equality, as well as using a unified and compact format to represent fuzzy rules and fuzzy facts.

The semantics of such fuzzy LP is an extension of classical LP semantics. Let B_R stand for the Herbrand base of a fuzzy knowledge base KB_{LP}. A fuzzy Herbrand interpretation H_I for KB_{LP} is defined as a mapping $B_R \rightarrow [0,1]$. It is a fuzzy subset of B_R under fuzzy semantics and can be specified by a function val with two arguments: a variable-free atom H (or $B_1, \cdots, B_n$) and a fuzzy Herbrand interpretation H_I. The returned result of the function val is the membership value of H (or $B_1, \cdots, B_n$) under H_I, denoted as $val(H, H_I)$ (or $val(B_i, H_I)$).

Therefore, if min is specified as the intersection operator and $\times$ is as the implication operator, a variable-free instance of a rule 2 is true under H_I iff $val(H, H_I) \geq c \times \min\{val(B_i, H_I)|i \in \{1, \cdots, n\}\}$ ($\min\{\}=1$ if $n = 0$). In other words, such an interpretation can be separated into the following two parts [12–14].

- The body of the rule consists of n atoms. Our confidence that all these atoms are true is interpreted under Gödel's semantics for fuzzy logic:
  $val((B_1, \cdots, B_n), H_I) = \min\{val(B_i, H_I)|i \in \{1, \cdots, n\}\}$
- The implication is interpreted as the product:
  $val(H, H_I) = c \times val((B_1, \cdots, B_n), H_I)$

Furthermore, a rule is true under H_I iff each variable-free instance of this rule is true under H_I and a fuzzy knowledge base KB_{LP} is true under H_I iff every rule in KB_{LP} is true under H_I. Such a Herbrand interpretation H_I is called a Herbrand model of KB_{LP}.

For a fuzzy knowledge base KB_{LP}, the reasoning task is a fuzzy entailment problem written as $KB_{LP} \models H \quad /c$ ($H \in B_R$, $c \in [0,1]$).

For simplicity, we take the min and $\times$ operators as default specifications in the examples presented hereafter.

Example 1. Consider the following fuzzy LP knowledge base:
$cheapFlight(x,y) \leftarrow affordableFlight(x,y) \quad /0.9 \quad (1)$
$affordableFlight(x,y) \quad /left\_shoulder0k4k1k3k(y) \quad (2)$

Figure 1 shows the left_shoulder membership function $left\_shoulder(0, 4000, 1000, 3000)$. We use the name $left\_shoulder0k4k1k3k$ for this parameterization. The function has the mathematical form

$$left\_shoulder0k4k1k3k(y)=\begin{cases} 1 & 0 \leq y \leq 1000 \\ -0.0005y + 1.5 & 1000 < y \leq 3000 \\ 0 & 3000 < y \leq 4000 \end{cases}$$

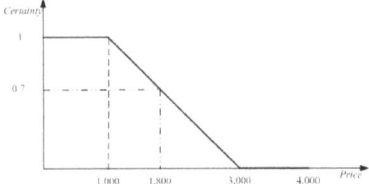

Fig. 1. A Left-shoulder Membership Function

For example, the certainty degree computed by this function for the fact $affordableFlight(flight0001, 1800)$ is 0.7.

Applying the semantics we discussed, $val(cheapFlight(flight0001, 1800), H_I)$ $= 0.9*0.7 = 0.63$, so we have that $KB_{LP} \models cheapFlight(flight0001, 1800) /0.63$.

Example 2. Consider the following fuzzy LP knowledge base:

$A(x) \leftarrow B(x), C(x) \quad /0.5 \qquad (1)$
$C(x) \leftarrow D(x) \quad /0.5 \qquad (2)$
$B(d) \quad /0.5 \qquad (3)$
$D(d) \quad /0.8 \qquad (4)$

We have that $KB_{LP} \models A(d) \qquad /0.2$. The reasoning steps of example 2 are described as follows:

$$\begin{aligned}
val(A(d), H_I) &= 0.5 \times \min(val(B(d), H_I), val(C(d), H_I)) \quad according to (1) \\
&= 0.5 \times \min(val(B(d), H_I), 0.5 \times val(D(d), H_I)) \quad according to (2) \\
&= 0.5 \times \min(0.5, 0.5 \times val(D(d), H_I)) \quad according to (3) \\
&= 0.5 \times \min(0.5, 0.5 \times 0.8) \quad according to (4) \\
&= 0.5 \times 0.4 \\
&= 0.2
\end{aligned}$$

3.2   Encoding Uncertainty Using RIF Functions

RIFs main logic dialect is RIF-BLD [3]. RIF-BLD corresponds to the language of definite Horn rules with equality and a standard first-order semantics. Syntactically, RIF-BLD has a number of extensions to support features such as objects and frames as in F-logic, internationalized resource identifiers (IRIs) as identifiers for concepts, and a rich set of datatypes and built-ins. RIF-BLD uses a standard first order semantics. For example, there is a rule in English describes that *A buyer buys an item from a seller if the seller sells the item to the buyer* and a fact *John sells LeRif to Mary*. Assuming Web IRIs for the predicates buy and sell, as well as for the individuals John, Mary, and LeRif, the above English text can be represented in the RIF-BLD Presentation Syntax as follows.

```
Document(
Base(<http://example.com/people#>)
Prefix(cpt <http://example.com/concepts#>)
Prefix(bks <http://example.com/books#>)
Group
( Forall ?Buyer ?Item ?Seller (
    cpt:buy(?Buyer ?Item ?Seller) :- cpt:sell(?Seller ?Item ?Buyer)
)
cpt:sell(<John> bks:LeRif "Mary" 'rif:iri)
))
```

One technique to encode uncertainty in logics with equality such as the current RIF-BLD (where equality in the head is "At Risk") is mapping all predicates to functions and using equality for letting them return uncertainty values [15]. We assume that H, B_i of the fuzzy rule of equation 2 contain variables in $\{?x_1, \cdots, ?x_k\}$ and that the head and body predicates are applied to terms $t_1, \cdots, t_r$ and $t_{j,1}, \cdots, t_{j,sj}$ ($1 \le j \le n$) respectively, which can all be variables, constants or complex terms. A fuzzy rule in the form of equation 2 can then be represented in RIF-BLD as (for simplicity, we will omit prefix declarations)

```
Document(
Group
( Forall ?x_1 ... ?x_k (
    h(t_1 ... t_r)=?c_h :- And(b_1(t_{1,1} ... t_{1,s1})=?c_1 ... b_n(t_{n,1} ... t_{n,sn})=?c_n
    ?c_t =External(F_IN(?c_1 ... ?c_n))
    ?c_h=External(F_IM(c ?c_t))) ))
```

Each predicate in the fuzzy rule thus becomes a function. Body predicates b_i ($1 \le i \le n$) in the fuzzy rule has uncertainty values between 0 and 1 by definition. The semantics of the fuzzy rules is then defined by the norm parameters: the intersection operator F_{IN} and the implication operator F_{IM}. For example, if F_{IN} and F_{IM} are specified using the minimum membership function and the multiply membership function respectively, the semantics of the fuzzy rules can be encoded in RIF-BLD using the built-in functions numeric-multiply from RIF Datatypes and Built-Ins (RIF-DTB) [25] and an aggregate function numeric-minimum proposed here as an addition to RIF-DTB (this could also be defined using rules). Based on the properties of the functions, it is fairly obvious that the uncertainty value for the variable $?c_t$ is a positive number less than 1 and the value for the variable $?c_h$ (i.e., the value returned for the head predicate function)is between 0 and 1. Therefore, each predicate in the fuzzy rule returns a uncertainty value between 0 and 1.

A fact of the form H /c can be represented in RIF-BLD presentation syntax as

```
h(t₁ ... tᵣ)=c
```

Example 3. We can rewrite example 1 using RIF functions as follows:

```
(* <http://example.org/fuzzy/membershipfunction > *)
Document(
 Group
 (
(*"Definition of membership function left_shoulder(0, 4000, 1000, 3000)" []*)
   Forall ?y(
     left_shoulder0k4k1k3k(?y)=1 :-
       And(External(numeric-less-than-or-equal(0 ?y))
       External(numeric-less-than-or-equal(?y 1000))))
   Forall ?y(
     left_shoulder0k4k1k3k(?y)=External(numeric-add(
       External(numeric-multiply(-0.0005 ?y)) 1.5))
       :- and(External(numeric-less-than(1000 ?y))
       External(numeric-less-than-or-equal(?y 3000))))
   Forall ?y(
     left_shoulder0k4k1e3k(?y)=0 :-
       And(External(numeric-less-than(3000 ?y))
       External(numeric-less-than-or-equal(?y 4000))))
. . .
) )
```

Note that membership function $left\_shoulder(0, 4000, 1000, 3000)$ is encoded as three rules.

```
Document(
 Import(<http://cxample.org/fuzzy/membershipfunction >)
 Group
 ( Forall ?x ?y(
     cheapFlight(?x ?y)  ?cₕ :- And(affordableFlight(?x ?y)=?c₁
       ?cₕ−External(numeric-multiply(0.9 ?c₁))))
   Forall ?x ?y(affordableFlight(?x ?y)=left_shaulder0k4k1k3k(?y))
) )
```

The Import statement loads the left_shoulder0k4k1k3k function defined at the given "< ... >" IRI.

Example 4. We can rewrite example 2 in RIF functions as follows:

```
Document(
 Group
 ( Forall ?x(
     A(?x)=?xₕ :- And(B(?c)=?c₁ C(?x)=?c₂
       ?cₜ =External(numeric-minimum(?c₁ ?c₂))
       ?cₕ=External(numeric-multiply(0.5 ?cₜ))))
   Forall ?x(
     C(?x)= ?cₕ :- And(D(?x)=?c₁ ?cₕ=External(numeric-multiply(0.5 ?c₁)))
   )
   B(d)=0.5
   D(d)=0.8
) )
```

3.3   Encoding Uncertainty Using RIF Predicates

Another encoding technique is making all n-ary predicates into (1+n)-ary predicates, each being functional in the first argument which captures the certainty factor of predicate applications. A fuzzy rule in the form of equation 2 can then be represented in RIF-BLD as

```
Document(
 Group
 ( Forall ?x₁ ... ?xₖ (
     h(?cₕ t₁ ... tᵣ) :- And(b₁(?c₁ t₁,₁ ... t₁,ₛ₁) ... bₙ(?cₙ tₙ,₁ ... t_{c,sn})
       ?cₜ =Exetrnal(F_{IN}(?c₁ ... ?cₙ))
       ?cₕ=Exetrnal(F_{IM}(c ?cₜ))    )
) )
```

Likewise, a fact of the form $H\ /c$ can be represented in RIF-BLD as

```
h(c t₁ ... tᵣ)
```

Example 5. We can rewrite example 1 in RIF predicates as follows:

```
Document(
 Import (<http://example.org/fuzzy/membershipfunction>)
 Group
 ( Forall ?x ?y(
     cheapFlight(?cₕ ?x ?y) :- And(affordabldFlight(?c₁ ?x ?y)
       ?cₕ=External(numberic-multiply(0.9 ?c₁)))
   )
   Forall ?x ?y(affordableFlight(?c₁?x ?y) :- ?c₁=left_shoulder0k4k1k3k(?y))
) )
```

4   Uncertainty Extension of RIF

In this section, we adapt the definition of the set of truth values from RIF-FLD and its semantic structure. We then propose a RIF extension for directly representing uncertain knowledge.

4.1   Definition of Truth Values and Truth Valuation

In previous sections, we showed how to represent the semantics of fuzzy LP with RIF functions and predicates in RIF presentation syntax. We now propose to introduce a new dialect for RIF, RIF Uncertainty Rule Dialect (RIF-URD), so as to directly represent uncertain knowledge and extend the expressive power of RIF.

The set TV of truth values in RIF-BLD consists of just two values, t and f. This set forms a two-element Boolean algebra with $t = 1$ and $f = 0$. However, in order to represent uncertain knowledge, all intermediate truth values must be allowed. Therefore, the set TV of truth values is extended to a set with infinitely many truth values ranging between 0 and 1. Our uncertain knowledge representation is specifically based on Fuzzy Logic, thus a member function maps a variable to a truth value in the 0 to 1 range.

Definition 1. *(Set of truth values as a specialization of the set in RIF-FLD) In RIF-FLD, $\leq_t$ denotes the truth order, a binary relation on the set of truth values TV. Instantiating RIF-FLD, which just requires a partial order, the set of truth values in RIF-URD is equipped with a total order over the 0 to 1 range. In RIF-URD, we specialize $\leq_t$ to $\leq$, denoting the numerical truth order. Thus, we observe that the following statements hold for any element e_i, e_j or e_k in the set of truth values TV in the 0 to 1 range, justifying to write it as the interval [0,1].*

1. *The set TV is a complete lattice with respect to $\leq$, i.e., the least upper bound (lub) and the greatest lower bound (glb) exist for any subset of $\leq$.*
2. *Antisymmetry. If $e_i \leq e_j$ and $e_j \leq e_i$ then $e_i = e_j$.*
3. *Transitivity. If $e_i \leq e_j$ and $e_j \leq e_k$ then $e_i \leq e_k$.*
4. *Totality. Any two elements should satisfy one of these two relations: $e_i \leq e_j$ or $e_j \leq e_i$.*
5. *The set TV has an operator of negation, $\sim: TV \to TV$, such that*
   (a) *$\sim e_i = 1 - e_i$*
   (b) *$\sim$ is self inverse, i.e., $\sim \sim e_i = e_i$.*

Let $TVal(\varphi)$ denote the truth value of a non-document formula, φ, in RIF-BLD. Here a non-document formula could be a well-formed term whose signature is formula, or a group formula, but not a document formula. $TVal(\varphi)$ is a mapping from the set of all non-document formulas to TV, I denotes an interpretation, and c is a real number in the interval [0,1].

Definition 2. *(Truth valuation adapted from RIF-FLD) Truth valuation for well-formed formulas in RIF-URD is determined as in RIF-FLD, adapting the following cases.*

(1) Conjunction (glb_t becomes F_{IN}): $TVal_I(And(B_1 \cdots B_n)) = F_{IN}(TVal(B_1) \cdots TVal(B_n))$.

(2) Disjunction (lub_t becomes F_U): $TVal_I(Or(B_1 \cdots B_n)) = F_U(TVal(B_1) \cdots TVal(B_n))$.

(3) Rule implication (t becomes 1, f becomes 0, condition valuation is multiplied with c):
$TVal_I(conclusion : -condition /c) = 1$ if $TVal_I(conclusion) \geq F_{IM}(c, TVal_I(condition))$
$TVal_I(conclusion : -condition /c) = 0$ if $TVal_I(conclusion) < F_{IM}(c, TVal_I(condition))$

4.2 Using RIF-URD to Represent Uncertain Knowledge

A fuzzy rule in the form of equation 2 can be directly represented in RIF-URD as

```
Document(
 Group
 ( Forall ?x₁ ... ?xₖ (
     h(t₁ ... tᵣ) :- And(b₁(t₁,₁ ... t₁,ₛ₁) ... bₙ(tₙ,₁ ... tₙ,ₛₙ))
 ) / c
)
```

Likewise, a fact of the form H $/c$ can be represented in RIF-URD as

```
h(t₁ ... tᵣ) / c
```

Such a RIF-URD document of course cannot be executed by an ordinary RIF-compliant reasoner. RIF-URD-compliant reasoners will need to be extended to support the above semantics and the reasoning process shown in Section 3.

Example 6. We can directly represent example 1 in RIF predicates as follows:

```
Document(
 Import (<http://example.hog/fuzzy/membershipfunction >)
 Group
 ( Forall ?x ?y(
     cheapFlight(?x ?y) :- affordableFlight(?x ?y)
     ) / 0.9
   Forall ?x ?y(affordableFlight(?x ?y)) / left_shoulder0k4k1k3k(?y)
) )
```

5   Fuzzy Description Logic Programs and Their Representation in RIF

In this section, we extend Description Logic Programs (DLP) [11] to Fuzzy DLP by fuzzifying each axiom in DLP and studying the semantics and the mappings in Fuzzy DLP; we also show how to represent such mappings in RIF-BLD and RIF-URD based on the three uncertainty treatment methods addressed in previous sections.

Since DL is a subset of FOL, it can also be seen in terms of that subset of FOL, where individuals are equivalent to FOL constants, concepts and concept descriptions are equivalent to FOL formulas with one free variable, and roles and role descriptions are equivalent to FOL formulas with two free variables.

A concept inclusion axiom of the form $C \sqsubseteq D$ is equivalent to an FOL sentence of the form $\forall x.C(x) \rightarrow D(x)$, i.e. an FOL implication. In uncertainty representation and reasoning, it is important to represent and compute the degree of subsumption between two fuzzy concepts, i.e., the degree of overlap, in addition to crisp subsumption. Therefore, we consider fuzzy axioms of the form $C \sqsubseteq D = c$ generalizing the crisp $C \sqsubseteq D$. The above equivalence leads to a straightforward mapping from a fuzzy concept inclusion axiom of the form $C \sqsubseteq D = c(c \in [0,1])$ to an LP rule as follows: $D(x) \leftarrow C(x) \; /c$.

Similarly, a role inclusion axiom of the form $R \sqsubseteq P$ is equivalent to an FOL sentence consisting of an implication between two roles. Thus we map a fuzzy role inclusion axiom of the form $R \sqsubseteq P = c(c \in [0,1])$ to a fuzzy LP rule as $P(x,y) \leftarrow R(x,y) \; /c$. Moreover, $\bigcap_{i=1}^{n} R_i \sqsubseteq P = c$ can be transformed to $P(x,y) \leftarrow R_1(x,y), \cdots, R_n(x,y) \; /c$.

A DL assertion $C(a)$ (respectively, $R(a,b)$) is equivalent to an FOL atom of the form $C(a)$ (respectively, $R(a,b)$), where a and b are individuals. Therefore, a fuzzy DL concept-individual assertion of the form corresponds to a ground fuzzy atom $C(a) \; /c$ in fuzzy LP, while a fuzzy DL role-individual assertion of the form $R(a,b) = c$ corresponds to a ground fuzzy fact $R(a,b) \; /c$.

The intersection of two fuzzy concepts in fuzzy DL is defined as $(C_1 \sqcap C_2)^I(x) = F_{IN}(C_1^I(x), C_2^I(x))$. Therefore, a fuzzy concept inclusion axiom of the form $C_1 \sqcap C_2 \sqsubseteq D = c$ including the intersection of C_1 and C_2 can be transformed to an LP rule $D(x) \leftarrow C_1(x), C_2(x) \; /c$. Here the certainty degree of (variable-free) instantiations of the atom $D(x)$ is defined by the valuation $val(D, H_I) \geq F_{IM}(c, F_{IN}(val(C_i, H_I)|i \in \{1,2\}))$. If the intersection connective is within the *Superclass* description, that is, $C \sqsubseteq D_1 \sqcap D_2 = c$, it can be transformed to LP rules $D_1(x) \leftarrow C(x) \; /c$ and $D_2(x) \leftarrow C(x) \; /c$. Instantiations of the atoms D_1 and D_2 as well as the conjunctive query of the two atoms have a certainty degree defined by the valuation $F_{IM}(c, val(C, H_I))$. It is easy to see that such fuzzy concept inclusion axioms can be extended to include the intersection of n concepts $(n > 2)$. Furthermore, when the *Union* connective is adopted in the subclass descriptions of a fuzzy concept inclusion axiom, $C_1 \sqcup C_2 \sqsubseteq D = c$, it can be transformed to two LP rules $D(x) \leftarrow C_1(x) \; /c$ and $D(x) \leftarrow C_2(x) \; /c$. Semantically, the certainty degree of the atom $D(x)$ is defined by the valuation $val(D, H_I) \geq F_{IM}(c, F_U(val(C_i, H_I)|i \in \{1,2\})) = F_U(F_{IM}(c, val(C_1, H_I)), F_{IM}(c, val(C_2, H_I)))$.

For an axiom stating that the *Domain* of a property R is a class C is true to some degree, $\top \sqsubseteq \forall R^-.C = c$, it can be mapped to a fuzzy LP rule $C(x) \leftarrow R(x,y) \; /c$ with the valuation $val(C, H_I) \geq F_{IM}(c, val(R, H_I))$; an axiom stating that the

Range of a property R is a class C, $\top \sqsubseteq \forall R.C = c$, can be mapped to a fuzzy LP rule $C(y) \leftarrow R(x,y)$ /c with the valuation $val(C, H_I) \geq F_{IM}(c, val(R, H_I))$. As in DLP, Fuzzy DLP allows the *Universal quantification* within superclass descriptions, $C \sqsubseteq \forall R.D = c$. Such an axiom is mapped to the following fuzzy LP rule $D(y) \leftarrow C(x), R(x,y)$ /c. Next, a fuzzy axiom using the *Existential quantification* within subclass descriptions in the form of $\exists R.C \sqsubseteq D = c$ can be mapped to the fuzzy LP rule $D(x) \leftarrow C(y), R(x,y)$ /c.

In classic logics, a role R is symmetric iff for all $x, y \in H_I$, $val(R^-, H_I) = val(R, H_I)$, where R^- defines the inverse of a role. The same property holds for a fuzzy symmetric role. Therefore, in Fuzzy DLP, the axiom stating that a property R is the Inverse of a property P has the same syntax as in DLP.

In classic logics, a role R is transitive iff for all $x, y, z \in H_I$, $R(x,y)$ and $R(y,z)$ imply $R(x,z)$. While in Fuzzy Logic, a fuzzy role R is transitive iff for all $x, y, z \in H_I$, it satisfies the following inequality [8]:

$$R(x,z) \geq \sup_{y \in H_I} F_{IN}(R(x,y), R(y,z)) \qquad (3)$$

where F_{IN} denotes the intersection operator. For example, in the case of Zadeh Logic, a transitive role satisfies:

$$R(x,z) \geq \sup_{y \in H_I} \min(R(x,y), R(y,z)) \qquad (4)$$

Therefore, in Fuzzy DLP, we define the axiom stating that a property R is *Transitive* use the following syntax $R^+ \sqsubseteq R$. Table 2 summarizes all the mappings in Fuzzy DLP. In summary, Fuzzy DLP is an extension of Description Logic Programs supporting the following concept and role inclusion axioms, range and domain axioms, concept and role assertion axioms to build a knowledge base: $\cap_{i=1}^n C_i \sqsubseteq D = c$, $\top \sqsubseteq \forall R.C = c$, $\top \sqsubseteq \forall R^-.C = c$, $\cap_{i=1}^n R_i \sqsubseteq P = c$, $P \equiv R^-$, $R^+ \sqsubseteq R$, $C(a) = c$, and $R(a,b) = c$, where $C, D, C_1, \cdots, C_n$ are concepts, P, R are roles, a, b are individuals, $c \in [0, 1]$ and $n \geq 1$. Notice that the crisp DLP axioms in DLP are special cases of their counterparts in Fuzzy DLP. For example, $C \sqsubseteq D$ is equal to its fuzzy version $\cap_{i=1}^n C_i \sqsubseteq D = c$ for $n = 1$ and $c = 1$.

Table 2. Representing Fuzzy DLP in RIF

LP syntax	$D(x) \leftarrow C_1(x), \cdots, C_n(x)$ /c
DL syntax	$\cap_{i=1}^n C_i \sqsubseteq D = c$
RIF function	$Forall \ ?x(\ D(?x) = ?c_h \ : -$ $And(C_1(?x) = ?c_1 \cdots C_n(?x) = ?c_n$ $?c_t = External(F_{IN}(?c_1 \cdots ?c_n))$ $?c_h = External(F_{IM}(c \ ?c_t)))$
RIF predicate	$Forall \ ?x(\ D(?c_h ?x) \ : -$ $And(C_1(?c_1 \ ?x) \cdots C_n(?c_n \ ?x)$ $?c_t = External(F_{IN}(?c_1 \cdots ?c_n))$ $?c_h = External(F_{IM}(c \ ?c_t)))$
	Continued on next page

RIF-URD	$Forall\ ?x($ $\quad D(?x)\ \ :-\ And(C_1(?x)\cdots C_n(?x))$ $\quad)\ \ /c$
LP syntax	$P(x,y) \leftarrow R_1(x,y),\cdots,R_n(x,y)\ \ \ /c$
DL syntax	$\cap_{i=1}^n R_i \sqsubseteq P = c$
RIF function	$Forall\ ?x\ ?y(\ P(?x\ ?y) =?c_h\ \ :-$ $\quad And(R_1(?x\ ?y)=?c_1 \cdots R_n(?x\ ?y)=?c_n$ $\quad ?c_t = External(F_{IN}(?c_1\cdots ?c_n))$ $\quad ?c_h = External(F_{IM}(c\ ?c_t)))$
RIF predicate	$Forall\ ?x\ ?y(\ P(?c_h\ ?x\ ?y)\ \ :-$ $\quad And(R_1(?c_1\ ?x\ ?y)\cdots R_n(?c_n\ ?x\ ?y)$ $\quad ?c_t = External(F_{IN}(?c_1\cdots ?c_n))$ $\quad ?c_h = External(F_{IM}(c\ ?c_t)))$
RIF-URD	$Forall\ ?x\ ?y($ $\quad P(?x\ ?y)\ \ :-\ And(R_1(?x\ ?y)\cdots R_n(?x\ ?y))$ $\quad)\ \ /c$
LP syntax	$C(y) \leftarrow R(x,y)\ \ /c$
DL syntax	$\top \sqsubseteq \forall R.C\ \ = c$
RIF function	$Forall\ ?x?y(\ C(?y)=?c_h\ \ :-$ $\quad And(R(?x\ ?y)=?c_1\ \ ?c_h = External(F_{IM}(c\ ?c_1)))$
RIF predicate	$Forall\ ?x?y(\ C(?c_h\ ?y)\ \ :-$ $\quad And(R(?c_1\ ?x\ ?y)\ \ ?c_h = External(F_{IM}(c\ ?c_1)))$
RIF-URD	$Forall\ ?x?y(\ C(?y)\ \ :-\ R(?x\ ?y))\ /c$
LP syntax	$C(x) \leftarrow R(x,y)\ \ /c$
DL syntax	$\top \sqsubseteq \forall R^-.C\ \ = c$
RIF function	$Forall\ ?x?y(\ C(?x)=?c_h\ \ :-$ $\quad And(R(?x\ ?y)=?c_1\ \ ?c_h = External(F_{IM}(c\ ?c_1)))$
RIF predicate	$Forall\ ?x?y(\ C(?c_h\ ?x)\ \ :-$ $\quad And(R(?c_1\ ?x\ ?y)\ \ ?c_h = External(F_{IM}(c\ ?c_1)))$
RIF-URD	$Forall\ ?x?y(\ C(?x)\ \ :-\ R(?x\ ?y))\ /c$
LP syntax	$C(x) \leftarrow R(x,y)\ \ /c$
DL syntax	$\top \sqsubseteq \forall R^-.C\ \ = c$
RIF function	$Forall\ ?x?y(\ C(?x)=?c_h\ \ :-$ $\quad And(R(?x\ ?y)=?c_1\ \ ?c_h = External(F_{IM}(c\ ?c_1)))$
RIF predicate	$Forall\ ?x?y(\ C(?c_h\ ?x)\ \ :-$ $\quad And(R(?c_1\ ?x\ ?y)\ \ ?c_h = External(F_{IM}(c\ ?c_1)))$
RIF-URD	$Forall\ ?x?y(\ C(?x)\ \ :-\ R(?x\ ?y))\ /c$
LP syntax	$D_1(x) \leftarrow C(x)\ /c, D_2(x) \leftarrow C(x)\ /c$
DL syntax	$C \sqsubseteq D_1 \sqcap D_2\ \ = c$
RIF function	$Forall\ ?x(\ D_1(?x)=?c_h\ \ :-$ $\quad And(C(?x)=?c_1\ \ ?c_h = External(F_{IM}(c\ ?c_1)))$ $Forall\ ?x(\ D_2(?x)=?c_h\ \ :-$ $\quad And(C(?x)=?c_1\ \ ?c_h = External(F_{IM}(c\ ?c_1)))$
RIF predicate	$Forall\ ?x(\ D_1(?c_h\ ?x)\ \ :-$ $\quad And(C(?c_1\ ?x)\ ?c_h = External(F_{IM}(c\ ?c_1)))$ $Forall\ ?x(\ D_2(?c_h\ ?x)\ \ :-$
	Continued on next page

	$And(C(?c_1\ ?x)\ ?c_h = External(F_{IM}(c\ ?c_1)))$
RIF-URD	$Forall\ ?x(\ D_1(?x)\ :-\ C(?x))\ /c$
	$Forall\ ?x(\ D_2(?x)\ :-\ C(?x))\ /c$
LP syntax	$D(x) \leftarrow C_1(x)\ /c, D(x) \leftarrow C_2(x)\ /c$
DL syntax	$C_1 \sqcup C_2 \sqsubseteq D\ = c$
RIF function	$Forall\ ?x(\ D(?x) =?c_h\ :-$
	$And(C_1(?x) =?c_1\ C_2(?x) =?c_2$
	$?c_h = External(F_U(F_{IM}(c\ ?c_1), F_{IM}(c\ ?c_2)))$
RIF predicate	$Forall\ ?x(\ D(?c_h\ ?x)\ :-$
	$And(C_1(?c_1\ ?x)\ C_2(?c_2\ ?x)$
	$?c_h = External(F_U(F_{IM}(c\ ?c_1), F_{IM}(c\ ?c_2)))$
RIF-URD	$Forall\ ?x(\ D(?x)\ :-\ C_1(?x))\ /c$
	$Forall\ ?x(\ D(?x)\ :-\ C_2(?x))\ /c$
LP syntax	$D(y) \leftarrow C(x), R(x,y)\ /c$
DL syntax	$C \sqsubseteq \forall R.D\ = c$
RIF function	$Forall\ ?x?y(\ D(?y) =?c_h\ :-$
	$And(C(?x) =?c_1\ R(?x\ ?y) =?c_2\ ?c_h$
	$= External(F_{IM}(c\ F_{IN}(?c_1\ ?c_2))))$
RIF predicate	$Forall\ ?x?y(\ D(?c_h\ ?y)\ :-$
	$And(C(?c_1\ ?x)\ R(?c_2\ ?x\ ?y)\ ?c_h$
	$= External(F_{IM}(c\ F_{IN}(?c_1\ ?c_2))))$
RIF-URD	$Forall\ ?x?y(\ D(?y)\ :-\ And(C(?x)\ R(?x\ ?y)))\ /c$
LP syntax	$D(x) \leftarrow C(y), R(x,y)\ /c$
DL syntax	$\exists R.C \sqsubseteq D\ = c$
RIF function	$Forall\ ?x?y(\ D(?x) =?c_h\ :-$
	$And(C(?y) =?c_1\ R(?x\ ?y) =?c_2\ ?c_h$
	$= External(F_{IM}(c\ F_{IN}(?c_1\ ?c_2))))$
RIF predicate	$Forall\ ?x?y(\ D(?c_h\ ?x)\ :-$
	$And(C(?c_1\ ?y)\ R(?c_2\ ?x\ ?y)\ ?c_h$
	$= External(F_{IM}(c\ F_{IN}(?c_1\ ?c_2))))$
RIF-URD	$Forall\ ?x?y(\ D(?x)\ :-\ And(C(?y)\ R(?x\ ?y)))\ /c$
LP syntax	$R(x,y) \leftarrow P(y,x), P(y,x) \leftarrow R(x,y)$
DL syntax	$R^- \equiv P$
RIF function	$Forall\ ?x\ ?y(\ R(?x\ ?y) =?c_h\ :-$
	$And(P(?y\ ?x) =?c_1\ ?c_h =?c_1)$
RIF predicate	$Forall\ ?x\ ?y(\ R(?c_h\ ?x\ ?y)\ :-$
	$And(P(?c_1\ ?y\ ?x)\ ?c_h =?c_1)$
RIF-URD	$Forall\ ?x\ ?y(\ R(?x\ ?y)\ :-\ P(?y\ ?x))$

LP syntax	$C(a)\ /c$	$R(a,b)\ /c$
DL syntax	$C(a)\ = c$	$R(a,b)\ = c$
RIF function	$C(a)\ = c$	$R(a\ b)\ = c$
RIF predicate	$C(c\ a)$	$R(c\ a\ b)$
RIF-URD	$C(a)\ /c$	$R(a\ b)\ /c$

In previous sections, we presented two techniques of encoding uncertainty in RIF and proposed a method based on an extension of RIF for uncertainty representation. Subsequently, we also showed how to represent Fuzzy DLP in RIF-BLD and RIF-URD in Table 2.

Layered on Fuzzy DLP, we can build fuzzy hybrid knowledge bases in order to build fuzzy rules on top of ontologies for the Semantic Web and reason on such KBs.

Definition 3. *A fuzzy hybrid knowledge base KB_{hf} is a pair $< K_{DL}, K_{LP} >$, where K_{DL} is the finite set of (fuzzy) concept inclusion axioms, role inclusion axioms, and concept and role assertions of a decidable DL defining an ontology. K_{LP} consists of a finite set of (fuzzy) hybrid rules and (fuzzy) facts.*

A hybrid rule r in K_{LP} is of the following generalized form (we use the BNF choice bar, |):

$$(H(\overrightarrow{y})|\&H(\overrightarrow{z})) \leftarrow B_1(\overrightarrow{y1}), \cdots, B_l(\overrightarrow{yl}), \&Q_1(\overrightarrow{zl}), \cdots, \&Q_n(\overrightarrow{zn}) \quad /c \quad (5)$$

Here, $H(\overrightarrow{y})$, $H(\overrightarrow{z})$, $B_i(\overrightarrow{yi})$, $Q_j(\overrightarrow{zj})$ are atoms, $\&$ precedes a DL atom, $\overrightarrow{y}$, $\overrightarrow{z}$, $\overrightarrow{yi}$, $\overrightarrow{zj}$ are vectors of variables or constants, where $\overrightarrow{y}$ and each $\overrightarrow{yi}$ have arbitrary lengths, $\overrightarrow{z}$ and each $\overrightarrow{zj}$ have length 1 or 2, and $c \in [0, 1]$. Also, $\&$ atoms and $/c$ degrees are optional (if all $\&$ atoms and $/c$ degrees are missing from a rule, it becomes a classical rule of Horn Logic).

Such a fuzzy hybrid rule must satisfy the following constraints:

(1) H is either a DL predicate or a rule predicate ($H \in \sum T \bigcup \sum R$). H is a DL predicate with the form $\&H$, while it is a rule predicate without the $\&$ operator.

(2) Each B_i ($1 < i \leq l$) is a rule predicate ($B_i \in \sum R$), and $B_i(y_i)$ is an LP atom.

(3) Each Q_j ($1 < j \leq n$) is a DL predicate ($Q_j \in \sum T$), and $Q_j(z_j)$ is a DL atom.

(4, pure DL rule) If a hybrid rule has head $\&H$, then each atom in the body must be of the form $\&Q_j$ ($1 < j \leq n$); in other words, there is no B_i ($l = 0$). A head $\&H$ without a body ($l = 0$, $n = 0$) constitutes the special case of a pure DL fact.

Example 7. The rule $\&CheapFlight(x, y) \leftarrow AffordableFlight(x, y)$ $/c$ is not a pure DL rule according to (4), hence not allowed in our hybrid knowledge base, while $CheapFlight(x, y) \leftarrow \&AffordableFlight(x, y)$ $/c$ is allowed.

A hybrid rule of the form $\&CheapFlight(x, y) \leftarrow \&AffordableFlight(x, y)$ $/c$ can be mapped to a fuzzy DL role subsumption axiom $AffordableFlight \sqsubseteq CheapFlight = c$.

Our approach thus allows DL atoms in the head of hybrid rules which satisfy the constraint (4, pure DL rule), supporting the mapping of DL subsumption axioms to rules. We also deal with fuzzy subsumption of fuzzy concepts of the form $C \sqsubseteq D = c$ as shown in Example 7.

An arbitrary hybrid knowledge base cannot be fully embedded into the knowledge representation formalism of RIF with uncertainty extensions. However, in the proposed Fuzzy DLP subset, DL components (DL axioms in LP syntax) can be mapped to LP rules and facts in RIF. A RIF-compliant reasoning engine can be extended to do reasoning on a hybrid knowledge base on top of Fuzzy DLP by adding a module that first maps atoms in rules to DL atoms, and then derives the reasoning answers with a DL reasoner, e.g. Racer or Pellet, or with a fuzzy DL reasoner, e.g. fuzzyDL [2]. The specification of such a reasoning algorithm for a fuzzy hybrid knowledge base KB_{hf} based on Fuzzy DLP and a query q is treated in a companion paper [38].

6 Conclusion

In this paper, we propose two different principles of representing uncertain knowledge, encodings in RIF-BLD and an extension leading to RIF-URD. We also present a fuzzy extension to Description Logic Programs, namely Fuzzy DLP. We address the mappings between fuzzy DL and fuzzy LP within Fuzzy DLP, and give Fuzzy DLP representations in RIF. Since handling uncertain information, such as with fuzzy logic, was listed as a RIF extension in the RIF Working Group Charter [3] and RIF-URD is a manageable extension to RIF-BLD, we propose here a version of URD as a RIF dialect, realizing a fuzzy rule sublanguage for the RIF standard.

The paper is an extended version of our previous work with the same title [39]. Here we presented a unified framework for uncertainty representation in RIF. Our fuzzy extension directly relates to the semantics of fuzzy sets and fuzzy logic, allowing the parameterization of RIF-URD to support Lotfi Zadeh's, Jan Lukasiewicz's, and other classes in the family of fuzzy logics. We do not yet cover here cover other uncertainty formalisms, based on probability theory, possibilities, or rough sets. Future work will include generalizing our fuzzy extension of hybrid knowledge bases to some of these different kinds of uncertainty.

The combination strategy presented in this paper is based on resolving some atoms in the hybrid knowledge base to DL queries. Therefore, another direction of future work would be the extension of uncertain knowledge to various combination strategies of DL and LP without being limited to DL queries.

References

1. Baaz, M.: Infinite-valued gödel logic with 0-1-projections and relativisations. In: Hájek, P. (ed.) Gödel 1996: Logical Foundations of Mathematics, Computer Science, and Physics. Lecture Notes in Logic, vol. 6, pp. 23–33. Springer, Brno (1996)
2. Bobillo, F., Straccia, U.: fuzzyDL: An expressive fuzzy description logic reasoner. In: Proceedings of the 2008 International Conference on Fuzzy Systems, FUZZ 2008 (2008)
3. Boley, H., Kifer, M.: A Guide to the Basic Logic Dialect for Rule Interchange on the Web. IEEE Transactions on Knowledge and Data Engineering 22(11), 1593–1608 (2010)

4. Boley, H., Kifer, M.: RIF Basic Logic Dialect. Tech. Rep. W3C Recommendation (June 30, 2010), `http://www.w3.org/TR/rif-bld/`
5. Boley, H., Kifer, M.: RIF Framework for Logic Dialects. Tech. Rep. W3C Recommendation (June 30, 2010), `http://www.w3.org/TR/rif-fld/`
6. Damásio, C.V., Pan, J.Z., Stoilos, G., Straccia, U.: Representing uncertainty in ruleml. Fundamenta Informaticae 82, 1–24 (2008)
7. Damásio, C.V., Pereira, L.M.: Monotonic and Residuated Logic Programs. In: Benferhat, S., Besnard, P. (eds.) ECSQARU 2001. LNCS (LNAI), vol. 2143, pp. 748–759. Springer, Heidelberg (2001)
8. Díaz, S., De Baets, B., Montes, S.: General results on the decomposition of transitive fuzzy relations. Fuzzy Optimization and Decision Making 9(1), 1–29 (2010)
9. Donini, F.M., Lenzerini, M., Nardi, D., Schaerf, A.: Al-log: Integrating datalog and description logics. Journal of Intelligent Information Systems 10(3), 227–252 (1998)
10. Dummett, M.: A propositional calculus with denumerable matrix. Journal of Symbolic Logic 24(2), 97–106 (1959)
11. Grosof, B.N., Horrocks, I., Volz, R., Decker, S.: Description logic programs: Combining logic programs with description logic. In: Proceedings of the 12th International Conference on World Wide Web, pp. 48–57 (2003)
12. Hájek, P.: Fuzzy Logic from the Logical Point of View. In: Bartosek, M., Staudek, J., Wiedermann, J. (eds.) SOFSEM 1995. LNCS, vol. 1012, pp. 31–49. Springer, Heidelberg (1995)
13. Hájek, P.: Metamathematics of fuzzy logic. Kluwer (1998)
14. Hájek, P.: Fuzzy logic and arithmetical hierarchy iii. Studia Logica 68(1), 129–142 (2001)
15. Hall, V.: Uncertainty-valued horn clauses. Tech. Rep. (1994), `http://www.dfki.uni-kl.de/~vega/relfun+/fuzzy/fuzzy.ps`
16. Hay, L.S.: Axiomatization of the infinite-valued predicate calculus. Journal of Symbolic Logic 28(1), 77–86 (1963)
17. Horrocks, I., Patel-Schneider, P.F., Bechhofer, S., Tsarkov, D.: Owl rules: A proposal and prototype implementation. Journal of Web Semantics 3(1), 23–40 (2005)
18. Jaeger, M.: Probabilistic reasoning in terminological logics. In: Proc. of the 4th Int. Conf. on the Principles of Knowledge Representation and Reasoning, KR 1994, pp. 305–316 (1994)
19. Koller, D., Levy, A., Pfeffer, A.: P-classic: A tractable probabilistic description logic. In: Proceedings of the Fourteenth National Conference on Artificial Intelligence, AAAI 1997, pp. 390–397 (1997)
20. Laskey, K.J., Laskey, K.B., Costa, P.C.G., Kokar, M.M., Martin, T., Lukasiewicz, T.: W3C incubator group report. Tech. Rep. W3C (March 05, 2008), `http://www.w3.org/2005/Incubator/urw3/wiki/DraftFinalReport`
21. Lukasiewicz, T., Straccia, U.: Managing uncertainty and vagueness in description logics for the semantic web. Journal of Web Semantics (2008)
22. Mei, J., Boley, H., Li, J., Bharvsar, V.C.: The datalogdl combination of deduction rules and description logics. Computational Intelligence 23(3), 356–372 (2007)
23. Motik, B., Sattler, U., Studer, R.: Query answering for owl dl with rules. Web Semantics: Science, Services and Agents on the World Wide Web 3(1), 41–60 (2005)
24. Novák, V., Perfilieva, I., Mockor, J.: Mathematical principles of fuzzy logic. Kluwer Academic, Dodrecht (1999)
25. Polleres, A., Boley, H., Kifer, M.: RIF Datatypes and Built-ins 1.0, W3C Recommendation (June 2010), `http://www.w3.org/TR/rif-dtb`

26. Stoilos, G., Stamou, G., Pan, J.Z., Tzouvaras, V., Horrocks, I.: Reasoning with very expressive fuzzy description logics. Journal of Artificial Intelligence Research 30, 273–320 (2007)
27. Straccia, U.: A fuzzy description logic. In: Proceedings of the 15th National Conference on Artificial Intelligence, AAAI 1998, pp. 594–599 (1998)
28. Straccia, U.: Reasoning within fuzzy description logics. Journal of Artificial Intelligence Research 14, 137–166 (2001)
29. Straccia, U.: Fuzzy description logic with concrete domains. Tech. Rep. Technical Report 2005-TR-03, Istituto di Elaborazione dell'Informazione (2005)
30. Straccia, U.: Fuzzy description logic programs. In: Proceedings of the 11th International Conference on Information Processing and Management of Uncertainty in Knowledge-Based Systems, IPMU 2006, pp. 1818–1825 (2006)
31. van Emden, M.: Quantitative deduction and its fixpoint theory. Journal of Logic Programming 30(1), 37–53 (1986)
32. Venetis, T., Stoilos, G., Stamou, G., Kollias, S.: f-dlps: Extending description logic programs with fuzzy sets and fuzzy logic. In: IEEE International Fuzzy Systems Conference, FUZZ-IEEE 2007, pp. 1–6 (2007) ID: 1
33. Vojtás, P.: Fuzzy logic programming. Fuzzy Sets and Systems 124, 361–370 (2004)
34. Vojtás, P., Paulík, L.: Soundness and Completeness of Non-Classical Extended SLD-Resolution. In: Herre, H., Dyckhoff, R., Schroeder-Heister, P. (eds.) ELP 1996. LNCS, vol. 1050, pp. 289–301. Springer, Heidelberg (1996)
35. Wang, H.-F.: Comparative studies on fuzzy t-norm operators. BUSFEL 50, 16–24 (1992)
36. Yager, R.R., Filev, D.P.: Generalizing the modeling of fuzzy logic controllers by parameterized aggregation operators. Fuzzy Sets and Systems 70, 303–313 (1995)
37. Zadeh, L.A.: Fuzzy sets. Information and Control 8(3), 338–353 (1965)
38. Zhao, J., Boley, H.: Combining fuzzy description logics and fuzzy logic programs. In: Web Intelligence/IAT Workshops, pp. 273–278. IEEE (2008)
39. Zhao, J., Boley, H.: Uncertainty Treatment in the Rule Interchange Format: From Encoding to Extension. In: URSW, vol. 423. CEUR-WS.org (2008)
40. Zhao, J., Boley, H.: Knowledge Representation and Reasoning in Norm-Parameterized Fuzzy Description Logics. In: Canadian Semantic Web: Technologies and Applications, pp. 27–54. Springer (2010)

PrOntoLearn: Unsupervised Lexico-Semantic Ontology Generation Using Probabilistic Methods

Saminda Abeyruwan[1], Ubbo Visser[1], Vance Lemmon[2], and Stephan Schürer[3]

[1] Department of Computer Science, University of Miami, Florida, USA
{saminda,visser}@cs.miami.edu
[2] The Miami Project to Cure Paralysis, University of Miami Miller School of Medicine, Florida, USA
vlemmon@miami.edu
[3] Department of Molecular and Cellular Pharmacology, University of Miami Miller School of Medicine, Florida, USA
sschuerer@med.miami.edu

Abstract. It is well known that manually formalizing a domain is a tedious and cumbersome process. It is constrained by the knowledge acquisition bottleneck. Therefore, many researchers have developed algorithms and systems to help automate the process. Among them are systems that incorporate text corpora in the knowledge acquisition process. Here, we provide a novel method for unsupervised bottom-up ontology generation. It is based on lexico-semantic structures and Bayesian reasoning to expedite the ontology generation process. To illustrate our approach, we provide three examples generating ontologies in diverse domains and validate them using qualitative and quantitative measures. The examples include the description of high-throughput screening data relevant to drug discovery and two custom text corpora. Our unsupervised method produces viable results with sometimes unexpected content. It is complementary to the typical top-down ontology development process. Our approach may therefore also be useful to domain experts.

Keywords: Ontology Modeling, Ontology Learning, Probabilistic Methods.

1 Introduction

An ontology is a formal, explicit specification of a shared conceptualization [15], [36]. Formalizing an ontology for a given domain with the supervision of domain experts is a tedious and cumbersome process. The identification of the structures and the characteristics of the domain knowledge through an ontology is a demanding task. This problem is known as the knowledge acquisition bottleneck (KAB) and a suitable solution presently does not exist.

There exists a large number of text corpora available from different domains (e.g., the BioAssay high throughput screening assays[1]) that need to be

---

[1] http://bioassayontology.org/

F. Bobillo et al. (Eds.): URSW 2008-2010/UniDL 2010, LNAI 7123, pp. 217–236, 2013.
© Springer-Verlag Berlin Heidelberg 2013

classified into ontologies to facilitate the discovery of new knowledge. A domain of discourse (i.e., sequential number of sentences) shows characteristics such as 1) redundancy 2) structured and unstructured text 3) noisy and uncertain data that provide a degree of belief 4) lexical ambiguity, and 5) semantic heterogeneity problems. We discuss in depth the importance of these characteristics in section 3. Our goal in this research is to provide a novel method to construct an ontology from the evidence collected from the corpus. In order to achieve our goal, we use the lexico-semantic features of the lexicon and probabilistic reasoning to handle the uncertainty of features. Since our method is applied to build an ontology for a corpus without domain experts, this method can be seen as an unsupervised learning technique. Since the method starts from the evidence present in the corpus, it can be seen as a reverse engineering technique. We use WordNet[2] to handle lexico-semantic structures, and the Bayesian reasoning to handle degree of belief of an uncertain event. We implement a Java based application to serialize the learned conceptualization to OWL DL[3] format.

The rest of the paper is organized as follows: Section 2 provides a broad investigation of the related work. Section 3 provides details of our research approach. Section 4 provides a brief overview of the implementation details. Section 5 provides a detailed description of the experiments based on three different text corpora and the detailed discussion. Finally, Section 6 provides the summary and the future work.

2   Related Work

The problem of learning a conceptualization from a corpus has been studied in many disciplines such as machine learning, text mining, information retrieval, natural language processing, and Semantic Web. Knowledge representation languages such as first-order logic and description logics reflect a major trade-off between expressivity and tractability [22]. Even though description logic, the logic that we are most interested in this work, has made compromises that allows to be a more successful one, it is limited in ability to express uncertainty. P-CLASSIC probabilistic version of description logic [22] has used Bayesian networks to express probabilistic subsumption, which computes the probability that a random individual in class C is subsumed by class D. Zig & Ping [14] present an ongoing research on probabilistic extension to OWL using Bayesian networks. Additional cases are used to tag existing classes with prior probabilities. Using these prior probabilities and set of predefined translation rules OWL T-Box is converted into a Bayesian network to do reasoning of the conceptualization covered by the T-Box. Lukasiewicz [24,25] presents expressive probabilistic description logic P-SHIF(D) and P-SHOIN(D) which are probabilistic extensions to description logics SHIF(D) and SHOIN(D) that OWL-Lite and OWL-DL is based on. This allows to express probabilistic knowledge about concepts and roles as well as assertional probabilistic knowledge about instances of concepts

---

[2] http://wordnet.princeton.edu/
[3] http://www.w3.org/TR/owl-guide/

and roles. Bergamaschi et al. [3] has used probabilistic world sense disambiguation (PWSD) to annotate relationships among elements in a data source. They
have used predefine relationships encoded in an ontology to annotate these elements, thus, providing an automatic generation of relationships subject to a
probability measure.

Machine learning approaches, especially semi-supervised learning models, have
shown a great improvement in extracting information from structured and unstructured text. This is mainly due to the fact that supervised training is very
expensive and fully categorized examples are very hard to overcome. Semi-
supervised learning methods use a small number of examples to bootstrap the
learning process. The most recent work on learning a conceptualization is addressed by Never Ending Language Learning (NELL) research that is being done
at CMU in the "Read the Web" project. The goal of the system is to run 24×7
and to learn good classification of the concepts and relations in next iteration
compared to the current iteration. The project has been focused on several semantic categories and uses 200 million web pages for the classification [6].

Knowledge acquisition is the transformation of knowledge from the form in
which it is available in the world into machine readable forms that can infer
useful results. It is not a trivial task to transfer domain knowledge to a machine
readable form because the interpretations should not be ambiguous. Hence, it
is required to model the domain in a way that it does not lose the underline
interpretation. Due to these intricacies, knowledge acquisition and representation is a hard problem. Natural language processing deals with the problem of
knowledge acquisition with part-of-speech tagging. This endeavor has been thoroughly investigated in knowledge acquisition from text. Many integrated tasks
in natural language processing require a large amount of world knowledge to
create expectations, assess plausibility and guide disambiguation. Building on
ideas by Schubert, a system called DART (Discovery and Aggregation of Relations in Text) has been developed that extracts simple, semi-formal statements
of world knowledge (e.g., air-planes can fly, people can drive cars) from text and
this has used it to create a database of 23 million propositions of this kind [11].
Recognize textual entailment (RTE) is the task to find out whether some text $\mathcal{T}$
entails a hypothesis $\mathcal{H}$. The recognition of textual entailment is without doubt
one of the ultimate challenges for any NLP system: if it is able to do so with
reasonable accuracy, it is clearly an indication that it has some thorough under
standing of how language works. Logical inference is the most common method
used in RTE. Boss & Markert [4] uses model building techniques borrowed from
automated reasoning to approximate entailment.

A common hypothesis is that there exists a large collection of general knowledge in texts, lying at a level beneath the explicit assertional content. This
knowledge is used to infer logical consequences that are possible in the world or
under certain conditions infer to be normal or commonplace in the world. Marinho et al. [1] focuses on deriving general propositions from noun phrase clauses
and then to fortify stronger generalization based on the nature and statistical
distribution of the propositions obtained in the first phase. Cankaya & Moldovan

[5] presents a semi-automatic method for generating commonsense axioms. The ultimate goal of knowledge capturing is to build systems that automatically construct a knowledge base by reading texts. This requires solving the problem of Natural Language Understanding (NLU) [34]. The main objective of NLU is to read texts to build a formal representation of their content in order to support a variety of tasks such as answering a query or RTE. Ambiguity is largely inherent to any natural language. This will cause a considerable challenge in full NLU understanding. The Learning-by-Reading system [21] focuses to answer the prior problem by integrating snippets of knowledge drawn from multiple texts to build a single coherent knowledge base, which has shown both feasible and promising. Some of the methods we have shown so far use either supervised or unsupervised machine learning algorithms to find a solution. The other aspect of the spectrum is active learning, a promising solution for named entity recognition [38]. Hearst [17] describes six lexico-syntactic patterns for automatic acquisition of the hyponym lexical relations from unrestricted text.

Text2Onto [10] represents the learned ontological structures at a meta-level in form of so called modeling primitives rather than in a concrete knowledge representation language, which gives the flexibility on handling the most prevalent representation languages currently used within the Semantic Web such as RDFS, OWL and F-Logic. Text2Onto uses Probabilistic Ontology Models (POMs) where the results of the system are attached by probabilities. In addition to this, it uses a data-driven change discovery, which is responsible for detecting changes in the corpus, calculating POM deltas with respect to the changes and accordingly modifying the POM without recalculating it for the whole document collection. Many reasoning-based applications in domains such as bioinformatics or medicine rely on much more complex conceptualizations than subsumption hierarchies [4]. Some of these conceptualizations are constructed by pure manual efforts. Hence, methods for the automatic or semi-automatic construction of expressive ontologies could help to overcome the knowledge acquisition bottleneck. The amount of post-processing for complex learned ontologies can be relaxed if proper integration of ontology evaluation and debugging approaches are introduced. Particularly, the treatment of logical inconsistencies, mostly neglected by existing ontology learning frameworks, becomes a great challenge as soon as we start to learn huge and expressive conceptualizations. LexO [4] is such an implementation supporting the automatic generation of complex class descriptions from lexical resources. The learned ontologies may represent uncertain and possibly contradicting knowledge [16]. RoundTrip Ontology Authoring (ROA) by Davis et al. [12] uses control language for ontology engineering. This method generates a controlled language using a given ontology. The text may be edited/modified by a user to their requirements and parsed back as an ontology. This process iterates until the user is satisfied with the logical consequences. There are many ways to represent lexical resources. Lemmon [29] is such a model that represents the lexical resources in a domain that can be coupled with NLP tools to generate satisfiable lexica for ontologies.

Formal Concept Analysis (FCA) [9] is a method that automatically constructs taxonomies or concept hierarchies from a corpus. In FCA paradigm, a concept consist of two parts, which is known as extension and intention. Extension covers all objects belonging to this concept and intention comprises all attributes valid for all those objects. These objects and attributes are used in deriving the subconcept-superconcept relations between concepts with respect to a formal context. This formal context later translates into a lattice from which the partial order of the concept hierarchy is learned. Since objects and attributes describe the extent and intent, it is important to extract theses from the text using NLP concepts. There exists a great demand for simpler and less costly methods for ontology refining, cleansing, completing, improving, and generation [18]. We believe that the research proposed in this paper is a step towards the fulfillment of methods that needs to generate domain independent less costly ontologies.

Table 1. The summary of the related work. Probabilistic learning (PR), probabilistic word sense disambiguation (PWSD), never ending language learning (NELL), discovery and aggregation of relations in text (DART), recognizing textual entailment (RTE), automated theorem proving (ATP), natural language understanding (NLU), formal concept analysis (FCA), RoundTrip Ontology Authoring (ROA), and ontology population (OP)

Work	Purpose	T-Box	A-Box	Method
PR [14,24,25,31]	reasoning	available	available	prob. theory
PWSD [3]	relationships	$\sqrt{}$	×	prob. theory
NELL [6]	24 × 7 learning	fixed	dynamic	ML techniques
DART [11]	world knowledge	×	×	semi-automated
RTE [4,23]	entailment	×	×	ATP
NLU [34]	commonsense rules	×	×	semi-supervised
Text2Onto [10]	ontology learning	$\sqrt{}$	$\sqrt{}$	semi-supervised
ROA [12]	ontology engineering	available	×	NLP
LexO [39]	complex classes	$\sqrt{}$	×	semi-supervised
FCA [9]	taxonomy	$\sqrt{}$	×	FCA
OP [7,37]	ontology population	available	available	semi-supervised

The other side of ontology learning is ontology population. Ontology population means finding instances of conceptualization [8]. Human-defined concepts are fundamental building-blocks in constructing knowledge bases such as ontologies. Statistical learning techniques provide an alternative automated approach to concept definition, driven by data rather than prior knowledge. Chemudugunta et al. [7] performs automatically tagging of Web pages with concepts from a known set of concepts without any need for labeled documents using latent Dirichlet allocation models. NLP is used in ontology population, using a combination of rule-based approaches and machine learning [28]. Linguistic and statistical technique methods are used for contextual information to bootstrap learning. Named entities are populated using a weakly supervised automatic approach in [37]. A syntactic model is learned for categories using an ontology. Then, this model is populated using the corpus. Pantel & Pennacchiotti [30] presents algorithms for harvesting semantic relations from text and then automatically linking the knowledge into existing semantic repositories.

Table 1 summarizes the pros and cons of different techniques to solve the problem of *ontology learning*. Each method covers some portion of the problem and each method learns the conceptualization from terms, and present it as taxonomies and axioms to an ontology. On the other hand, most of the methods use a top-down approach, i.e., an initial classification of an ontology is given. The uncertainty inherited from the domain is usually dealt with by a domain expert, and the conceptualization is normally defined using predefined rules or templates. These methods show the characteristics of a semi-supervised and a semi-automated learning paradigm.

3   Research Approach

Our research focuses on an unsupervised method to quantify the degree of belief that a grouping of words in the corpus will provide a substantial conceptualization of the domain of interest. The degree of belief in world states influences the uncertainty of the conceptualization. The uncertainty arises from partial observability, non-determinism, laziness and theoretical and practical ignorance [33]. The partial observability arises from the size of the corpus. Even though a corpus could be large, it might not contain all the necessary evidence of an event of interest. A corpus contains ambiguous statements about an event that leads to a non-determinism of the state of the event. The laziness arises from plethora of work that needs to be done in order to learn exceptionless rules and it is difficult to learn such rules. The theoretical and practical ignorance arises from lack of complete evidence and it is not possible to conduct all the necessary tests to learn a particular event. Hence, the domain knowledge, and in our case the domain conceptualization, can at best provide only a degree of belief of the relevant groups of words. We use probability theory to deal with the degrees of belief. As mentioned in [33], the probability theory has the same ontological commitment as the formal logic, though the epistemological commitment differs. The process of learning and presenting a probabilistic conceptualization is divided into four phases as shown in Figure 1. They are, 1) pre-processing 2) syntactic analysis 3) semantic analysis, and 4) representation.

3.1   Pre-processing

A corpus contains a plethora of structured and unstructured sentences built from a lexicon. A lexicon of a language is its vocabulary built from lexemes [20], [26]. A lexicon contains words belonging to a language and in our work individual words from the corpus will be treated as the vocabulary, thus, the lexicon of the corpus. In pure form, the lexicon may contain words that appear frequently in the corpus but have little value in formalizing a meaningful criterion. These words are called stop words or in our terminology: negated lexicon, and they are excluded from the vocabulary. The definition of the lexicon of our work is given as follows.

Definition 1. *A lexicon $\mathcal{L}_\mathcal{O}$ is the set that contains words belonging to the English vocabulary, which is part-of-speech (POS) type tagged with the Penn Treebank English POS tag set [27]. The set $\mathcal{L}_\mathcal{O}$ is built from the tag set: NN (noun, singular or mass), NNP (proper Noun, singular), NNS (noun, plural), NNPS (proper Noun, plural), JJ (adjective), JJR (adjective, comparative), JJS (adjective, superlative), VB (verb, base form), VBD (verb, past tense), VBG (verb, gerund or present participle), VBN (verb, past participle), VBP (verb, non-3rd person singular present), and VBZ (verb, 3rd person singular present)*

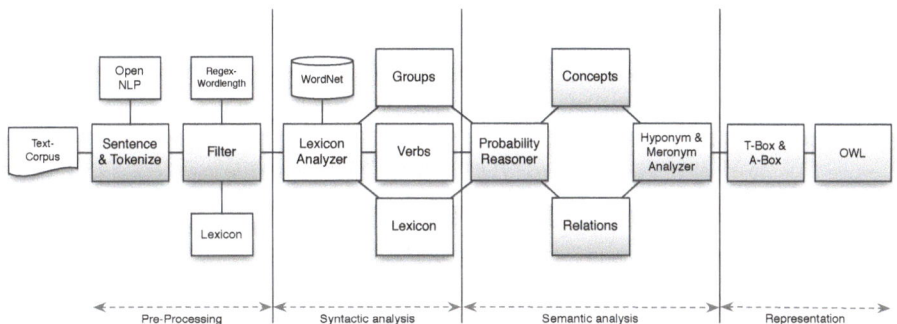

Fig. 1. Overall process: process categorizes into four phases; pre-processing, syntactic analysis, semantic analysis & representation

Definition 1 implies that negated lexicon $\overline{\mathcal{L}_\mathcal{O}}$ is the set that contains English words that are POS tagged with the Penn Treebank English POS tag set, other than the tags given in Definition 1. In addition, the word length W_L above some threshold W_{L_t} is also considered when building $\overline{\mathcal{L}_\mathcal{O}}$. The length of a word, with respect to POS context, is the sequence of characters or symbols that made up the word (e.g., the word "mika" has a word length of four $W_L = 4$). By default, we consider that a word with $W_L > 2$ sufficiently formalizes to some criterion.

Building up the pure lexicon at this stage, excluding the negated lexicon of the pre-processing, is known as tokenization from sentences [26]. Here, the pure form of the lexicon might contain words that need to be further purified according to some criterion. Words of the corpus contain many standard and constructed words. As mentioned, some words do not provide useful information (e.g., on, off, and at). In order to filter out these words, in the next phase of the pre-processing, each word is processed through a regular expression filter. The regular expression filter is a parameter of the system. The default regular expression is given as $[a - zA - Z]+$ if this parameter is not specified by the user. e.g., a word such as *du-145* will be filtered out from this regular expression. We also try to do token normalization to some extent. This is the process of canonicalizing the tokens so that matches occur despite superficial differences in the character sequences of the tokens [26]. In the next step, the vocabulary learned from the corpus is subjected to case-folding by reducing all letters to lower case. e.g., *Protocol* case-folds to *protocol*. Generally, documents use different forms of a word such as *organize*, *organizes* and *organizing* for grammatical reasons. In addition to this there are families of derivationally related words with similar meanings. We use stemming and lemmatization to reduce the inflectional forms and derivational forms of a word to a common base form [26]. We achieve this with the aid of WordNets' stemming algorithms. We couple the knowledge of POS tag of the lexicon to get the correct context of the word.

3.2 Syntactic Analysis

The pre-processing phase eliminates the noise of the corpus and tags the $\mathcal{L}_\mathcal{O}$ according to Definition 1. The primary focus on this phase is to look at the structure of the sentences and learn the associations among the words in $\mathcal{L}_\mathcal{O}$. We assume that each sentence of the corpus follows the POS pattern:

$$(Subject_{NounPhrase}+)(Verb+)(Object_{NounPhrase}+). \tag{1}$$

We hypothesize that the associations learned from this phase of the lexicon $\mathcal{L}_\mathcal{O}$ provide the potential candidates for concepts and relations of the ontology. But the words in the $\mathcal{L}_\mathcal{O}$ itself do not provide sufficient ontology concepts. We use a notion of grouping of consecutive sequence of words to form an OWL concept. This grouping is done using an appropriate N-gram model [2]. We illustrate this idea using Figure 2.

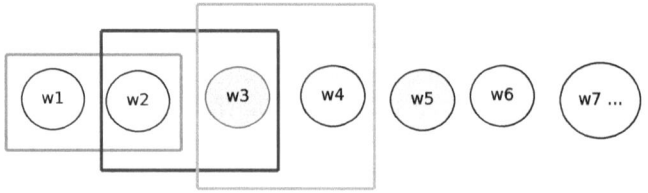

Fig. 2. An example three-gram model

According to Figure 2, group $w_1 \circ w_2$ forms a potential concept in the conceptualization. We use the notation $x \circ y$ to show that the word y is appended to the word x. The groups $w_2 \circ w_3$, $w_3 \circ w_4$ etc. form other potential concepts in the conceptualization. Word w_3 comes after group $w_1 \circ w_2$. According to the Bayes viewpoint, we collect information to estimate the probability $P(w_3|\{w_1 \circ w_2\})$, which will be used to form IS-A relationships, $w_1 \circ w_2 \sqsubseteq w_3$ using an independent Bayesian network with conditional probability $P(\{w_1 \circ w_2\}|w_3)$. In addition to this, we count the groups appearing in the left hand side and the right hand side of the expression 1 and the association of these groups given the verbs of lexicon $\mathcal{L}_\mathcal{O}$. These counts are used in the third phase to create the relations among concepts.

3.3 Semantic Analysis

This phase conducts the semantic analysis with probabilistic reasoning, which constitutes the most important operation of our work. This phase determines the conceptualization of the domain using a probability distribution for IS-A relations and relations among the concepts. In addition to this, and in order to provide a useful taxonomy, we induce concepts from clustered concepts. Our definition of concept learning is given in Definition 2.

Definition 2. *The set $W = \{w_1, \ldots, w_n\}$ represents independent words of the $\mathcal{L}_\mathcal{O}$ and each w_i has a prior probability θ_i. The set $G = \{g_1, \ldots, g_m\}$ represents independent N-gram groups learned from the corpus and each g_j has a prior probability η_j. When $w \in W$ and $g \in G$, $P(w|g)$ is the likelihood probability π learned from the corpus. The entities w and g represent the potential concepts of the conceptualization. Within this environment, an IS-A relationship between w and g is given by the posterior probability $P(g|w)$ and this is represented with a Bayesian network having two nodes w and g as shown in the Figure 3 and,*

$$P(g|w) = \frac{\pi \times \eta}{\sum_i p(w|g_i) \times p(g_i)}. \tag{2}$$

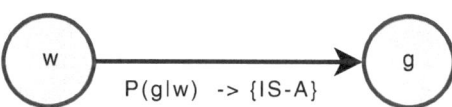

Fig. 3. Probabilistic IS-A relationship representation of the conceptualization(2). w and g are defined as the concepts of the conceptualization.

Lets define the *knowledge factor*: the lower-bound that select the super-concepts of the conceptualization.

Definition 3. $W = \{w_1, \ldots, w_n\}$ *represents independent words of the $\mathcal{L}_\mathcal{O}$ and each w_i has a prior probability θ_i. Lets define the knowledge factor (KF) as the lower-bound: if $\theta_i \geq \tau$ with $0 \leq \tau \leq 1$ then w_i is considered as a super-concept of the conceptualization.*

Definition 3 states that W of Definition 2 is considered as a super-concept of the conceptualization.

Definition 4. *The probabilistic conceptualization of the domain is represented by an n-number of independent Bayesian networks sharing groups (Figure 4).*

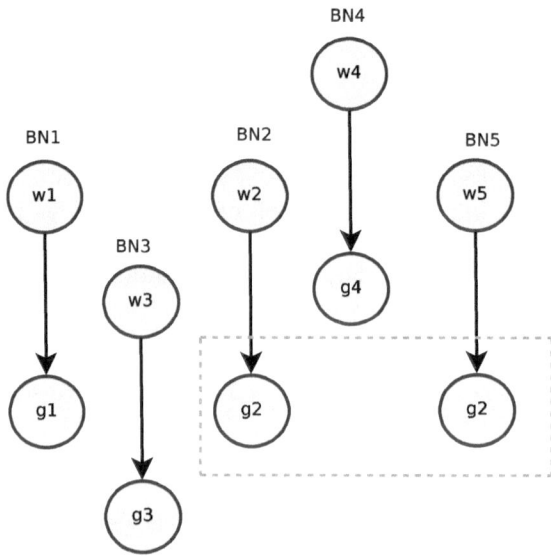

Fig. 4. w_1, w_2, w_3, w_4 and w_5 are super-concepts. g_1, g_2, g_3 and g_4 are candidate sub-concepts. There are 5 independent Bayesian networks. Bayesian networks 2 and 5 share the group g_2 when representing the concepts of the conceptualization.

Figure 4 shows multiple Bayesian networks that share a common group g_2. The interpretation of Definition 4 is: Let a set G contain an n-number of finite random variables $\{g_1, \ldots, g_n\}$. There exist a group g_i, which is shared by m words $\{w_1, \ldots, w_m\}$. Then, with respect to the Bayesian framework, BN_i of $P(g_i|w_i)$ is calculated and $max(P(g_i|m_i))$ is selected for the construction of the ontology. This means that if there exists two Bayesian networks and the Bayesian network one is given by the pair w_1, g_1 and the Bayesian network two is given by the pair $\{w_2, g_1\}$ then the Bayesian network that has the most substantial IS-A relationship is obtained through $max_{BN_i}(P(g_1|w_1))$ and this network is retained and other Bayesian networks will be ignored when building the ontology. If all $P(g_1|w_1)$ remains equal, then the Bayesian network with the highest super-concept probability will be retained. These two conditions will resolve any naming issues.

Definition 5. *Given a subset of concepts $G_S = \{g_1, \ldots, g_n\}$, $G_S \subset G$, with size n, for a given super-concept w, when $P(g_1|w), \ldots, P(g_n|w)$ holds, the prefixes of the concepts are extracted and known as an induced concepts. For a m-gram model, at most up to $m-1$ concepts can be induced. For all induced concepts c, the concepts name collision will be avoided by assigning different namespaces. The induced concept will be given a prior probability of 0.*

Definition 5 gives an efficient way to represent the taxonomy of the conceptualization. Newly induced concepts contain words up to at most $m - 1$. These concepts induction lead to concepts collision in the given namespace. This situation is avoided according to Definition 6.

Definition 6. *When a concept is induced from a group of concepts, the induced concept is assigned to a different namespace in order to avoid possible concept name conflicts. The namespace assignment is forced, if and only if there exist a concept with the same name in the system, otherwise induced concepts will be subjected to the default namespace of the system.*

The next step is to induce the relationships to complete the conceptualization. In order to do this, we need to find semantics associated with each *verb*. The relations are as important as the concepts in a conceptualization. The relations exist among the concept of the conceptualization. We hypothesize that relations are generated by the verbs in the corpus.

Definition 7. *The relationships of the conceptualization are learned from the syntactic structure model by the expression 1 and the semantic structure model by the lambda expression $\lambda obj.\lambda sub.Verb(sub, obj)$, where β-reduction is applied for obj and sub of the expression 1.*

Definition 8. *If there exists a verb V between two groups of concepts C_1 and C_2, the relationship of the triple (V, C_1, C_2) is written as $V(C_1, C_2)$ and model with conditional probability $P(C_1, C_2|V)$. The Bayesian network for relationship is and the model semantic relationship is given by,*

$$P(C_1, C_2|V) = p(C_1|V)p(C_2|V) \rightarrow V(C_1, C_2)$$

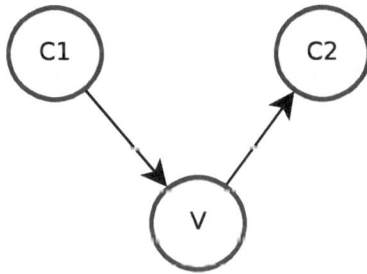

Fig. 5. Bayesian networks for relations modeling. C_1 and C_2 are groups and V is a verb

Using definitions 7 and 8, the relationship among multiple concepts are defined in 9. We define the relations in terms of groups of words in $\mathcal{L_O}$. These groups are clustered around the most probable words found in the corpus.

Definition 9. *Let $S_p \subset S$ be a part of co-occurrence sentence of the corpus, which can be transformed into $\{G_i \, v_j \, G_k\}$ groups and a verb. The sizes of G_i and G_k are $|G_i|$, $|G_k|$ and $G_i = \{g_1, \ldots, g_m\}$ and $G_k = \{g_{m+1}, \ldots, g_n\}$, $n > m$. Then, the relationships among G_i and G_k are build from the combinations of the elements from G_i and G_k with respect to v_j in accordance with the Bayesian model $p(G_i, G_k | V_j)$. There will be $|G_i| \times |G_k|$ relations,*

$$v_j(g_1, g_{m+1}) \leftarrow p(g_1, g_{m+1}|v_j)$$
$$v_j(g_1, g_{m+2}) \leftarrow p(g_1, g_{m+2}|v_j)$$
$$\ldots$$
$$v_j(g_1, g_n) \leftarrow p(g_1, g_n|v_j)$$
$$v_j(g_2, g_{m+1}) \leftarrow p(g_2, g_{m+1}|v_j)$$
$$\ldots$$
$$v_j(g_p, g_{m+1}) \leftarrow p(g_p, g_{m+1}|v_j)$$
$$\ldots$$
$$v_j(g_m, g_n) \leftarrow p(g_n, g_m|v_j)$$

The relations learned from Definitions 7 and 8 sometimes needs to be subjected to a lower bound. The lower bound is known as the *relations factor*, and it is used as an input parameter to the semantic analysis phase to set this lower bound.

Definition 10. *Let set $R = \{v_1(C_1, C_2), \ldots, v_m(C_k, C_r)\}$ be the relations that are learned from the corpus. Relations $v_i(C_j, C_k)$ are assigned a probability using a Bayesian model $p(C_j, C_k | V_i)$. When these relations are ordered based on their probability, a threshold φ is defined as the Relations Factor (RF) of the system.*

Definition 10 allows the user to limit the number of relations learned from the system. When the corpus is substantially large, the number of relations is proportional to the number of verbs in $\mathcal{L}_\mathcal{O}$. Not all relations may relevant and the RF is used as the limiting factor.

Definition 11. *Let v_i be a verb and v_j is the antonym verb of v_j learned from WordNet ($v_i \bowtie v_j$). Let there be relations $v_i(G_m, G_n)$ and $v_j(G_m, G_n)$ modeled by $p(G_m, G_n | v_r)$ ($v_r = i, j$). Since, $v_i \bowtie v_j$ for G_m and G_n, the relationship with the highest $p(G_m, G_n | v_r)$ value will be selected and the other relationship will be removed.*

We use verbs in $\mathcal{L}_\mathcal{O}$ as the key elements in forming relationships among concepts. Verbs have opposite verbs. Thus, according to Definition 11, if a verb is associated with some concepts and these concepts happen to be associated with a opposite verb, the verb with the highest Bayesian probability value is selected for the relations map and the other relationship will be removed from the system. Finally, the probabilistic conceptualization is serialized as an OWL DL ontology in the representation phase.

4   Implementation

The implementation of our approach uses several open source projects to populate the required contexts at different phases as we introduce in section 3. The bootstrapping algorithm requires tokenizing sentences and stemming or lemmatizing of tokens to produce $\mathcal{L}_\mathcal{O}$ of the corpus. According to Definition 1, $\mathcal{L}_\mathcal{O}$ is defined based on the Penn Treebank English POS tag set. We use the Stanford log-linear POS tagger[4], which uses the standard Penn Treebank tag set. We use the OpenNLP[5] tools to analyze sentences and tokens and the WordNet project to lookup for the type, stem and lemma of a word. In order to access the WordNet electronic library, we use the JWI[6] project. The BioAssayOntology corpus contains XHTML documents. We use the HTML parser[7] library to extract text from these documents. One of our other corpora contains PDF documents. We use the Apache PDFBox[8] library to extract the contents from the PDF documents. Finally, we use the Jena API[9] to serialize the probabilistic conceptualization model into OWL DL. Our implementation is based on Java 6 and it is named as `PrOntoLearn` (Probabilistic Ontology Learning).

5   Experiments

We have conducted experiments on three main data corpora, 1) the BioAssay Ontology (BAO) dataset 2) a sample collection of 38 PDF files from ISWC 2009 proceedings, and 3) a substantial portion of the web pages extracted from the University of Miami, Department of Computer Science[10] domain . We have constructed ontologies for all three corpora with different parameter settings. One of the key problems we have encountered is the ontology evaluation. The BAO and the PDF dataset were hard to evaluate as there are no existing reference ontologies or no ground truth that we could find of. Therefore, we use the third dataset from the University of Miami, Department of Computer Science domain and we measure recall and precision given a reference ontology.

The BAO dataset contains high throughput screening (HTS) assays performed on various screening centers (e.g., the Molecular Libraries Probe Production Centers Network (MLPCN)). HTS is the most common approach to initiate the development of novel drugs as therapeutics Increasingly complex biological systems and processes can be interrogated using HTS, leveraging innovative assay designs and new detection technologies. Driven by the NIH Molecular Libraries Initiative, HTS has become available to public research sector along with a public HTS data repository, PubChem [40]. However, HTS data are rarely used

---

[1] http://nlp.stanford.edu/software/tagger.shtml
[5] http://opennlp.sourceforge.net/
[6] http://projects.csail.mit.edu/jwi/
[7] http://htmlparser.sourceforge.net/samples.html
[8] http://pdfbox.apache.org/
[9] http://jena.sourceforge.net/
[10] http://www.cs.miami.edu

beyond one drug development project. This is due to the lack of standardized descriptions for HTS experiments and lacking standards to report the HTS results. The available data are therefore not used to their fullest potential [19]. The motivation for developing BAO was to address this problem and to enable categorization of assays by concepts that are relevant to interpret screening results, which would then facilitate meaningful data retrieval and analysis across diverse HTS assays. We specifically limited our dataset to assays available on the 1st of January 2010. Table 2 provides the statistics of the corpus. We extract the vocabulary generated from $[a\text{-}zA\text{-}Z]+[\text{-}\_]?\backslash w^*$ regular expression, and normalized them to create $\mathcal{L}_\mathcal{O}$ of the corpus.

Table 2. The PCAssay (the BioAssay Ontology project) corpus statistics

Title	Statistics	Description
Documents	1,759	All documents are XHTML formated with a given template
Unique $ConceptWords$	13,017	Normalized candidate concept words from NN, NNP, NNS, JJ, JJR & JJS using $[a\text{-zA-Z}]+[\text{-}\_]?\backslash w^*$
Unique $Verbs$	1,337	Normalized verbs from VB, VBD, VBG, VBN, VBP & VBZ using $[a\text{-zA-Z}]+[\text{-}\_]?\backslash w^*$
Total $ConceptWords$	631,623	
Total $Verbs$	109,421	
Total Lexicon	741,044	$Lexicon = ConceptWords \bigcap Verbs$
Total $Groups$	631,623	

One of the other obstacles we have encountered in terms of time complexity is in the representation layer. We use the Jena API to serialize the probabilistic conceptualization into OWL DL. When the system produced more than 1,000 concepts and relations, it is found that the Jena API takes a considerable amount of time to serialize the model. We use different architectural schemes to improve its performance. With all optimization, the presentation layer requires approximately 3.2 hours to serialize the model for the BAO dataset contains 1,758 documents. In order to provide a fast visualization of the conceptualization, we have written a simple yet flexible Java swing graphical user interface (GUI). This GUI has provided us visualizing and debugging the code as smoothly as possible. One of the other advantages of using a GUI is that it also provides the probabilities of the joint probability distribution $P(X, G)$, which is the representation of our probabilistic conceptualization.

The idea of our work is to generate an ontology without the supervision of a domain expert (unsupervised) for any given corpus. The user has to set system parameters such as KF, RF and regular expression of $\mathcal{L}_O$. Since we use corpora from the bio medical domain, a collection of research papers and set of documents collected from computer science web site, the evaluation of the created ontology using standard techniques such as precision and recall is not easy. We evaluate the generated ontologies with human domain experts. We obtain the comments and recommendations from the domain expert on the importance of the generated ontology. The ontology that is generated is too large to show in here.Instead, we provide a few distinct snapshots of the ontology with the help of Protégé OWLViz plugin. Figures 6 and 7 show snapshots of the ontology created from the BioAssay Ontology corpus for input parameters $KF = 0.5$, N-gram = 3, and $RF = 0.9$. Figure 6 shows the IS-A relationships and Figure 7 shows the binary relationships.

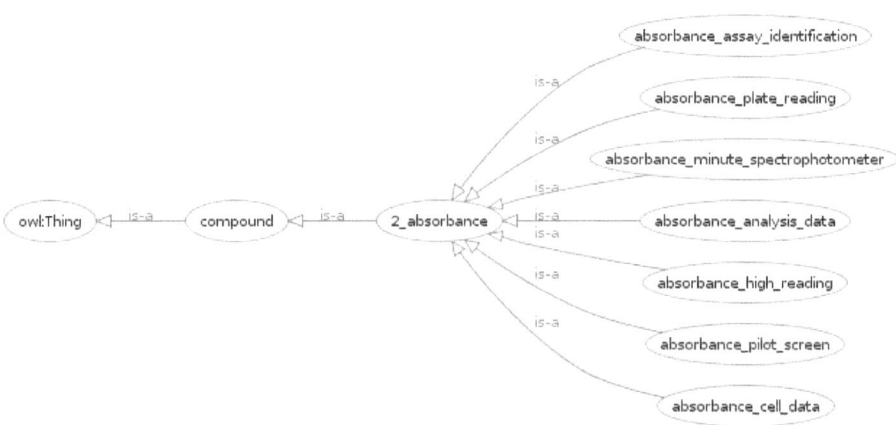

Fig. 6. An example snapshot of the BioAssay Ontology corpus with IS-A relations

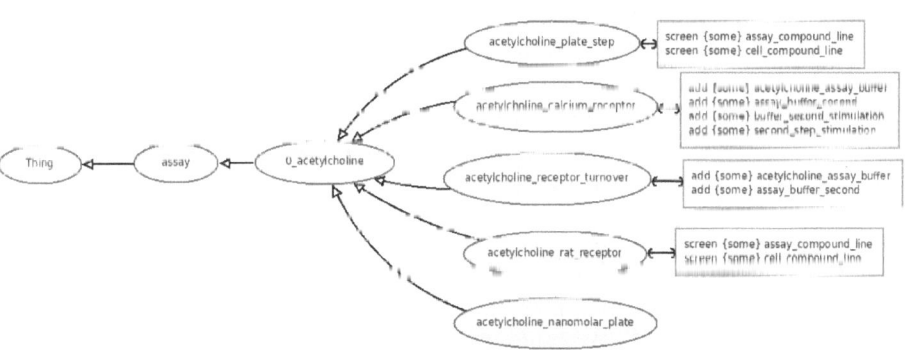

Fig. 7. An example snapshot of the BioAssay Ontology corpus with binary relations

Table 3. Precision, recall and F1 measurement for N-gram=4 and RF=1 using extended reference ontology

KF	Precision	Recall	F1
0.1	0.424	1	0.596
0.2	0.388	1	0.559
0.3	0.445	1	0.616
0.4	0.438	1	0.609
0.5	0.438	1	0.609
0.6	0.424	1	0.595
0.7	0.415	1	0.587
0.8	0.412	1	0.583
0.9	0.405	1	0.576
1.0	0.309	1	0.472

The generation of an ontology such as BAO from the PubChem corpus using the described methodology is very challenging. BAO concepts labels in many cases are not present in the text of assay descriptions, but are assigned by domain experts. This makes it even difficult to automatically annotate assays given a finished ontology and certainly it is a barrier to construct an ontology from scratch. It is therefore not possible to quantify the quality of such the ontology. However, we can qualitatively evaluate the ontology via domain exports how meaningful the extracted concepts and their relationships are. We have consulted one expert and the recommendation is that the ontology contains rich set of vocabulary, which is very useful for top-down ontology construction. But expert also mentioned that the ontology have a flat structure. The main reason for this observation is that we use a 3-gram generator to create the ontology. Therefore, the maximum levels this model achieve is at most 3.

The *www.cs.miami.edu* corpus is used to calculate quantitative measurements. The gold standard based approaches such as precision (P), recall (R) and F-measure (F_1) are used to evaluate ontologies [13]. We use a slightly modified version of [35] as our reference ontology. Table 3 shows the results. The average precision of the constructed ontology is approximately 42%. It is to be noted that we use only one reference ontology. If we use another reference ontology the precision values varies. This means that the precision value depends on the available ground truth.

The results show that our method creates an ontology for any given domain with acceptable results. This is shown in the precision value, if the ground truth is available. On the other hand, if the domain does not have ground truth the results are subject to domain expert evaluation of the ontology. One of the potential problems we have seen in our approach is search space. Since our method is unsupervised, it tends to search the entire space for results, which is computationally costly. We thus need a better method to prune the search space so that out method provide better results. According to domain experts, our method extracts good vocabulary but provides a flat structure. We have proposed a semi-supervised approach to correct these problems by combining the knowledge from domain experts and results produced by our system. We left the detailed investigation for future work.

Since our method is based on the Bayesian reasoning (which uses N-gram probabilities), it is paramount that the corpus contains enough evidence of the redundant information. This condition requires that the corpus to be large enough so that we can hypothesize that the corpus provides enough evidence to build the ontology.

We hypothesize that a sentence of the corpus would generally be subjected to the grammar rule given in expression 1. This constituent is the main factor that uses to build the relationships among concepts. In NLP, there are many other finer grained grammar rules that specifically fit for given sentences. If these grammar rules are used, we believe we can build a better relationship model. We have left this for future work.

At the moment our system does not distinguish between concepts and the individuals of the concepts. The learned A-Box primarily consists of the probabilities of each concepts. This is one area where we are eager to work on. Using the state-of-the art NLP techniques, we plan to fill this gap in a future work. Since our method has the potential to be used in any corpus, it could be seen that the lemmatizing and stemming algorithms that are available in WordNet would not recognize some of the words. Specially in the BioAssay corpus, we observe that some of the domain specific words are not recognized by WordNet. We use the Porter stemming algorithm [32] to get the word form and it shows that this algorithm constructs peculiar word forms. Therefore, we deliberately remove it from the processing pipeline.

The complexity of our algorithms is as follows. The bootstrapping algorithm available in the syntactic layer has a worst case running time of $O(M \times max(s_j) \times max(w_k))$, where M is the number of documents, s_j is a the number of sentences in a document, and w_k is the number of words in a sentence. The probabilistic reasoning algorithm has the worst case running time of $O(|\mathcal{L}| \times |SuperConcepts|)$, where $|\mathcal{L}|$ is the size of the lexicon and $|SuperConcepts|$ is the size of the super concepts set. The ontologies generated from the system are consistent with Pellet[11] and FaCT++[12] reasoners.

Finally, our method provides a process to create a lexico-semantic ontology for any domain. For our knowledge, this is a very first research on this line of work. So we continue our research along this line and to provide better results for future use.

6   Conclusion

We have introduced a novel process to generate an ontology for any random text corpus. We have shown that our process constructs a flexible ontology. It is also shown that in order to achieve high precision, it is paramount that the corpus

---

[11] http://clarkparsia.com/pellet
[12] http://owl.man.ac.uk/factplusplus/

should be large enough to extract important evidence. Our research has also shown that probabilistic reasoning on lexico-semantic structures is a powerful solution to overcome or at least mitigate the knowledge acquisition bottleneck. Our method also provides evidence to domain experts to build ontologies using a top-down approach. Though we have introduced a powerful technique to construct ontologies, we believe that there is a lot of work that can be done to improve the performance of our system. One of the areas our method lacks is the separation between concepts and individuals. We would like to use the generated ontology as a seed ontology to generate instances for the concepts and extract the individuals already classified as concepts. Finally, we would like to increase the lexicon of the system with more tags available from the Penn Treebank tag set. We believe that if we introduce more tags into the system, our system can be trained to construct human readable (friendly) concepts and relations names.

Acknowledgements. This work was partially funded by the NIH grant RC2 HG005668. We thank all anonymous reviewers for their comments that greatly improve the structure and the presentation of the paper.

References

1. Balby Marinho, L., Buza, K., Schmidt-Thieme, L.: Folksonomy-Based Collabulary Learning. In: Sheth, A.P., Staab, S., Dean, M., Paolucci, M., Maynard, D., Finin, T., Thirunarayan, K. (eds.) ISWC 2008. LNCS, vol. 5318, pp. 261–276. Springer, Heidelberg (2008)
2. Banerjee, S., Pedersen, T.: The Design, Implementation and Use of the N-Gram Statistics Package. In: Gelbukh, A. (ed.) CICLing 2003. LNCS, vol. 2588, pp. 370–381. Springer, Heidelberg (2003)
3. Bergamaschi, S., Po, L., Sorrentino, S., Corni, A.: Uncertainty in data integration systems: automatic generation of probabilistic relationships. In: Management of the Interconnected World: ItAIS: the Italian Association for Information Systems, p. 221 (2010)
4. Bos, J., Markert, K.: Recognising textual entailment with logical inference. In: HLT 2005: Proceedings of the Conference on Human Language Technology and Empirical Methods in Natural Language Processing, pp. 628–635. Association for Computational Linguistics, Morristown (2005)
5. Cankaya, H.C., Moldovan, D.: Method for extracting commonsense knowledge. In: K-CAP 2009: Proceedings of the Fifth International Conference on Knowledge Capture, pp. 57–64. ACM, New York (2009)
6. Carlson, A., Betteridge, J., Wang, R.C., Hruschka Jr., E.R., Mitchell, T.M.: Coupled semi-supervised learning for information extraction. In: WSDM 2010: Proceedings of the Third ACM International Conference on Web Search and Data Mining, pp. 101–110. ACM, New York (2010)
7. Chemudugunta, C., Holloway, A., Smyth, P., Steyvers, M.: Modeling Documents by Combining Semantic Concepts with Unsupervised Statistical Learning. In: Sheth, A.P., Staab, S., Dean, M., Paolucci, M., Maynard, D., Finin, T., Thirunarayan, K. (eds.) ISWC 2008. LNCS, vol. 5318, pp. 229–244. Springer, Heidelberg (2008)
8. Cimiano, P.: Ontology Learning and Population from Text: Algorithms, Evaluation and Applications. Springer-Verlag New York, Inc., Secaucus (2006)

9. Cimiano, P., Hotho, A., Staab, S.: Learning concept hierarchies from text corpora using formal concept analysis. Journal of Artificial Intelligence Research 24, 305–339 (2005)

10. Cimiano, P., Völker, J.: Text2Onto - A Framework for Ontology Learning and Data-driven Change Discovery (2005)

11. Clark, P., Harrison, P.: Large-scale extraction and use of knowledge from text. In: K-CAP 2009: Proceedings of the Fifth International Conference on Knowledge Capture, pp. 153–160. ACM, New York (2009)

12. Davis, B., Iqbal, A.A., Funk, A., Tablan, V., Bontcheva, K., Cunningham, H., Handschuh, S.: RoundTrip Ontology Authoring. In: Sheth, A.P., Staab, S., Dean, M., Paolucci, M., Maynard, D., Finin, T., Thirunarayan, K. (eds.) ISWC 2008. LNCS, vol. 5318, pp. 50–65. Springer, Heidelberg (2008)

13. Dellschaft, K., Staab, S.: Strategies for the evaluation of ontology learning. In: Proceeding of the 2008 Conference on Ontology Learning and Population: Bridging the Gap between Text and Knowledge, pp. 253–272. IOS Press, Amsterdam (2008)

14. Ding, Z., Peng, Y.: A Probabilistic Extension to Ontology Language OWL. In: HICSS 2004: The Proceedings of the 37th Annual Hawaii International Conference on System Sciences (HICSS 2004) - Track 4, p. 40111.1. IEEE Computer Society, Washington, DC (2004)

15. Gruber, T.R.: A translation approach to portable ontology specifications. Knowledge Acquisition 5(2), 199–220 (1993)

16. Haase, P., Völker, J.: Ontology Learning and Reasoning — Dealing with Uncertainty and Inconsistency. In: da Costa, P.C.G., d'Amato, C., Fanizzi, N., Laskey, K.B., Laskey, K.J., Lukasiewicz, T., Nickles, M., Pool, M. (eds.) URSW 2005-2007. LNCS (LNAI), vol. 5327, pp. 366–384. Springer, Heidelberg (2008)

17. Hearst, M.A.: Automatic acquisition of hyponyms from large text corpora. In: Proceedings of the 14th Conference on Computational Linguistics, pp. 539–545. Association for Computational Linguistics, Morristown (1992)

18. Hitzler, P.: What's Happening in Semantic Web ... and What FCA Could Have to Do with It. In: Valtchev, P., Jäschke, R. (eds.) ICFCA 2011. LNCS, vol. 6628, pp. 18–23. Springer, Heidelberg (2011)

19. Inglese, J., Shamu, C.E., Guy, R.K.:

20. Jurafsky, D., Martin, J.H.: Speech and Language Processing: An Introduction to Natural Language Processing. Computational Linguistics and Speech Recognition, 2nd edn. Prentice Hall, Pearson Education International (2009)

21. Kim, D.S., Barker, K., Porter, B.: Knowledge integration across multiple texts. In: K-CAP 2009: Proceedings of the Fifth International Conference on Knowledge Capture, pp. 49–56. ACM, New York (2009)

22. Koller, D., Levy, A., Pfeffer, A.: P-CLASSIC: A tractable probabilistic description logic. In: Proceedings of AAAI 1997, pp. 390–397 (1997)

23. Lin, D., Pantel, P.: Discovery of inference rules for question-answering. Natural Language Engineering 7(4), 343–360 (2001)

24. Lukasiewicz, T.: Expressive probabilistic description logics. Artif. Intell. 172(6-7), 852–883 (2008)

25. Lukasiewicz, T., Straccia, U.: Managing uncertainty and vagueness in description logics for the semantic web. J. Web Sem. 6(4), 291–308 (2008)

26. Manning, C.D., Raghavan, P., Schütze, H.: Introduction to Information Retrieval. Cambridge University Press, New York (2008)

27. Marcus, M.P., Marcinkiewicz, M.A., Santorini, B.: Building a large annotated corpus of English: the penn treebank. Comput. Linguist. 19(2), 313–330 (1993)

28. Maynard, D., Li, Y., Peters, W.: Nlp techniques for term extraction and ontology population. In: Proceeding of the 2008 conference on Ontology Learning and Population: Bridging the Gap between Text and Knowledge, pp. 107–127. IOS Press, Amsterdam (2008)
29. McCrae, J., Spohr, D., Cimiano, P.: Linking Lexical Resources and Ontologies on the Semantic Web with Lemon. In: Antoniou, G., Grobelnik, M., Simperl, E., Parsia, B., Plexousakis, D., De Leenheer, P., Pan, J. (eds.) ESWC 2011, Part I. LNCS, vol. 6643, pp. 245–259. Springer, Heidelberg (2011)
30. Pantel, P., Pennacchiotti, M.: Automatically harvesting and ontologizing semantic relations. In: Proceeding of the 2008 Conference on Ontology Learning and Population: Bridging the Gap between Text and Knowledge, pp. 171–195. IOS Press, Amsterdam (2008)
31. Poon, H., Domingos, P.: Unsupervised ontology induction from text. In: Proceedings of the 48th Annual Meeting of the Association for Computational Linguistics, ACL 2010, pp. 296–305. Association for Computational Linguistics, Stroudsburg (2010)
32. Porter, M.F.: An algorithm for suffix stripping. Program 14(3), 130–137 (1980)
33. Russell, S.J., Norvig, P.: Artificial Intelligence: A Modern Approach, 3rd edn. Prentice Hall (2009)
34. Salloum, W.: A question answering system based on conceptual graph formalism. In: KAM 2009: Proceedings of the 2009 Second International Symposium on Knowledge Acquisition and Modeling, pp. 383–386. IEEE Computer Society, Washington, DC (2009)
35. SHOE: Example computer science department ontology, http://www.cs.umd.edu/projects/plus/SHOE/cs.html (last visited on June 2, 2011)
36. Studer, R., Benjamins, V.R., Fensel, D.: Knowledge engineering: Principles and methods. Data and Knowledge Engineering 25(1-2), 161–197 (1998)
37. Tanev, H., Magnini, B.: Weakly supervised approaches for ontology population. In: Proceeding of the 2008 Conference on Ontology Learning and Population: Bridging the Gap between Text and Knowledge, pp. 129–143. IOS Press, Amsterdam (2008)
38. Tomanek, K., Hahn, U.: Reducing class imbalance during active learning for named entity annotation. In: K-CAP 2009: Proceedings of the Fifth International Conference on Knowledge Capture, pp. 105–112. ACM, New York (2009)
39. Völker, J., Hitzler, P., Cimiano, P.: Acquisition of OWL DL Axioms from Lexical Resources. In: Franconi, E., Kifer, M., May, W. (eds.) ESWC 2007. LNCS, vol. 4519, pp. 670–685. Springer, Heidelberg (2007)
40. Wang, Y., Xiao, J., Suzek, T.O., Zhang, J., Wang, J., Bryant, S.H.: Pubchem: a public information system for analyzing bioactivities of small molecules. Nucleic Acids Research, 623–633 (2009)

Semantic Web Search and Inductive Reasoning

Claudia d'Amato[1], Nicola Fanizzi[1], Bettina Fazzinga[2],
Georg Gottlob[3,4], and Thomas Lukasiewicz[3]

[1] Dipartimento di Informatica, Università degli Studi di Bari, Italy
{claudia.damato,fanizzi}@di.uniba.it
[2] Dipartimento di Elettronica, Informatica e Sistemistica, Università della Calabria, Italy
bfazzinga@deis.unical.it
[3] Department of Computer Science, University of Oxford, UK
Georg.Gottlob@cs.ox.ac.uk
[4] Oxford-Man Institute of Quantitative Finance, University of Oxford, UK

Abstract. Extensive research activities are recently directed towards the Semantic Web as a future form of the Web. Consequently, Web search as the key technology of the Web is evolving towards some novel form of Semantic Web search. A very promising recent such approach is based on combining standard Web pages and search queries with ontological background knowledge, and using standard Web search engines as the main inference motor of Semantic Web search. In this paper, we further enhance this approach to Semantic Web search by the use of inductive reasoning techniques. This adds especially the important ability to handle inconsistencies, noise, and incompleteness, which are all very likely to occur in distributed and heterogeneous environments, such as the Web. We report on a prototype implementation of the new approach and experimental results.

1 Introduction

Web search [6] as the key technology of the Web is about to change radically with the development of the *Semantic Web* [3]. As a consequence, the elaboration of a new search technology for the Semantic Web, called *Semantic Web search* [18], is currently an extremely hot topic, both in Web-related companies and in academic research. In particular, there is a fast growing number of commercial and academic Semantic Web search engines. The research can be roughly divided into two main directions. The first (most common) one is to develop a new form of search for searching the pieces of data and knowledge that are encoded in the new representation formalisms of the Semantic Web (e.g., [18]), while the second (less explored) direction is to use the data and knowledge of the Semantic Web to add some semantics to Web search (e.g., [27]).

A very promising recent representative of the second direction to Semantic Web search has been presented in [22]. The approach is based on (i) using ontological (unions of) conjunctive queries (which may contain negated subqueries) as Semantic Web search queries, (ii) combining standard Web pages and search queries with ontological background knowledge, (iii) using the power of Semantic Web formalisms and technologies, and (iv) using standard Web search engines as the main inference motor of Semantic Web search. It consists of an offline ontology compilation step, based on deductive reasoning techniques, and an online query processing step.

F. Bobillo et al. (Eds.): URSW 2008-2010/UniDL 2010, LNAI 7123, pp. 237–261, 2013.
© Springer-Verlag Berlin Heidelberg 2013

In this paper, we propose to further enhance this approach to Semantic Web search by the use of inductive reasoning techniques for the offline ontology compilation step. This allows to cope with inconsistencies, noise, and incompleteness as forms of uncertainty. The main idea behind combining Semantic Web search with inductive reasoning is also closely related to the idea of using probabilistic ontologies to increase the precision and the recall of querying databases and of information retrieval in general, but rather than learning probabilistic ontologies from data, representing them, and reasoning with them, we directly use the data in the inductive inference step. To our knowledge, this is the first combination of Semantic Web search with inductive reasoning. The main contributions of this paper are briefly summarized as follows:

- We develop a combination of Semantic Web search as presented in [22] with an inductive reasoning technique (based on similarity search [52] for retrieving the resources that likely belong to a query concept [14]). The latter serves in an offline ontology compilation step to compute completed semantic annotations.
- Importantly, the new approach to Semantic Web search can handle inconsistencies, noise, and incompleteness in Semantic Web knowledge bases, which are all very likely to occur in distributed and heterogeneous environments, such as the Web. We provide several examples illustrating this important advantage of the new approach.
- We report on a prototype implementation of the new approach in the context of desktop search. We also provide very positive experimental results for the precision and the recall of the new approach, comparing it to the deductive approach in [22].

The rest of this paper is organized as follows. In Sections 2 and 3, we give a brief overview of our Semantic Web search system and the underlying theoretical model, respectively. Section 4 proposes to use inductive rather than deductive reasoning in Semantic Web search. In Section 5, we describe the main advantages of using inductive reasoning in this context. Section 6 reports on a prototype implementation in desktop search along with experimental results. In Section 7, we discuss related work. Section 8 finally summarizes the main results and gives an outlook on future research.

2   System Overview

The overall architecture of our Semantic Web search system is shown in Fig. 1. It consists of the *Interface*, the *Query Evaluator*, and the *Inference Engine*, where the Query Evaluator is implemented on top of standard Web *Search Engines*. Standard *Web* pages and their objects are enriched by *Annotation* pages, based on an underlying *Ontology*.

Ontology. Our approach to Semantic Web search is done relative to a fixed underlying *ontology*, which defines an alphabet of elementary ontological ingredients, as well as terminological relationships between these ingredients. The ontology may either describe fully general knowledge (such as the knowledge encoded in Wikipedia) for general ontology-based search on the Web, or it may describe some specific knowledge (such as biomedical knowledge) for vertical ontology-based search on the Web. The former results into a general ontology-based interface to the Web similar to Google, while the latter produces different vertical ontology-based interfaces to the Web. There

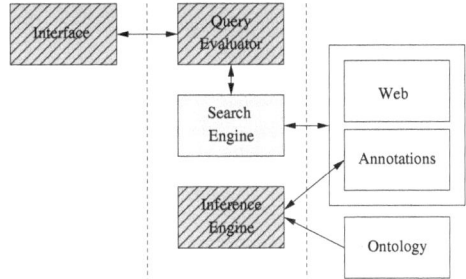

Fig. 1. System architecture

are many existing ontologies that can be used, which have especially been developed for the Semantic Web, but also in biomedical and technical areas. They are generally created and updated by human experts in a knowledge engineering process. Recent research attempts are also directed towards an automatic generation of ontologies from text documents, eventually coming along with existing ontological knowledge [7, 21].

For example, an ontology may contain the knowledge that (i) conference and journal papers are articles, (ii) conference papers are not journal papers, (iii) *isAuthorOf* relates scientists and articles, (iv) *isAuthorOf* is the inverse of *hasAuthor*, and (v) *hasFirstAuthor* is a functional binary relationship, which is formalized as follows:

$$ConferencePaper \sqsubseteq Article, \ JournalPaper \sqsubseteq Article, \ ConferencePaper \sqsubseteq \neg JournalPaper,$$
$$\exists isAuthorOf \sqsubseteq Scientist, \ \exists isAuthorOf^- \sqsubseteq Article, \ isAuthorOf^- \sqsubseteq hasAuthor, \qquad (1)$$
$$hasAuthor^- \sqsubseteq isAuthorOf, \ (\text{funct } hasFirstAuthor) \ .$$

Annotations. As a second ingredient of our Semantic Web search, we assume the existence of assertional pieces of knowledge about Web pages and their objects, also called *(semantic) annotations*, which are defined relative to the terminological relationships of the underlying ontology. Such annotations are starting to be widely available for a large class of Web resources, especially user-defined annotations with the Web 2.0. They may also be automatically learned from Web pages and their objects (e.g., [9]). As a midway between such fully user-defined and fully automatically generated annotations, one can also automatically extract annotations from Web pages using user-defined rules [22].

For example, in a very simple scenario relative to the ontology in Eq (1), a Web page i_1 (Fig. 2, left side) may contain information about a Ph.D. student i_2, called Mary, and two of her papers, a conference paper i_3 with title "*Semantic Web search*" and a journal paper i_4 entitled "*Semantic Web search engines*" and published in 2008. There may now exist one semantic annotation each for the Web page, the Ph.D. student Mary, the journal paper, and the conference paper. The annotation for the Web page may simply encode that it mentions Mary and the two papers, while the one for Mary may encode that she is a Ph.D. student with the name Mary and the author of the papers i_3 and i_4. The annotation for i_3 may encode that i_3 is a conference paper and has the title "*Semantic Web search*", while the one for i_4 may encode that i_4 is a journal paper, authored by Mary, has the title "*Semantic Web search engines*", was published in 2008, and has the keyword "RDF". The annotations of i_1, i_2, i_3, and i_4 are formally expressed

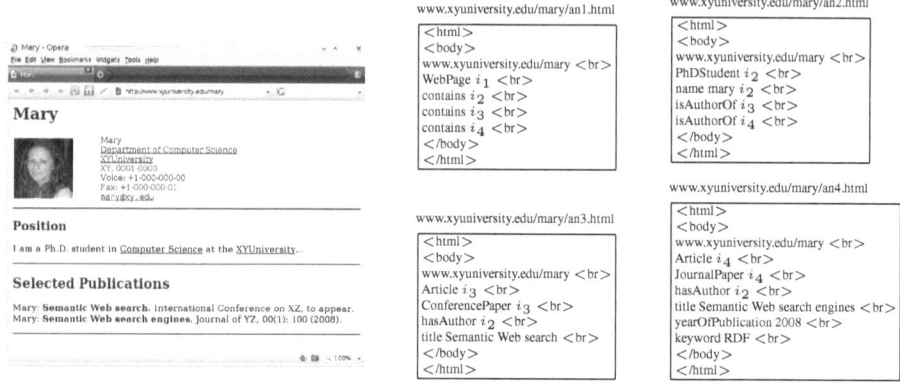

Fig. 2. Left side: HTML page p; right side: four HTML pages p_1, p_2, p_3, and p_4, which encode (completed) semantic annotations for p and the objects on p

as the following sets of ontological axioms $\mathcal{A}_{i_1}$, $\mathcal{A}_{i_2}$, $\mathcal{A}_{i_3}$, and $\mathcal{A}_{i_4}$, respectively:

$$\mathcal{A}_{i_1} = \{contains(i_1, i_2), \, contains(i_1, i_3), contains(i_1, i_4)\},$$
$$\mathcal{A}_{i_2} = \{PhDStudent(i_2), \, name(i_2, \text{``mary''}), \, isAuthorOf(i_2, i_3), \, isAuthorOf(i_2, i_4)\},$$
$$\mathcal{A}_{i_3} = \{ConferencePaper(i_3), \, title(i_3, \text{``Semantic Web search''})\}, \qquad (2)$$
$$\mathcal{A}_{i_4} = \{JournalPaper(i_4), \, hasAuthor(i_4, i_2), \, title(i_4, \text{``Semantic Web search engines''}),$$
$$yearOfPublication(i_4, 2008), \, keyword(i_4, \text{``RDF''})\}.$$

Inference Engine. Differently from the ontology, the semantic annotations can be directly published on the Web and searched via standard Web search engines. To also make the ontology visible to standard Web search engines, it is compiled into the semantic annotations: all semantic annotations are completed in an offline ontology compilation step, where the *Inference Engine* adds all properties (that is, ground atoms) that can be derived (deductively in [22] and inductively here) from the ontology and the semantic annotations. The resulting (*completed*) semantic annotations are then published as Web pages, so that they can be searched by standard Web search engines.

For example, considering again the running scenario, using the ontology in Eq. (1), in particular, we can derive from the semantic annotations in Eq. (2) that the two papers i_3 and i_4 are also articles, and both authored by Mary.

HTML Encoding of Annotations. The above searchable (completed) semantic annotations of (objects on) standard Web pages are published as HTML Web pages with pointers to the respective object pages, so that they (in addition to the standard Web pages) can be searched by standard search engines. For example, the HTML pages for the completed semantic annotations of the above $\mathcal{A}_{i_1}$, $\mathcal{A}_{i_2}$, $\mathcal{A}_{i_3}$, and $\mathcal{A}_{i_4}$ are shown in Fig. 2, right side. We here use the HTML address of the Web page/object's annotation page as an identifier for that Web page/object. The plain textual representation of the completed semantic annotations allows their processing by existing standard search engines for the Web. It is important to point out that this textual representation is simply a list of properties, each eventually along with an identifier or a data value as attribute value, and it can thus immediately be encoded as a list of RDF triples. Similarly, the completed semantic annotations can easily be encoded in RDFa or microformats.

Query Evaluator. The *Query Evaluator* reduces each Semantic Web search query of the user in an online query processing step to a sequence of standard Web search queries on standard Web and annotation pages, which are then processed by a standard Web *Search Engine*. The Query Evaluator also collects the results and re-transforms them into a single answer which is returned to the user. As an example of a Semantic Web search query, one may ask for all Ph.D. students who have published an article in 2008 with RDF as a keyword, which is formally expressed as follows:

$$Q(x) = \exists y \, (PhDStudent(x) \wedge isAuthorOf(x, y) \wedge Article(y) \wedge \\ yearOfPublication(y, 2008) \wedge keyword(y, \text{``}RDF\text{''})) \,. \tag{3}$$

This query Q is transformed into the two queries $Q_1 = PhDStudent$ AND *isAuthorOf* and $Q_2 = Article$ AND *"yearOfPublication* 2008" AND *"keyword* RDF", which can both be submitted to a standard Web search engine. The result of the original query Q is then built from the results of the two queries Q_1 and Q_2. Note that a graphical user interface, such as the one of Google's advanced search, and ultimately a natural language interface (for queries in written or spoken natural language) can help to hide the conceptual complexity of ontological queries to the user.

3   Semantic Web Search

We now introduce Semantic Web knowledge bases and the syntax and semantics of Semantic Web search queries to such knowledge bases. We then generalize the PageRank technique to our approach. We assume the reader is familiar with the syntax and semantics of description logics (DLs) [2], which we use as underlying ontology languages.

3.1   Semantic Web Knowledge Bases

Intuitively, a Semantic Web knowledge base consists of a background TBox and a collection of ABoxes, one for every concrete Web page and for every object on a Web page. For example, the homepage of a scientist may be such a concrete Web page and be associated with an ABox, while the publications on the homepage may be such objects, which are also associated with one ABox each.

We assume pairwise disjoint sets $\mathbf{D}$, $\mathbf{A}$, $\mathbf{R}_A$, $\mathbf{R}_D$, $\mathbf{I}$, and $\mathbf{V}$ of atomic datatypes, atomic concepts, atomic roles, atomic attributes, individuals, and data values, respectively. Let $\mathbf{I}$ be the disjoint union of two sets $\mathbf{P}$ and $\mathbf{O}$ of *Web pages* and *Web objects*, respectively. Informally, every $p \in \mathbf{P}$ is an identifier for a concrete Web page, while every $o \in \mathbf{O}$ is an identifier for a concrete object on a concrete Web page. We assume the atomic roles *links_to* between Web pages and *contains* between Web pages and Web objects. The former represents the link structure between concrete Web pages, while the latter encodes the occurrences of concrete Web objects on concrete Web pages.

Definition 1. A *semantic annotation* $\mathcal{A}_a$ for a Web page or object $a \in \mathbf{P} \cup \mathbf{O}$ is a finite set of concept membership axioms $A(a)$, role membership axioms $P(a, b)$, and attribute membership axioms $U(a, v)$, where $A \in \mathbf{A}$, $P \in \mathbf{R}_A$, $U \in \mathbf{R}_D$, $b \in \mathbf{I}$, and $v \in \mathbf{V}$. A *Semantic Web knowledge base* $KB = (\mathcal{T}, (\mathcal{A}_a)_{a \in \mathbf{P} \cup \mathbf{O}})$ consists of a TBox $\mathcal{T}$ and one semantic annotation $\mathcal{A}_a$ for every Web page or object $a \in \mathbf{P} \cup \mathbf{O}$.

Informally, a Semantic Web knowledge base consists of some background termino-logical knowledge and some assertional knowledge for every concrete Web page and for every concrete object on a Web page. The background terminological knowledge may be an ontology from some global Semantic Web repository or an ontology defined locally by the user site. In contrast to the background terminological knowledge, the assertional knowledge will be directly stored on the Web (on annotation pages like the described standard Web pages) and is thus accessible via Web search engines.

Example 1. (Scientific Database). We use a Semantic Web knowledge base $KB = (\mathcal{T}, \mathcal{A})$ to specify some simple information about scientists and their publications. The sets of atomic concepts, atomic roles, atomic attributes, and data values are as follows:

$$\mathbf{A} = \{Scientist, PhDStudent, Article, ConferencePaper, JournalPaper\},$$
$$\mathbf{R}_A = \{hasAuthor, isAuthorOf, hasFirstAuthor, links\_to, contains\},$$
$$\mathbf{R}_D = \{name, title, yearOfPublication, keyword\},$$
$$\mathbf{V} = \{\text{"mary"}, \text{"Semantic Web search"}, 2008, \text{"Semantic Web search engines"}, \text{"RDF"}\}.$$

Let $\mathbf{I} = \mathbf{P} \cup \mathbf{O}$ be the set of individuals, where $\mathbf{P} = \{i_1\}$ is the set of Web pages, and $\mathbf{O} = \{i_2, i_3, i_4\}$ is the set of Web objects on the Web page i_1. The TBox $\mathcal{T}$ contains the axioms in Eq. (1). Then, a Semantic Web knowledge base is given by $KB = (\mathcal{T}, (\mathcal{A}_a)_{a \in \mathbf{P} \cup \mathbf{O}})$, where the semantic annotations of all $a \in \mathbf{P} \cup \mathbf{O}$ are the ones in Eq. (2).

3.2 Semantic Web Search Queries

We use unions of conjunctive queries with negated conjunctive subqueries as Semantic Web search queries to Semantic Web knowledge bases. We now first define the syntax and then the semantics of positive and general Semantic Web search queries.

Syntax. Let $\mathbf{X}$ be a finite set of variables. A *term* is either a Web page $p \in \mathbf{P}$, a Web object $o \in \mathbf{O}$, a data value $v \in \mathbf{V}$, or a variable $x \in \mathbf{X}$. An *atomic formula* (or *atom*) α is of one of the following forms: (i) $d(t)$, where d is an atomic datatype, and t is a term; (ii) $A(t)$, where A is an atomic concept, and t is a term; (iii) $P(t, t')$, where P is an atomic role, and t, t' are terms; and (iv) $U(t, t')$, where U is an atomic attribute, and t, t' are terms. An *equality* has the form $=(t, t')$, where t and t' are terms. A *conjunctive formula* $\exists \mathbf{y}\, \phi(\mathbf{x}, \mathbf{y})$ is an existentially quantified conjunction of atoms α and equalities $=(t, t')$, which have free variables among $\mathbf{x}$ and $\mathbf{y}$.

Definition 2. A *Semantic Web search query* $Q(\mathbf{x})$ is an expression $\bigvee_{i=1}^{n} \exists \mathbf{y}_i\, \phi_i(\mathbf{x}, \mathbf{y}_i)$, where each ϕ_i, $i \in \{1, \ldots, n\}$, is a conjunction of atoms α (also called *positive atoms*), negated conjunctive formulas *not* ψ, and equalities $=(t, t')$, which have free variables among $\mathbf{x}$ and $\mathbf{y}_i$, and the $\mathbf{x}$'s are exactly the free variables of $\bigvee_{i=1}^{n} \exists \mathbf{y}_i\, \phi_i(\mathbf{x}, \mathbf{y}_i)$.

Intuitively, Semantic Web search queries are unions of conjunctive queries with negated conjunctive queries in addition to atoms and equalities as conjuncts. Note that the negation "*not*" in Semantic Web search queries is a default negation and thus differs from the classical negation "¬" used in concepts in Semantic Web knowledge bases.

Example 2. (Scientific Database cont'd). Two Semantic Web search queries are:

$$Q_1(x) = (Scientist(x) \land not\ doctoralDegree(x, \text{``oxford university''}) \land worksFor(x,$$
$$\text{``oxford university''})) \lor (Scientist(x) \land doctoralDegree(x, \text{``oxford university''}) \land$$
$$not\ worksFor(x, \text{``oxford university''}));$$
$$Q_2(x) = \exists y\ (Scientist(x) \land worksFor(x, \text{``oxford university''}) \land isAuthorOf(x, y) \land$$
$$not\ ConferencePaper(y) \land not\ \exists z\ yearOfPublication(y, z)).$$

Informally, $Q_1(x)$ asks for scientists who are either working for *oxford university* and did not receive their Ph.D. from that university, or who received their Ph.D. from *oxford university* but do not work for it. Whereas query $Q_2(x)$ asks for scientists of *oxford university* who are authors of at least one unpublished non-conference paper. Note that when searching for scientists, the system automatically searches for all subconcepts (known according to the ontology), such as, e.g., Ph.D. students or computer scientists.

Semantics of Positive Search Queries. We now define the semantics of positive Semantic Web search queries, which are free of negations, in terms of ground substitutions via the notion of logical consequence.

A search query $Q(\mathbf{x})$ is *positive* iff it contains no negated conjunctive subqueries. A *(variable) substitution* θ maps variables from $\mathbf{X}$ to terms. A substitution θ is *ground* iff it maps to Web pages $p \in \mathbf{P}$, Web objects $o \in \mathbf{O}$, and data values $v \in \mathbf{V}$. A closed first-order formula ϕ is a *logical consequence* of a knowledge base $KB = (\mathcal{T}, (\mathcal{A}_a)_{a \in \mathbf{P} \cup \mathbf{O}})$, denoted $KB \models \phi$, iff every first-order model $\mathcal{I}$ of $\mathcal{T} \cup \bigcup_{a \in \mathbf{P} \cup \mathbf{O}} \mathcal{A}_a$ also satisfies ϕ.

Definition 3. Given a Semantic Web knowledge base KB and a positive Semantic Web search query $Q(\mathbf{x})$, an *answer* for $Q(\mathbf{x})$ to KB is a ground substitution θ for the variables $\mathbf{x}$ (which are exactly the free variables of $Q(\mathbf{x})$) such that $KB \models Q(\mathbf{x}\theta)$.

Example 3. (Scientific Database cont'd). Consider the Semantic Web knowledge base KB of Example 1. The search query $Q(x)$ of Eq. (3) is positive, and an answer for $Q(x)$ to KB is $\theta = \{x/i_2\}$. Recall that i_2 represents the Ph.D. student Mary.

Semantics of General Search Queries. We next define the semantics of general Semantic Web search queries by reduction to the semantics of positive ones, interpreting negated conjunctive subqueries *not* ψ as the lack of evidence about the truth of ψ. That is, negations are interpreted by a closed-world semantics on top of the open world semantics of DLs (we refer to [22] for more motivation and background).

Definition 4. Given a Semantic Web knowledge base KB and search query

$$Q(\mathbf{x}) = \bigvee_{i=1}^{n} \exists \mathbf{y}_i\ \phi_{i,1}(\mathbf{x}, \mathbf{y}_i) \land \cdots \land \phi_{i,l_i}(\mathbf{x}, \mathbf{y}_i) \land not\ \phi_{i,l_i+1}(\mathbf{x}, \mathbf{y}_i) \land \cdots \land not\ \phi_{i,m_i}(\mathbf{x}, \mathbf{y}_i),$$

an *answer* for $Q(\mathbf{x})$ to KB is a ground substitution θ for the variables $\mathbf{x}$ such that $KB \models Q^+(\mathbf{x}\theta)$ and $KB \not\models Q^-(\mathbf{x}\theta)$, where $Q^+(\mathbf{x})$ and $Q^-(\mathbf{x})$ are defined as follows:

$Q^+(\mathbf{x}) = \bigvee_{i=1}^{n} \exists \mathbf{y}_i\, \phi_{i,1}(\mathbf{x}, \mathbf{y}_i) \wedge \cdots \wedge \phi_{i,l_i}(\mathbf{x}, \mathbf{y}_i)$ and

$Q^-(\mathbf{x}) = \bigvee_{i=1}^{n} \exists \mathbf{y}_i\, \phi_{i,1}(\mathbf{x}, \mathbf{y}_i) \wedge \cdots \wedge \phi_{i,l_i}(\mathbf{x}, \mathbf{y}_i) \wedge (\phi_{i,l_i+1}(\mathbf{x}, \mathbf{y}_i) \vee \cdots \vee \phi_{i,m_i}(\mathbf{x}, \mathbf{y}_i))$.

Roughly, a ground substitution θ is an answer for $Q(\mathbf{x})$ to KB iff (i) θ is an answer for $Q^+(\mathbf{x})$ to KB, and (ii) θ is not an answer for $Q^-(\mathbf{x})$ to KB, where $Q^+(\mathbf{x})$ is the positive part of $Q(\mathbf{x})$, while $Q^-(\mathbf{x})$ is the positive part of $Q(\mathbf{x})$ combined with the complement of the negative one. Note that both $Q^+(\mathbf{x})$ and $Q^-(\mathbf{x})$ are positive queries.

Example 4. (Scientific Database cont'd). Consider the Semantic Web knowledge base $KB = (\mathcal{T}, (\mathcal{A}_a)_{a \in \mathbf{P} \cup \mathbf{O}})$ of Example 1 and the following general Semantic Web search query, asking for Mary's unpublished non-journal papers:

$Q(x) = \exists y\, (Article(x) \wedge hasAuthor(x, y) \wedge name(y, \text{"mary"}) \wedge not\, JournalPaper(x) \wedge$
$\qquad not\, \exists z\, yearOfPublication(x, z)).$

An answer for $Q(x)$ to KB is given by $\theta = \{x/i_3\}$. Recall that i_3 represents an unpublished conference paper entitled *"Semantic Web search"*. Observe that the membership axioms $Article(i_3)$ and $hasAuthor(i_2, i_3)$ do not appear in the semantic annotations $\mathcal{A}_a$ with $a \in \mathbf{P} \cup \mathbf{O}$, but they can be inferred from them using the background ontology $\mathcal{T}$.

Ranking Answers. As for the ranking of all answers for a Semantic Web search query Q to a Semantic Web knowledge base KB (i.e., ground substitutions for all free variables in Q, which correspond to tuples of Web pages, Web objects, and data values), we use a generalization of the PageRank technique: rather than considering only Web pages and the link structure between Web pages (expressed through the role *links_to* here), we also consider Web objects, which may occur on Web pages (expressed through the role *contains*), and which may also be related to other Web objects via other roles. More concretely, we define the *ObjectRank* of a Web page or an object a as follows:

$$R(a) = d \cdot \sum_{b \in B_a} R(b) / N_b + (1 - d) \cdot E(a),$$

where (i) B_a is the set of all Web pages and Web objects that relate to a, (ii) N_b is the number of Web pages and Web objects that relate from b, (iii) d is a damping factor, and (iv) E associates with every Web page and every Web object a source of rank. Note that ObjectRank can be computed by reduction to the computation of PageRank [22].

3.3   Realizing Semantic Web Search

Processing Semantic Web search queries Q is divided into

- an offline ontology reasoning step, where the TBox $\mathcal{T}$ of a Semantic Web knowledge base KB is compiled into KB's ABox $\mathcal{A}$ via completing all semantic annotations of Web pages and objects by membership axioms entailed from KB, and
- an online reduction to standard Web search, where Q is transformed into standard Web search queries whose answers are used to construct the answer for Q.

In the offline ontology reasoning step, we check whether the Semantic Web knowledge base is satisfiable, and we compute the completion of all semantic annotations, i.e., we augment the semantic annotations with all concept, role, and attribute membership axioms that can be derived (deductively in [22] and inductively here) from the semantic annotations and the ontology. In the online reduction to standard Web search, we decompose a given Semantic Web search query Q into a collection of standard Web search queries, of which the answers are then used to construct the answer for Q. These standard Web search queries are processed with existing search engines on the Web.

Note that such online query processing on the data resulting from an offline ontology inference step is very close to current Web search techniques, which also include the offline construction of a search index, which is then used for rather efficiently performing online query processing. In a sense, offline ontology inference can be considered as the offline construction of an ontological index, in addition to the standard index for Web search. That is, our approach to Semantic Web search can perhaps be best realized by existing search engine companies as an extension of their standard Web search.

3.4   Deductive Offline Ontology Compilation

In this section, we describe the (deductive) offline ontology reasoning step, which compiles the implicit terminological knowledge in the TBox of a Semantic Web knowledge base into explicit membership axioms in the ABox, i.e., in the semantic annotations of Web pages / objects, so that it (in addition to the standard Web pages) can be searched by standard Web search engines. For the online query processing step, see [22].

The compilation of TBox knowledge into ABox knowledge is formalized as follows. Given a satisfiable Semantic Web knowledge base $KB = (\mathcal{T}, (\mathcal{A}_a)_{a \in \mathbf{P} \cup \mathbf{O}})$, the *simple completion* of KB is the Semantic Web knowledge base $KB' = (\emptyset, (\mathcal{A}_a')_{a \in \mathbf{P} \cup \mathbf{O}})$ such that every $\mathcal{A}_a'$ is the set of all concept memberships $A(a)$, role memberships $P(a, b)$, and attribute memberships $U(a, v)$ that logically follow from $\mathcal{T} \cup \bigcup_{a \in \mathbf{P} \cup \mathbf{O}} \mathcal{A}_a$, where $A \in \mathbf{A}$, $P \in \mathbf{R}_A$, $U \in \mathbf{R}_D$, $b \in \mathbf{I}$, and $v \in \mathbf{V}$. Informally, for every Web page and object, the simple completion collects all available and deducible facts (whose predicate symbols shall be usable in search queries) in a completed semantic annotation.

Example 5. Consider again the TBox $\mathcal{T}$ and the semantic annotations $(\mathcal{A}_a)_{a \in \mathbf{P} \cup \mathbf{O}}$ of Example 1. The simple completion contains in particular the new axioms $Article(i_3)$, $hasAuthor(i_3, i_2)$, and $Article(i_1)$. The first two are added to $\mathcal{A}_{i_3}$ and the last one to $\mathcal{A}_{i_4}$.

Semantic Web search queries can be evaluated on the simple completion of KB (which contains only compiled but no explicit TBox knowledge anymore). This is always sound, and in many cases also complete [22], including (i) the case of general quantifier-free queries to a Semantic Web knowledge base KB over $DL\text{-}Lite_{\mathcal{A}}$ [15] as underlying DL, and (ii) the case where the TBox of KB is equivalent to a Datalog program, and the query is fully general. For this reason, and since completeness of query processing is actually not that much an issue in the inherently incomplete Web, we propose to use the simple completion as the basis of our Semantic Web search.

Once the completed semantic annotations are computed, we encode them as HTML pages, so that they are searchable via standard keyword search. Specifically, we build one HTML page for the semantic annotation $\mathcal{A}_a$ of each individual $a \in \mathbf{P} \cup \mathbf{O}$. That is, for each such a, we build a page p containing all the atomic concepts whose argument is a and all the atomic roles/attributes where the first argument is a (see Section 2).

4   Inductive Offline Ontology Compilation

In this section, we propose to use inductive inference based on a notion of similarity as an alternative to deductive inference for offline ontology compilation in our approach to Semantic Web search. Hence, rather than obtaining the simple completion of a semantic annotation by adding all logically entailed membership axioms, we now obtain it by adding all inductively entailed membership axioms. Section 5 then summarizes the central advantages of this proposal, namely, an increased robustness due to the additional ability to handle inconsistencies, noise, and incompleteness.

4.1   Inductive Inference Based on Similarity Search

The *inductive inference* (or *classification*) problem here can be briefly described as follows. Given a Semantic Web knowledge base $KB = (\mathcal{T}, (\mathcal{A}_a)_{a \in \mathbf{P} \cup \mathbf{O}})$, a set of training individuals $TrExs \subseteq IS = \mathbf{P} \cup \mathbf{O}$, a Web page or object a, and a query property $Q(x)$, decide whether KB and $TrExs$ inductively entail $Q(a)$. Here, (i) a *property* $Q(x)$ is either a concept membership $A(x)$, a role membership $P(x, b)$, or an attribute membership $U(x, v)$, where $A \in \mathbf{A}$, $P \in \mathbf{R}_A$, $U \in \mathbf{R}_D$, $b \in \mathbf{I}$, and $v \in \mathbf{V}$, and (ii) *inductive entailment* is defined using a notion of similarity between individuals as follows.

We now review the basics of the k-*nearest-neighbor* (k-NN) method in the context of the Semantic Web [14]. Informally, the k-NN method is a classification method that assigns an individual a to the class that is most common among the k nearest (most similar) neighbors of a in the training set. A notion of nearness, i.e., a similarity (or dissimilarity) measure [52], is exploited for selecting the k most similar training examples with respect to the individual a to be classified. Formally, the method aims at inducing an approximation for a discrete-valued target hypothesis function $h \colon IS \to V$ from a space of individuals IS to a set of values $V = \{v_1, \ldots, v_s\}$, standing for the properties that have to be predicted. The approximation moves from the availability of training individuals $TrExs \subseteq IS$, which is a subset of all prototypical individuals whose correct classification $h(\cdot)$ is known.

Let x_q be the query individual whose property is to be determined. Using a dissimilarity measure $d \colon IS \times IS \mapsto \mathbb{R}$, we select the set of the k-nearest training individuals (neighbors) of $TrExs$ relative to x_q, denoted $NN(x_q) = \{x_1, \ldots, x_k\}$. Hence, the k-NN procedure approximates h for classifying x_q on the grounds of the values that h assumes for the neighbor training individuals in $NN(x_q)$. Precisely, the value is decided by means of a weighted majority voting procedure: it is the most *voted* value by the neighbor individuals in $NN(x_q)$ weighted by their similarity. The estimate of the hypothesis function for the query individual is as follows:

$$\hat{h}(x_q) = \mathrm{argmax}_{v \in V} \sum_{i=1}^{k} w_i \cdot \delta(v, h(x_i)), \tag{4}$$

where the indicator function δ returns 1 in case of matching arguments, and 0 otherwise, and the weights w_i are determined by $w_i = 1 \, / \, d(x_i, x_q)$. Note that for the case $d(x_i, x_q) = 0$, a constant ϵ such that $\epsilon \simeq 0$ and $\epsilon \neq 0$ is considered as an approximation of the null dissimilarity value.

But this approximation determines a value that stands for one in a set of disjoint properties. Indeed, this is intended for simple settings with attribute-value representations [41]. In a multi-relational context, like with typical representations of the Semantic Web, this is no longer valid, since one deals with multiple properties, which are generally not implicitly disjoint. A further problem is related to the open-world assumption (OWA) generally adopted with Semantic Web representations; the absence of information of an individual relative to some query property should not be interpreted negatively, as in knowledge discovery from databases, where the closed-world assumption (CWA) is adopted; rather, this case should count as neutral (uncertain) information. Therefore, under the OWA, the multi-classification problem is transformed into a number of ternary problems (one per property), adopting $V = \{-1, 0, +1\}$ as the set of classification values relative to each query property Q, where the values denote explicitly membership $(+1)$, non-membership (-1), and uncertainty (0) relative to Q.

Hence, inductive inference can be restated as follows: given a Semantic Web knowledge base $KB = (\mathcal{T}, (\mathcal{A}_a)_{a \in \mathbf{P} \cup \mathbf{O}})$, a set of training individuals $TrExs \subseteq IS = \mathbf{P} \cup \mathbf{O}$, and a query property $Q(x)$, find an approximation $\hat{h}_Q$ (on IS) of the hypothesis function h_Q, whose value $h_Q(x)$ for every training individual $x \in TrExs$ is as follows:

$$
h_Q(x) = \begin{cases} +1 & KB \models Q(x) \\ -1 & KB \not\models Q(x),\ KB \models \neg Q(x) \\ 0 & \textit{otherwise.} \end{cases}
$$

That is, the value of h_Q for the training individuals is determined by logical entailment. Alternatively, a mere look-up for the assertions $(\neg)Q(x)$ in $(\mathcal{A}_a)_{a \in \mathbf{P} \cup \mathbf{O}}$ could be considered, to simplify the inductive process, but also adding a further approximation.

Once the set of training individuals $TrExs$ has been constructed, the inductive classification $\hat{h}_Q(x_q)$ of an individual x_q through the k-NN procedure is done via Eq. (4).

To assess the similarity between individuals, a totally semantic and language-independent family of dissimilarity measures is used [14]. They are based on the idea of comparing the semantics of the input individuals along a number of dimensions represented by a committee of concepts $\mathbf{F} = \{F_1, \ldots, F_m\}$, which stands as a context of discriminating *features* expressed in the considered DL. Possible candidates for the feature set $\mathbf{F}$ are the concepts already defined in the knowledge base of reference or concepts that are learned starting from the knowledge base of reference (see Section 4.2 for more details). The family of dissimilarity measures is defined as follows [14].

Definition 5 (family of measures). Let $KB = (\mathcal{T}, (\mathcal{A}_a)_{a \in \mathbf{P} \cup \mathbf{O}})$ be a Semantic Web knowledge base. Given a set of concepts $\mathbf{F} = \{F_1, \ldots, F_m\}$, $m \geq 1$, weights $w_1, \ldots,$ w_m, and $p > 0$, a family of dissimilarity functions $d_p^{\mathbf{F}} : \mathbf{P} \cup \mathbf{O} \times \mathbf{P} \cup \mathbf{O} \mapsto [0, 1]$ is defined as follows (where $\mathcal{A} = \bigcup_{a \in \mathbf{P} \cup \mathbf{O}} \mathcal{A}_a$):

$$
\forall a, b \in \mathbf{P} \cup \mathbf{O} \qquad d_p^{\mathbf{F}}(a, b) \quad \frac{1}{m} \left[\sum_{i=1}^{m} w_i \mid \delta_i(a, b) \mid^p \right]^{1/p},
$$

where the dissimilarity function δ_i $(i \in \{1, \ldots, m\})$ is defined as follows:

$$
\delta_i(a, b) = \begin{cases} 0 & (F_i(a) \in \mathcal{A} \wedge F_i(b) \in \mathcal{A}) \vee (\neg F_i(a) \in \mathcal{A} \wedge \neg F_i(b) \in \mathcal{A}) \\ 1 & (F_i(a) \in \mathcal{A} \wedge \neg F_i(b) \in \mathcal{A}) \vee (\neg F_i(a) \in \mathcal{A} \wedge F_i(b) \in \mathcal{A}) \\ \frac{1}{2} & \textit{otherwise.} \end{cases}
$$

An alternative definition for the functions δ_i requires the logical entailment of the assertions $(\neg)F_i(x)$, rather than their simple ABox look-up; this makes the measure more accurate, but also more complex to compute. Moreover, using logical entailment, induction is done on top of deduction, thus making it a kind of completion of deduction.

The weights w_i in the family of measures should reflect the impact of the single feature F_i relative to the overall dissimilarity. This is determined by the quantity of information conveyed by the feature, which is measured in terms of its entropy. Namely, the probability of belonging to F_i may be quantified in terms of a measure of the extension of F_i relative to the whole domain of objects (relative to the canonical interpretation $\mathcal{I}$): $P_{F_i} = \mu(F_i^{\mathcal{I}})/\mu(\Delta^{\mathcal{I}})$. This can be roughly approximated by $|\{x \in \mathbf{P} \cup \mathbf{O} \mid F_i(x) \in \mathcal{A}\}| / |\mathbf{P} \cup \mathbf{O}|$. Hence, considering also the probability related to the complement of F_i, denoted $P_{\neg F_i}$, and the one related to the unclassified individuals (relative to F_i), denoted P_{U_i}, one obtains an entropic measure for the feature:

$$H(F_i) = -[P_{F_i} \log(P_{F_i}) + P_{\neg F_i} \log(P_{\neg F_i}) + P_{U_i} \log(P_{U_i})].$$

Alternatively, these weights may be based on the variance related to each feature [21].

4.2   Optimizing the Feature Set

The underlying idea in the measure definition is that similar individuals should exhibit the same behavior relative to the concepts in $\mathbf{F}$. We assume that the feature set $\mathbf{F}$ represents a sufficient number of (possibly redundant) features that are able to discriminate really different individuals. Preliminary experiments, where the measure has been exploited for instance-based classification (*nearest-neighbor* algorithm) and similarity search [52], demonstrated the effectiveness of the measure using even the very set of both primitive and defined concepts in the knowledge base [14]. However, the choice of the concepts to be included in the committee $\mathbf{F}$ is crucial and may be the object of a preliminary learning problem to be solved (*feature selection for metric learning*).

Before introducing any approach for learning a suitable feature sets $\mathbf{F}$, we introduce some criteria for defining what a *good feature set* is. Among the possible feature sets, we prefer those that can discriminate the individuals in the ABox.

Definition 6 (good feature set). Let $\mathbf{F} = \{F_1, \ldots, F_m\}$, $m \geq 1$, be a set of concepts over an underlying DL. Then, $\mathbf{F}$ is a *good feature set* for the Semantic Web knowledge base $KB = (\mathcal{T}, (\mathcal{A}_a)_{a \in \mathbf{P} \cup \mathbf{O}})$ iff for any two different individuals $a, b \in \mathbf{P} \cup \mathbf{O}$, either (a) $KB \models F_i(a)$ and $KB \not\models F_i(b)$, or (b) $KB \not\models F_i(a)$ and $KB \models F_i(b)$ for some $i \in \{1, \ldots, m\}$. Alternatively, the simple look-up in KB can be considered.

Hence, when the previously defined function (see Def. 5) is parameterized on a good feature set, it has the property of a metric function.

Since the function is strictly dependent on the feature set $\mathbf{F}$, two immediate heuristics arise: (1) the *number* of concepts of the feature set; (2) their discriminating power in terms of a *discernibility factor*. Indeed, the number of features in $\mathbf{F}$ should be controlled in order to avoid high computational costs. At the same time, the considered features in $\mathbf{F}$ should really discriminate the considered individuals. Furthermore, finding optimal sets of discriminating features should profit also by their composition, by employing the specific constructors made available by the DL of choice.

These objectives can be accomplished by means of randomized optimization search, especially for knowledge bases with large sets of individuals [20, 19]. Namely, part of the entire data can be drawn to learn optimal feature sets F, in advance with respect to the successive usage for all other purposes. The space of the feature sets (with a definite maximal cardinality) may be explored by means of *refinement operators* [33, 37]. The optimization of a fitness function based on the (finite) available dataset ensures that this process does not follow infinite refinement chains, as a candidate refinement step is only made when a better solution is reached in terms of the fitness function. In the following, two solutions for learning optimal discriminating feature sets by exploiting the refinement operators introduced in [33] are presented.

Optimization through Genetic Programming. We have cast the problem solution as an optimization algorithm in *genetic programming* [41]. Essentially, this algorithm encodes the traversal of the search space as a result of simple operations carried out on a representation of the problem solutions (*genomes*). Such operations mimic modifications of the solutions that may lead to better ones in terms of a fitness function, which is here based on the discernibility of the individuals. The resulting algorithm is shown in Fig. 3. It essentially searches the space of all possible feature sets, starting from an initial guess (determined by the call to MAKEINITIALFS) based on the concepts (both primitive and defined) in the knowledge base $KB = (\mathcal{T}, (\mathcal{A}_a)_{a \in \mathbf{P} \cup \mathbf{O}})$. The algorithm starts with a feature set, made by atomic concepts randomly chosen from KB, of a given initial cardinality (INIT_CARD), which may be determined as a function of $\lceil \log_3(N) \rceil$, where $N = |\mathbf{P} \cup \mathbf{O}|$, since each feature projection can categorize the individuals in three sets.

The outer loop gradually augments the cardinality of the candidate feature sets. It is repeated until the threshold fitness is reached or the algorithm detects some fixpoint: employing larger feature sets would not yield a better feature set with respect to the best fitness recorded in the previous iteration (with fewer features). Otherwise, the EX-TENDFS procedure extends the current feature sets for the next generations by including a newly generated random concept. The inner while-loop is repeated for a number of generations until a stop criterion is met, based on the maximal number of generations maxGenerations or, alternatively, when a minimal fitness threshold fitnessThr is crossed by some feature set in the population, which can be returned. As regards the BESTFITNESS routine, it computes the best fitness of the feature sets in the input vector, namely it determines the feature sets that maximize the fitness function. The fitness function is determined as the *discernibility factor* yielded by the feature set, as computed on the whole set of individuals or on a smaller sample. Specifically, given the fixed set of individuals $IS \subseteq \mathbf{P} \cup \mathbf{O}$, the fitness function is defined as follows:

$$\text{DISCERNIBILITY}(\mathsf{F}) :- \nu \cdot \sum\nolimits_{(a,b) \subseteq IS^2} \sum\nolimits_{i=1}^{|F|} |\pi_i(a) - \pi_i(b)|,$$

where ν is a normalizing factor depending on the overall number of couples involved, and $\pi_i(a)$ is defined as follows, for all $a \in \mathbf{P} \cup \mathbf{O}$:

$$\pi_i(a) = \begin{cases} 1 & KB \models F_i(a) \\ 0 & KB \models \neg F_i(a) \\ \frac{1}{2} & otherwise. \end{cases}$$

```
FeatureSet GP_OPTIMIZATION(KB, maxGenerations, fitnessThr)
input:      KB: current knowledge base;
            maxGenerations: maximal number of generations;
            fitnessThr: minimal required fitness threshold.
output:     FeatureSet: set of concept descriptions.
static:     currentFSs, formerFSs; arrays of current/previous feature sets;
            currentBestFitness, formerBestFitness = 0; current/previous best fitness values;
            offsprings; array of generated feature sets;
            fitnessImproved; improvement flag;
            generationNo = 0: number of current generation.
begin
currentFSs = MAKEINITIALFS(KB, INIT_CARD);
formerFSs = currentFSs;
repeat
       currentBestFitness = BESTFITNESS(currentFSs);
       while (currentBestFitness < fitnessThr) and (generationNo < maxGenerations)
             begin
             offsprings = GENERATEOFFSPRINGS(currentFSs);
             currentFSs = SELECTFROMPOPULATION(offsprings);
             currentBestFitness = BESTFITNESS(currentFSs);
             ++generationNo
             end;
       if (currentBestFitness > formerBestFitness) and (currentBestFitness < fitnessThr) then
             begin
             formerFSs = currentFSs;
             formerBestFitness = currentBestFitness;
             currentFSs = EXTENDFS(currentFSs);
             fitnessImproved = true
             end
       else
             fitnessImproved = false
       end
until not fitnessImproved;
return SELECTBEST(formerFSs)
end.
```

Fig. 3. Feature set optimization based on genetic programming

Finding candidate feature sets to replace the current feature set (in GENERATEOFF-SPRINGS) is based on some transformations of the current best feature sets as follows:

- choose $F \in$ currentFSs;
- randomly select $F_i \in F$;
  - replace F_i with a randomly generated $F_i' \in$ RANDOMMUTATION(F_i), where RANDOMMUTATION, for instance, performs the negation of a feature concept F_i, removes the negation from a negated concept, or transforms a concept conjunction into a concept disjunction; alternatively,
  - replace F_i with one of its refinements $F_i' \in$ REF(F_i) (that are generated by adopting the refinement operators presented in [33]).

The possible refinements of feature concepts are language-specific. For example, for the DL $\mathcal{ALC}$, refinement operators have been proposed in [37, 33].

This is iterated until a suitable number of offsprings is generated. Then these off-spring feature sets are evaluated (by the use of the fitness function) and the best ones (maximizing the fitness function) are included in the new version of currentFSs.

Once the while-loop is terminated, the current best fitness is compared with the best one computed for the former feature set length; if an improvement is detected, then the outer repeat-loop is continued, otherwise (one of) the former best feature set(s) (having the best fitness value) is selected and returned as the result of the algorithm.

Optimization through Simulated Annealing. The above randomized optimization algorithm based on genetic programming may suffer from being possibly caught in plateaux or local minima if a limited number of generations are explored before check-ing for an improvement. This is likely due to the extent of the search space, which, in turn, depends on the language of choice. Moreover, maintaining a single best genome for the next generation may slow down the search process.

FeatureSet SA_OPTIMIZATION(KB, ΔT)
input: KB: knowledge base; $\Delta T()$: cooling function.
output: FeatureSet: set of concept descriptions.
static: currentFS: current Feature Set;   nextFS: new Feature Set;
          time: time controlling variable; ΔE: energy increment;
          temp: temperature (probability of replacement).
begin
currentFS = MAKEINITIALFS(KB);
for time = 1 **to** ∞ **do**
          temp = temp $- \Delta T$(time);
          if (temp == 0) **then**
                    return currentFS;
          nextFS = RANDOMSUCCESSOR(currentFS, KB);
          ΔE = FITNESS(nextFS) $-$ FITNESS(currentFS);
          if ($\Delta E > 0$) **then**
                    // replacement
                    currentFS = nextFS
          else
                    // conditional replacement with given probability
                    currentFS = REPLACE(nextFS, $e^{\Delta E/\text{temp}}$)
end

Fig. 4. Feature set optimization based on simulated annealing

To prevent such cases, different randomized search procedures that aim at global op-timization can be adopted. In particular, an algorithm based on *simulated annealing* [1] has been proposed [19], which is shown in Fig. 4. The algorithm searches the space of feature sets starting from an initial guess (determined by MAKEINITIALFS(KB)) based on the concepts (both primitive and defined) in the knowledge base, which can be freely combined to form new concepts. The loop controlling the search is repeated for a number of times that depends on the temperature temp controlled by the cooling function ΔT, which gradually decays to 0, when the current feature set can be returned. In this cycle, the current feature set is iteratively refined by calling the procedure RAN-DOMSUCCESSOR, which, by the adoption of the refinement operators defined in [33], makes a step in the space by refining the current feature set. Then, the fitness of the new feature set is computed (as shown above) and compared to that of the current one determining the increment of energy ΔE. If this is positive, then the candidate feature set replaces the current one. Otherwise, it is (less likely) replaced with a probability that depends on ΔE and on the current temperature.

The energy increase ΔE is determined by the FITNESS of the new and current feature sets, which can be computed as the average *discernibility* factor, as defined above.

As for finding candidates to replace the current feature set, RANDOMSUCCESSOR can be implemented by recurring to simple transformations of the feature set:

- add (resp., remove) a concept C:
  nextFS $\leftarrow$ currentFS $\cup$ $\{C\}$ (resp., nextFS $\leftarrow$ currentFS $\setminus$ $\{C\}$);
- randomly choose one of the current concepts from currentFS, say C;
  replace it with one of its refinements $C' \in \text{REF}(C)$.

Note that these transformation may change the cardinality of the current feature set. As mentioned before, refining feature concepts is language-dependent. Complete operators are to be preferred, to ensure exploring the whole search space.

Given a suitable cooling schedule, the algorithm finds an optimal solution. More practically, to control the complexity of the process, alternate schedules may be preferred that guarantee the construction of suboptimal solutions in polynomial time [1].

4.3 Measuring the Likelihood of an Answer

The inductive inference made by the procedure presented above is not guaranteed to be deductively valid. Indeed, it naturally yields a certain degree of uncertainty. So, from a more general perspective, the main idea behind the above inductive inference for Semantic Web search is closely related to the idea of using probabilistic ontologies to increase the precision and the recall of querying databases and of information retrieval in general. However, rather than learning probabilistic ontologies from data, representing them, and reasoning with them, we directly use the data in the inductive inference step.

To measure the likelihood of the inductive decision (x_q has the query property Q denoted by the value v, maximizing the argmax argument in Eq. (4), given $NN(x_q) = \{x_1, \ldots, x_k\}$), the quantity that determined the decision should be normalized:

$$l(Q(x_q) = v | NN(x_q)) = \frac{\sum_{i=1}^{k} w_i \cdot \delta(v, h_Q(x_i))}{\sum_{v' \in V} \sum_{i=1}^{k} w_i \cdot \delta(v', h_Q(x_i))} . \qquad (5)$$

Hence, the likelihood of $Q(x_q)$ corresponds to the case when $v = +1$. The computed likelihood can be used for building a probabilistic ABox, which is a collection of pairs, each consisting of a classical ABox axiom and a probability value $(Q(x_q), \ell)$.

5   Inconsistencies, Noise, and Incompleteness

We now illustrate the main advantages of using inductive rather than deductive inference in Semantic Web search. In detail, inductive inference can better handle cases of inconsistency, noise, and incompleteness in Semantic Web knowledge bases than deductive inference. These cases are all very likely to occur when knowledge bases are fed by multiple heterogeneous sources and maintained on distributed peers on the Web.

Inconsistency. Since our inductive inference is triggered by factual knowledge (assertions concerning prototypical neighboring individuals in the presented algorithm), it can provide a correct classification even in the case of knowledge bases that are inconsistent due to wrong assertions. This is illustrated by the following example. Note that for an inconsistent knowledge base, the measure evaluation (see Section 4) for the case provoking the inconsistency can be done by the use of one of the following criteria: (a) short-circuit evaluation; or (b) prior probability, if available.

Example 6. Consider the following DL knowledge base $KB = (\mathcal{T}, \mathcal{A})$:

$\mathcal{T} = \{$ *Professor* $\equiv$ *Graduate* $\sqcap$ $\exists worksAt.University$ $\sqcap$ $\exists teaches.Course$;
　　　　 Researcher $\equiv$ *Graduate* $\sqcap$ $\exists worksAt.Institution$ $\sqcap$ $\neg \exists teaches.Course$; . . .$\}$;
$\mathcal{A} = \{$ *Professor(franz)*; *teaches(franz, course$_1$)*; *Professor(jim)*;
　　　　 teaches(jim, course$_2$); *Professor(flo)*; *teaches(flo, course$_3$)*;
　　　　 Researcher(nick); *Researcher(ann)*; *teaches(nick, course$_4$)*; . . .$\}$.

Suppose that Nick is actually a professor, and he is indeed asserted to be a lecturer of some course. However, by mistake, he is also asserted to be a researcher, and because of the definition of researcher in KB, he cannot teach any course. Hence, KB is inconsistent, and thus logically entails anything under deductive inference. Under inductive inference as described above, in contrast, Nick turns out to be a professor, because of the similarity of Nick to other individuals known to be professors (Franz, Jim, and Flo).

Noise. In the former case, noisy assertions may be pinpointed as the very source of inconsistency. An even trickier case is when noisy assertions do not produce any inconsistency, but are indeed wrong relative to the intended true models. Inductive reasoning can also provide a correct classification in such a presence of incorrect assertions on concepts, roles, and/or attributes relative to the intended true models.

Example 7. Consider the DL knowledge base $KB = (\mathcal{T}', \mathcal{A})$, where the ABox $\mathcal{A}$ does not change relative to Example 6 and the TBox $\mathcal{T}'$ is obtained from $\mathcal{T}$ of Example 6 by simplifying the definition of *Researcher* dropping the negative restriction:

$$Researcher = Graduate \sqcap \exists worksAt.Institution .$$

Again, suppose that Nick is actually a professor, but by mistake asserted to be a researcher. Due to the new definition of researcher in KB, there is no inconsistency anymore. However, by deductive inference, Nick turns out to be a researcher, while by inductive inference, the returned classification result is that Nick is a professor, as above, because the most similar individuals (Franz, Jim, and Flo) are all professors.

Incompleteness. Clearly, inductive reasoning may also be able to give a correct clas-sification in the presence of incompleteness in a knowledge base. That is, inductive reasoning is not necessarily deductively valid, and can suggest new knowledge.

Example 8. Consider yet another slightly different DL knowledge base $KB = (\mathcal{T}', \mathcal{A}')$, where the TBox $\mathcal{T}'$ is as in Example 7 and the ABox $\mathcal{A}'$ is obtained from the ABox $\mathcal{A}$ of Example 6 by removing the axiom *Researcher(nick)*. Then, KB is neither inconsistent nor noisy, but we know less about Nick. Nonetheless, by the same line of argumentation as in the previous examples, Nick is inductively entailed to be a professor.

6   Implementation and Experiments

In this section, we describe our prototype implementation for a semantic desktop search engine. Furthermore, we report on very positive experimental results on the precision and the recall under inductively vs. deductively completed semantic annotations. Fur-ther experimental results in [22] (for the deductive case) show that the completed se-mantic annotations are rather small in practice, that the online query processing step po-tentially scales to Web search, and that, compared to standard Web search, our approach to Semantic Web search results in a very high precision and recall for the query result.

Implementation. We have implemented a prototype for a semantic desktop search engine. We have realized both a deductive and an inductive version of the offline in-ference step for generating the completed semantic annotation for every considered resource. The deductive version uses PELLET[1], while the inductive one is based on the k-NN technique, integrated with an entropic measure, as proposed in Section 4, without any feature set optimization. Specifically, each individual i of a Semantic Web knowl-edge base KB is classified relative to all atomic concepts and all restrictions $\exists R^-.\{i\}$ with roles R. The parameter k was set to $\log(|\mathbf{P} \cup \mathbf{O}|)$, where $\mathbf{P} \cup \mathbf{O}$ is the set of all individuals in KB. The simpler distances d_1^{F} were employed, using all the atomic concepts in KB for determining F.

Precision and Recall of Inductive Semantic Web Search. We next give an experi-mental comparison between Semantic Web search under inductive and under deductive reasoning, by providing the precision and the recall of the latter vs. the former.

   The experiments have been performed on a standard laptop (ASUS PRO31 series, with 2.20 GHz Intel Core Duo processor and 2 GB RAM). Two ontologies have been considered: the FINITE-STATE-MACHINE (FSM) and the SURFACE-WATER-MODEL (SWM) ontology from the Protégé Ontology Library[2]. The knowledge base relative to the FSM (resp., SWM) ontology consists of 37 (resp., 115) annotations with 130 (resp., 621) facts. We evaluated 9 queries on the FSM annotations, and 7 queries on the SWM annotations. The queries vary from single atoms to conjunctive formulas, possibly with negations. All the queries, along with the experimental results are sum-marized in Table 1. For example, Query (8) asks for all transitions having no target

---

[1] http://www.mindswap.org
[2] http://protegewiki.stanford.edu/index.php/
   Protege_Ontology_Library

Table 1. Precision and recall of inductive vs. deductive Semantic Web search

	Onto-logy	Query	No. Results Deduction	No. Results Induction	No. Correct Results Induction	Precision Induction	Recall Induction
1	FSM	$State(x)$	11	11	11	1	1
2	FSM	$StateMachineElement(x)$	37	37	37	1	1
3	FSM	$Composite(x) \wedge hasStateMachineElement(x, accountDetails)$	1	1	1	1	1
4	FSM	$State(y) \wedge StateMachineElement(x) \wedge hasStateMachineElement(x, y)$	3	3	3	1	1
5	FSM	$Action(x) \vee Guard(x)$	12	12	12	1	1
6	FSM	$\exists y, z\,(State(y) \wedge State(z) \wedge Transition(x) \wedge source(x, y) \wedge target(x, z))$	11	2	2	1	0.18
7	FSM	$StateMachineElement(x) \wedge not\ \exists y\,(StateMachineElement(y) \wedge hasStateMachineElement(x, y))$	34	34	34	1	1
8	FSM	$Transition(x) \wedge not\ \exists y\,(State(y) \wedge target(x, y))$	0	5	0	0	1
9	FSM	$\exists y\,(StateMachineElement(x) \wedge not\ hasStateMachineElement(x, accountDetails) \wedge hasStateMachineElement(x, y) \wedge State(y))$	2	2	2	1	1
10	SWM	$Model(x)$	56	56	56	1	1
11	SWM	$Mathematical(x)$	64	64	64	1	1
12	SWM	$Model(x) \wedge hasDomain(x, lake) \wedge hasDomain(x, river)$	9	9	9	1	1
13	SWM	$Model(x) \wedge not\ \exists y\,(Availability(y) \wedge hasAvailability(x, y))$	11	11	11	1	1
14	SWM	$Model(x) \wedge hasDomain(x, river) \wedge not\ hasAvailability(x, public)$	2	8	0	0	0
15	SWM	$\exists y\,(Model(x) \wedge hasDeveloper(x, y) \wedge University(y))$	1	1	1	1	1
16	SWM	$Numerical(x) \wedge hasDomain(x, lake) \wedge hasAvailability(x, public) \vee Numerical(x) \wedge hasDomain(x, coastalArea) \wedge hasAvailability(x, commercial)$	12	9	9	1	0.75

state, while Query (16) asks for all numerical models having either the domain "lake" and public availability, or the domain "coastalArea" and commercial availability.

The experimental results in Table 1 essentially show that the answer sets under inductive reasoning are very close to the ones under deductive reasoning.

7   Related Work

In this section, we discuss related work on (i) state-of-the-art systems for Semantic Web search (see especially [23] for a more detailed recent survey), focusing on the most closely related to ours, and (ii) inductive reasoning from ontologies.

Semantic Web Search. Related approaches to Semantic Web search can roughly be divided into (1) those based on structured query languages, such as [12, 25, 30, 35, 43, 44, 48], keyword-based approaches, such as [8, 27, 29, 38, 49, 50, 51], where queries consist of lists of keywords, and natural-language-based approaches, such as [10, 16, 24, 26, 39, 40], where users can express queries in natural language. To evaluate user queries on Semantic Web documents, both keyword-based and natural-language-based approaches need a reformulation phase, where user queries are transformed into "semantic" queries. In keyword-based approaches, query processing generally starts with the assignment of a semantic meaning to the keywords, i.e., each keyword is mapped to an ontological concept (property, entity, class, etc.). Since each keyword can match a class, a property, or an instance, several combinations of semantic matchings of the keywords are considered, and, in some cases, the user is asked for choosing the right assignment. Similarly, natural-language-based approaches focus mainly on the translation of queries from natural language to structured languages, by directly mapping query terms to ontological concepts or by using some ad-hoc translation techniques.

The approaches based on structured query languages which are most closely related to ours are [12, 30, 35], in that they aim at providing general semantic search facilities. The Corese system [12] is an ontology-based search engine for the Semantic

Web, which retrieves Web resources that are annotated in RDF(S) via a query language based on RDF(S). It is the system that is perhaps closest in spirit to our approach. In a first phase, Corese translates annotations into conceptual graphs, it then applies proper inference rules to augment the information contained in the graphs, and finally evaluates a user query by projecting it onto the annotation graphs. The Corese query language is based on RDF, and it allows variables and operators.

SHOE [30] is one of the first attempts to semantically query the Web. It provides the following: a tool for annotating Web pages, allowing users to add SHOE markup to a page by selecting ontologies, classes, and properties from a list; a Web crawler, which searches for Web pages with SHOE markup and stores the information in a knowledge base (KB); an inference engine, which provides new markups by means of inference rules (basically, Horn clauses); and several query tools, which allow users to pose structured queries against an ontology. One of the query tools allows users to draw a graph in which nodes represent constant or variable instances, and arcs represent relations. To answer the query, the system retrieves subgraphs matching the user graph. The SHOE search tool allows users to pose queries by choosing first an ontology from a drop-down list and next classes and properties from another list. Finally, the system builds a conjunctive query, issues the query to the KB, and presents the results in a tabular form.

NAGA [35] provides a graph-based query language to query the underlying knowledge base (KB) encoded as a graph. The KB is built automatically by a tool that extends the approach proposed in [47] and extracts knowledge from three Web sources: Wordnet, Wikipedia, and IMDB. The nodes and edges in the knowledge graph represent entities and relationships between entities, respectively. The query language is based on SPARQL, and adds the possibility of formulating graph queries with regular expressions on edge labels, but the language does not allow queries with negation. Answers to a query are subgraphs of the knowledge graph matching the query graph and are ranked using a specific scoring model for weighted labeled graphs.

Comparing the above three approaches to ours, in addition to the differences in the adopted query languages (in particular, SHOE and NAGA do not allow complex queries with negation) and underlying ontology languages, and to the fact that all above three approaches are based on deductive rather than inductive reasoning, there is a strong difference in the query-processing strategy. Indeed, Corese, SHOE, and NAGA all rely on building a unique KB, which collects the information disseminated among the data sources, and which is suitably organized for query processing via the adopted query language. However, this has a strong limitations. First, representing the whole information spread across the Web in a unique KB and efficiently processing each user query on the thus obtained huge amount of data is a rather challenging task. This makes these approaches more suitable for specific domains, where the amount of data to be dealt with is usually much smaller. In contrast, our approach allows the query processing task to be supported by well-established Web search technologies. In fact, we do not evaluate user queries on a single KB, but we represent the information implied by the annotations on different Web pages, and evaluate queries in a distributed way. Specifically, user queries are processed as Web searches over completed annotations. We thus realize Semantic Web search by using standard Web search technologies as well-established solutions to the problem of querying huge amounts of data. Second, a closely related limitation of query processing in Corese, SHOE, and NAGA is its tight connection to the underlying ontology language, while our approach is actually independent from the ontology language and works in the same way for other underlying ontology languages.

Note that besides being a widely used keyword search engine, Google [26] is recently also evolving towards a natural-language-based search engine, and starting to incorporate ideas from the Semantic Web. In fact, it has recently been augmented with a new functionality, which provides more precise answers to queries: instead of returning Web page links as query results, Google now tries to build query answers, collecting information from several Web pages. As an example, the simple query "barack obama date of birth" gets the answer "4 August, 1961". Next to the answer, the link *Show sources* is shown, that leads to the Web pages from which the answer has been obtained. As an important example of an initiative towards adding structure and/or semantics to Web contents in practice, Google's *Rich Snippets*[3] highlight useful information from Web pages via structured data standards such as microformats and RDFa. Differently from our approach, in particular, Google does not allow for complex structured queries, which are evaluated via reasoning over the Web relative to a background ontology.

Inductive Reasoning from Ontologies. Most research on formal ontologies focuses on methods based on deductive reasoning. However, these methods may fail on large-scale and/or noisy data, coming from heterogeneous sources. In order to overcome these limitations, other forms of reasoning are being investigated, such as *nonmonotonic*, *paraconsistent* [28], and *approximate* reasoning [31]. However, most of them may fail in the presence of data inconsistencies, which can easily happen in the context of heterogeneous and distributed sources of information, such as the (Semantic) Web. Inductive (instance-based) learning methods can effectively be employed to overcome this weakness, since they are known to be both very efficient and fault-tolerant compared to classic logic-based methods. Nonetheless, research on inductive methods and knowledge discovery applied to ontological representations have received less attention [11, 36, 17, 4]. The most widely investigated reasoning service to be solved by the use of inductive learning methods is concept retrieval. By casting concept retrieval as a classification problem, the goal is assessing the memberships of individuals to query concepts. One of the first proposals that exploits inductive learning methods for concept retrieval has been presented in [14]. As summarized above, it is based on an extension of the *k-nearest neighbor* algorithm for OWL ontologies, with the goal of classifying individuals relative to query concepts. Successively, alternative classification methods have been considered. In particular, due to their efficiency, kernel methods [46] for the induction of classifiers have been taken into account [20, 4].

Both the k-NN approach and kernel methods are based on the exploitation of a notion of similarity. Specifically, kernel methods represent a family of statistical learning algorithms, including the *support vector machines* (SVMs) [13], which can be very efficient, since they map, by means of a kernel function, the original feature space into a higher-dimensional space, where the learning task is simplified, and where the kernel function implements a dissimilarity notion.

---

[3] http://knol.google.com/k/google-rich-snippets-tips-and-tricks

Various other attempts to define semantic similarity (or dissimilarity) measures for ontology languages have been made, but they still have a limited applicability to simple languages [5] or they are not completely semantic, depending also on the structure of concepts [15, 34]. Very few works deal with the comparison of individuals rather than concepts [14, 32]. In the context of clausal logics, a metric was defined [42] for Herbrand interpretations of logic clauses as induced from a distance defined on the space of ground atoms. Such measures may be used to assess similarity in *deductive databases*. Although it represents a form of fully semantic measure, different assumptions are made compared to those that are standard for knowledge bases in the Semantic Web. Thus, the transposition to the context of the Semantic Web is not straightforward.

8   Summary and Outlook

We have presented a combination of Semantic Web search as presented in [22] with an inductive reasoning technique, based on similarity search [52] for retrieving the resources that likely belong to a query concept [14]. As a crucial advantage, the new approach to Semantic Web search has an increased robustness, as it allows for handling inconsistencies, noise, and incompleteness, which are all very likely in distributed and heterogeneous environments, such as the Web. We have also reported on a prototype implementation and very positive experimental results on the precision and the recall of the new inductive approach to Semantic Web search.

As for future research, we aim especially at extending the desktop implementation to a real Web implementation, using existing search engines, such as Google. Another interesting topic is to explore how search expressions that are formulated as plain natural language sentences can be translated into the ontological conjunctive queries of our approach. Furthermore, it would also be interesting to investigate the use of probabilistic ontologies rather than classical ones.

Acknowledgments. This work was supported by the European Research Council under the EU's 7th Framework Programme (FP7/2007-2013)/ERC grant 246858 – DIADEM, by the EPSRC grant EP/J008346/1 "PrOQAW: Probabilistic Ontological Query Answering on the Web", a Yahoo! Research Fellowship, and a Google Research Award. Georg Gottlob is a James Martin Senior Fellow, and also gratefully acknowledges a Royal Society Wolfson Research Merit Award. The work was carried out in the context of the James Martin Institute for the Future of Computing. We thank the reviewers of this paper and its URSW-2009 abstract for their useful and constructive comments, which have helped to improve this work.

References

[1] Aarts, E., Korst, J., Michiels, W.: Simulated annealing. In: Burke, E.K., Kendall, G. (eds.) Search Methodologies, ch. 7, pp. 187–210. Springer (2005)
[2] Baader, F., Calvanese, D., McGuinness, D.L., Nardi, D., Patel-Schneider, P.F. (eds.): The Description Logic Handbook. Cambridge University Press (2003)
[3] Berners-Lee, T., Hendler, J., Lassila, O.: The Semantic Web. Sci. Am. 284, 34–43 (2001)

[4] Bloehdorn, S., Sure, Y.: Kernel Methods for Mining Instance Data in Ontologies. In: Aberer, K., Choi, K.-S., Noy, N., Allemang, D., Lee, K.-I., Nixon, L.J.B., Golbeck, J., Mika, P., Maynard, D., Mizoguchi, R., Schreiber, G., Cudré-Mauroux, P. (eds.) ISWC/ASWC 2007. LNCS, vol. 4825, pp. 58–71. Springer, Heidelberg (2007)

[5] Borgida, A., Walsh, T.J., Hirsh, H.: Towards measuring similarity in description logics. In: Proc. DL 2005. CEUR Workshop Proceedings, vol. 147. CEUR-WS.org (2005)

[6] Brin, S., Page, L.: The anatomy of a large-scale hypertextual web search engine. Comput. Netw. 30(1-7), 107–117 (1998)

[7] Buitelaar, P., Cimiano, P.: Ontology Learning and Population: Bridging the Gap Between Text and Knowledge. IOS Press (2008)

[8] Cheng, G., Ge, W., Qu, Y.: Falcons: Searching and browsing entities on the Semantic Web. In: Proc. WWW 2008, pp. 1101–1102. ACM Press (2008)

[9] Chirita, P.-A., Costache, S., Nejdl, W., Handschuh, S.: P-TAG: Large scale automatic generation of personalized annotation TAGs for the Web. In: Proc. WWW 2007, pp. 845–854. ACM Press (2007)

[10] Cimiano, P., Haase, P., Heizmann, J., Mantel, M., Studer, R.: Towards portable natural language interfaces to knowledge bases — The case of the ORAKEL system. Data Knowl. Eng. 65(2), 325–354 (2008)

[11] Cohen, W.W., Hirsh, H.: Learning the CLASSIC description logic. In: Proc. KR 1994, pp. 121–133. Morgan Kaufmann (1994)

[12] Corby, O., Dieng-Kuntz, R., Faron-Zucker, C.: Querying the Semantic Web with Corese search engine. In: Proc. ECAI 2004, pp. 705–709. IOS Press (2004)

[13] Cristianini, N., Shawe-Taylor, J.: An Introduction to Support Vector Machines. Cambridge University Press (2000)

[14] d'Amato, C., Fanizzi, N., Esposito, F.: Query Answering and Ontology Population: An Inductive Approach. In: Bechhofer, S., Hauswirth, M., Hoffmann, J., Koubarakis, M. (eds.) ESWC 2008. LNCS, vol. 5021, pp. 288–302. Springer, Heidelberg (2008)

[15] d'Amato, C., Staab, S., Fanizzi, N.: On the Influence of Description Logics Ontologies on Conceptual Similarity. In: Gangemi, A., Euzenat, J. (eds.) EKAW 2008. LNCS (LNAI), vol. 5268, pp. 48–63. Springer, Heidelberg (2008)

[16] Damljanovic, D., Agatonovic, M., Cunningham, H.: Natural Language Interfaces to Ontologies: Combining Syntactic Analysis and Ontology-Based Lookup through the User Interaction. In: Aroyo, L., Antoniou, G., Hyvönen, E., ten Teije, A., Stuckenschmidt, H., Cabral, L., Tudorache, T. (eds.) ESWC 2010, Part I. LNCS, vol. 6088, pp. 106–120. Springer, Heidelberg (2010)

[17] d'Aquin, M., Lieber, J., Napoli, A.: Decentralized Case-Based Reasoning for the Semantic Web. In: Gil, Y., Motta, E., Benjamins, V.R., Musen, M.A. (eds.) ISWC 2005. LNCS, vol. 3729, pp. 142–155. Springer, Heidelberg (2005)

[18] Ding, L., Finin, T.W., Joshi, A., Peng, Y., Pan, R., Reddivari, P.: Search on the Semantic Web. IEEE Computer 38(10), 62–69 (2005)

[19] Fanizzi, N., d'Amato, C., Esposito, F.: Evolutionary conceptual clustering based on induced pseudo-metrics. Int. J. Semantic Web Inf. Syst. 4(3), 44–67 (2008)

[20] Fanizzi, N., d'Amato, C., Esposito, F.: Induction of classifiers through non-parametric methods for approximate classification and retrieval with ontologies. Int. J. Semant. Comput. 2(3), 403–423 (2008)

[21] Fanizzi, N., d'Amato, C., Esposito, F.: Metric-based stochastic conceptual clustering for ontologies. Inform. Syst. 34(8), 725–739 (2009)

[22] Fazzinga, B., Gianforme, G., Gottlob, G., Lukasiewicz, T.: Semantic Web search based on ontological conjunctive queries. J. Web Sem. 9(4), 453–473 (2011)

[23] Fazzinga, B., Lukasiewicz, T.: Semantic search on the Web. Sem. Web 1(1/2), 89–96 (2010)

[24] Fernández, M., Lopez, V., Sabou, M., Uren, V.S., Vallet, D., Motta, E., Castells, P.: Semantic search meets the Web. In: Proc. ICSC 2008, pp. 253–260. IEEE Computer Society (2008)

[25] Finin, T.W., Ding, L., Pan, R., Joshi, A., Kolari, P., Java, A., Peng, Y.: Swoogle: Searching for knowledge on the Semantic Web. In: Proc. AAAI 2005, pp. 1682–1683. AAAI Press/MIT Press (2005)

[26] Google, http://www.google.com

[27] Guha, R.V., McCool, R., Miller, E.: Semantic search. In: Proc. WWW 2003, pp. 700–709. ACM Press (2003)

[28] Haase, P., van Harmelen, F., Huang, Z., Stuckenschmidt, H., Sure, Y.: A Framework for Handling Inconsistency in Changing Ontologies. In: Gil, Y., Motta, E., Benjamins, V.R., Musen, M.A. (eds.) ISWC 2005. LNCS, vol. 3729, pp. 353–367. Springer, Heidelberg (2005)

[29] Harth, A., Hogan, A., Delbru, R., Umbrich, J., O'Riain, S., Decker, S.: SWSE: Answers before links! In: Proc. Semantic Web Challenge 2007. CEUR Workshop Proceedings, vol. 295. CEUR-WS.org (2007)

[30] Heflin, J., Hendler, J.A., Luke, S.: SHOE: A blueprint for the Semantic Web. In: Fensel, D., Wahlster, W., Lieberman, H. (eds.) Spinning the Semantic Web: Bringing the World Wide Web to Its Full Potential, pp. 29–63. MIT Press (2003)

[31] Hitzler, P., Vrandečić, D.: Resolution-Based Approximate Reasoning for OWL DL. In: Gil, Y., Motta, E., Benjamins, V.R., Musen, M.A. (eds.) ISWC 2005. LNCS, vol. 3729, pp. 383–397. Springer, Heidelberg (2005)

[32] Hu, B., Kalfoglou, Y., Alani, H., Dupplaw, D., Lewis, P.H., Shadbolt, N.: Semantic Metrics. In: Staab, S., Svátek, V. (eds.) EKAW 2006. LNCS (LNAI), vol. 4248, pp. 166–181. Springer, Heidelberg (2006)

[33] Iannone, L., Palmisano, I., Fanizzi, N.: An algorithm based on counterfactuals for concept learning in the Semantic Web. Int. J. Appl. Intell. 26(2), 139–159 (2007)

[34] Janowicz, K., Wilkes, M.: SIM-DL$_A$: A Novel Semantic Similarity Measure for Description Logics Reducing Inter-concept to Inter-instance Similarity. In: Aroyo, L., Traverso, P., Ciravegna, F., Cimiano, P., Heath, T., Hyvönen, E., Mizoguchi, R., Oren, E., Sabou, M., Simperl, E. (eds.) ESWC 2009. LNCS, vol. 5554, pp. 353–367. Springer, Heidelberg (2009)

[35] Kasneci, G., Suchanek, F.M., Ifrim, G., Ramanath, M., Weikum, G.: NAGA: Searching and ranking knowledge. In: Proc. ICDE 2008, pp. 953–962. IEEE Computer Society (2008)

[36] Kietz, J.U., Morik, K.: A polynomial approach to the constructive induction of structural knowledge. Mach. Learn. 14, 193–218 (1994)

[37] Lehmann, J., Hitzler, P.: Foundations of Refinement Operators for Description Logics. In: Blockeel, H., Ramon, J., Shavlik, J., Tadepalli, P. (eds.) ILP 2007. LNCS (LNAI), vol. 4894, pp. 161–174. Springer, Heidelberg (2008)

[38] Lei, Y., Uren, V.S., Motta, E.: SemSearch: A Search Engine for the Semantic Web. In: Staab, S., Svátek, V. (eds.) EKAW 2006. LNCS (LNAI), vol. 4248, pp. 238–245. Springer, Heidelberg (2006)

[39] Lopez, V., Pasin, M., Motta, E.: AquaLog: An Ontology-Portable Question Answering System for the Semantic Web. In: Gómez-Pérez, A., Euzenat, J. (eds.) ESWC 2005. LNCS, vol. 3532, pp. 546–562. Springer, Heidelberg (2005)

[40] Lopez, V., Sabou, M., Motta, E.: PowerMap: Mapping the Real Semantic Web on the Fly. In: Cruz, I., Decker, S., Allemang, D., Preist, C., Schwabe, D., Mika, P., Uschold, M., Aroyo, L.M. (eds.) ISWC 2006. LNCS, vol. 4273, pp. 414–427. Springer, Heidelberg (2006)

[41] Mitchell, T.: Machine Learning. McGraw Hill (1997)

[42] Nienhuys-Cheng, S.-H.: Distances and Limits on Herbrand Interpretations. In: Page, D.L. (ed.) ILP 1998. LNCS, vol. 1446, pp. 250–260. Springer, Heidelberg (1998)

[43] Nováček, V., Groza, T., Handschuh, S.: CORAAL – Towards Deep Exploitation of Textual Resources in Life Sciences. In: Combi, C., Shahar, Y., Abu-Hanna, A. (eds.) AIME 2009. LNCS, vol. 5651, pp. 206–215. Springer, Heidelberg (2009)

[44] Oren, E., Guéret, C., Schlobach, S.: Anytime Query Answering in RDF through Evolutionary Algorithms. In: Sheth, A.P., Staab, S., Dean, M., Paolucci, M., Maynard, D., Finin, T., Thirunarayan, K. (eds.) ISWC 2008. LNCS, vol. 5318, pp. 98–113. Springer, Heidelberg (2008)

[45] Poggi, A., Lembo, D., Calvanese, D., De Giacomo, G., Lenzerini, M., Rosati, R.: Linking Data to Ontologies. In: Spaccapietra, S. (ed.) Journal on Data Semantics X. LNCS, vol. 4900, pp. 133–173. Springer, Heidelberg (2008)

[46] Schölkopf, B., Smola, A.J.: Learning with Kernels. MIT Press (2002)

[47] Suchanek, F.M., Kasneci, G., Weikum, G.: Yago: A core of semantic knowledge. In: Proc. WWW 2007, pp. 697–706. ACM Press (2007)

[48] Thomas, E., Pan, J.Z., Sleeman, D.H.: ONTOSEARCH2: Searching ontologies semantically. In: Proc. OWLED 2007. CEUR Workshop Proceedings, vol. 258. CEUR-WS.org (2007)

[49] Tran, T., Cimiano, P., Rudolph, S., Studer, R.: Ontology-Based Interpretation of Keywords for Semantic Search. In: Aberer, K., Choi, K.-S., Noy, N., Allemang, D., Lee, K.-I., Nixon, L.J.B., Golbeck, J., Mika, P., Maynard, D., Mizoguchi, R., Schreiber, G., Cudré-Mauroux, P. (eds.) ISWC/ASWC 2007. LNCS, vol. 4825, pp. 523–536. Springer, Heidelberg (2007)

[50] Tummarello, G., Cyganiak, R., Catasta, M., Danielczyk, S., Delbru, R., Decker, S.: Sig.ma: Live views on the Web of data. In: Proc. WWW 2010, pp. 1301–1304. ACM Press (2010)

[51] Zenz, G., Zhou, X., Minack, E., Siberski, W., Nejdl, W.: From keywords to semantic queries — Incremental query construction on the Semantic Web. J. Web Sem. 7(3), 166–176 (2009)

[52] Zezula, P., Amato, G., Dohnal, V., Batko, M.: Similarity Search — The Metric Space Approach. Advances in Database Systems, vol. 32. Springer (2006)

Ontology Enhancement
through Inductive Decision Trees

Bart Gajderowicz, Alireza Sadeghian, and Mikhail Soutchanski

Ryerson University, Computer Science Department,
250 Victoria Street, Toronto, Ontario, Canada
{bgajdero,asadeghi,mes}@ryerson.ca
http://www.scs.ryerson.ca

Abstract. The popularity of ontologies for representing the semantics behind many real-world domains has created a growing pool of ontologies on various topics. Different ontologists, experts, and organizations create a great variety of ontologies, often for narrow application domains. Some of the created ontologies frequently overlap with other ontologies in broader domains if they pertain to the Semantic Web. Sometimes, they model similar or matching theories that may be inconsistent. To assist in the reuse of these ontologies, this paper describes a technique for enriching manually created ontologies by supplementing them with inductively derived rules, and reducing the number of inconsistencies. The derived rules are translated from decision trees with probability measures, created by executing a tree based data mining algorithm over the data being modelled. These rules can be used to revise an ontology in order to extend the ontology with definitions grounded in empirical data, and identify possible similarities between complementary ontologies. We demonstrate the application of our technique by presenting an example, and discuss how various data types may be treated to generalize the semantics of an ontology for a broader application domain.

Keywords: probabilistic ontology, extending ontologies, decision trees.

1 Introduction

This paper proposes an algorithm for extending an existing ontology with decision trees (DT) obtained from executing a tree learning algorithm on an external dataset of data related to the ontology's domain[1]. The resulting decision trees refine the ontology's definitions and terms, by grounding them with empirical facts inferred from the external dataset.

A possible domain where this is applicable is in scientific research, where the results are only as accurate as their underlying data. For example, when qualifying collected specimens or observed phenomena as a concept in a geoscientific ontology, the researcher often relies on a combination of data-driven and theory-driven information [4]. In fields such as geology, qualifying various types of rock

---

[1] This paper targets ontologies which can be represented by a direct acyclic graph (DAG) and compatible languages.

F. Bobillo et al. (Eds.): URSW 2008-2010/UniDL 2010, LNAI 7123, pp. 262–281, 2013.
© Springer-Verlag Berlin Heidelberg 2013

depends greatly on the specimens found and the geologist's knowledge about the region, rock types, and properties which are infrequently observed but theoretically important. Due to personal bias, some theoretical knowledge may be used incorrectly due to incorrect classification of the location, for example as a *lake* instead of *stream*. Brodaric et al. [4] observed that more consistent, and presumed correct, qualifications were exhibited using data-driven information, versus theory-based. In other work, Kieler et. al. match specific geospatial locations [20] and specific rivers [21] based on rules derived from records of different systems. These records were captured by different sensors, and most importantly at different scales. A set of derived rules was needed to successfully match the same concepts represented differently by different systems.

Biological and chemical information is increasingly being published and shared using semantic technologies [1][5]. Much of the analysis on this type of information has not yet begun to use the latest representation languages such as RDF and OWL. For example, the toxicity of chemical products is often analyzed using statistical analysis of chemical features. These features focus on a chemical's structure and function. A popular method to achieve this is the development of *decision trees* by mining empirical toxicology data. It is beneficial to use compatible languages to represent domain knowledge and to formulate problems for analysis. Chepelev et al. [5] have created such decision trees represented in the OWL language specifically for toxicity classification. The result are OWL rules which classify toxicity features. An OWL reasoner was then used to characterize the toxicity of various chemical products. Datasets were compared semantically by examining logical equivalences between the OWL decision trees. However, the underlying decision trees differentiating between toxic and non-toxic classes were not easily created due to significant overlap. The lack of uncertainty measures associated with the generated decision trees made it difficult to differentiate between concrete rules. The addition of chemical product structure was required to disambiguate the various classification rules.

The field of *ontology matching* consists of matching a concept from one ontology to another. Several issues have been brought up as obstacles in the manual matching process [12,27], specifically inconsistency, incompletness and redundancy. This results in incorrectly defined relationships, missing information, or simply human error. Various methods have been identified by Euzenat et al. [11], for automated and semi-automated matching techniques. Specifically *instance identification techniques*, such as comparing data values of instance data, are described to determine data correspondences, especially when ID keys are not available. When datasets are not similar to each other, disjoint *extension comparison techniques* are described, which can be based on statistical measures of class member features matched between entity sets [11]. The information created by our algorithm is targeted at datasets for such matchings. BayesOWL has been proposed to perform automatic ontology mapping [9] by associating probabilities with text based information, and using Jeffrey's Rule to propagate those probabilities. Text documents are classified using a classifier such as Rainbow[2]. Tools such as OWL-CM [4] have begun looking at how similarity

---

[2] http://www-2.cs.cmu.edu/~mccallum/bow/rainbow

measures and uncertainties in the mapping process can be improved to increase access correspondences between lexical ontology entities.

As an example, the classification of *cat*, *tiger*, and *panther* as subclasses of *felinae* does not have sufficient non-lexical information to differentiate them from each other. The addition of physical attributes such as weight ranges or geographical habitats may provide information which allows successful differentiation. Further, attribute level information may be consistent amongst the instances observed by other ontologists, even when it does not apply to their domain. If so, it may be used to match these *classes*[3] at a more detailed level based on a model *learned* from instance data [11], presented in the form of decision trees. These trees may then be associated with edges representing relations between concepts in the ontologies, creating extensions refining the ontology. As will be expanded on in Section 4, the consistency demonstrated between clusters in Figure 1 may be used to match the *classified* concepts from one ontology to another. In Section 2 we give relevant background information on the covered topics. Section 3 gives a detailed definition of our contribution, the extension algorithm, and describes how such extensions may be used for ontology matching[4]. In Section 4 we expanded on the applicability of the algorithm, and summarize our findings in Section 5.

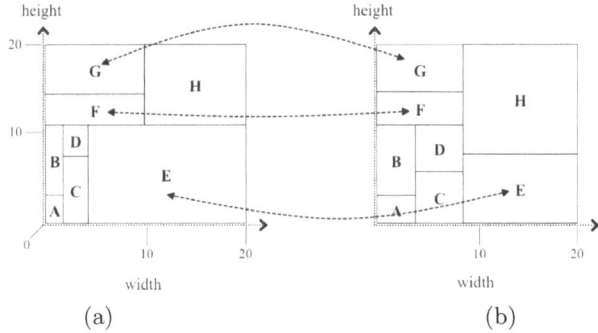

(a)   (b)

Fig. 1. Classifying instances using concepts of different ontologies based on a pair of attributes *width* and *height*, reveal similarity correlation between the same pair of attributes in separate ontologies

2   Background Knowledge

2.1   Description Logic and Uncertainty

The current research on including inductively derived information has focused on classification of assertions (ABox) in a Description Logic (DL) knowledge

---

[3] The word *class* in the context of data-mining is used here in order to differentiate it from the word *concept* used as the label for an ontological entity.

[4] We make a distinction between matching as the alignment between entities of different ontologies, and mapping as the directed version alignment of entities in one ontology to at most one entity in another, as in [11].

base, by associating uncertainty to its terminology (TBox). Description Logic provides constructors to build complex concepts out of atomic ones [10], with various extensions derived to handle different types of constructs [18,10]. In recent years, much attention has been placed on the SH family of extensions because it provides sufficient expressivity, useful for intended application domains. More recently, the $SHOQ$(D) extension has added the capability to specify qualified number restrictions, and the $SHOIN$(D) extension has combined singleton classes, inverse roles and unqualified number restrictions. Further, $SHOIN$(D) has been used to create the Web Ontology Language (OWL), which has been adopted as the web ontology standard by W3C [18]. Most recently, the $SROIQ$ [17] DL has extended $SHOIN$(D) by adding complex role inclusion, reflexive roles, transitive roles, irreflexive roles, disjoint roles, universal role, and a new construct $\exists R.Self$. To ensure greater tractability of OWL, $SROIQ$ has been proposed as the basis for OWL2 which allows strict partial order on roles in the form of role-hierarchies [23]. OWL2 is comprised of three semantic subsets called profiles[5], where each has a set of properties suitable for different tasks.

In the past several years, significant contributions have been made in introducing uncertainty to DL. Some notable ones have been the introduction of P-$SHOQ$(D)[16], a probabilistic extension to $SHOQ$(D) [19,25], fuzzy $SHOIN$(D) [30], a fuzzy extension to $SHOIN$(D) [18], as well as BayesOWL [8] and PR-OWL [7], probabilistic extensions to OWL. These techniques offer new ways of querying, modelling, and reasoning with DL ontologies. P-$SHOQ$(D) has provided a sound, complete, and decidable reasoning technique for probabilistic Description Logics. Fuzzy $SHOIN$(D) demonstrates subsumption and entailment relationship to hold to a certain degree, with the use of fuzzy modifiers, fuzzy concrete domain predicates, and fuzzy axioms. Fuzzy $SHOIN$(D) is an extension to work done on extending the ALC DL with fuzzy operators [28,29] (see Straccia et. al. [30] for a more complete list of extensions). PR-OWL is a language as well as a framework which allows ontologists to add probabilistic measures and reasoning to OWL ontologies. In order to incorporate uncertainty to an ontology based on a data mining algorithm, Gajderowicz [15] has derived decision trees from an external dataset associated with that ontology, and represented the trees in OWL2 syntax and RL profile[6] semantics.

As Fanizzi et. al [13] demonstrate, it is not only desirable to incorporate uncertainty with ontologies, but also to learn new rules from them that contain uncertainty. In their work, an existing OWL ontology is used to generate decision trees called *terminological decision trees* which are represented as OWL DL classes. Like their traditional *data-based* decision tree counterparts, *terminological decision trees* are based on frequent patterns in the ontology's defined OWL concepts. Unlike traditional decision trees that use conditions such as $wa{:}Direction = 'North'$ or $wa{:}Temp = 30$, these rules, called *concept descriptions*, use the OWL *concepts* defined in the ontology, such as $\exists hasPart.Worn$ and $\exists hasPart.(\neg Replaceable)$. Such *concept descriptions* are in the form:

---

[5] http://www.w3.org/TR/owl2-overview/#Profiles
[6] http://www.w3.org/TR/owl2-new-features/#OWL_2_RL

$$SendBack \equiv \exists hasPart.(Worn \sqcap \neg Replaceable).$$

The remaining sections demonstrate how to extend existing ontologies with information inferred from empirical data in the form of decision trees with associated probabilities.

2.2   Decision Trees

As a data structure, *decision trees* (DT) are used to classify a particular object based on patterns in data records associated with that object. These patterns are represented as logical structures representing classification rules. Each *rule* is a sequence of *nodes* and *edges* that make up a *branch*. Within a *branch*, each *edge* represents a single *condition* that differentiates concepts on a particular *attribute*. Any objects classified by sub-branches for this *edge* are members of the set of objects where this *condition* is true. In Figure 2, we present two decision trees utilizing (a) ordinal (numeric) attributes *height* and *weight*, and (b) a combination of ordinal and nominal (categorical) attributes *height* and *habitat*. These trees classify object models A through H using the three attributes. As Figure 2 illustrates, *nodes* can represent either a data attribute (e.g. *height, weight, habitat*) or a classified object (e.g. *Model A, Model B*).

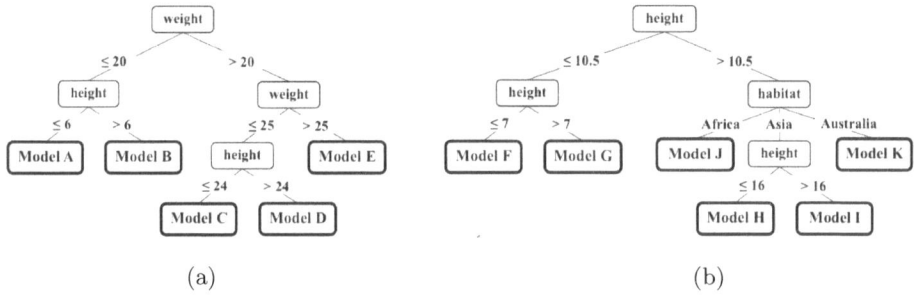

(a)   (b)

Fig. 2. Decision Trees with (a) ordinal attributes (*height* and *width*), and a combination of ordinal (*height*) and nominal (*habitat*) attributes

A *condition* which creates branching could be made up of a *weight* parent node with two sub-nodes where *weight* < 20 kg represents, for example, *servals* or the common *house cat*, while *weight* ≥ 20 kg represents larger felidae such as a *lions* or *tigers*. Nominal attributes are treated as categorically disjoint sets, which can result in as many branches as there are values. For example, various types of wild felidae may be identified by their geographical *habitat*. A possible branching scenario could be made up of the *habitat* parent node with three sub-nodes, *Africa, Asia* and *Australia*. Each sub-node could then show that *habitat* = *Africa* represents *lions* and *servals*, *habitat* = *Asia* represents *tigers*, and *habitat* = *Australia* represents *feral cats*. A branch bounding a nominal attribute can also represent a list of values resulting in less branches than possible values. For example, a single branch could represent *Asia* and *Australia*, and a second one *Africa*. Regardless of how many branches are produced or whether the branch bounds an ordinal attribute, any sub-node could be further split on additional attributes to represent further subsets of felidae. These subsets would be smaller in cardinality, but more exact in precision when classifying a particular *class* of felidae.

The key factor in the classifying process is the attribute and value combinations which identify concepts best and make up the branching conditions (classification rules). An advantage in using ontology concepts as the classification objects is that this attribute/value factor is guided by the attribute's semantic relation to a particular concept. As described further in Section 3.3, this advantage is utilized in our algorithm to build decision trees that select the most appropriate attributes and values that identify a semantic relationship deductively from the data.

2.3   Motivation for Extending Concepts

In the next section we describe how DT rules can extend ontology concepts and add a higher level of precision to concept definitions. In this chapter we discuss the motivation behind our approach.

With the use of DTs, a single concept is defined as clusters of data-points that uniquely identify that concept based on empirical data. The consistency between different sets of empirical data is then proposed in Section 4 as benefitting ontology matching. Defining individual concepts with several clusters is related to the field of granular computing [34] which views elements as parts of groups. The goal of this field is to study the reasons why elements are grouped together by indistinguishability, similarity, proximity, and functionality [33]. It takes advantage of rough and fuzzy sets to gain a level of granularity through inductive means, by defining crisp sets from fuzzy or possibilistic scoring models [32,22], and similar to DTs, are non-parametric [26]. By inductively reducing the dimensionality of an ontology concept, both rough sets and DTs are able to provide discrete partitions, required to identify and distinguish instances. Bittner et al. [2] identifies the requirements for crisp and vague boundaries, which are provided by rough and fuzzy sets, respectively. This paper presents our approach for achieving granularity of ontology concepts through decision trees.

3   Ontology Extension

In this section, we use a commerce use case to describe our algorithm for extending ontology concepts. Again, we will utilize DTs generated from external data that contain complementary information to the original ontology. Our work differs from ÖDT [35] and SemTree [3], in that while they use an ontology to build a DT, we use DTs to extend an existing ontology. The deductively derived DTs will hold classification rules which may overlap with another set of rules for similar concepts in a different ontology. The simple ontology in Figure 3 is a small hierarchy of objects, with a breakdown on physical objects, and further broken down to grains and animals. Notice the categories used to identify different ontology levels (described in Definition 1 in the next section). Target ontologies are ones which can be represented by a directed acyclic graph (DAG).

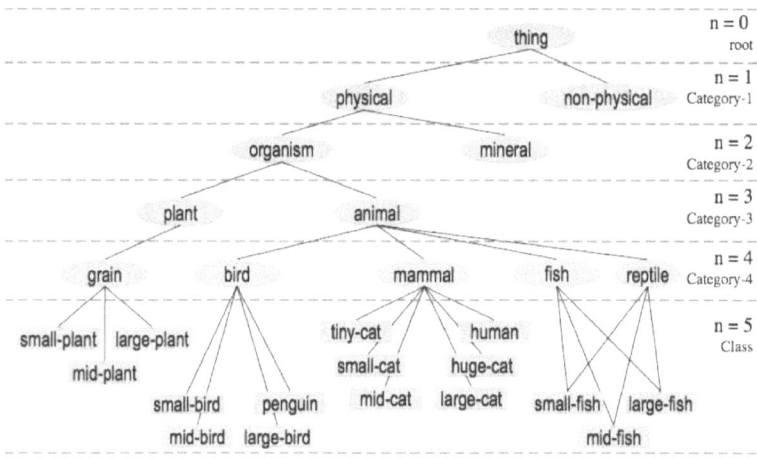

Fig. 3. An ontology, split by levels n, which are used for iterating edges in our algorithm in Section 3.3.

3.1   Database Preparation

Definition 1. (**Data preparation**) *Given the ontology O to be extended, the related database DB has:*

f := *number of attributes in normalized version of DB*

a_i := *attribute; where $i = \{\, 0, \ldots, f \,\}$*

$v_i := \begin{cases} value\ of\ a_i : if\ attribute\ a_i\ is\ defined; \\ null \qquad\quad : otherwise. \end{cases}$

C_n := *concepts at level n; i.e. $\{C_1, \ldots, C_n\}$*

Our algorithm uses supervised learning to build *decision trees* that classify data records in the database DB. Traditionally, there is a single static class that is associated with a record, such as *Siamese* identifying a record as a siamese cat. The algorithm presented in Section 3.3 uses a bottom concept such as *Siamese* as well as its parent concepts such as *HouseCat* and *Felinae* including the top most concept *Physical* at level $n = 1$ to classify a single record in DB[7].

$$Siamese \sqsubseteq HouseCat(tiny\text{-}cat) \sqsubseteq Felinae \sqsubseteq Felidae \sqsubseteq Mammal$$
$$\sqsubseteq Animal \sqsubseteq Organism \sqsubseteq Physical \sqsubseteq Thing \qquad (1)$$

---

[7] The root concept *owl:Thing* is implied, and not included in the record.

Consider again ontology O in Figure 3 which identifies various types of cats, in addition to different animals, plants, organisms, and minerals. Now imagine an external database DB (similar to Table 2) that supplements the O ontology by containing data on different species of cats around the world; specifically hierarchy (1) and a record in DB classified as *Siamese*. The *Siamese* class differentiates the record from other breeds of *tiny cats*. The same record is also classified as a *HouseCat* to differentiate it from other smaller cats in the wild, such as the *Cougar* and *Cheetah*. All *HouseCats* and small cats in the wild are classified as *Felinae* to differentiate them from other larger cats belonging to the *Panthera* class, such as the *Tiger* and *Lion*. Finally, the *Felinae* and *Panther* classes are both sub-classes of the *Felidae* family, which itself is a sub-class of *Mammals*. The algorithm generates individual decision trees for the same records for each class in the hierarchy defined in the O ontology. Specifically, any record classified as *Siamese* will contain a new attribute associated with each superclass in Equation 1.

Table 1. Conversion of ontology concepts to database columns

Concept	Level (n)	DB Column
physical	1	C_1
organism	2	C_2
animal	3	C_3
mammal	4	C_4
tiny-cat	5	C_5

In order to use ontology concepts as classes in a decision tree algorithm, O must first be represented in a suitable format. For this reason each bottom concept and its super-classes are presented as columns in a database, as illustrated in Figure 3 and listed in Table 1. For example *tiny-cat* is the value for column $C_5(Class)$, its super-class *mammal* is the value for column C_4, and so on until each class is associated with the record it classifies. It is important to represent concepts at equivalent levels by the same column C_n, with different classes as separate values[8]. This is depicted in Figure 3, with all nodes at level $n = 4$, for example, representing possible values for the column $C_4 = \{mammal, bird, fish, reptile, grain\}$. A value of "?" identifies an unknown value.

Table 2 demonstrates this hierarchy as a denormalized table with all other attributes. Note that an animal can be classified as both a *fish* and a *reptile* which means it has two parent nodes. Multiple parent nodes are represented by a duplication of records with different category values, as illustrated by instances 10 to 14. These are represented by a different parent in C_4, mainly *reptile* and *fish*, but the same *Class* value of *small-fish*.

---

[8] It is not required for ontology levels to align when matching concept characteristics (see Definition 6) across ontologies. Each target concept is compared to each local concept, regardless of its ontology level.

Table 2. Normalized Data Sample

Instance #	habitat	length	width	height	weight	fly	walk	swim	move	growth	ID	Size	C_1	C_2	C_3	C_4	C_5 (Class)
1	Algeria	12	4	6	115	N	Y	N	Y	Y	63	small	physical	organism	animal	mammal	small-cat
2	Amrcn-Samoa	4	1	3	4	N	Y	N	Y	Y	353	tiny	physical	organism	animal	mammal	tiny-cat
3	Armenia	51	14	29	8282	N	Y	?	Y	Y	354	?	physical	organism	animal	mammal	huge-cat
4	New-Zealand	7	1	3	2	Y	Y	N	Y	Y	469	small	physical	organism	animal	bird	small-bird
5	New-Zealand	14	6	6	50	Y	Y	N	?	Y	617	?	physical	organism	animal	bird	mid-bird
6	land-Islands	17	10	17	289	Y	?	N	Y	Y	767	large	physical	organism	animal	bird	large-bird
7	Antarctia	5	5	28	560	N	Y	Y	Y	?	841	?	physical	organism	animal	bird	penguin
8	Antig&Brbda	89	58	99	255519	N	Y	N	Y	Y	909	mid	physical	organism	animal	mammal	human
9	Aruba	75	55	43	88688	N	Y	N	Y	Y	912	mid	physical	organism	animal	mammal	human
10	New-Zealand	8	1	3	7.2	N	N	Y	Y	Y	1183	small	physical	organism	animal	fish	small-fish
11	New-Zealand	8	1	3	7.2	N	N	Y	Y	Y	1183	small	physical	organism	animal	reptile	small-fish
12	New-Zealand	7	1	4	8.4	N	N	Y	Y	Y	1185	?	physical	organism	animal	fish	small-fish
13	New-Zealand	7	1	4	8.4	N	N	Y	Y	Y	1185	?	physical	organism	animal	fish	small-fish
14	New-Zealand	7	1	4	8.4	N	N	Y	Y	Y	1186	?	physical	organism	animal	reptile	small-fish
15	Bahrain	0.001	0.001	0.001	0.000	?	?	?	N	Y	945	small	physical	organism	plant	grain	small-plant
16	Anguilla	1.001	0.001	3.001	0.000	?	?	?	N	Y	1100	mid	physical	organism	plant	grain	mid-plant
17	Bahamas	4.000	3.000	10.00	1.200	?	?	?	N	Y	1164	?	physical	organism	plant	grain	large-plant

3.2   Rule Insertion and Enhancement

In Section 2.1, we presented current work on introducing uncertainty to Description Logic, and in Section 2.3 we presented our motivation for utilizing decision trees. These ideas allow for a more accurate representation of real-world occurrences. In this section we introduce the notion of a *Region*, as per Definition 2, which is a 2-dimensional plane representing a decision tree branch with two attributes in a database DB.

Definition 2. *(Region) A Region (Reg) is a 2-dimensional space representing a decision tree branch that utilizes 2 attributes. The branch defines ranges of values for the attributes that fall within the values covered by the 2-dimensional region.*

Generating rules by inductive means allows us to extend existing axioms which define an ontology and its concepts. These types of extensions may introduce exceptions for a particular axiom that defines a concept, by splitting that axiom into two or more variations, which more accurately covers a broader range of observations. Such extensions have been described by Kwok [24] as Ripple Down Rules (RDR) that add knowledge to existing axioms through these exceptions. This prolongs the usability and maintainability of existing rules, while they are refined and added to [24]. RDR exceptions can also introduce closed world defaults [24].

To present the algorithm, we describe a typical commerce use case, where a manufacturer sets out to find customers interested in purchasing their product. Our manufacturer Mats for Cats (MAC) has a set of criteria identifying the size and weight of cats on which they base their product design. What they need now is a way to find a target market to advertise their product to, and the types of felines potential customers own. As part of the Semantic Web and

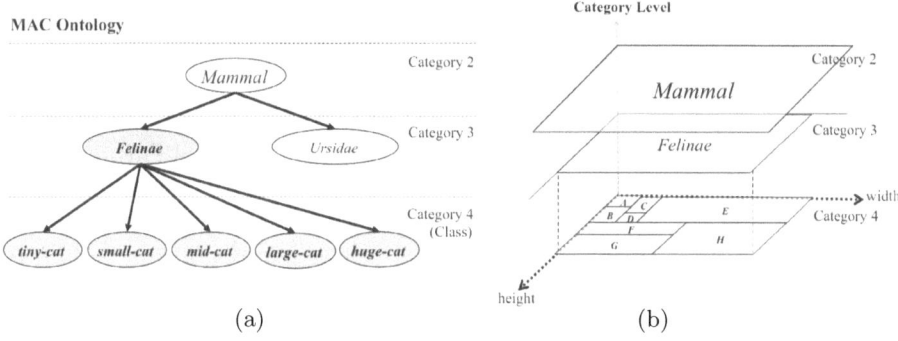

Fig. 4. (a) MAC Ontology separated into Category levels C_2 - C_4 and (b) the same ontology with each Category level on a separate *width* × *height* plane

a contribution to Open Data, another group Cats as Pets (CAP) has opened up their database and associated ontology of their member's cats, with various types of *felinae*. CAP stores easily obtainable information about their cats, such as *height, length, weight, colour*, and *habitat*, and does not store a full ontology like the one stored by the Animal Diversity Web[9] (AWD) database. Also, because this is a world wide group, the pets range from house cats to large felines such as tigers. As a result, the stored information will vary, but correlation between attributes will classify various types of MAC *felinae*. The following sections describe how to create and match regions between the MAC and CAP ontologies.

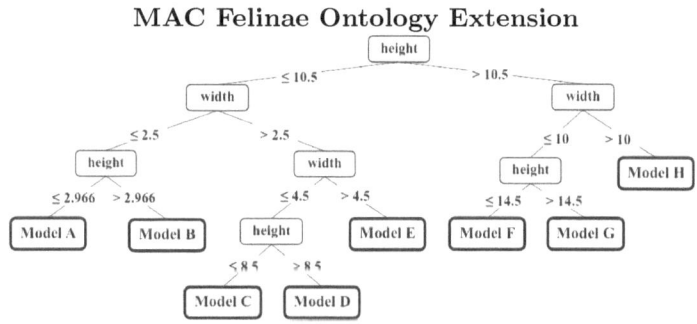

Fig. 5. NBTree classifying MAC Felinae model based on *height* and *width*

Consider the decision tree in Figure 5 that classified various types of MAC Felinae. As part of our commerce use case, Figure 4 (a) contains these Felinae as concepts of the MAC Ontology. Figure 4 (b) contains the *regions* that represent these concepts. The lowest plane (C_4) represents regions A - H defined by the DT in Figure 5.

---

[9] Animal Diversity Web: http://animaldiversity.ummz.umich.edu

We were able to successfully represent ontology concepts as regions by the database preparation procedure in Definition 1. Once the concepts have been associated with data points, the decision tree algorithm generates the 2-dimensional regions at various levels of the ontology. To achieve this, consider Figures 4 (a) and (b) that demonstrates how the MAC ontology's Felinae sub-branch can be represented on 2-dimensional planes. *Mammal* is a super-class of *Felinae* which in turn is a super-class of various cats. The data-points for *width* and *height* are used to represent these concepts at their own category levels. In Figure 4, each super-class subsumes its sub-classes. It follows then that the area in 4 (b) representing the super-class equally subsumes the area representing its sub-classes.

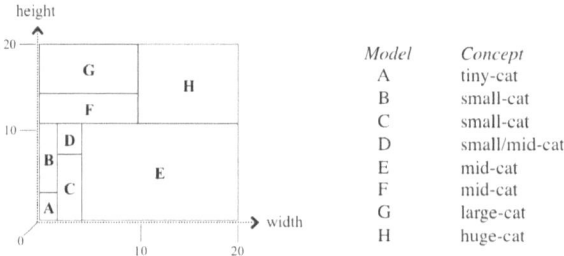

Model	Concept
A	tiny-cat
B	small-cat
C	small-cat
D	small/mid-cat
E	mid-cat
F	mid-cat
G	large-cat
H	huge-cat

Fig. 6. C_4 values of MAC Ontology on a single $width \times height$ plane. Each *region* A - H represents a decision tree branch in Figure 5.

Looking at the C_4 plane in Figure 4 (b), we show how the subclasses of *Felinae* are clustered into a *region* on the $width \times height$ plane. Each *region* represents a type of cat class, as demonstrated in Figure 6. This plane and its regions correspond to the DT in Figure 5. For example, the region $Reg_B(small\text{-}cat)$ may differentiate small cats from other larger felines with the rule $(height < 10.5) \wedge (width \leq 2.5) \wedge (height > 2.996)$.

3.3   Ontology Extension Algorithm

The extension process involves deriving decision trees from a database that classify ontology concepts. The trees are rules made up of ordinal number ranges and nominal category identifiers. We begin by listing elements needed to prepare the ontology for classification and the creation of *regions*.

Attributes of DB, mainly, $A = \{a_0, a_1, \ldots, a_f\}$, are selected into the subset $A_n : A_n \subseteq A$, based on their ability to classify concepts at level n, and construct a decision tree. When constructing trees however, only attributes which are required to differentiate between different models are included in the final tree. This subset $A_m : A_m \subseteq A_n$, is chosen to extend an ontology concept at level n with the class $c_{n,j}$, where j is the unique index of classification model c, as defined in Definition 3.

Definition 3. *(Ontology hierarchy) A given ontology O can be represented as a hierarchy (see Figure 3) with the following properties:*

$$O_h := a\ hierarchical\ representation\ O_h\ of\ the\ given\ ontology\ O$$
$$that\ contains\ the\ following\ properties:$$

$$levels(O_h) := the\ total\ number\ of\ levels\ in\ O_h$$

$$n := 0 \le n \le levels(O_h);\ where\ level\ n = 0\ represents\ the$$
$$tree\ root$$

$$c_{n,j} := a\ model\ classifying\ an\ ontology\ concept\ at\ a\ level\ n;$$
$$where\ j \in \{0,\ldots,|\ c_n\ |\}$$

$$|c| := number\ of\ instances\ classified\ as\ class\ c$$

$$edge(c_{n,j}, c_{n-1,k}) := edge\ between\ node\ c_{n,j}\ and\ its\ parent\ node\ c_{n-1,k}$$

Definition 4. (*Attribute relevance*) The attributes chosen to build a decision tree extending an ontology concept at level n, mainly $c_{n,j}$, depend on
$rank(c_{n,j}, a_i)$, which is the relevance of a_i in classifying $c_{n,j}$ and can be chosen
by an expert or automatically through a attribute ranking criterion.

When choosing an attribute automatically based on its contribution to classification, various rankings can be used. The data mining tool we are using is an
open source package called WEKA [31], which provides several modules, such
as *information gain, entropy*, and *principal component*. The *information gain*[10]
module has produced the best results for our dataset.

When selecting an attribute to build decision trees for $c_{n,j}$, our experience has
indicated that attributes which ranked significantly less (see Equation 2) than
attributes representing the parent node of $c_{n,j}$, will prevent creating a tree which
resembles a decision tree classifying the parent node. For example, the *weight*
attribute can be successfully used to differentiate different types of Felinae (wilds
cats vs house cats). In this case, the *weight* attribute would be ranked high.
When moving to the child nodes of house cats, the *weight* attribute would not
be a good indicator because many house cats are of the same *weight*. If the
information gain method chose *weight* to differentiate house cats, all house cats
would be grouped into a single class due to the similar *weight* values, producing
a less meaningful classification rule. In the same sense, attributes ranked closely
to ones used to classify child nodes or which are close to 0 should be avoided
(Equation 3), otherwise they will have a relatively high level of misclassification.
While the information gain method should disregard low ranking attributes,
Equations 2 and 3 should be considered when creating custom configurations for
decision trees.

$$rank(a_i) \ll rank(c_{n-1,j}), \tag{2}$$

$$0 \ll rank(c_{n+1,j}) \ll rank(a_i). \tag{3}$$

---

[10] The WEKA 3.6.0 module *weka.attributeSelection.InfoGainAttributeEval* was used
in our tests.

Traditional decision trees classify records in a binary fashion where a record either belongs in a class or not. For this paper, we chose a tree learning algorithm that may select several trees to classify a record, with Bayesian models indicating how well that tree and its branches classifies a particular record.

Definition 5. *(Concept extension) Given the set A_m, attributes utilized by the DT, we use the NBTree[11] module which produces several Bayesian models of the class $c_{n,j}$, as leaf nodes of the decision tree. Each leaf node, which we call a region Reg, contains the following elements:*

  $\sigma :=$ *Bayesian probability of classifying $c_{n,j}$ correctly with a Reg.*

  $\varphi :=$ *coverage (number of instances in a Reg classifying $c_{n,j}$ out of $|c|$).*

  $P := \sigma(\varphi/|c|)$ *: probability of Reg being correct and its accuracy covering entire set of c_n instances.*

  where the k-th region Reg_k is comprised of a DT branch, producing an associated clause with

  $P_k Reg_k(c_{n,j}) \leftarrow (a_0 \Diamond_0 v_0) \wedge (a_1 \Diamond_1 v_1) \wedge \ldots \wedge (a_m \Diamond_m v_m).$
  where $\Diamond \in \{\leq, >, =\}$.

Definition 5 describes the properties each Bayesian model possesses. The resulting clause $P_k Reg_k$ represents a single branch of a decision tree, and the conditions that represent a particular region. It should be noted that this clause is sensitive to the distribution of data points in DB, which is a shortfall of the NBTree module and other machine learning algorithms based on supervised learning. As Chien et al. [6] point out, there are many algorithms which handle missing data in a principle way. The configuration of NBTree module, however, is not sufficient for our needs. As a result, any missing values v_i for an attribute a_i cause a_i to act as a wild card and increases the probability (P_k) of the associated region Reg_k, while decreasing the accuracy. Also, if the number of instances representing each *class* does not have a relatively equal distribution, NBTree introduces a bias that skews the generated DT to classify only the best represented *classes*. For example, if 95% of observations are of *class A* and 5% of *class B*, B will not be represented by the DT, as the probability of incorrectly choosing A is negligible at only 5%. For the DTs and associated probabilities to be meaningful, the number of instances of *classes* should be approximately equal [15]. This ensures each concept has equal representation in the DT.

Definition 6. *(Concept Characteristic) Given a set of regions Reg_k used to classify $c_{n,j}$, we create the clause*

  $Ch_j(\Sigma P_j, c_{n,j}) \leftarrow (P_x Reg_x) \vee (P_y Reg_y) \vee \ldots \vee (P_z Reg_z).$

where ΣP_j is the probability for class $c_{n,j}$, calculated from summing all probabilities (P_k) with an associated coverage $|c|$.

---

[11] WEKA Naïve Bayes classifiers weka.classifiers.trees.NBTree

Definition 6 defines the entire rule for classifying a particular ontology concept class $c_{n,j}$. It combines all the clauses from all trees that classify $c_{n,j}$ into a disjunction of those clauses. The resulting composite condition creates a *concept characteristic* Ch_j which states that if a record's values fall within the range of any clauses in Ch_j, that record is classified as that ontology concept class $c_{n,j}$, with the probability ΣP_j.

The algorithm described below combines Definitions 1 to 6 to create decision trees which are used to build ontology extensions with probabilist classification models. First, features important to the identification and differentiation of a set of classes (steps 1 - 3) are identified. Those features are used to build a DT (step 4), which results in a set of rules that identify the classes with different levels of coverage, accuracy, and probability. Each concept is then associated with a concept characteristic and a probability identifying its confidence (steps 5 - 7). The derived rules are used to build the characteristic clause Ch (step 10) and probability ΣP (step 11). The concept characteristic is then associated with a concept in Cn as class $c_{n,j}$ in the ontology hierarchy O_h (step 13).

Extension Algorithm

```
1)   Denormalize DB, applying ontology classes as attributes
     (see Section 3.1 for a discussion and Table 2 for an example).
2)   For each n ∈ levels(Oₕ)
3)      Select attribute set Aₙ using rank(aᵢ), to best classify Cₙ
        with an automated attribute selection method such as WEKA's
        information gain.
4)      Execute the classification algorithm  as defined in Definition 5,
        to produce a decision trees classifying the concept Cₙ.
5)      For each cₙ,ⱼ
6)         Initialize Chⱼ to an empty clause.
7)         Initialize probability ΣPⱼ to 0.
8)      For each k ∈ z; where z is the number of regions classifying c.
9)         Capture entire branch of a DT model for cₙⱼ, giving Regₖ
           and associated Pₖ.
10)        Append Regₖ(cₙ,ⱼ) to the Ch(cₙ,ⱼ) clause with the OR operator.
11)        Calculate new ΣPⱼ as: ΣPⱼ = ΣPⱼ + Pₖ.
12)     End
13)     Associate ΣPⱼCh(cₙ,ⱼ) with edge(cₙ,ⱼ, c₍ₙ₋₁₎,ₖ).
14) End
```

To increase the quality of decision trees, attribute selection in step (3) can be performed manually by the ontology author or a subject matter expert (SME).

4   Commerce Scenario In Detail

4.1   Extending MAC and CAP Ontologies

As a continuation of the MAC and CAP use case introduced in Section 3.2, this section uses specific examples generated by our algorithm using the WEKA package. Figure 7 contains the CAP decision tree generated using the *height*, *width*, and *weight* attributes. The database *DB* as well as the MAC and CAP ontologies are simulated, but suffer from real-world issues such as incomplete and incorrect data. We test the attribute ranking and classification algorithm for their ability to handle such cases.

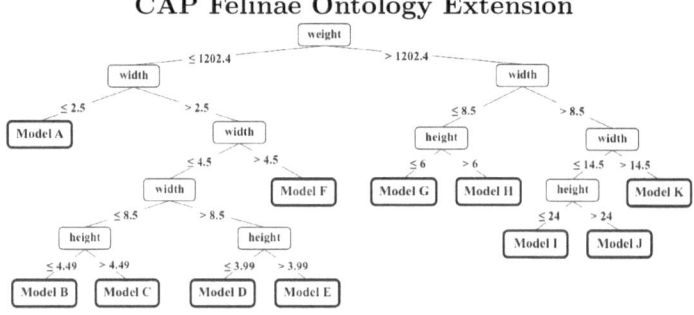

Fig. 7. NBTree classifying CAP Felinae based on height, width, weight

The simulated ontologies have related concepts with various differences. The hypothesis we made is that even though individual ontologies covering the same domain may differ (possibly due to the ontology author's bias), the associated empirical dataset will remain somewhat consistent [11], and the resulting decision trees will retain some of that consistency.

4.2   Matching CAP And MAC Regions

Using the NBTree classifier in WEKA to classify different sizes of felines, we classify *Felinae* as $F = \{tiny\text{-}cat, small\text{-}cat, mid\text{-}cat, large\text{-}cat, huge\text{-}cat\}$, and derive the DT in Figure 5. The corresponding 2-dimensional regions are illustrated in Figures 8 and 9. For a decision tree, each leaf node represents a Bayesian model for each concept, with various degrees of probability σ and coverage φ.

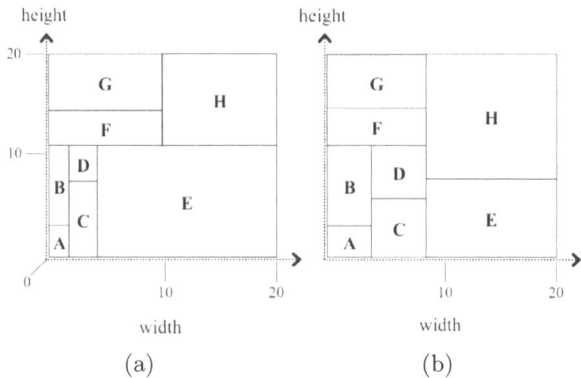

Fig. 8. Regions for attributes *width* × *height* for (a) MAC and (b) CAP ontologies

Each leaf node also represents a single region *Reg*. Here, regions are derived using the *width* × *height* and *weight* × *height* attribute combinations.

Looking at the regions in Figure 8 using *height* and *width*, we see overlapping clusters between (a) MAC and (b) CAP regions, specifically regions $Reg_A(tiny\text{-}cat)$, $Reg_E(mid\text{-}cat)$, $Reg_F(mid\text{-}cat)$, $Reg_G(large\text{-}cat)$, and a partial overlap on $Reg_H(huge\text{-}cat)$. We can begin to infer not only a match between the concepts represented by these regions (*tiny-cat*, *small-cat*, etc), but also between the attributes *height* and *weight* themselves.

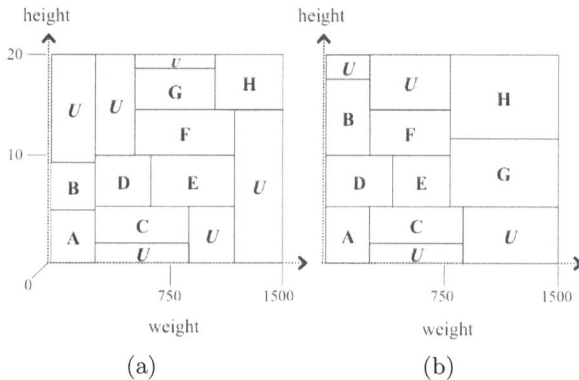

Fig. 9. Regions for attributes *weight* × *height* for (a) MAC and (b) CAP ontologies. The *U* regions are *unknown* due to a low probability *P*.

Unfortunately, not all databases are this well aligned, and various measures of similarity must be considered. In Figure 9, the correlation between the *height* and *weight* attributes lack the definite majority of correlation and overlapping between regions as was observed in Figure 8. As a result, a mix of similarities would need to be considered as classification characteristics. As with Figure 8, Figure also 9 contains a correlation between $Reg_A(small\text{-}cat)$, $Reg_C(small\text{-}cat)$, and $Reg_F(mid\text{-}cat)$ in the centre, and to a lesser degree $Reg_H(huge\text{-}cat)$ regions. Unlike Figure 8 however, no significant correlation exists between other regions. A series of decision trees with various permutations of attributes would produce the best regions to match against regions from other ontologies and datasets.

4.3 MAC And CAP Characteristics

Throughout this paper we have described how to built *regions* and concept *characteristics* from decision tress, and used these to match concepts in two complementary ontologies. As a reference, this section presents the resulting clauses representing regions and charactersistics for the commerce use case MAC and CAP systems.

The MAC decision tree in Figure 5, results in regions presented in Figure 8. These regions represent the clauses in Table 3, and *charactersistics* in Table 4. The CAP decision tree in Figure 7, results in regions presented in Figure 9. These regions represent the clauses in Table 5, and *charactersistics* in Table 6.

Table 3. MAC regions build from the decision tree in Figure 3, using *height* (x) and *width* (y)

Model	P		Region
	σ	φ	
A	0.89	101	$Reg_0(tiny\text{-}cat) \leftarrow (x \leq 10.5) \wedge (y \leq 2.5) \wedge (x \leq 2.99)$
	0.09	9	$Reg_1(small\text{-}cat) \leftarrow (x \leq 10.5) \wedge (y \leq 2.5) \wedge (x \leq 2.99)$
B	0.92	44	$Reg_2(small\text{-}cat) \leftarrow (x \leq 10.5) \wedge (y > 2.5) \wedge (x > 2.99)$
C	0.90	34	$Reg_3(small\text{-}cat) \leftarrow (x \leq 10.5) \wedge (y > 2.5) \wedge (y \leq 4.5) \wedge (x \leq 8.5)$
D	0.58	13	$Reg_4(small\text{-}cat) \leftarrow (x \leq 10.5) \wedge (y > 2.5) \wedge (y \leq 4.5) \wedge (x > 8.5)$
	0.29	6	$Reg_5(mid\text{-}cat) \leftarrow (x \leq 10.5) \wedge (y > 2.5) \wedge (y \leq 4.5) \wedge (x > 8.5)$
E	0.64	6	$Reg_6(mid\text{-}cat) \leftarrow (x \leq 10.5) \wedge (y > 2.5) \wedge (y > 4.5)$
F	0.87	26	$Reg_7(mid\text{-}cat) \leftarrow (x > 10.5) \wedge (y \leq 10) \wedge (y \leq 14.5)$
G	0.78	93	$Reg_8(large\text{-}cat) \leftarrow (x > 10.5) \wedge (y \leq 10) \wedge (y > 14.5)$
H	0.96	105	$Reg_{10}(huge\text{-}cat) \leftarrow (x > 10.5) \wedge (y > 10)$

Table 4. MAC Characteristics classifying Felinae built from regions in Table 3

ΣP		Characteristic
ΣP	$|c|$	
0.89	101	$Ch_0(tiny\text{-}cat) \leftarrow (P_0\,Reg_0)$
0.78	100	$Ch_1(small\text{-}cat) \leftarrow (P_1\,Reg_1) \vee (P_2\,Reg_2) \vee (P_3\,Reg_3) \vee (P_4\,Reg_4)$
0.78	60	$Ch_2(mid\text{-}cat) \leftarrow (P_5\,Reg_5) \vee (P_6\,Reg_6) \vee (P_7\,Reg_7) \vee (P_9\,Reg_9)$
0.78	93	$Ch_3(large\text{-}cat) \leftarrow (P_8\,Reg_8)$
0.96	105	$Ch_4(huge\text{-}cat) \leftarrow (P_{10}\,Reg_{10})$

Table 5. CAP regions built from the decision tree in Figure 7, using *height* (x), *width* (y) and *weight* (z)

Model	P		Region
	σ	φ	
A	0.51	124	$Reg_0(small\text{-}cat) \leftarrow (z \leq 1202.4) \wedge (y \leq 1.5)$
	0.43	20	$Reg_1(mid\text{-}cat) \leftarrow (z \leq 1202.4) \wedge (y \leq 1.5)$
B	0.09	4	$Reg_2(small\text{-}cat) \leftarrow (z \leq 1202.4) \wedge (y > 1.5) \wedge (y \leq 3.5) \wedge (y \leq 2.5) \wedge (x \leq 4.5)$
	0.85	45	$Reg_3(tiny\text{-}cat) \leftarrow (z \leq 1202.4) \wedge (y > 1.5) \wedge (y \leq 3.5) \wedge (y \leq 2.5) \wedge (x \leq 4.5)$
C	0.38	13	$Reg_4(small\text{-}cat) \leftarrow (z \leq 1202.4) \wedge (y > 1.5) \wedge (y \leq 3.5) \wedge (y \leq 2.5) \wedge (x > 4.5)$
	0.54	19	$Reg_5(mid\text{-}cat) \leftarrow (z \leq 1202.4) \wedge (y > 1.5) \wedge (y \leq 3.5) \wedge (y \leq 2.5) \wedge (x > 4.5)$
D	0.15	10	$Reg_6(small\text{-}cat) \leftarrow (z \leq 1202.4) \wedge (y > 1.5) \wedge (y \leq 3.5) \wedge (y > 2.5) \wedge (x \leq 4)$
	0.80	56	$Reg_7(tiny\text{-}cat) \leftarrow (z \leq 1202.4) \wedge (y > 1.5) \wedge (y \leq 3.5) \wedge (y > 2.5) \wedge (x \leq 4)$
E	0.40	15	$Reg_8(small\text{-}cat) \leftarrow (z \leq 1202.4) \wedge (y > 1.5) \wedge (y \leq 3.5) \wedge (y > 2.5) \wedge (x > 4)$
	0.53	20	$Reg_9(mid\text{-}cat) \leftarrow (z \leq 1202.4) \wedge (y > 1.5) \wedge (y \leq 3.5) \wedge (y > 2.5) \wedge (x > 4)$
F	0.48	19	$Reg_{10}(small\text{-}cat) \leftarrow (z \leq 1202.4) \wedge (y > 1.5) \wedge (y > 3.5)$
	0.45	18	$Reg_{11}(mid\text{-}cat) \leftarrow (z \leq 1202.4) \wedge (y > 1.5) \wedge (y > 3.5)$
G	0.67	7	$Reg_{12}(mid\text{-}cat) \leftarrow (z > 1202.4) \wedge (y \leq 8.5) \wedge (y \leq 6)$
H	0.87	26	$Reg_{13}(large\text{-}cat) \leftarrow (z > 1202.4) \wedge (y \leq 8.5) \wedge (y > 6)$
I	0.96	97	$Reg_{14}(large\text{-}cat) \leftarrow (z > 1202.4) \wedge (y > 8.5) \wedge (y \leq 11.5) \wedge (x \leq 24)$
J	0.95	78	$Reg_{15}(huge\text{-}cat) \leftarrow (z > 1202.4) \wedge (y > 8.5) \wedge (y \leq 11.5) \wedge (x > 24)$
K	0.87	26	$Reg_{16}(huge\text{-}cat) \leftarrow (z > 1202.4) \wedge (y > 8.5) \wedge (y > 11.5)$

Table 6. CAP Characteristics classifying Felinae built from regions in Table 5

ΣP		Characteristic
ΣP	$|c|$	
0.82	101	$Ch_0(tiny\text{-}cat) \leftarrow (P_3\,Reg_3) \wedge (P_7\,Reg_7)$
0.40	85	$Ch_1(small\text{-}cat) \leftarrow (P_0\,Reg_0) \wedge (P_2\,Reg_2) \wedge (P_4\,Reg_4) \wedge (P_6\,Reg_6) \wedge (P_8\,Reg_8) \wedge (P_{10}\,Reg_{10})$
0.50	84	$Ch_2(mid\text{-}cat) \leftarrow (P_1\,Reg_1) \wedge (P_5\,Reg_5) \wedge (P_9\,Reg_9) \wedge (P_{11}\,Reg_{11}) \wedge (P_{12}\,Reg_{12})$
0.94	123	$Ch_3(large\text{-}cat) \leftarrow (P_{13}\,Reg_{13}) \wedge (P_{14}\,Reg_{14})$
0.93	104	$Ch_4(huge\text{-}cat) \leftarrow (P_{15}\,Reg_{15}) \wedge (P_{16}\,Reg_{16})$

5   Conclusion

In this paper, we presented an algorithm for enhancing ontologies with inductively derived decision trees, in order to enhance concepts with empirical data. The concept extension process aims to produce partitions of characteristics of ontology concepts, based on the ontology's observed instances, such that the concepts are represented by 2-dimensional regions, as per Definition 2. We then describe how these regions can be used to match concepts of different but similar ontologies with each other. We apply our algorithm to a simulated dataset of Felines, with a matching scenario in the commerce domain. This paper describes potential benefits of data that describes similar concepts, and how the similarities can be utilized. The simulated database for MAC and CAP contained key real-life database features, positive and negative, required to demonstrate our algorithm.

In our research, we have identified several key ontology matching observations and issues. It is important to find attributes in one ontology which are subsumed by a hybrid attribute derived from multiple attributes in the other. Relevant work has been done in the field of Object Based Representation Systems (OBRS) [3], where looking at subsumptions made about classified instances can lead to deducing new information about those instances. Our *regions* and *characteristics* represent ranges and clusters which identify some class. For ordinal values, units of measure may be less relevant then ratios of values and their ranges, specifically when matching concepts at higher levels. For example, identifying traits in objects may depend on a correlation between two or more attributes. A long life span for one animal is short for another, so when grouping long life span factors, for example, it would make most sense to use the "relative" life span (in the form of ratios) of a particular species, when comparing life expectancy factors across multiple species.

When dealing with concrete values, such as those found in a database, it would be unrealistic to assume exact matches between these values exist in the local and target databases. For this reason, buffers must be introduced in order to make the derived rules more inclusive [15]. Numeric values can be expanded with varying degrees, depending on the strictness of a particular domain. Nominal values can be extended using resource such as WordNet [14] or translation tools.

Acknowledgement. Bart Gajderowicz gratefully acknowledges the discussions with Michael Gruninger of the University of Toronto, which benefited this research. The authors would like to thank the numerous reviewers for their suggestions and comments.

References

1. Belleau, F., Nolin, M.-A., Tourigny, N., Rigault, P., Morissette, J.: Bio2rdf: towards a mashup to build bioinformatics knowledge systems. Journal of Biomedical Informatics 41(5), 706–716 (2008)

2. Bittner, T., Smith, B.: Granular partitions and vagueness. In: FOIS 2001: Proceedings of the International Conference on Formal Ontology in Information Systems, pp. 309–320. ACM, New York (2001)

3. Bouza, A., Reif, G., Bernstein, A., Gall, H.: Semtree: Ontology-based decision tree algorithm for recommender systems. In: 7th International Semantic Web Conference, ISWC 2008, Germany (October 2008)

4. Brodaric, M., Gahegan, M.: Experiments to examine the situated nature of geoscientific concepts. Spatial Cognition and Computation: An Interdisciplinary Journal 7(1), 61–95 (2007)

5. Chepelev, L., Klassen, D., Dumontier, M.: Chemical hazard estimation and method comparison with owl-encoded toxicity decision trees. In: OWLED 2011 OWL: Experiences and Directions (June 2011)

6. Chien, B.-C., Lu, C.-F., Hsu, S.-J.: Handling incomplete quantitative data for supervised learning based on fuzzy entropy. In: 2005 IEEE International Conference on Granular Computing, vol. 1, pp. 135–140 (July 2005)

7. da Costa, P.C.G., Laskey, K.B., Laskey, K.J.: PR-OWL: A Bayesian Ontology Language for the Semantic Web. In: da Costa, P.C.G., d'Amato, C., Fanizzi, N., Laskey, K.B., Laskey, K.J., Lukasiewicz, T., Nickles, M., Pool, M. (eds.) URSW 2005-2007. LNCS (LNAI), vol. 5327, pp. 88–107. Springer, Heidelberg (2008)

8. Ding, Z., Peng, Y., Pan, R.: BayesOWL: Uncertainty Modeling in Semantic Web Ontologies. In: Ma, Z. (ed.) Soft Computing in Ontologies and Semantic Web. STUDFUZZ, vol. 204, pp. 3–29. Springer, Heidelberg (2005)

9. Ding, Z., Peng, Y., Pan, R., Yu, Y.: A bayesian methodology towards automatic ontology mapping. In: Proceedings of the AAAI 2005 C&O Workshop on Contexts and Ontologies: Theory, Practice and Applications, pp. 72–79. AAAI Press (July 2005)

10. Erdur, C., Seylan, I.: The design of a semantic web compatible content language for agent communication. Expert Systems 25(3), 268–294 (2008)

11. Euzenat, J., Shvaiko, P.: Ontology Matching. Springer-Verlag New York, Secaucus (2007)

12. Falconer, S.M., Noy, N.F., Storey, M.A.: Ontology mapping - a user survey. In: Proceedings of the Workshop on Ontology Matching, OM 2007, pp. 113–125. ISWC/ASWC (2007)

13. Fanizzi, N., d'Amato, C., Esposito, F.: Towards the induction of terminological decision trees. In: Proceedings of the 2010 ACM Symposium on Applied Computing, SAC 2010, pp. 1423–1427. ACM, New York (2010)

14. Fellbaum, C.: Wordnet: An electronic lexical database (1998)

15. Gajderowicz, B.: Using decision trees for inductively driven semantic integration and ontology matching. Master's thesis, Ryerson University, 250 Victoria Street, Toronto, Ontario, Canada (2011)

16. Giugno, R., Lukasiewicz, T.: P-$\mathcal{SHOQ}$(**D**): A Probabilistic Extension of $\mathcal{SHOQ}$(**D**) for Probabilistic Ontologies in the Semantic Web. In: Flesca, S., Greco, S., Leone, N., Ianni, G. (eds.) JELIA 2002. LNCS (LNAI), vol. 2424, pp. 86–97. Springer, Heidelberg (2002)

17. Horrocks, I., Kutz, O., Sattler, U.: The even more irresistible sroiq. In: Proc. of the 10th Int. Conf. on Principles of Knowledge Representation and Reasoning, KR 2006, pp. 57–67. AAAI Press (June 2006)

18. Horrocks, I., Patel-Schneider, P.F., Harmelen, F.V.: From shiq and rdf to owl: The making of a web ontology language. Journal of Web Semantics 1, 7–26 (2003)

19. Horrocks, I., Sattler, U.: Ontology reasoning in the shoq(d) description logic. In: Proc. of the 17th Int. Joint Conf. on Artificial Intelligence, IJCAI 2001, pp. 199–204. Morgan Kaufmann (2001)
20. Kieler, B.: Semantic data integration across different scales: Automatic learning generalization rules. In: International Archives of Photogrammetry, Remote Sensing and Spatial Information Sciences, vol. 37 (2008)
21. Kieler, B., Huang, W., Haunert, J.H., Jiang, J.: Matching river datasets of different scales. In: AGILE Conf. Lecture Notes in Geoinformation and Cartography, pp. 135–154. Springer (2009)
22. Klinov, P., Mazlack, L.J.: Granulating semantic web ontologies. In: 2006 IEEE International Conference on Granular Computing, pp. 431–434 (2006)
23. Krotzsch, M., Rudolph, S., Hitzler, P.: Description logic rules. In: Proceeding of the 2008 Conference on ECAI 2008, pp. 80–84. IOS Press, Amsterdam (2008)
24. Kwok, R.B.H.: Translations of Ripple Down Rules into Logic Formalisms. In: Dieng, R., Corby, O. (eds.) EKAW 2000. LNCS (LNAI), vol. 1937, pp. 366–379. Springer, Heidelberg (2000)
25. Pan, J.Z., Horrocks, I.: Reasoning in the shoq(dn) description logic. In: Proc. of the 2002 Int. Workshop on Description Logics, DL 2002 (April 2002)
26. Sikder, I.U., Munakata, T.: Application of rough set and decision tree for characterization of premonitory factors of low seismic activity. Expert Systems with Applications 36(1), 102–110 (2009)
27. Staab, S., Studer, R.: Handbook on Ontologies. International Handbooks on Information Systems. Springer (2004)
28. Straccia, U.: A fuzzy description logic. In: 15th National Conference on Artificial Intelligence, Madison, USA, vol. 4, pp. 594–599 (1998)
29. Straccia, U.: Reasoning within fuzzy description logics. Journal of Artificial Intelligence Research 14, 137–166 (2001)
30. Straccia, U.: A fuzzy description logic for the semantic web. In: Fuzzy Logic and the Semantic Web, Capturing Intelligence, vol. 4, pp. 167–181. Elsevier (2005)
31. Witten, I., Frank, E.: Data Mining: Practical machine learning tools and techniques, 2nd edn. Morgan Kaufmann Publishers, San Francisco (2005)
32. Yao, Y.Y.: Granular computing: basic issues and possible solutions. In: Proceedings of the 5th Joint Conference on Information Sciences, pp. 186–189 (2000)
33. Zadeh, L.A.: Toward a theory of fuzzy information granulation and its centrality in human reasoning and fuzzy logic. Fuzzy Sets Syst. 90, 111–127 (1997)
34. Zadeh, L.A.: Fuzzy sets, fuzzy logic, and fuzzy systems. World Scientific Publishing Co., Inc., River Edge (1996)
35. Zhang, J., Silvescu, A., Honavar, V.: Ontology Driven Induction of Decision Trees at Multiple Levels of Abstraction. In: Koenig, S., Holte, R.C. (eds.) SARA 2002. LNCS (LNAI), vol. 2371, pp. 316–323. Springer, Heidelberg (2002)

Assertion Prediction with Ontologies through Evidence Combination

Giuseppe Rizzo, Claudia d'Amato, Nicola Fanizzi, and Floriana Esposito

Dipartimento di Informatica
Università degli studi di Bari, Italy
rizzo.giuseppe@unibari.net,
firstname.lastname@uniba.it

Abstract. Following previous works on inductive methods for ABox reasoning, we propose an alternative method for predicting assertions based on the available evidence and the analogical criterion. Once neighbors of a test individual are selected through some distance measures, a combination rule descending from the Dempster-Shafer theory can join together the evidence provided by the various neighbor individuals in order to predict unknown values in a learning problem. We show how to exploit the procedure in the problems of determining unknown class- and role-memberships or fillers for datatype properties which may be the basis for many further ABox inductive reasoning algorithms. This work presents also an empirical evaluation of the method on real ontologies.

1 Introduction

In the context of reasoning in the Semantic Web, a growing interest is being shown to alternative procedures extending the standard methods so that they can deal with the various facets of uncertainty related with Web reasoning [21].

Extensions of the classic probability measures [20] offer alternative ways to deal with inherent uncertainty of the knowledge bases in the Semantic Web. Particularly, belief and plausibility measures adopted in the *Dempster-Shafer Theory of Evidence* [20] have been exploited as means for dealing with incompleteness [11] and also inconsistency [22], which may arise from the aggregation of data and metadata on a large and distributed scale. The Dempster-Shafer theory is a generalization of the Bayesian theory of subjective probability in which functions base degrees of belief on the probability of a proposition. Belief functions represent the probability that a given proposition is provable from a set of other propositions, to which probabilities are assigned.

In this work we undertake again the inductive point of view. Indeed, in many Semantic Web domains a very large number of assertions can potentially be true but often only a small number of them is known to be true or can be inferred to be true. So far the application of combination rules related to the Dempster-Shafer theory in this field has regarded the induction of metrics which are essential for all similarity-based reasoning methods [11]. One of the applications of such measures was related to the prediction of assertions through nearest neighbor

F. Bobillo et al. (Eds.): URSW 2008-2010/UniDL 2010, LNAI 7123, pp. 282–299, 2013.

procedures. A general-purpose evidential nearest neighbor procedure based on the Dempster-Shafer combination rule (a generalization of the classic Bayes rule) has been proposed [8]. In this work this method is extended to the specific case of semantic knowledge bases through a more epistemically appropriate combination procedure [29], exploiting specific metrics to assess the similarity of the individuals involved.

In the perspective of inductive methods, the need for a definition of a semantic similarity measure for *individuals* arises, that is a problem that so far received less attention in the literature compared to the measures for concepts. Recently proposed dissimilarity measures for individuals in specific languages founded in *Description Logics* [1] turned out to be practically effective for the targeted inductive tasks [5], however they are still based on structural criteria so that they can hardly scale to more complex languages. We devised families of dissimilarity measures for semantically annotated resources, which can overcome the aforementioned limitations [6,13]. Our measures are mainly based on the Minkowski's norms for Euclidean spaces induced by means of a method developed in the context of *multi-relational learning* [26]. Namely, the measures are based on the degree of discernibility of the input individuals with respect to a given context [15] (or committee of features), which are represented by concept descriptions expressed in the language of choice.

The main contributions of this paper regard the extension of a framework for the classification of individuals through a prediction procedure based on evidence theory and similarity. In [14,25] we investigate the use of alternative rules of combination and exploiting the mentioned families of metrics defined for individuals in ontologies. This allows for measuring the confirmation of the truth of candidate assertions. The prediction of the values (related to class-membership or datatype and object properties) may have plenty of applications in uncertainty reasoning with ontologies.

The remainder of the paper is organized as follows. In the next section (Sect. 2), we recall the definition of distance measures that shall be utilized for selecting neighbor individuals. Then (Sect. 3), the basics of the Dempster-Shafer theory and a nearest-neighbor procedure based on an alternative rule of combination are recalled. Hence (Sect. 4) we present the applications of the method to the problems of determining the class- or role-membership of individuals w.r.t. given query concepts / roles as well as the prediction of fillers for datatype properties. An experimental evaluation of the method in three prediction tasks is showed in (Sect. 5). Relevant related work are discussed in (Sect. 6) and we conclude (Sect. 7) by proposing extensions and applications of these methods in further works.

2 Dissimilarity Measures between Individuals

Since the reasoning method to be presented in the following is intended to be general purpose, no specific language, will be assumed in the following for resources, concepts (classes) and their properties. It suffices to consider a generic representation that can be mapped to some Description Logic language with the standard model-theoretic semantics (see [1] for a thorough reference).

A *knowledge base* $\mathcal{K} = \langle \mathcal{T}, \mathcal{A} \rangle$ comprises a *TBox* $\mathcal{T}$ and an *ABox* $\mathcal{A}$. $\mathcal{T}$ is a set of axioms concerning the (partial) definition of concepts (and roles) through class (role) expressions. $\mathcal{A}$ contains assertions (ground facts) concerning the world state. The set of the individuals occurring in $\mathcal{A}$ will be denoted with $\mathsf{Ind}(\mathcal{A})$. Each individual can be assumed to be identified by its own URI (it is useful in this context to make the *unique names assumption*).

Similarity-based tasks, such as individual classification, retrieval, and clustering require language-independent measures for individuals whose definition can capture semantic aspects of their occurrence in the knowledge base [6,13].

For our purposes, we need functions to assess the similarity of individuals. However individuals do not have an explicit syntactic (or algebraic) structure that can be compared (unless one resorts to language-specific notions [5], such as the *most specific concept* [1]). Hence it turns out to be difficult to adapt measures for concepts that have been recently proposed, e.g. see [4,10,19].

Focusing on the semantic level, the fundamental idea is that similar individuals should behave similarly with respect to the same concepts. A way for assessing the similarity of individuals in a knowledge base can be based on the comparison of their semantics along a number of dimensions represented by a set of concept descriptions (henceforth referred to as the *committee* or *context* [15]). Specifically, the measure may compare individuals on the grounds of their behavior w.r.t. a given context, say $\mathsf{C} = \{C_1, C_2, \ldots, C_m\}$, which stands as a group of discriminating relevant concepts (*features*) expressed in the considered language.

We begin with defining the behavior of an individual w.r.t. a certain concept in terms of projecting it in this dimension: Given a concept $C_i \in \mathsf{C}$, the related *projection function* $\pi_i : \mathsf{Ind}(\mathcal{A}) \mapsto [0,1]$ is defined:

$$\forall a \in \mathsf{Ind}(\mathcal{A}) \qquad \pi_i(a) = \begin{cases} 1 & \mathcal{K} \models C_i(a) \\ 0 & \mathcal{K} \models \neg C_i(a) \\ \pi_i & \text{otherwise} \end{cases}$$

The intermediate value π_i corresponds to the case when a reasoner cannot assign the truth value for a certain membership query. This is due to the *Open World Assumption* normally made in Semantic Web reasoning. Hence, as in the classic probabilistic models, the prior membership probability π_i w.r.t. C_i may be considered, if known. Otherwise a uniform distribution is assumed ($\pi_i = \frac{1}{2}$). Priors may be determined as measures of the concept extension w.r.t. the domain of objects to be approximated by the ratio between the cardinality of the retrieval of the concept and the number of individuals occurring in the knowledge base $|\mathsf{Ind}(\mathcal{A})|$. Further ways to approximate these values in case of uncertainty are investigated in [11]. A further degree of approximation (for densely populated ontologies) can be introduced by replacing reasoning on individuals ($\mathcal{K} \models (\neg)C_i(a)$) with ABox lookup (($(\neg)C_i(a) \in \mathcal{A}$)) when feature concepts C_i are chosen among those that.

Similarly to [6,13], a family of dissimilarity measures for individuals inspired to the Minkowski's metrics can be defined as follows: Let $\mathcal{K} = \langle \mathcal{T}, \mathcal{A} \rangle$ be a knowledge base. Given a context C and a related vector of weights $\boldsymbol{w}$, a family of dissimilarity measures $\{d_p^C\}_{p \in \mathbb{N}}$, is made up of the functions

$$d_p^C : \mathsf{Ind}(\mathcal{A}) \times \mathsf{Ind}(\mathcal{A}) \mapsto [0, 1]$$

defined as follows:

$$\forall (a, b) \in \mathsf{Ind}(\mathcal{A}) \times \mathsf{Ind}(\mathcal{A}) \qquad d_p^C(a, b) = \sqrt[p]{\sum_{C_i \in C} w_i \left[1 - \pi_i(a) \pi_i(b) \right]^p}$$

Note that there is a sort of assumption of independence here that is somewhat similar to the Naïve Bayesian approach [17]. The effect of the weights is to normalize w.r.t. the other features involved. Obviously these measures are not absolute and they should be also be considered w.r.t. the context of choice, hence comparisons across different contexts may not be meaningful. Larger contexts are likely to decrease the measures because of the normalizing factor yet these values are affected also by the degree of redundancy of the features employed. In other works the choice of the weights is done according to variance or entropy associated to the various concepts in the context [6,13].

Compared to other proposed measures [4,5,7], the presented functions do not depend on the constructors of a specific language, rather they require only instance-checking inferences for computing the projections through class-membership queries to the knowledge base.

The complexity of measuring the dissimilarity of two individuals depends on the complexity of such inferences (see [1], Ch. 3). Note also that the projections that determine the measure can be computed (or derived from statistics maintained on the knowledge base) before the actual distance application, thus determining a speed-up in the computation of the measure. This is very important for algorithms that massively use this distance, such as all instance-based methods.

One should assume that C represents a set of (possibly redundant) features that are able to discriminate individuals that are actually different. The choice of the concepts to be included (a *feature selection* problem [17]) may be crucial. Therefore, specific optimization algorithms founded in *randomized search* have been devised which are able to find optimal choices of discriminating contexts [6,13]. However, the results obtained so far with knowledge bases drawn from ontology libraries showed that (a selection) of the primitive and defined concepts are often sufficient to induce sufficiently discriminating measures.

3   Evidence-Theoretic Prediction

In this section the basics of the theory of evidence and combination rules [20] are recalled. Then a nearest neighbor classification procedure based on the rule of combination [8] is extended in order to perform prediction of unobserved values (related to datatype properties or also class-membership).

3.1   Basics of the Evidence Theory

In the Dempster-Shafer theory, a *frame of discernment* Ω is defined as the set of all hypotheses in a certain domain. Particularly, in a classification problem it is the set of all possible classes. A *basic belief assignment* (BBA) is a function m that defines a mapping $m : 2^\Omega \mapsto [0,1]$ verifying: $\sum_{A \in \Omega} m(A) = 1$. Given a certain piece of evidence, the value of the BBA for a given set A expresses a measure of belief that is committed exactly to A. The quantity $m(A)$ pertains only to A and does not imply any additional claims about any of its subsets. If $m(A) > 0$, then A is called a *focal element* for m.

The BBA m cannot be considered a proper probability measure: it is defined over 2^Ω instead of Ω and it does not require the properties of monotone measures [20]. The BBA m and its associated focal elements define a *body of evidence*, from which a *belief function Bel* and a *plausibility function Pl* can be derived as mappings from 2^Ω to $[0,1]$. For a given $A \subseteq \Omega$, the *belief* in A, denoted $Bel(A)$, represents a measure of the total belief committed to A given the available evidence. *Bel* is defined as follows:

$$\forall A \in 2^\Omega \qquad Bel(A) = \sum_{\emptyset \neq B \subseteq A} m(B) \in [0,1] \qquad (1)$$

Analogously, the plausibility of A, denoted $Pl(A)$, represents the amount of belief that could be placed in A, if further information became available. *Pl* is defined as follows:

$$\forall A \in 2^\Omega \qquad Pl(A) = \sum_{B \cap A \neq \emptyset} m(B) \in [0,1] \qquad (2)$$

It is easy to see that: $Pl(A) = Bel(\Omega) - Bel(\bar{A})$. Moreover $m(\emptyset) = 1 - Bel(\Omega)$ and for each $A \neq \emptyset$: $m(A) = \sum_{B \subseteq A} (-1)^{|A \setminus B|} Bel(B)$. Using these equations, knowing just one function among m, *Bel*, and *Pl* allows to derive the others.

The Dempster-Shafer rule of combination [20] is an operation for pooling evidence from a variety of sources. This rule aggregates independent bodies of evidence defined within the same frame of discernment into one body of evidence. Let m_1 and m_2 be two BBAs. The new BBA obtained by combining m_1 and m_2 using the rule of combination, m_{12} is the orthogonal sum of m_1 and m_2 defined

$$\forall A \in 2^\Omega \qquad m_{12}(A) = (m_1 \oplus m_2)(A) = \sum_{B \cap C = A} m_1(B)\, m_2(C)$$

Generally, the normalized version of the rule is used:
$\forall A \in 2^\Omega \setminus \{\emptyset\}$

$$m_{12}(A) = (m_1 \oplus m_2)(A) = \frac{\sum_{B \cap C = A} m_1(B)\, m_2(C)}{1 - \sum_{B \cap C = \emptyset} m_1(B)\, m_2(C)}$$

(and $m_{12}(\emptyset) = 0$) where the numerator $(1 - c)$ normalizes the values of the combined BBA w.r.t. the amount of conflict c between m_1 and m_2.

Alternative rules of combination used in the experiments will be discussed in the Appendix.

3.2   An Evidential Nearest-Neighbors Procedure Applied to DL

Let us consider the finite set of instances X and a finite set of integers $V \subseteq \mathbb{Z}$ to be used as labels (which may correspond to disjoint classes or distinct attribute values). The available information is assumed to consist in a training set $\mathsf{TrSet} = \{(x_1, v_1), \ldots, (x_M, v_M)\} \subseteq \mathsf{Ind} \times V$ of single-labeled instances (*examples*). In our case, $X = \mathsf{Ind}(\mathcal{A})$, the set of individual names occurring in the ontology.

Let x_q be a new individual to be classified on the basis of its nearest neighbors in TrSet. Let $N_k(x_q) = \{(x_{o(j)}, v_{o(j)}) \mid j = 1, \ldots, k\}$ be the set of the k nearest neighbors of x_q in TrSet sorted by a function $o(\cdot)$ depending on an appropriate metric d which can be applied to ontology individuals (e.g., one of the measures in the family defined in the previous section §2).

Each pair $(x_i, v_i) \in N_k(x_q)$ constitutes a distinct item of evidence regarding the value to be predicted for x_q. If x_q is close to x_i according to d, then one will be inclined to believe that both instances are associated to the same value, while when $d(x_q, x_i)$ increases, this belief decreases and that leads to a situation of almost complete ignorance concerning the value to be predicted for x_q.

Consequently, each $(x_i, v_i) \in N_k(x_q)$ may induce a BBA m_i over V which can be defined as follows [8]:
$\forall A \in 2^V$

$$m_i(A) = \begin{cases} \lambda\sigma(d(x_q, x_i)) & A = \{v_i\} \\ 1 - \lambda\sigma(d(x_q, x_i)) & A = V \\ 0 & \text{otherwise} \end{cases} \tag{3}$$

where $\lambda \in\,]0, 1[$ is a parameter and $\sigma(\cdot)$ is a decreasing function such that $\sigma(0) = 1$ and $\lim_{d \to \infty} \sigma(d) = 0$ (e.g., $\sigma(d) = \exp(-\gamma d^n)$ with $\gamma > 0$ and $n \in \mathbb{N}$). The values of the parameters can be determined heuristically.

Considering each training individual in $N_k(x_q)$ as a separate source of evidence, k BBAs m_j are obtained. These can be pooled by means of the rule of combination leading to the aggregated BBA m that synthesizes the final belief:

$$\bar{m} = \bigoplus_{j=1}^{k} m_j = m_1 \oplus \cdots \oplus m_k \tag{4}$$

In order to predict a value, functions $\overline{Bel}$ and $\overline{Pl}$ can be derived from $\bar{m}$ using the equations seen above, and the query individual x_q is assigned the value in V that maximizes the belief or plausibility:

$$v^* = \operatorname*{argmax}_{(x_i, v_i) \subseteq N_k(x_q)} \overline{Bel}(\{v_i\})$$

or

$$v^* = \operatorname*{argmax}_{(x_i, v_i) \in N_k(x_q)} \overline{Pl}(\{v_i\})$$

The former choice (select the hypothesis with the greatest degree of belief – the most credible) corresponds to a *skeptical* viewpoint while the latter (select the hypothesis with the lowest degree of doubt – the most plausible) is more *credulous.* The degree belief (or plausibility) of the predicted value provides also a way to compare the answers of an algorithm built on top of such analogical procedure. This is useful for tasks such as ranking, matchmaking, etc..

Finally, analogously to necessity and possibility in *Possibility Theory* (which can be considered a special case[1] of Dempster-Shafer theory) it is possible to combine the two measures *Bel* and *Pl*, defining a measure of *confirmation C*, ranging in $[-1, +1]$, by means of a simple one-to-one transformation [20]:

$$\forall A \subseteq \Omega \qquad C(A) = Bel(A) + Pl(A) - 1 \tag{5}$$

Hence, denoted with $\overline{C}$ the combination of $\overline{Bel}$ and $\overline{Pl}$, the resulting rule for predicting the uncertain value for the test individual can be written as follows:

$$v^* = \operatorname*{argmax}_{(x_i, v_i) \in N_k(x_q)} \overline{C}(\{v_i\}) \tag{6}$$

Summing up, the procedure is as reported as Algorithm 1:

Algorithm 1. The Evidential Nearest-Neighbor procedure

$ENN_k(x_q, \mathsf{TrSet}, V) \to (v^*, c^*)$

Input:
    x_q: query individual, TrSet: training set, V value set

Output:
    $v^* \in V$: predicted value, c: corresponding confirmation value

1: Compute the neighbor set $N_k(x_q) \subseteq \mathsf{TrSet}$.
2: **for all** $i \leftarrow 1$ **to** k **do**
3:    Compute m_i (using Eq. 3)
4: **end for**
5: **for all** $v \in V$ **do**
6:    Compute $\bar{m}(\{v\})$ (using Eq. 4)
7:    Compute $\overline{Bel}(\{v\})$ and $\overline{Pl}(\{v\})$ (using Eqs. 1–2) based on $\bar{m}(\{v\})$
8:    Compute the confirmation $\overline{C}(\{v\})$ (using Eq. 5) based on $\overline{Bel}$ and $\overline{Pl}$
9: **end for**
10: **return** (v^*, c^*),
    where v^* is the label that maximizes $\overline{C}$ (Eq. 6) and c^* be this maximal value.

It is worthwhile to note that the complexity of the method is polynomial in the number of instances in the TrSet. If this set is compact and contains very prototypical individuals with plenty of related assertions, then the resulting predictions are likely to be accurate. Another source of complexity in the computations may be the number of values in V which may yield a large number of subsets $2^{|V|}$ for which BBAs are to be computed. However this depends also on the kind of problem that is to be solved (e.g., in class membership detection $|V| = 2$). Moreover what really matters in the number of focal sets for each BBA which may be much less than $2^{|V|}$.

---

[1] Precisely, the body of evidence must contain *consonant* focal sets, i.e. when the set of focal elements is a nested family [20].

4 Predicting Assertions

The utility of the presented procedure when applied to ABox reasoning can be manifold. In the following we propose its employment in the inductive prediction of unknown values related to class-membership and datatype / object property fillers. This feature may be easily embedded in an ontology management system in order to help the knowledge engineers elicit assertions which may be not be derived from the knowledge base, rather they can be made in analogy with the others [5].

In the following, the symbol $\approx\!\!\!\mid$ in expressions like $\mathcal{K} \approx\!\!\!\mid \alpha$ will denote the derivation of the assertion α from the knowledge base $\mathcal{K}$ obtained through an approximate procedure (like the evidence nearest neighbor presented in the previous section).

4.1 Class-Membership

Let us suppose that a target (query) concept Q is given. In this case one may consider only examples made up of individuals with a definite class-membership leading to a binary problem with a set of values $V_Q = \{+1, -1\}$ denoting, resp., membership and non-membership w.r.t. the query concept. Alternatively, one may admit ternary problems with a further label 0 to explicitly denote an indefinite (uncertain) class-membership [5,6]. We shall also consider the related training set $\mathsf{TrSet}_Q \subseteq \mathsf{Ind}(\mathcal{A}) \times V_Q$. The values of the labels v_i for the training examples can be obtained through deductive reasoning (instance-checking) or specific facilities made available by instance stores and similar knowledge management systems [18,3].

In order to predict the class-membership value v^* for some individual x_q w.r.t. Q, it suffices to call the procedure $ENN_k(x_q, \mathsf{TrSet}_Q, V_Q)$ and decide on the grounds of the returned value. Thus in a binary setting ($V_Q = \{+1, -1\}$), one will either conclude that $\mathcal{K} \approx\!\!\!\mid Q(x_q)$ or $\mathcal{K} \approx\!\!\!\mid \neg Q(x_q)$ depending on the value that maximizes $\overline{C}$ in Eq. 6 (resp., $v^* = +1$ or $v^* = -1$). Moreover the value c^* of the confirmation function which determined the returned value v^* can be exploited for ranking the hits by comparing the strength of the inductive conclusions.

Adopting a ternary setting, it may turn out that the most likely value is $v^* = 0$ resulting in an uncertain case. One may force the choice among the values of $\overline{C}$ for $v^* = -1$ and $v^* = +1$ e.g., when the confirmation degree exceeds some given threshold.

The inductive procedure described above can be trivially exploited for performing the retrieval of a certain concept inductively. Given a certain concept Q, it would suffice to find all individuals $a \in \mathsf{Ind}(\mathcal{A})$ that are such that $\mathcal{K} \approx\!\!\!\mid Q(a)$. The hits could be returned ranked by the respective confirmation value $\overline{C}(\{+1\})$.

4.2 Datatype Fillers

Extending the setting to the case datatype properties, we suppose that a certain (functional) datatype property P is given and the problem is to predict its

value for a certain test individual a (belonging to the domain of the property).
The set of values V_P may correspond to the (discrete and finite) range of the
property or to its restriction to the observed values for the training instances:
$V_P = \{v \in range(P) \mid \exists P(a, v) \in \mathcal{A}\}$. Different settings may be devised to
consider further special values denoting the case of a yet unobserved values for
the property.

The related training set will be some $\mathsf{TrSet}_P \subseteq domain(P) \times V_P$, where
$domain(P) \subseteq \mathsf{Ind}(\mathcal{A})$ is the set of individual names that have a known value
for P in the knowledge base. Differently from the previous problem, datatype
properties generally do not have a specific intensional definition in the knowledge
base (except for the specification of domain and range), hence a mere look-up
in the ABox should suffice to determine the TrSet.

Now to predict the value $v^* \in V_P$ of the datatype property P for some individual a, the method requires calling the procedure with $ENN_k(a, \mathsf{TrSet}_P, V_P)$.
Thus in this setting we can write $\mathcal{K} \models P(a, v^*)$. Also in this case the value of the
confirmation function which determined choice of the value v^* can be exploited
for comparing the strength of an inductive conclusion to others.

In case of special settings with dummy values indicating unobserved values,
when these are found to be the most credible among the others, a knowledge
engineer should be contacted for the necessary changes to the ontology.

Extensions. The inductive procedure described above can be trivially exploited
for performing alternate forms of retrieval e.g., finding all individuals with a
certain value for the given property. Given a certain value v, it would suffice to
find all individuals $a \in \mathsf{Ind}(\mathcal{A})$ that are such that $\mathcal{K} \models P(a, v)$. Again, the hits
could be returned ranked according to the respective confirmation value $\overline{C}(\{v\})$.

For datatypes ranging on continuous numerical sets, such as intervals, or the
whole $\mathbb{R}$, it is quite straightforward to adopt solutions for typical *regression*
problems. Instead of a weighted majority vote, the k-NN procedure can be modified to produce an average value mediated by their similarity:

The limitation of treating only functional datatype properties may be overcome by considering a different way to assign the probability mass to BBAs than
Eq. 3, including subsets of all possible values. Examples are to be constructed
accordingly (labels will be chosen in 2^{V_P}). Alternatively, more complex frames
of discernment e.g., $\Omega' = 2^{\Omega}$, so consider sets of values as possible fillers of
the property. In all such settings the computation of the BBAs and descending
measures would become of course much more complex and expensive, yet clever
solutions (or approximations) proposed in the literature [8] may contribute to
mitigate this problem.

4.3   Relationships among Individuals

In principle, a very similar setting may be used in order to establish the possibility that a certain test individual is related through some object property with
some other individual.

Since the set $\mathsf{Ind}(\mathcal{A})$ is finite (the target is not discovering relations with unseen individuals), one may want to find all individuals that are related to a test one through some object property, say R. The problem can be decomposed into smaller ones aiming at verifying whether $\mathcal{K} \mathrel{\vertiii{\approx}} R(a,b)$ holds, as shown in Algorithm 2.

Algorithm 2. Finding relationships among individuals

REL_PREDICTION($\mathsf{Ind}(\mathcal{A})$)

Input: $\mathsf{Ind}(\mathcal{A})$ set of individuals
Output: $\mathrel{\vertiii{\approx}}$ for the individuals in $\mathsf{Ind}(\mathcal{A})$
 1: **for all** $b \in \mathsf{Ind}(\mathcal{A})$ **do**
 2:     **for all** $a \in \mathsf{Ind}(\mathcal{A})$ **do**
 3:         $\mathsf{TrSet} \leftarrow \{(x,v) \mid x \in \mathsf{Ind}(\mathcal{A}) \setminus \{a\}, \text{ if } \mathcal{K} \models R(x,b) \text{ then } v \leftarrow +1 \text{ else } v \leftarrow -1\}$
 4:         $v_b^R \leftarrow ENN_k(a, \mathsf{TrSet}, \{+1, -1\})$
 5:         **if** $v_b^R = +1$ **then**
 6:             Predict $\mathcal{K} \mathrel{\vertiii{\approx}} R(a,b)$
 7:         **else**
 8:             Predict $\mathcal{K} \mathrel{\vertiii{\not\approx}} R(a,b)$
 9:         **end if**
10:     **end for**
11: **end for**

Note that, in the construction of the training sets, the inference $\mathcal{K} \models R(x,b)$ may turn out to be merely an ABox lookup operation for the given assertions (when roles are not intensionally defined in a proper RBox). Conversely, if an RBox is available (sometimes as a subset of the TBox) the values of the label for the training examples can be obtained through deductive reasoning (instance-checking) or the mentioned facilities made available by advanced reasoners or knowledge management systems [18].

This simple setting makes a sort of *closed-world assumption* in the decision of the induced assertions descending from the adoption of the binary value set and the composition of the TrSet. A more cautious setting would involve a ternary value set $V_R = \{-1, 0, +1\}$ which allows for an explicit treatment of those individuals a for which $R(a,b)$ is not derivable (or just absent from the ABox). The final decision on the induced conclusion has to consider also this new possibility (e.g., using a threshold of confirmation for accepting likely assertions).

5   Empirical Evaluation

5.1   Experiment Design

In order to test the algorithms on real ontologies, the resulting system prototype was applied to the three kinds of prediction problems involving the individuals therein.

Table 1. Facts concerning the ontologies employed in the experiments

Ontology	DL Language	#Concepts	#Object Properties	#Datatype Properties	#Individuals
FSM	$\mathcal{SOF}(\mathcal{D})$	20	10	7	37
BCO	$\mathcal{ALCROF}(\mathcal{D})$	196	22	3	112
IMDB	$\mathcal{ALIN}(\mathcal{D})$	7	5	13	302
BioPax	$\mathcal{ALCIF}(\mathcal{D})$	74	70	40	323
HDis	$\mathcal{ALCIF}(\mathcal{D})$	1498	10	15	639

To this purpose, a number of OWL ontologies from different domains have been selected[2], namely: FINITESTATEMACHINES (FSM) concerning finite state machines, NEWTESTAMENTNAMES (NTN) accounting for characters and places mentioned in that book, the *Internet Movie DataBase* ontology (IMDB), the BioPax glycolysis ontology (BioPax) describing the glycolysis pathway from the EcoCyc database, translated into BioPax format, and two medical knowledge bases, the *Breast Clinical Ontology* (BCO) and *Human Diseases* HDis. Tab. 1 summarizes important details concerning these ontologies, in terms of the numbers of concepts, object and datatype properties and individuals.

The experiments have been replicated adopting four different rules of evidence of combination, namely: Dempster-Shafer's, Dubois-Prade's, Yager's and the Mixing rule [20]. A 10-fold cross validation design was adopted to determine the average performance indices (described below). Given the training set TS in each experiment run, the *size of the neighborhood* was set to $k = \log |TS|$. As regards the settings for the other parameters $\lambda = .95$, and the other are determined heuristically (see [8]): n is a small integer (typically $n = 2$) and has little impact on the function applied to the distance among examples, while we set $\gamma = 1/d^*$, where d^* is the average distance among training instances belonging to the target class. All atomic concepts defined in each ontology have been considered to be included in the set of features for the contexts C.

The experiments concerning the three specific tasks required suitable ontologies to create suitable training and test sets for the task. As in previous experiments on the same task (e.g. see [6,12]) random classification problems have been generated by constructing concept descriptions using the concepts and relations offered by each ontology and the operators of the $\mathcal{ALC}$ languages. The role-filling tasks have been tested by randomly selecting 5 properties from the ontologies. Finally, only ontologies containing functional properties were usable for the problem of predicting the values of datatype properties for given individuals. Besides only ranges with small cardinalities are currently tractable by the system prototype.

---

[2] The ontologies can be found in standard repositories: the Protégé library (http://protege.stanford.edu/plugins/owl/owl-library) and TONES (http://owl.cs.manchester.ac.uk/repository).

Table 2. Outcomes of the experiments of class-membership prediction with the four combination rules: average values of the indices and standard deviations

Ontology		Dempster	Dubois-Prade	Mixing	Yager
FSM	M%	86.60 ± 04.42	84.75 ± 04.49	85.80 ± 03.90	89.00 ± 04.65
	C%	04.69 ± 03.05	06.65 ± 03.06	05.49 ± 02.33	02.29 ± 02.76
	O%	00.00 ± 00.00	00.00 ± 00.00	00.00 ± 00.00	00.00 ± 00.00
	I%	08.71 ± 00.29	08.71 ± 00.29	08.71 ± 00.29	08.71 ± 00.29
BIOPAX	M%	94.93 ± 00.32	94.76 ± 00.32	94.93 ± 00.32	94.93 ± 00.32
	C%	00.15 ± 00.00	00.32 ± 00.00	00.15 ± 00.00	00.15 ± 00.00
	O%	00.00 ± 00.00	00.00 ± 00.00	00.00 ± 00.00	00.00 ± 00.00
	I%	04.91 ± 00.29	04.91 ± 00.29	04.91 ± 00.29	04.91 ± 00.29
BCO	M%	85.21 ± 04.04	84.54 ± 04.83	85.21 ± 04.04	85.45 ± 04.18
	C%	00.81 ± 00.56	01.47 ± 01.54	00.81 ± 00.56	00.57 ± 00.70
	O%	00.05 ± 00.14	00.14 ± 00.23	00.05 ± 00.14	00.05 ± 00.14
	I%	13.93 ± 03.72	13.95 ± 03.64	13.93 ± 03.72	13.93 ± 03.72

5.2 Outcomes

Due to the open-world semantics, a situation may occur where the membership of an individual w.r.t. a query cannot be determined by a reasoner, since it can build models for the membership w.r.t. both the concept $C(a)$ and its negation $\neg C(a)$. Then a three-way classification was adopted and evaluated using the following indices already adopted in previous works [6,11,12]. Essentially they measure the correspondence between the deductive and inductive classification for the instances w.r.t. the query concept provided, resp., by the reasoner and the inductive algorithm:

- *match rate* (M%), i.e. number of cases of individuals that got the same classification with both modes;
- *omission error rate* (O%), amount of individuals for which the membership w.r.t. the given query could not be determined using the inductive method, while they can be proven to belong to the query concept or to its complement;
- *commission error rate* (C%), amount of individuals found to belong to the query concept according to the inductive classification, while they can be proven to belong to its complement and vice-versa;
- *induction rate* (I%), amount of individuals found to belong to the query concept or its complement according to the inductive classifier, while either case is not logically derivable from the knowledge base.

For each index, average value and standard deviation over the various folds is reported in the following tables.

Table 2 presents the outcomes of the experiments where the method aimed at determining the class membership of the test individuals. The table shows a good performance of the inductive classification in terms of the average match rates. It also shows that omission errors occurred quite rarely, while commission rates are slightly higher (especially with the smallest ontology). Finally the average

Table 3. Outcomes of the experiments of datatype property filler prediction with the four combination rules: average values of the indices and standard deviations

Ontology		Dempster	Dubois-Prade	Mixing	Yager
BCO	M%	64.15 ± 13.53	33.79 ± 11.64	63.52 ± 15.08	71.14 ± 10.00
	C%	35.85 ± 13.53	13.61 ± 10.52	36.48 ± 15.08	28.86 ± 10.00
	O%	00.00 ± 00.00	52.60 ± 15.95	00.00 ± 00.00	00.00 ± 00.00
	I%	00.00 ± 00.00	00.00 ± 00.00	00.00 ± 00.00	00.00 ± 00.00
IMDB	M%	65.60 ± 06.38	39.73 ± 14.19	66.25 ± 05.94	61.34 ± 08.28
	C%	30.74 ± 06.57	13.62 ± 10.52	30.09 ± 06.13	35.00 ± 09.78
	O%	03.66 ± 03.74	43.01 ± 19.99	03.66 ± 03.74	03.66 ± 03.74
	I%	00.00 ± 00.00	00.00 ± 00.00	00.00 ± 00.00	00.00 ± 00.00
HDIS	M%	61.00 ± 19.15	61.00 ± 19.15	61.00 ± 19.15	61.00 ± 19.15
	C%	35.62 ± 17.32	35.62 ± 17.32	35.62 ± 17.32	35.62 ± 17.32
	O%	03.38 ± 04.94	03.38 ± 04.94	03.38 ± 04.94	03.38 ± 04.94
	I%	00.00 ± 00.00	00.00 ± 00.00	00.00 ± 00.00	00.00 ± 00.00

induction rates are sensible (see especially the outcomes of the BCO case). This means that the system can actually help determining a class-membership by analogy when the reasoner cannot.

Comparing the outcomes in terms of the various evidence combination rules, no significant differences were observed not in terms of average values and of their variance (standard deviation). These outcomes are on average in line with those obtained through more sophisticated methods such as those exploiting kernels (e.g. SVMs in [12]).

In the experiments that aimed at predicting object property fillers the ontologies FSM, BIOPAX, and BCO were involved. An optimal performance was observed of the inductive method in terms of the average match rates (from 99.64 to 100% with each methods). Consequently the other cases (omission and commission error, induction) were not so frequent to be statistically significant. Also in terms of the evidence combination rules, no significant difference was observed. The careful consideration of the ontologies shows that they are likely too easy for the task for the primitive roles in the ontology used. Having a chance of combining different roles would probably provide harder learning problems.

Finally, Table 3 shows the results of the experiments where the method aimed at determining the value of a functional datatype property for the test individuals. Preliminarily, note that the induction rate is null because the considered properties had a value for almost all of the considered individuals. Similarly, omission errors occurred quite rarely. The table shows a fair performance of the inductive classification in terms of the average match rates. However the commission rates are considerably high. Another noticeable feature is the higher variance w.r.t. the outcomes of the previous experiments. Comparing the outcomes in terms of the various evidence combination rules, while they coincide for the case of BCO, in the other cases sensible differences where observed in terms of average values and of their variance (standard deviation). Specifically, Yager's rule seemed to perform slightly better, while with Dubois-Prade's rule the performance decreased.

6   Related Work

The proposed method is related to those approaches devised to offer alternative ways of reasoning with ABoxes for eliciting hidden knowledge (regularities) in order to complete and populate the ontology with likely assertions even in the occurrence of incorrect parts, supposing this kind of *noise* is not systematic.

The tasks of ontology completion and population have often been tackled through formal methods (such as *formal concept analysis* [2]). Discovering new assertions (and related probabilities in a classical setting) is another related task for eliciting hidden knowledge in the ontologies. In [28] a machine learning method is proposed to estimate the truth of statements by exploiting regularities in the data. In [24] another statistical learning method for OWL-DL ontologies is proposed, combining a latent relational graphical model with Description Logic inference in a modular fashion. The probability of unknown role-assertions can be inductively inferred and known concept-assertions can be analyzed by clustering individuals.

Similarity-based reasoning with ontologies is the primary aim of this work which follows a number of related methods founded on dissimilarity measures for individuals in knowledge bases expressed in Description Logics [5,6]. Mostly, they adopt some alternate form of the classic Nearest-Neighbor learning scheme [17] in order to draw inductive conclusions that often cannot be deductively entailed by the knowledge bases.

Similar approaches based on lazy learning have been proposed that adopt generalized probability theories such as the Dempster-Shafer. In [30], which was a source of inspiration for this paper, the standard rule of combination is exploited in an evidence-theoretic classification procedure where labels were not assumed to be mutually exclusive. Rules of combination had been used in [11] in order to learn precise metrics to be exploited in a lazy learning setting like those mentioned above.

One of the most appreciated advantages of performing inductive ABox reasoning through these methods is that they can naturally handle inconsistent (and inherently incomplete) knowledge bases, especially when inconsistency is not systematic. In [22] a method for dealing with inconsistent ABoxes populated through information extraction is proposed: it constructs ad hoc belief networks for the conflicting parts in an ontology and adopts the Dempster-Shafer theory for assessing the confidence of the resulting assertions.

7   Conclusions and Extensions

In line with our investigation of inductive methods for Semantic Web reasoning, we have proposed an alternative way for approximate ABox reasoning based on the nearest-neighbors analogical principle. Once neighbors of a test individual are selected through some distance measures, a combination rule descending from the Dempster-Shafer theory can fuse the evidence provided by the various neighbor individuals. We have shown how to exploit the procedure for assertion prediction problems such as determining unknown class- or role-memberships as well as attribute-values which may be the basis for many ABox inductive reasoning algorithms. The method has been implemented so to allow an empirical evaluation on real ontologies.

Special settings to accommodate cases of uncertain or unobserved values are to be investigated. One promising extension of the method concerns the possibility of considering infinite sets of values V following the studies [16,20]. This would allow dealing with domains where the total amount of values is unknown (also due to the inherent nature of the Semantic Web). Moreover the predicted values often need not be exclusive. Hence the prediction procedure would require an extension towards the consideration of sets of values instead of singletons.

As necessity and possibility measures are related to the belief measures (see note 1), a natural extension may be towards the possibilistic theory and its calculus which is, in general, different from the Dempster-Shafer theory and calculus. Further possible extensions concern all other monotone measures such as the Sugeno λ-measures [20]. The extension towards the Possibility Theory is also interesting because of its parallelism with modal logics [16] and possibilistic extensions of Description Logics [23].

References

1. Baader, F., Calvanese, D., McGuinness, D., Nardi, D., Patel-Schneider, P. (eds.): The Description Logic Handbook. Cambridge University Press (2003)
2. Baader, F., Ganter, B., Sertkaya, B., Sattler, U.: Completing description logic knowledge bases using formal concept analysis. In: Veloso, M. (ed.) Proceedings of the 20th International Joint Conference on Artificial Intelligence, IJCAI 2007, Hyderabad, India, pp. 230–235 (2007)
3. Bishop, B., Kiryakov, A., Ognyanoff, D., Peikov, I., Tashev, Z., Velkov, R.: OWLIM: A family of scalable semantic repositories. Semantic Web 2(1), 33–42 (2011)
4. Borgida, A., Walsh, T., Hirsh, H.: Towards measuring similarity in description logics. In: Horrocks, I., Sattler, U., Wolter, F. (eds.) Working Notes of the International Description Logics Workshop, Edinburgh, UK. CEUR Workshop Proceedings, vol. 147 (2005)
5. d'Amato, C., Fanizzi, N., Esposito, F.: Analogical Reasoning in Description Logics. In: da Costa, P.C.G., d'Amato, C., Fanizzi, N., Laskey, K.B., Laskey, K.J., Lukasiewicz, T., Nickles, M., Pool, M. (eds.) URSW 2005-2007. LNCS (LNAI), vol. 5327, pp. 330–347. Springer, Heidelberg (2008)
6. d'Amato, C., Fanizzi, N., Esposito, F.: Query Answering and Ontology Population: An Inductive Approach. In: Bechhofer, S., Hauswirth, M., Hoffmann, J., Koubarakis, M. (eds.) ESWC 2008. LNCS, vol. 5021, pp. 288–302. Springer, Heidelberg (2008)

7. d'Amato, C., Staab, S., Fanizzi, N.: On the Influence of Description Logics Ontologies on Conceptual Similarity. In: Gangemi, A., Euzenat, J. (eds.) EKAW 2008. LNCS (LNAI), vol. 5268, pp. 48–63. Springer, Heidelberg (2008)
8. Denoeux, T.: A k-nearest neighbor classification rule based on Dempster-Shafer theory. IEEE Transactions on Systems, Man and Cybernetics 25(5), 804–813 (1995)
9. Dubois, D., Prade, H.: On the combination of evidence in various mathematical frameworks. In: Flamm, J., Luisi, T. (eds.) Reliability Data Collection and Analysis, pp. 213–241. Springer (1992)
10. Euzenat, J., Shvaiko, P.: Ontology matching. Springer (2007)
11. Fanizzi, N., d'Amato, C., Esposito, F.: Approximate Measures of Semantic Dissimilarity under Uncertainty. In: da Costa, P.C.G., d'Amato, C., Fanizzi, N., Laskey, K.B., Laskey, K.J., Lukasiewicz, T., Nickles, M., Pool, M. (eds.) URSW 2005-2007. LNCS (LNAI), vol. 5327, pp. 348–365. Springer, Heidelberg (2008)
12. Fanizzi, N., d'Amato, C., Esposito, F.: Statistical Learning for Inductive Query Answering on OWL Ontologies. In: Sheth, A., Staab, S., Dean, M., Paolucci, M., Maynard, D., Finin, T., Thirunarayan, K. (eds.) ISWC 2008. LNCS, vol. 5318, pp. 195–212. Springer, Heidelberg (2008)
13. Fanizzi, N., d'Amato, C., Esposito, F.: Metric-based stochastic conceptual clustering for ontologies. Information Systems 34(8), 725–739 (2009)
14. Fanizzi, N., d'Amato, C., Esposito, F.: Evidential nearest-neighbors classification for inductive abox reasoning. In: Bobillo, F., et al. (eds.) Proceedings of the 5th International Workshop on Uncertainty Reasoning for the Semantic Web, URSW 2009. CEUR Workshop Proceedings, vol. 527, pp. 27–38. CEUR-WS.org (2009)
15. Goldstone, R., Medin, D., Halberstadt, J.: Similarity in context. Memory and Cognition 25(2), 237–255 (1997)
16. Harmanec, D., Klir, G., Wang, Z.: Modal logic interpretation of Dempster-Shafer theory: An infinite case. International Journal of Approximate Reasoning 14(2-3), 81–93 (1996)
17. Hastie, T., Tibshirani, R., Friedman, J.: The Elements of Statistical Learning – Data Mining, Inference, and Prediction, 2nd edn. Springer (2009)
18. Horrocks, I., Li, L., Turi, D., Bechhofer, S.: The Instance Store: DL reasoning with large numbers of individuals. In: Haarslev, V., Möller, R. (eds.) Proceedings of the 2004 Description Logic Workshop, DL 2004. CEUR Workshop Proceedings, vol. 104, pp. 31–40. CEUR (2004)
19. Janowicz, K., Wilkes, M.: SIM-DL$_A$: A Novel Semantic Similarity Measure for Description Logics Reducing Inter-concept to Inter-instance Similarity. In: Aroyo, L., Traverso, P., Ciravegna, F., Cimiano, P., Heath, T., Hyvönen, E., Mizoguchi, R., Oren, E., Sabou, M., Simperl, E. (eds.) ESWC 2009. LNCS, vol. 5554, pp. 353–367. Springer, Heidelberg (2009)
20. Klir, G.: Uncertainty and Information. Wiley (2006)
21. Laskey, K., Laskey, K., Costa, P., Kokar, M., Martin, T., Lukasiewicz, T.: Uncertainty Reasoning for the World Wide Web. W3C Incubator Group (2008), http://www.w3.org/2005/Incubator/urw3/XGR-urw3-20080331/
22. Nikolov, A., Uren, V., Motta, E., de Roeck, A.: Using the Dempster-Shafer Theory of Evidence to Resolve ABox Inconsistencies. In: da Costa, P.C.G., d'Amato, C., Fanizzi, N., Laskey, K.B., Laskey, K.J., Lukasiewicz, T., Nickles, M., Pool, M. (eds.) URSW 2005-2007. LNCS (LNAI), vol. 5327, pp. 143–160. Springer, Heidelberg (2008)

23. Qi, G., Pan, J., Ji, Q.: A possibilistic extension of description logics. In: Calvanese, D., et al. (eds.) Working Notes of the 20th International Description Logics Workshop, DL 2007, Bressanone, Italy. CEUR Workshop Proceedings, vol. 250, pp. 435–442 (2007)
24. Rettinger, A., Nickles, M.: Infinite hidden semantic models for learning with OWL DL. In: d'Amato, C., et al. (eds.) Proceedings of 1st ESWC Workshop on Inductive Reasoning and Machine Learning for the Semantic Web, IRMLeS 2009, Heraklion, Greece. CEUR Workshop Proceedings, vol. 474 (2009)
25. Rizzo, G., Fanizzi, N., d'Amato, C., Esposito, F.: Prediction of class and property assertions on OWL ontologies through evidence combination. In: Akerkar, R. (ed.) Proceedings of the International Conference on Web Intelligence, Mining and Semantics, WIMS 2011, p. 45. ACM (2011)
26. Sebag, M.: Distance Induction in First Order Logic. In: Džeroski, S., Lavrač, N. (eds.) ILP 1997. LNCS (LNAI), vol. 1297, pp. 264–272. Springer, Heidelberg (1997)
27. Sentz, K., Ferson, S.: Combination of evidence in Dempster-Shafer theory. Technical Report SAND2002-0835, SANDIATech (April 2002)
28. Tresp, V., Huang, Y., Bundschus, M., Rettinger, A.: Materializing and querying learned knowledge. In: d'Amato, C., et al. (eds.) Proceedings of 1st ESWC Workshop on Inductive Reasoning and Machine Learning for the Semantic Web, IRMLeS 2009, Heraklion, Greece. CEUR Workshop Proceedings, vol. 474 (2009)
29. Yager, R.: On the Dempster-Shafer framework and new combination rules. Information Sciences 41, 93–137 (1987)
30. Younes, Z., Abdallah, F., Denœux, T.: An Evidence-Theoretic k-Nearest Neighbor Rule for Multi-label Classification. In: Godo, L., Pugliese, A. (eds.) SUM 2009. LNCS, vol. 5785, pp. 297–308. Springer, Heidelberg (2009)

A   Further Evidence Combination Rules

Combination rules are the special types of aggregation methods for information obtained from multiple sources. These sources provide different assessments for the same frame of discernment and Dempster-Shafer theory is based on the assumption that these sources are independent. These rules can potentially occupy a continuum between conjunction (based on set intersection) and disjunction (based on set union). In the situation where all sources are considered reliable, a conjunctive operation is appropriate. Conversely, when only one or some reliable sources are given, a disjunctive combination would be justified [27].

Many combination operations lie between these two extremes [9]. The original combination rule of multiple BBAs known as the Dempster's rule is a generalization of Bayes rule. This rule emphasizes the agreement between multiple sources and ignores all the conflicting evidence through a normalization factor. This rule has come under serious criticism when significant conflict is encountered. Consequently, other rules have been proposed that attempt to represent the degree of conflict In the following, we briefly survey the combination rules considered for the experiments.

A.1   Yager's Rule

Yager [29] pointed out that an important feature of combination rules is the ability to update an already combined structure when new information becomes available. However in other cases a non-associative operator may be necessary, for example the arithmetic average. In these cases the *quasi-associativity* property may be pursued, i.e. the operator can be broken down into associative sub-operations.

Coming back to the problem of placing the conflicting evidence, an epistemologically sound combination rule places the probability mass related to the conflict between the BBAs to the case of maximal ignorance:

$\forall A \in 2^{\Omega}$

$$m_{12}(A) = \begin{cases} \sum_{B \cap C = A} m_1(B)\, m_2(C) & A \neq \Omega \wedge A \neq \emptyset \\ m_1(\Omega)\, m_2(\Omega) + c & A = \Omega \\ 0 & A = \emptyset \end{cases}$$

This means that the conflict between the two sources of evidence is not hidden, but it is explicitly recognized as a contributor to ignorance.

Due to the associativity and commutativity of the operations involved, it is easy to prove that the resulting combination operator is associative and commutative, and admits the vacuous BBA (Ω unique focal set) as neutral element.

A.2   Dubois and Prade's Disjunctive Pooling Rule

This rule takes into account the union of the probability masses (disjunctive rule) this allows avoiding conflict generation as there no rejection of information coming from the various sources.

The combination rule can be defined as follows:

$\forall A \in 2^{\Omega}$

$$m_{12}(A) = \sum_{B \cup C = A} m_1(B)\, m_2(C)$$

The union does not generate any conflict and does not reject any information asserted by the sources. As such, no normalization procedure is required. The drawback of this rule is that it may yield a more imprecise result than desirable. It is easy to see that this rule is commutative and associative.

A.3   Mixing Rule

This rule (also known as *averaging*) represents an extension of the average for probability distributions computed on the BBAs and describes the frequency of the various values withing a range of possible values. Formally it is merely a weighted average of the masses according to the various features:

$\forall A \in 2^{\Omega}$

$$m_{1 \dots n}(A) = \frac{1}{n} \sum_{i=1}^{n} w_i m_i(A)$$

where a normalized weight vector $\boldsymbol{w}$ is generally considered. The values of the weights should reflect a degree of confidence in the sources.

This rule is commutative, idempotent and quasi-associative.

Representing Uncertain Concepts in Rough Description Logics via Contextual Indiscernibility Relations

Claudia d'Amato[1], Nicola Fanizzi[1],
Floriana Esposito[1], and Thomas Lukasiewicz[2]

[1] LACAM, Dipartimento di Informatica, Università degli Studi di Bari, Italy
firstname.lastname@uniba.it
[2] Department of Computer Science, University of Oxford, UK
thomas.lukasiewicz@cs.ox.ac.uk

Abstract. We investigate the modeling of uncertain concepts via *rough description logics (RDLs)*, which are an extension of traditional description logics (DLs) by a mechanism to handle approximate concept definitions via lower and upper approximations of concepts based on a rough-set semantics. This allows to apply RDLs to modeling uncertain knowledge. Since these approximations are ultimately grounded on an indiscernibility relation, we explore possible logical and numerical ways for defining such relations based on the considered knowledge. In particular, we introduce the notion of context, allowing for the definition of specific equivalence relations, which are directly used for lower and upper approximations of concepts. The notion of context also allows for defining similarity measures, which are used for introducing a notion of tolerance in the indiscernibility. Finally, we describe several learning problems in our RDL framework.

1 Introduction

Uncertainty is an intrinsic characteristic of the current Web, which, being a heterogeneous and distributed source of information, naturally contains uncertain as well as incomplete and/or contradictory information. Managing uncertainty is thus a highly important topic also for the extension of the Web to the Semantic Web (SW).

Particularly, modeling uncertain concepts in description logics (DLs) [1] is generally done via numerical approaches, such as probabilistic and possibilistic ones [17]. A drawback of these approaches is that uncertainty is introduced in the model (e.g., by specifying a set of uncertainty measures, such as probability and possibility measures, respectively), which often has the consequence that the approach becomes conceptually and/or computationally more complex. An alternative (simpler) approach is based on the theory of *rough sets* [22], which gives rise to new representations and *ad hoc* reasoning procedures [4]. These languages are based on the idea of *indiscernibility*.

Among these recent developments, *rough description logics* (RDLs) [23] have introduced a complementary mechanism that allows for modeling uncertain knowledge by means of crisp approximations of concepts. RDLs extend classical DLs with two modal-like operators, the lower and the upper approximation. In the spirit of rough-set theory, two concepts approximate an underspecified (uncertain) concept C as particular

F. Bobillo et al. (Eds.): URSW 2008-2010/UniDL 2010, LNAI 7123, pp. 300–314, 2013.

sub- and superconcepts, describing which elements are definitely and possibly elements of the concept, respectively.

The approximations are based on capturing uncertainty as an indiscernibility relation R among individuals, and then formally defining the upper approximation of a concept C as the set of individuals that are indiscernible from at least one that is known to belong to the concept (where $(\Delta^{\mathcal{I}}, \cdot^{\mathcal{I}})$ is a standard first-order interpretation):

$$(\overline{C})^{\mathcal{I}} := \{a \in \Delta^{\mathcal{I}} \mid \exists b : (a,b) \in R^{\mathcal{I}} \wedge b \in C^{\mathcal{I}}\}.$$

Similarly, one can define the lower approximation as

$$(\underline{C})^{\mathcal{I}} := \{a \in \Delta^{\mathcal{I}} \mid \forall b : (a,b) \in R^{\mathcal{I}} \to b \in C^{\mathcal{I}}\}.$$

Intuitively, the upper approximation of a concept C covers the elements of a domain with the typical properties of C, whereas the lower approximation contains the proto-typical elements of C. This may be described in terms of necessity and possibility.

To avoid introducing uncertainty into the model (as for the approaches previously mentioned), these approximations are to be defined in a crisp way. In [23], a method-ology for representing approximations in a crisp way is introduced and it is also shown that RDLs can be simulated within standard DLs. Specifically, for any DL $\mathcal{DL}$ with uni-versal and existential quantification, and symmetric, transitive and reflexive roles, the rough extension of $\mathcal{DL}$ can be translated into $\mathcal{DL}$, and reasoning in the rough extension of $\mathcal{DL}$ can be performed by reduction to $\mathcal{DL}$, using a standard DL reasoner.

However, as shown in [19, 20], the representation of the upper and the lower ap-proximation of a concept C as crisp concepts may not be straightforward. A knowledge engineer or domain expert may not always be able to give intensional definitions of the approximated concepts, but only examples for such approximated concepts. To cope with these issues, the problem of representing concept approximations as crisp con-cepts can be seen as a learning problem, where one has a given set of examples (and counterexamples) for the lower (resp., upper) approximation of a given concept C, and the goal is to learn a crisp concept definition such that the examples and counterexam-ples are instances of the learned concept and its negation, respectively.

Looking at the semantics of the lower and upper approximations of a concept C (re-ported above), an important role is played by the indiscernibility relation. But to our knowledge, there are no existing works (different from [10] of which this paper is an extension) coping with the problem of defining an indiscernibility relation. Inspired by existing works on semantic metrics [2] and kernels [9], we propose to exploit semantic similarity measures, which can be optimized to maximize their capacity of distinguish-ing really different individuals, as indiscernibility relations. This naturally induces ways for defining an equivalence relation based on indiscernibility criteria.

The rest of this paper is organized as follows. Section 2 provides some preliminaries around DLs and RDLs. In Section 3, we introduce contextual indiscernibility relations Section 4 proposes a family of similarity measures based on such contexts along with a suggestion on their optimization. This also allows for the definition of tolerance degrees of indiscernibility. In Section 5, we introduce and discuss the problem of learning crisp descriptions of rough concepts. Section 6 finally summarizes the results of this paper and outlines further applications of ontology mining methods.

2   Preliminaries

In this section, we first recall the basic notions of description logics (DLs). We then describe the extension of DLs to rough DLs (RDLs).

2.1   Description Logics

We now briefly recall the syntax and the semantics of DLs. For ease of presentation, we consider only the DL $\mathcal{ALC}$; for further background and details on other DLs, we refer the reader to the standard textbook [1].

Basic elements of DLs are atomic concepts and roles. *Atomic concepts* from a set $N_C = \{C, D, \ldots\}$ are interpreted as subsets of a domain of objects (resources), while *atomic roles* from a set $N_R = \{R, S, \ldots\}$ are interpreted as binary relations on such a domain (properties). *Individuals* represent the objects through names chosen from a set $N_I = \{a, b, \ldots\}$. *Complex concepts* are built using atomic concepts and roles by means of specific concept constructors. The meaning of concepts and roles is defined by *interpretations* $\mathcal{I} = (\Delta^{\mathcal{I}}, \cdot^{\mathcal{I}})$, where $\Delta^{\mathcal{I}}$ is a set of objects, called *domain*, and $\cdot^{\mathcal{I}}$ is an *interpretation function*, mapping concepts and roles to subsets of the domain and to binary relations on the domain, respectively.

The *top* concept $\top$ is interpreted as the whole domain $\Delta^{\mathcal{I}}$, while the *bottom* concept $\bot$ corresponds to $\emptyset$. Complex concepts can be built in $\mathcal{ALC}$ using the following constructors. The *conjunction* of two concepts C and D, denoted $C \sqcap D$, is interpreted as $C^{\mathcal{I}} \cap D^{\mathcal{I}}$, while the *disjunction* of C and D, denoted $C \sqcup D$, is interpreted as $C^{\mathcal{I}} \cup D^{\mathcal{I}}$. Finally, there are two restrictions on roles, namely, the *existential restriction* on R relative to C, denoted $\exists R.C$, which is interpreted as the set $\{x \in \Delta^{\mathcal{I}} \mid \exists y \in \Delta^{\mathcal{I}} : (x, y) \in R^{\mathcal{I}} \land y \in C^{\mathcal{I}}\}$, and the *value restriction* on R relative to C, denoted $\forall R.C$, which is interpreted as $\{x \in \Delta^{\mathcal{I}} \mid \forall y \in \Delta^{\mathcal{I}} : (x, y) \in R^{\mathcal{I}} \to y \in C^{\mathcal{I}}\}$.

More expressive DLs allow for further constructors. The DL standing behind the ontology language OWL DL is $\mathcal{SHOIQ}(\mathbf{D})$, which extends $\mathcal{ALC}$ by transitive roles, role hierarchies, nominals, inverse roles, and qualified number restrictions, and which allows to deal with concrete domains $\mathbf{D}$ and their specific semantics.

A *knowledge base* $KB = (\mathcal{T}, \mathcal{A})$ consists of a TBox $\mathcal{T}$ and an ABox $\mathcal{A}$. The TBox $\mathcal{T}$ is a set of *subsumption axioms* $C \sqsubseteq D$ and *definition axioms* $A \equiv D$, where A is usually an atomic concept, and C and D are concepts. They are *satisfied* in an interpretation $\mathcal{I}$, or $\mathcal{I}$ is a *model* of them, denoted $\mathcal{I} \models C \sqsubseteq D$ and $\mathcal{I} \models A \equiv D$, respectively, iff $C^{\mathcal{I}} \subseteq D^{\mathcal{I}}$ and $A^{\mathcal{I}} = D^{\mathcal{I}}$, respectively. The ABox $\mathcal{A}$ contains *concept membership axioms* $C(a)$ and *role membership axioms* $R(a, b)$, where C is a concept, R is a role, and a and b are individuals. They are *satisfied* in $\mathcal{I}$, or $\mathcal{I}$ is a *model* of them, denoted $\mathcal{I} \models C(a)$ and $\mathcal{I} \models R(a, b)$, respectively, iff $a^{\mathcal{I}} \in C^{\mathcal{I}}$ and $(a^{\mathcal{I}}, b^{\mathcal{I}}) \in R^{\mathcal{I}}$, respectively. An interpretation $\mathcal{I}$ *satisfies* a knowledge base KB, or $\mathcal{I}$ is a *model* of KB, denoted $\mathcal{I} \models KB$, iff $\mathcal{I}$ satisfies all the axioms in KB. An axiom F is a *logical consequence* of KB, denoted $KB \models F$, iff every model of KB is also a model of F.

In DLs, one generally does not make the *unique name assumption (UNA)*, i.e., different individuals (which ultimately correspond to URIs in RDF/OWL) may be mapped to the same object (resource), if not explicitly forbidden. Furthermore, one usually adopts the *open-world assumption* (OWA). Thus, an object that cannot be proved to belong to a certain concept is not necessarily a counterexample for that concept. This is only interpreted as a case of insufficient (incomplete) knowledge for that assertion (i.e., models can be constructed for both the membership and non-membership case). This assumption is compatible with the typical scenario related to the Semantic Web, where new resources may continuously be made available (and unavailable) across the Web, and thus one generally cannot assume complete knowledge.

Some important inference problems in the context of DLs include subsumption checking, instance checking, and concept retrieval:

Subsumption Checking: Given a knowledge base KB and two concepts (or two roles, when role hierarchies are allowed) C and D, decide whether $KB \models C \sqsubseteq D$.

Instance Checking: Given a knowledge base KB, a concept C, and an individual a, decide whether $KB \models C(a)$.

Concept Retrieval: Given a knowledge base $KB = (\mathcal{T}, \mathcal{A})$ and a concept C, compute the set of all individuals $a \in \mathsf{Ind}(\mathcal{A})$ (among those in $\mathcal{A}$) such that $KB \models C(a)$.

2.2 Rough Description Logics

DLs are suitable for modeling crisp knowledge, but they cannot easily be used to model approximate information. For example, no explicit mechanism is provided when a definition is not commonly agreed upon, or when exceptions need to be captured. Rough DLs (RDLs) attempt to close this gap in a conceptually simple way.

The basic idea behind RDLs is to approximate an uncertain concept C by giving an upper and a lower bound. The upper approximation of C, denoted $\overline{C}$, is the set of all individuals that possibly belong to C, while the lower approximation of C, denoted $\underline{C}$, is the set of all individuals that definitely belong to C. Traditionally, this is modeled using subsumption axioms; in pure DL modeling, the relation between C and its approximations $\underline{C}$ and $\overline{C}$ is $\underline{C} \sqsubseteq C \sqsubseteq \overline{C}$.

RDLs are not restricted to particular DLs, and can be defined for an arbitrary DL $\mathcal{DL}$. Its RDL language $\mathcal{RDL}$ has the lower and upper approximation as additional unary concept constructors, i.e., if C is a concept in $\mathcal{RDL}$, then also $\overline{C}$ and $\underline{C}$ are concepts in $\mathcal{RDL}$. The notions of *rough TBox* and *ABox*, as well as *rough knowledge base* then canonically extend their classical counterparts.

Example 2.1 (Advertising Campaign). Suppose that we want to use some pieces of data collected from the Web to find a group of people to serve as addressees for the advertising campaign of a new product. Clearly, the collected pieces of data are in general highly incomplete and uncertain. The DL concept *Addressee* may now be approximated from below by all the definite addressees and from above by all the potential addressees. So, we can use a DL to specify the TBox knowledge about *Addressee*, and in the same time specify the ABox knowledge about which people are definite and potential addressees, i.e., belong to the two concepts $\underline{Addressee}$ and $\overline{Addressee}$, respectively. ∎

A *rough interpretation* is a triple $\mathcal{I} = (\Delta^{\mathcal{I}}, \cdot^{\mathcal{I}}, R^{\mathcal{I}})$, where $\Delta^{\mathcal{I}}$ is a domain of objects, $\cdot^{\mathcal{I}}$ is an interpretation function, and $R^{\mathcal{I}}$ is an equivalence (i.e., reflexive, symmetric, and transitive) relation over $\Delta^{\mathcal{I}}$. The function $\cdot^{\mathcal{I}}$ maps RDL concepts to subsets of the domain $\Delta^{\mathcal{I}}$, and atomic roles to binary relations over $\Delta^{\mathcal{I}}$. It interprets the classical DL constructs and atomic concepts as usual, and the new constructs as follows:

- $(\overline{C})^{\mathcal{I}} = \{a \in \Delta^{\mathcal{I}} \mid \exists b \in \Delta^{\mathcal{I}} : (a, b) \in R^{\mathcal{I}} \wedge b \in C^{\mathcal{I}}\},$
- $(\underline{C})^{\mathcal{I}} = \{a \in \Delta^{\mathcal{I}} \mid \forall b \in \Delta^{\mathcal{I}} : (a, b) \in R^{\mathcal{I}} \rightarrow b \in C^{\mathcal{I}}\}.$

Intuitively, the upper approximation of a concept C covers the elements of a domain with the *typical properties* of C, while the lower approximation of C contains the *prototypical elements* of C.

Example 2.2 (Advertising Campaign cont'd). To define the definite and potential addressees for the advertising campaign of a new product, we may exploit a classification of people into equivalence classes. For example, people with an income above 1 million dollars may be definite addressees for the advertising campaign of a new Porsche, while people with an income above 100 000 dollars may be potential addressees, and people with an income below 10 000 dollars may not be addressees. ∎

One of the advantages of this way of modeling uncertain concepts is that reasoning comes for free. Indeed, reasoning with approximations can be reduced to standard DL reasoning, by translating RDL concepts into classical DL concepts with a special reflexive and symmetric role.

A translation function for RDL concepts $\cdot^{t} : \mathcal{RDL} \mapsto \mathcal{DL}$ is defined as follows (introducing the new atomic role R for the indiscernibility relation): For every $\mathcal{RDL}$ concept C, the $\mathcal{DL}$ concept C^{t} is obtained from C by recursively (over the structure of C) replacing every $\overline{D}$ and $\underline{D}$ in C by $\exists R.D$ and $\forall R.D$, respectively, and using the identical mapping for all other constructs and atomic concepts. The translation function is naturally extended to axioms and knowledge bases (see [23]).

For any DL $\mathcal{DL}$ with universal and existential quantification, and reflexive, symmetric, and transitive roles, there is no increase in expressiveness, i.e., RDLs can be simulated in (almost) standard DLs: an $\mathcal{RDL}$ concept C is satisfiable in a rough interpretation relative to $\mathcal{T}$ iff the $\mathcal{DL}$ concept C^{t} is satisfiable relative to $\mathcal{T}^{t}$ [23]. In the presence of negation, other inference problems (such as subsumption checking) can be reduced to checking concept satisfiability (and finally to checking ABox satisfiability). Since the translation is linear, the complexity of reasoning in an RDL is the same as the one of reasoning in its DL counterpart with quantifiers as well as reflexive, symmetric, and transitive roles.

Since RDLs do not specify the nature of the indiscernibility relation, except prescribing its encoding as a (special) new equivalence relation, we introduce possible ways for defining it. The first one (see Section 3) makes it depend on a specific set of concepts determining the indiscernibility of the individuals relative to a specific context described by the concepts in the knowledge base. Then (see Section 4), we also define the indiscernibility relation in terms of a similarity measure (based on a context of features), which allows for relaxing the discernibility using a tolerance threshold. In case an indiscernibility relation cannot be specified (e.g., due to lack of knowledge), crisp descriptions of the concept approximations may be learned (see Section 5).

3 Contextual Indiscernibility Relations

In this section, we first define the notion of a context via a collection of DL concepts. We then introduce indiscernibility relations based on such contexts. We finally define upper and lower approximations of DL concepts using these notions, and we provide some theoretical results about them.

It is well known that classification by analogy cannot be really general-purpose, since the number of features on which the analogy is made may be very large [21]. The key point is that indiscernibility is not absolute, but, rather, an induced notion, which depends on the specific contexts of interest. Instead of modeling indiscernibility through a single relation in the interpretation, one may consider diverse contexts, each giving rise to a different relation, which determines also different ways of approximating uncertain concepts. We first recall the notion of projection function [6].

Definition 3.1 (projection). Let $\mathcal{I} = (\Delta^{\mathcal{I}}, \cdot^{\mathcal{I}})$ be an interpretation, and let F be a DL concept. Then, the projection function $\pi_F^{\mathcal{I}} : \Delta^{\mathcal{I}} \mapsto \{0, 1\}$ is defined as follows:

$$\forall a \in \Delta^{\mathcal{I}} : \quad \pi_F^{\mathcal{I}}(a) = \begin{cases} 1 & a \in F^{\mathcal{I}}; \\ 0 & \text{otherwise.} \end{cases}$$

We define a *context* as a finite set of relevant features in the form of DL concepts, which may encode context information for the similarity to be measured [12].

Definition 3.2 (context). A *context* is a set of DL concepts $\mathsf{C} = \{F_1, \ldots, F_m\}$.

Example 3.1 (Advertising Campaign cont'd). One possible context C for the advertising campaign of a new product is given as follows:

$$\mathsf{C} = \{SalaryAboveMillion, HouseOwner, Manager\},$$

where *SalaryAboveMillion*, *HouseOwner*, and *Manager* are DL concepts.   ∎

Two individuals a and b are *indiscernible* relative to the context $\mathsf{C} = \{F_1, \ldots, F_m\}$ iff $\pi_{F_i}(a) = \pi_{F_i}(b)$ for all $i \in \{1, \ldots, m\}$. This induces an equivalence relation. Note that one may define multiple such relations by considering different contexts.

Definition 3.3 (indiscernibility relation). Let $\mathcal{I} = (\Delta^{\mathcal{I}}, \cdot^{\mathcal{I}})$ be an interpretation, and let $\mathsf{C} = \{F_1, \ldots, F_m\}$ be a context. Then, the indiscernibility relation $R_{\mathsf{C}}^{\mathcal{I}}$ induced by C under $\mathcal{I}$ is defined as follows:

$$R_{\mathsf{C}}^{\mathcal{I}} = \{(a, b) \in \Delta^{\mathcal{I}} \times \Delta^{\mathcal{I}}) \mid \forall i \in \{1, \ldots, m\} : \pi_{F_i}^{\mathcal{I}}(a) = \pi_{F_i}^{\mathcal{I}}(b)\}.$$

Any indiscernibility relation splits $\Delta^{\mathcal{I}}$ in a partition of equivalence classes (also known as *elementary sets*) denoted $[a]_{\mathsf{C}}$, for a generic individual a. Each class naturally induces a concept, denoted C_a.

Example 3.2 (Advertising Campaign cont'd). Consider again the context C of Example 3.1. Observe that C defines an indiscernibility relation on the set of all people, which is given by the extensions of all atomic concepts constructed from C as its equivalence classes. For example, one such atomic concept is the conjunction of *SalaryAboveMillion*, *HouseOwner*, and *Manager*; another one is the conjunction of *SalaryAboveMillion*, *HouseOwner*, and $\neg$ *Manager*.   ∎

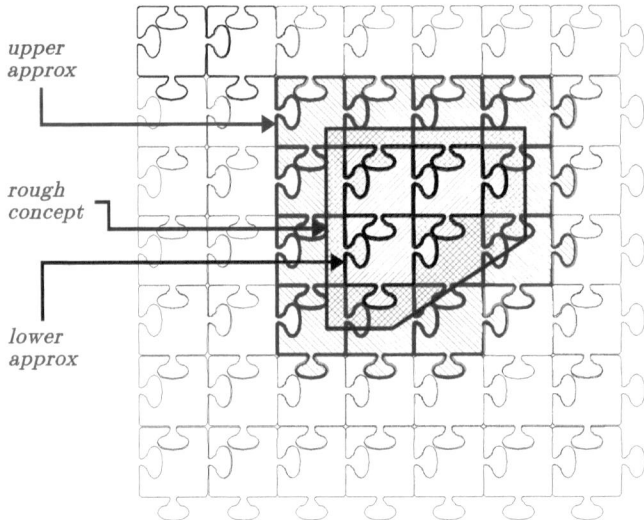

Fig. 1. Lower and upper approximations of rough concepts

Thus, a C-*definable* concept has an extension that corresponds to the union of elementary sets. The other concepts may be approximated as usual (we give a slightly different definition of the approximations relative to those in Section 2.2).

Definition 3.4 (contextual approximations). Let $C = \{F_1, \ldots, F_m\}$ be a context, let D be a DL concept, and let $\mathcal{I}$ be an interpretation. Then, the *contextual upper and lower approximations* of D relative to C, denoted $\overline{D}^C$ and $\underline{D}_C$, respectively, are defined as follows:

- $(\overline{D}^C)^{\mathcal{I}} = \{a \in \Delta^{\mathcal{I}} \mid \mathcal{I} \not\models C_a \sqcap D \sqsubseteq \bot\}$,
- $(\underline{D}_C)^{\mathcal{I}} = \{a \in \Delta^{\mathcal{I}} \mid \mathcal{I} \models C_a \sqsubseteq D\}$.

Figure 1 illustrates the contextual upper and lower approximations. The partition is determined by the feature concepts included in the context, each block standing for one of the C-definable concepts. The blocks inscribed in the concept polygon represent its lower approximation, while the blocks having a nonempty intersection with the concept polygon stand for its upper approximation.

These approximations can be encoded in a DL knowledge base through special indiscernibility relationships, as in [23], so to exploit standard reasoners for implementing inference services (with crisp answers). Alternatively, new constructors for contextual rough approximation may be defined to be added to the standard ones in the specific DL.

Following an analogous approach to the one presented in [18], it is easy to see that the following properties hold for these operators:

Proposition 3.1 (properties). *Let* $C = \{F_1, \ldots, F_m\}$ *be a context, let* D *and* E *be two DL concepts. Then:*

1. $\underline{\bot}_C = \overline{\top}^C = \bot$,

2. $\underline{\top}_C = \overline{\top}^C = \top$,

3. $\underline{D \sqcup E}_C \sqsupseteq \underline{D}_C \sqcup \underline{E}_C$,

4. $\overline{D \sqcup E}^C = \overline{D}^C \sqcup \overline{E}^C$,

5. $\underline{D \sqcap E}_C = \underline{D}_C \sqcap \underline{E}_C$,

6. $\overline{D \sqcap E}^C \sqsubseteq \overline{D}^C \sqcap \overline{E}^C$,

7. $\underline{\neg D}_C = \neg \overline{D}^C$,

8. $\overline{\neg D}^C = \neg \underline{D}_C$,

9. $\underline{D_C}_C = \underline{D}_C$,

10. $\overline{\overline{D}^C}^C = \overline{D}^C$.

4   Numerical Extensions

In this section, the indiscernibility relation is expressed in terms of a similarity measure. We introduce contextual similarity measures, and we discuss the aspect of finding optimal contexts. We finally describe how indiscernibility relations can be defined on top of tolerance functions.

4.1   Contextual Similarity Measures

Since indiscernibility can be graded in terms of the similarity between individuals, we propose a set of similarity functions, based on ideas that inspired a family of inductive distance measures [6, 2]:

Definition 4.1 (family of similarity functions). Let $KB = (\mathcal{T}, \mathcal{A})$ be a knowledge base. Given a context $C = \{F_1, F_2, \ldots, F_m\}$, a family of similarity functions

$$s_p^C : \mathsf{Ind}(\mathcal{A}) \times \mathsf{Ind}(\mathcal{A}) \mapsto [0, 1]$$

is defined as follows ($\forall a, b \in \mathsf{Ind}(\mathcal{A})$):

$$s_p^C(a, b) := \frac{1}{m} \left[\sum_{i=1}^{m} \sigma_i(a, b)^p \right]^{\frac{1}{p}}, \qquad (1)$$

where $p > 0$, and the basic similarity function σ_i ($\forall i \in \{1, \ldots, m\}$) is defined by:

$$\sigma_i(a, b) = \begin{cases} 1 & (KB \models F_i(a) \wedge KB \models F_i(b)) \vee (KB \models \neg F_i(a) \wedge KB \models \neg F_i(b)); \\ 0 & (KB \models \neg F_i(a) \wedge KB \models F_i(b)) \vee (KB \models F_i(a) \wedge KB \models \neg F_i(b)); \\ \frac{1}{2} & \text{otherwise.} \end{cases}$$

The rationale for these functions is that similarity between individuals is determined relative to a given context [12]. Two individuals are maximally similar relative to a given concept F_i if they exhibit the same behavior, i.e., both are instances of the concept or of its negation. Conversely, the minimal similarity holds when they belong to opposite concepts. By the open-world semantics, sometimes a reasoner cannot assess the concept-membership, hence, since both possibilities are open, an intermediate value is assigned to reflect such uncertainty.

As mentioned, instance-checking is used for assessing the value of the basic similarity functions. As this is known to be computationally expensive (also depending on the specific DL language), a simple look-up may be sufficient, especially for ontologies that are rich of explicit class-membership information (assertions). Hence, alternatively, for densely populated knowledge bases, the σ_i's can be efficiently approximated by defining them as follows ($\forall a, b \in \mathsf{Ind}(\mathcal{A})$):

$$
\sigma_i(a,b) = \begin{cases} 1 & (F_i(a) \in \mathcal{A} \wedge F_i(b) \in \mathcal{A}) \vee (\neg F_i(a) \in \mathcal{A} \wedge \neg F_i(b) \in \mathcal{A}); \\ 0 & (F_i(a) \in \mathcal{A} \wedge \neg F_i(b) \in \mathcal{A}) \vee (\neg F_i(a) \in \mathcal{A} \wedge F_i(b) \in \mathcal{A}); \\ \frac{1}{2} & \text{otherwise.} \end{cases}
$$

The parameter p in (1) was borrowed from the form of Minkowski's measures [24]. Once the context is fixed, the possible values for the similarity function are determined; hence, p has an impact on the granularity of the measure.

Furthermore, the uniform choice of the weights assigned to the similarity related to the various features in the sum ($1/m^p$) may be replaced by assigning different weights reflecting the importance of a certain feature in discerning the various instances. A good choice may be based on the amount of *entropy* related to each feature concept (then the weight vector has only to be normalized) [2].

4.2   Optimization of the Contexts

It is worthwhile to note that Definition 4.1 introduces a family of functions that are parameterized on the choice of features.

Preliminarily, the very set of both atomic and defined concepts found in the knowledge base can be used as a context.[1] But the choice of the concepts to be included in the context C is crucial, both for the effectiveness of the measure and for the computational efficiency itself. Specifically, the required computational effort grows with the size of the context C.

As performed for inducing the pseudo-metric that inspired the definition of the similarity function [6], a preliminary phase may concern finding optimal contexts. This may be carried out by means of randomized optimization procedures.

Since the underlying idea in the definition of the functions is that similar individuals should exhibit the same behavior relative to the concepts in C, the context C should represent a sufficient number of (possibly redundant) features that are able to discriminate different individuals.

---

[1] Preliminary experiments, reported in [2], demonstrated the effectiveness of the similarity function using the very set of both atomic and defined concepts found in the knowledge base.

The problem may be regarded as a learning problem having as a goal finding an optimal context (given the knowledge base) provided that two crucial factors are considered/optimized:

- the *number* of concepts of the context C,
- the discriminating power of the concepts in C in terms of a *discernibility factor*, i.e., a measure of the amount of difference between individuals.

The learned discriminating concepts in C may be complex concepts that are built via the specific constructors of the underlying DL.

A possible solution to the learning problem has been discussed in [6], where a randomized optimization procedure is proposed. This solution is particularly well-suited when knowledge bases with large sets of individuals are considered.

4.3   Approximation by Tolerance

In [4], a less strict type of approximation is introduced, based on the notion of *tolerance*. Exploiting the similarity functions that have been defined in Section 4.1, it is easy to extend this kind of (contextual) approximation to the case of RDLs.

Let a *tolerance function* on a set U be any function $\tau : U \times U \mapsto [0, 1]$ such that for all $a, b \in U$, it holds that $\tau(a, a) = 1$ and $\tau(a, b) = \tau(b, a)$.

Considering a tolerance function τ on a (*universal*) set U and a *tolerance threshold* $\theta \in [0, 1]$, a *neighborhood function* $\nu : U \mapsto 2^U$ is defined as follows:

$$\nu_\theta(a) = \{b \in U \mid \tau(a, b) \geq \theta\}.$$

For each element $a \in U$, the set $\nu_\theta(a)$ is the *neighborhood* of a.

Consider now the domain $\Delta^{\mathcal{I}}$ of an interpretation $\mathcal{I}$ as a universal set, a similarity function s_p^C on $\Delta^{\mathcal{I}}$ (for some context C) as a tolerance function, and a threshold $\theta \in [0, 1]$. It is then easy to derive a *tolerance relation*[2], i.e., a reflexive and symmetric relation on $\Delta^{\mathcal{I}}$, inducing tolerance classes that consist of individuals within a certain degree of similarity, indicated by the threshold: $[a]_{C,\theta} = \nu_\theta(a)$. The notions of upper and lower approximation relative to the tolerance relation induced by C and θ descend straightforwardly:

- $(\overline{D})^{\mathcal{I}} = \{a \in \Delta^{\mathcal{I}} \mid \exists b \in \Delta^{\mathcal{I}} : s_p^C(a, b) \geq \theta \wedge b \in D^{\mathcal{I}}\}$,
- $(\underline{D})^{\mathcal{I}} = \{a \in \Delta^{\mathcal{I}} \mid \forall b \in \Delta^{\mathcal{I}} : s_p^C(a, b) \geq \theta \rightarrow b \in D^{\mathcal{I}}\}$.

Given the similarity measure defined in Section 4.1 as a tolerance function, the approximation by tolerance allows a less strict approximation with respect to the adoption of the indiscernibility relation exploited for the case of the contextual approximation (see Section 3). The granularity of the approximation is specifically controlled by the threshold. Indeed, if θ is (very close to) 1, then we obtain almost the indiscernibility relation for the contextual approximation. Considering lower values for θ, additional individuals will be included in the neighborhood $\nu_\theta(a)$ of a given individual a. This aspect may result to be particularly useful when not enough information is available for defining a suitable context of interest C.

---

[2] Transitivity is not necessary, however, the case of an indiscernibility relation can be considered with the equivalence classes $[a]_\theta = \bigcap\{\nu_\theta(b) \mid a \in \nu_\theta(b)\}$.

Example 4.1 (Advertising Campaign cont'd). Given the context

$$\mathsf{C} = \{SalaryAboveMillion, HouseOwner, Manager\}$$

introduced in Example 3.1, $\theta = 0.8$, and the similarity function s_p^{C} defined in Section 4.1, the concepts $\underline{Addressee}$ and $\overline{Addressee}$ will be given by

- $(\overline{Addressee})^{\mathcal{I}} = \{a \in \Delta^{\mathcal{I}} \mid \exists b \in \Delta^{\mathcal{I}} : s_p^{\mathsf{C}}(a, b) \geq 0.8 \wedge b \in Addressee^{\mathcal{I}}\}$,
- $(\underline{Addressee})^{\mathcal{I}} = \{a \in \Delta^{\mathcal{I}} \mid \forall b \in \Delta^{\mathcal{I}} : s_p^{\mathsf{C}}(a, b) \geq 0.8 \rightarrow b \in Addressee^{\mathcal{I}}\}$. ∎

Note that these approximations depend on the threshold. Thus, we have a numerical way to control the degree of indiscernibility that is needed to model uncertain concepts. This applies both to the standard RDL setting and to the new contextual one presented in the previous section.

Alternatively, by learning an intensional concept description (see the next Section 5) for the neighborhood $\nu_\theta(a)$, lower and upper approximations of a given concept D may be defined as for the case of the contextual approximation (see Section 3).

5   Learning Crisp Definitions of Rough Concepts

There may be cases where modeling uncertain knowledge is a difficult task, even in the framework of RDLs. One main problem is that a domain expert may not always have a clear idea about the concepts to model via lower and upper concept approximations, thus having incomplete besides rough concept definitions. Another problem are the difficulties in defining a suitable indiscernibility relation. Even the indiscernibility relations in Sections 3 and 4 are based on the notion of a context, which for some cases may be difficult to define (see, e.g., the *Sepsis* example in [23]). Furthermore, even the translation function (presented at the end of Section 2.2) for transforming RDL concepts into crisp concepts, via an appropriate predicate for the indiscernibility relation, may be hard to apply in practice in DLs with low expressiveness. Furthermore, a domain expert often has a clear idea about counterexamples to a concept definition, but may not be able to give an explanation or a clear definition for them.

To cope with these problems, and to still be able to represent uncertain knowledge in RDLs, we propose an alternative way, grounded on DL concept learning methods, for describing lower and upper approximations of a given concept.

Since a domain expert often has a clear idea about counterexamples to a given concept definition, we assume that he/she is able to supply (a) a set of positive examples for the upper (resp., lower) approximation of a given concept C, i.e., a set of individuals standing as the possible (resp., certain) instances of the concept C, and (b) a set of negative examples for the upper (resp., lower) approximation of C, i.e., a set of individuals that are surely not instances of the upper (resp., lower) approximation of C.

Given these sets of examples, the problem now is to find a suitable crisp definition for them. Specifically, the problem can be formally defined as follows:

Definition 5.1 (learning problem). Let $KB = (\mathcal{T}, \mathcal{A})$ be a knowledge base. Then, given

- $\mathsf{Ind}(\mathcal{A})$ the set of all individuals occurring in $\mathcal{A}$,
- a set of positive and negative examples $\mathsf{Ind}_C^+(\mathcal{A}) \cup \mathsf{Ind}_C^-(\mathcal{A}) \subseteq \mathsf{Ind}(\mathcal{A})$ for the upper (resp., lower) approximation of a given concept C,

we build a concept definition $\widehat{C}$ such that

$$KB \models \widehat{C}(a) \quad \forall a \in \mathsf{Ind}_C^+(\mathcal{A}) \qquad \text{and} \qquad KB \models \neg\widehat{C}(b) \quad \forall b \in \mathsf{Ind}_C^-(\mathcal{A}).$$

Example 5.1 (Advertising Campaign cont'd). Consider again Examples 2.1 and 2.2, where the RDL concepts $\underline{Addressee}$ and $\overline{Addressee}$ are introduced, representing the definite and potential addressees, respectively, for an advertising campaign for a new product (i.e., Porsche). Now, a crisp DL definition for each of the two concepts has to be given. Suppose now that no indiscernibility function is adopted and/or specified, because the domain expert does not have enough knowledge (e.g., for defining a suitable context C), but the domain expert is able to identify some instances (i.e., individuals of the knowledge base, i.e., instances of some concepts in the knowledge base) that are definitely addressees (positive examples) and that are surely not addressees (negative examples). This information is exploited for learning an intensional concept description such that all positive examples are instances of the learned concept and that all negative examples are instances of the negation of the learned concept. The same process can be applied for the concept $\overline{Addressee}$. In this way, a crisp description for an RDL concept can be given without adopting any indiscernibility function. ∎

The definition given above can be interpreted as a generic supervised concept learning task. The problem consists of finding a DL concept definition $\widehat{C}$ such that all positive examples are instances of $\widehat{C}$, while all negative examples are instances of $\neg\widehat{C}$. This problem is well-studied in the literature, resulting in different inductive learning methods that are grounded on the (greedy) exploration of the search space by the adoption of suitable refinement operators for DL representations [14, 15]. Among the most well-known algorithms and systems, there are DL-FOIL [7], DL-LEARNER[3] [16], and TERMITIS [11]. Hence, given the set of the positive and negative examples for the upper (resp., lower) approximation of a concept C, the crisp definitions of the approximate concepts can be learned by adopting one of the systems cited above.

Alternatively/additionally, we may also be interested in assessing/learning the crisp description of the upper (resp., lower) approximation of a crisp concept D that is already existing in the knowledge base. In this case, finding a domain expert who provides the set of positive and negative examples for the upper (resp., lower) approximation of D may not always be possible. The problem of learning a crisp concept description for the upper (resp., lower) approximation of D is now shifted to the problem of determining the positive and negative examples for the upper (resp., lower) approximation of D. In the following, the possible solutions are illustrated.

Definition 5.2 (positive/negative examples for lower approximation). Let $KB = (\mathcal{T}, \mathcal{A})$ be a knowledge base. Then, given

- $\mathsf{Ind}(\mathcal{A})$ the set of all individuals occurring in $\mathcal{A}$,
- a target atomic concept D,

---

[3] http://dl-learner.org/Projects/DLLearner

we define

- the set of positive examples as $\mathsf{Ind}_{\underline{D}}^{+}(\mathcal{A}) = \{a \in \mathsf{Ind}(\mathcal{A}) \mid KB \models D(a)\}$,
- the set of negative examples as $\mathsf{Ind}_{\underline{D}}^{-}(\mathcal{A}) = \{a \in \mathsf{Ind}(\mathcal{A}) \mid KB \not\models D(a)\}$.

The set of the positive examples for the lower approximation of the concept D is given by all individuals of the knowledge base that are instances[4] of D, while the set of negative examples is given by all individuals for which it is not possible to prove that they are instances of D. This set includes both the individuals that are instances of $\neg D$ and the individuals for which the reasoner is not able to give any reply due to the OWA.

Definition 5.3 (positive/negative examples for upper approximation). Let $KB = (\mathcal{T}, \mathcal{A})$ be a knowledge base. Then, given

- $\mathsf{Ind}(\mathcal{A})$ the set of all individuals occurring in $\mathcal{A}$,
- a target atomic concept D,

we define

- the set of positive examples as $\mathsf{Ind}_{\overline{D}}^{+}(\mathcal{A}) = \{a \in \mathsf{Ind}(\mathcal{A}) \mid KB \not\models \neg D(a)\}$,
- the set of negative examples as $\mathsf{Ind}_{\overline{D}}^{-}(\mathcal{A}) = \{a \in \mathsf{Ind}(\mathcal{A}) \mid KB \models \neg D(a)\}$.

The set of the negative examples for the upper approximation of the concept D is given by all individuals of the knowledge base that are instances of $\neg D$, while the set of positive examples is given by all individuals for which it is not possible to prove that they are instances of $\neg D$ (e.g., because of the absence of disjointness axioms in the considered ontology). This set includes both the individuals that are instances of D and the individuals for which the reasoner is not able to give any reply due to the OWA.

Once the set of positive and negative examples for the lower (resp., upper) approximation of D have been determined, the crisp definition for the lower (resp., upper) approximation of D can be learned as illustrated above. Note, however, that the learned definitions may be noisy when a high percentage of unlabeled examples (due to the OWA) is included in the set of negative (resp., positive) examples. To cope with this problem, alternative learning methods such as methods for learning from positive (and unlabeled) examples [3, 25] only were investigated and can be exploited.

6   Summary and Outlook

Inspired by previous works on dissimilarity measures in DLs, we have defined a notion of context, which allows to extend the indiscernibility relation adopted by rough DLs, thus allowing for various kinds of approximations of uncertain concepts within the same knowledge base. It also saves the advantage of encoding the relation in the same DL language, thus allowing for reasoning with uncertain concepts through standard tools, obtaining crisp answers to queries.

Alternatively, these approximations can be implemented as new modal-like language operators. Some properties of the approximations deriving from the theory of rough sets have also been investigated.

---

[4] Here, *concept retrieval* may be adopted.

A novel family of semantic similarity functions for individuals has also been defined based on their behavior relative to a number of features (concepts). The functions are language-independent, being based on instance-checking (or ABox look-up). This allows for defining further kinds of graded approximations based on the notion of tolerance relative to a certain threshold.

Since data can be classified into indiscernible clusters, unsupervised learning methods for grouping individuals on the grounds of their similarity can be used for the definition of an equivalence relation [13, 6, 8]. Besides, it is also possible to learn rough DL concepts from the explicit definitions of the instances of particular concepts [14, 15, 7].

Acknowledgments. This work was partially supported by the Engineering and Physical Sciences Research Council (EPSRC) grant EP/J008346/1 (PrOQAW), the European Research Council under the EU's 7th Framework Programme (FP7/2007-2013/ERC) grant 246858 (DIADEM), a Google Research Award, and a Yahoo! Research Fellowship. Many thanks also to the reviewers of this paper and its URSW-2008 abstract for their useful and constructive comments, which have helped to improve this work.

References

[1] Baader, F., Calvanese, D., McGuinness, D., Nardi, D., Patel-Schneider, P. (eds.): The Description Logic Handbook. Cambridge University Press (2003)

[2] d'Amato, C., Fanizzi, N., Esposito, F.: Query Answering and Ontology Population: An Inductive Approach. In: Bechhofer, S., Hauswirth, M., Hoffmann, J., Koubarakis, M. (eds.) ESWC 2008. LNCS, vol. 5021, pp. 288–302. Springer, Heidelberg (2008)

[3] De Comité, F., Denis, F., Gilleron, R., Letouzey, F.: Positive and Unlabeled Examples Help Learning. In: Watanabe, O., Yokomori, T. (eds.) ALT 1999. LNCS (LNAI), vol. 1720, pp. 219–230. Springer, Heidelberg (1999)

[4] Doherty, P., Grabowski, M., Łukaszewicz, W., Szalas, A.: Towards a framework for approximate ontologies. Fundamenta Informaticae 57(2-4), 147–165 (2003)

[5] Donini, F., Lenzerini, M., Nardi, D., Nutt, W.: An epistemic operator for description logics. Artificial Intelligence 100(1/2), 225–274 (1998)

[6] Fanizzi, N., d'Amato, C., Esposito, F.: Randomized metric induction and evolutionary conceptual clustering for semantic knowledge bases. In: Proceedings of the 16th Conference on Information and Knowledge Management, CIKM 2007, pp. 51–60. ACM Press (2007)

[7] Fanizzi, N., d'Amato, C., Esposito, F.: DL-FOIL Concept Learning in Description Logics. In: Železný, F., Lavrač, N. (eds.) ILP 2008. LNCS (LNAI), vol. 5194, pp. 107–121. Springer, Heidelberg (2008)

[8] Fanizzi, N., d'Amato, C., Esposito, F.: Conceptual Clustering and Its Application to Concept Drift and Novelty Detection. In: Bechhofer, S., Hauswirth, M., Hoffmann, J., Koubarakis, M. (eds.) ESWC 2008. LNCS, vol. 5021, pp. 318–332. Springer, Heidelberg (2008)

[9] Fanizzi, N., d'Amato, C., Esposito, F.: Statistical Learning for Inductive Query Answering on OWL Ontologies. In: Sheth, A.P., Staab, S., Dean, M., Paolucci, M., Maynard, D., Finin, T., Thirunarayan, K. (eds.) ISWC 2008. LNCS, vol. 5318, pp. 195–212. Springer, Heidelberg (2008)

[10] Fanizzi, N., d'Amato, C., Esposito, F., Lukasiewicz, T.: Representing uncertain concepts in rough description logics via contextual indiscernibility relations. In: Proceedings of the 4th International Workshop on Uncertainty Reasoning for the Semantic Web, URSW 2008. CEUR Workshop Proceedings, vol. 423. CEUR-WS.org (2008)

[11] Fanizzi, N., d'Amato, C., Esposito, F.: Induction of Concepts in Web Ontologies through Terminological Decision Trees. In: Balcázar, J.L., Bonchi, F., Gionis, A., Sebag, M. (eds.) ECML PKDD 2010, Part I. LNCS, vol. 6321, pp. 442–457. Springer, Heidelberg (2010)

[12] Goldstone, R., Medin, D., Halberstadt, J.: Similarity in context. Memory and Cognition 25(3), 237–255 (1997)

[13] Hirano, S., Tsumoto, S.: An indiscernibility-based clustering method. In: Proceedings of the 2005 IEEE International Conference on Granular Computing, pp. 468–473. IEEE Computer Society (2005)

[14] Iannone, L., Palmisano, I., Fanizzi, N.: An algorithm based on counterfactuals for concept learning in the Semantic Web. Applied Intelligence 26(2), 139–159 (2007)

[15] Lehmann, J., Hitzler, P.: A Refinement Operator Based Learning Algorithm for the $\mathcal{ALC}$ Description Logic. In: Blockeel, H., Ramon, J., Shavlik, J., Tadepalli, P. (eds.) ILP 2007. LNCS (LNAI), vol. 4894, pp. 147–160. Springer, Heidelberg (2008)

[16] Lehmann, J.: DL-Learner: Learning concepts in description logics. Journal of Machine Learning Research 10, 2639–2642 (2009)

[17] Lukasiewicz, T., Straccia, U.: Managing uncertainty and vagueness in description logics for the Semantic Web. Journal of Web Semantics 6(4), 291–308 (2008)

[18] Jiang, Y., Wang, J., Tang, S., Xiao, B.: Reasoning with rough description logics: An approximate concepts approach. Information Sciences 179(5), 600–612 (2009)

[19] Keet, C.M.: On the feasibility of description logic knowledge bases with rough concepts and vague instances. In: Proceedings of the 23rd International Workshop on Description Logics, DL 2010. CEUR Workshop Proceedings, vol. 573. CEUR-WS.org (2010)

[20] Keet, C.M.: Ontology Engineering with Rough Concepts and Instances. In: Cimiano, P., Pinto, H.S. (eds.) EKAW 2010. LNCS, vol. 6317, pp. 503–513. Springer, Heidelberg (2010)

[21] Mitchell, T.: Machine Learning. McGraw-Hill (1997)

[22] Pawlak, Z.: Rough Sets: Theoretical Aspects of Reasoning About Data. Kluwer Academic Publishers (1991)

[23] Schlobach, S., Klein, M.C.A., Peelen, L.: Description logics with approximate definitions — precise modeling of vague concepts. In: Proceedings of the 20th International Joint Conference on Artificial Intelligence, IJCAI 2007, pp. 557–562 (2007)

[24] Zezula, P., Amato, G., Dohnal, V., Batko, M.: Similarity Search — The Metric Space Approach. Advances in Database Systems. Springer (2007)

[25] Zhang, B., Zuo, W.: Learning from positive and unlabeled examples: A survey. In: Proceeding of the International Symposium on Information Processing, ISP 2008, pp. 650–654 (2008)

Efficient Trust-Based Approximate SPARQL Querying of the Web of Linked Data

Kuldeep B.R. Reddy and P. Sreenivasa Kumar

Indian Institute of Technology Madras,
Chennai, India
{brkreddy,psk}@cse.iitm.ac.in

Abstract. The web of linked data represents a globally distributed dataspace, which can be queried using the SPARQL query language. However, with the growth in size and complexity of the web of linked data, it becomes impractical for the user to know enough about its structure and semantics for the user queries to produce enough answers. Moreover, there is a prevalence of unreliable data which can dominate the query results misleading the users and software agents. These problems are addressed in the paper by making use of ontologies available on the web of linked data to produce approximate results and also by presenting a trust model that associates RDF statements with trust values, which is used to give prominence to trustworthy data. Trustworthy approximate results can be generated by performing the relaxation steps at compile-time leading to the generation of multiple relaxed queries that are sorted in decreasing order of their similarity scores with the original query and executed. During their execution the trust scores of RDF data fetched are computed. However, the relaxed queries generated have conditions in common and we propose that by performing trust-based relaxations on-the-fly at runtime, the shared data between several relaxed queries need not be fetched repeatedly. Thus, the trust-based relaxation steps are integrated with the query execution itself resulting in performance benefits. Further opportunities for optimizations during query execution are identified and are used to prune relaxation steps which do not produce results. The implementation of our approach demonstrates its efficacy.

1   Introduction

The traditional World Wide Web has allowed sharing of documents among users on a global scale. The documents are generally represented in HTML, XHTML, DHTML formats and are accessed using URL and HTTP(S) protocols creating a global information space. However, in the recent years the web has evolved towards a web of data as the conventional web's data representation sacrifices much of its structure and semantics [1] and the links between documents are not expressive enough to establish the relationship between them. This has lead to the emergence of the global data space known as Linked Data [1].

F. Bobillo et al. (Eds.): URSW 2008-2010/UniDL 2010, LNAI 7123, pp. 315–330, 2013.

Linked data basically interconnects pieces of data from different sources utilizing the existing web infrastructure. The data published is machine readable that means it is explicitly defined. Instead of using HTML, linked data uses RDF format to represent data. The connection between data is made by typed statements in RDF which clearly defines the relationship between them resulting in a web of data. The Linked Data Principles outlined by Berners-Lee for publishing data on the web basically suggests using URIs for names of things which are described in RDF format and accessed using HTTP protocol in a way that all published data becomes part of a single global data space.

RDF is a W3C standard for modeling and sharing distributed knowledge based on a decentralized open-world assumption. Any knowledge about anything can be decomposed into triples (3-tuples) consisting of subject, predicate, and object; essentially, RDF is the lowest common denominator for exchanging data between systems. The subject and object of a triple can be both URIs that each identify an entity, or a URI and a string value respectively. The predicate denotes the relationship between the subject and object, and is also represented by a URI. SPARQL is the query language proposed by W3C recommendation to query RDF data. A SPARQL query basically consists of a set of triple patterns. It can have variables in the subject,object or predicate positions in each of the triple pattern. The solution consists of binding these variables to entities which are related with each other in the RDF model according to the query structure.

There have been a number of approaches proposed to query the web of linked data. One direction has been to crawl the web by following RDF links and build an index of discovered data. The queries are then executed against these indexes. This approach is followed by Sindice[2]. Another approach has been to follow the federated query processing concepts, as in DARQ[3], which decomposes a SPARQL query in subqueries, forwards these subqueries to multiple, distributed query services, and, finally, integrates the results of the subqueries. Another execution approach for evaluating SPARQL queries on linked data is proposed in [4]. It is basically a run-time approach which executes the query by asynchronously traversing RDF links to discover data sources at run-time. SPARQL query execution takes place by iteratively dereferencing URIs to fetch their RDF descriptions from the web and building solutions from the retrieved data. The SPARQL query execution according to [4] is explained with an example below.

```
SELECT ?prof ?publ WHERE
{
  <http://site//univ> univ:hasPublications ?publ
      ?publ  univ:authoredBy  ?prof
      ?prof  rdf:type  Professor
}
```

Fig. 1. Example SPARQL query

Example. The SPARQL query shown in Figure 1 searches for Professors employed by the university who have authored a publication. The query execution begins by fetching the RDF description of the university by dereferencing its URI. The fetched RDF description is then parsed to gather a list of all of its publications. Parsing is done by looking for triples that match the first pattern in the query. The object URIs in the matched triples form the list of publications in the university. Lets say `<http://site1/publ1.rdf>`, `<http://site2/publ2.rdf>`, `<http://site3/publ3.rdf>` were found to be the papers. The query execution proceeds by fetching the RDF descriptions corresponding to the three publications. Lets say first publ1's graph is retrieved. It is parsed to check for triples matching the second query pattern and it is found that publ1 was authored by John `<http://site4/John.rdf>`. John's details are again fetched and the third triple pattern in the query is searched in the graph to see whether he is of type Professor and if he is, the result of query is formed and displayed as output. Publ1's and Publ2's graphs and their author details would also be retrieved and the query execution proceeded in a way similar to Publ1's.

In the previous example, consider the situation where the retrieved list of publications authored by the professors may not meet the requirements of the user, in which case query conditions need to be relaxed to produce more results. For example, instead of looking for only Professors, the query can be generalized by searching for all kinds of people including lectures,graduate students etc. Moreover, considering the prevalence of unreliable data we would like trustworthy results be ranked higher. Therefore to meet the user needs, the relaxation steps can be performed before the execution of the query begins, similar to the approach in [5] to query centralized RDF repositories, which generates multiple relaxed SPARQL queries. They are sorted in the decreasing order of their semantic similarity scores with the original query and executed to produce approximate answers. But the approach in [5] was designed to work on centralized RDF repositories and it also does not consider the reliability of RDF information. In addition, we notice that the relaxed SPARQL queries formed share many query conditions in common, which are not utilized to optimize the queries. Especially in a distributed environment, like the web of linked data, avoiding repeated fetching of data shared across the queries results in significant performance benefits [6]. To ensure that trustworthy results are ranked higher, the RDF data as it is fetched from the web is assigned trust scores based on the user defined trust function. The trust values are then combined with the relaxation scores to arrive at a weighted score which is used to rank the results.

Example. Figure 2 gives the two relaxed queries formed after the query term Professor has been replaced by Faculty and Person terms using RDFS ontology. The first query involving the terms faculty is semantically closer to the original query and is executed first followed by the second query. During their execution, the fetched RDF triples are given trust values and used along with relaxation scores for ranking purposes. However, we notice that the first two predicates are common between the two queries, therefore the information corresponding to

SELECT ?prof ?publ WHERE	SELECT ?prof ?publ WHERE
{	{
<http://site//univ> univ:hasPublications ?publ	<http://site//univ> univ:hasPublications ?publ
?publ univ:authoredBy ?prof	?publ univ:authoredBy ?prof
?prof rdf:type Faculty	?prof rdf:type Person
}	}

Fig. 2. Relaxed SPARQL query

them can be fetched just once to gain efficiency. Hence, instead of generating the two queries, if the execution of the original query continues by dereferencing the URIs corresponding to Publ1 and its author and retrieving their RDF descriptions, and the check performed by the third predicate to see whether the author is a Professor or not only replaced by Faculty and Person at the last step, the shared data is fetched just once.

We introduce the idea of trust as an additional component in order to rate the results. The information consumer in this scheme gives each triple a score between [-1,1] based on his judgement of the information contained in the triple. These scores are used in addition to the scores associated with the relaxations, so that the results which are semantically closer to the original query and trustworthy appear higher.

The goal of this paper is therefore to perform trust-based approximate SPARQL querying of the web of linked data in an optimized way. We present an approach which uses the ontologies available on the web of linked data to relax the query conditions before its execution. This produces many relaxed queries whose execution taking into account the trust values of RDF triples produces trustworthy approximate results. We improve upon this approach by presenting the idea of delaying query relaxation steps to run-time [6]. To ensure that trustworthy results are given prominence, we introduce the idea of integrating the approach of performing run-time relaxations with the trust assignment steps to RDF triples that are fetched from the web during the query execution. We further only choose the relaxation steps during the query processing based on their semantic similarity with the original query and their trust scores, resulting in an optimized query execution. To reiterate, the difference between the two approaches is that the trust assignment mechanism is disassociated with the relaxation steps in the compile-time approach with the former taking place at run-time and the latter at compile-time. Whereas in the proposed trust-based run-time approach the trust assignment and relaxation steps are done together at run-time.

2   Trust Model

We present the trust model introduced in [7] whose purpose is to rate the RDF information. The RDF data is made up of a set of triples which inturn consists of subject, predicate and the object components. Each triple asserts the relationship between the subject and the object which is described by the predicate. The trust model assigns scores to each triple pattern that indicates the degree of trustworthiness of the relationship asserted by the triple. A high trust score means that the consumer has a high degree of faith in the information contained in the triple and vice-versa. The trust scores are assigned to the triples by the information consumer based on his subject belief after assessing the triples.

The trust model does not prescribe a specific way of determining trust values. Instead, each system is allowed to provide its own, application specific trust function. Determining trust values may be based on provenance information [8] as in the Inference Web trust component [9]; the TRELLIS system [10] additionally considers related data; the FilmTrust application [11] combines provenance information and recommendations from other consumers.

RDF statements are given scores in the range [-1,1]. A positive score indicates that the triple is trustworthy. The higher it is, that is the closer it is to the score of +1, more trustworthy it is. A score closer to 0 indicates that the consumer is not sure about his assessment on the trustworthiness of the triple. Whereas, a negative score indicates that the consumer believes that the triple is untrustworthy. More negative the score, that is closer it is to -1, more untrustworthy it is.

As each triple pattern is given a trust score the user can define an aggregation function which determines the trust score of the RDF graph as a whole from the trust score of its component RDF triple. There is no standard aggregation function, it is left to the applications to use the functions which suits their needs. The minimum, for instance, is a cautious choice; it assumes the trustworthiness of a set of triples is only as trustworthy as the least trusted triple. The median, a more optimistic choice, is another reasonable trust aggregation function.

A metric is presented which makes use of RDFS/OWL schemas to reveal inconsistent statements describing an entity. The presence of such statements decrease the likelihood of the knowledge about the entity being accurate.

For example, suppose the RDFS/OWL schema describes that a university employs only professors and not managers. The RDF graph retrieved for an entity describes that the person works in a university and is of type manager, then it is claimed that the other information about the entity too may not be correct.

The process of trust assignment works as follows. First, class of the entity is recorded. Then for each triple describing the entity, class of the object is recorded. From this a new triple is constructed which tries to establish the relationship among the classes involved. Then, we check for its confirmation in the RDFS/OWL schema to see if it too describes the relationship amongst the classes in the same way. If it does not, then we decrement the trust scores of the triples by a certain amount δ till it reaches 0.

3   Similarity Measures

The similarity measures were defined in [5] and describes their computation from the RDFS ontology. However, the measures were designed for centralized RDF repositories and considered only one ontology and therefore they cannot be directly applies in the context of web of linked data. That is because each user publishing data in the web of data has the freedom to define his own ontology. But according to the principles of linked data, if a user has defined his own ontology it has to be mapped to existing ontologies like FOAF and others, therefore we assume such mappings exist for the purposes of this paper.

A triple pattern can be replaced by terms in the ontology in a number of ways. Therefore, there is a need to attach a score to each relaxation which can be used to rank them to ensure the quality of results. The score given to each relaxation measures its similarity with the original triple pattern. Highest scoring relaxation are executed first followed by others in the decreasing order of the similarity score. For example, we would rank the relaxation from (?X,type,professor) to (?X,type,faculty) higher than (?X,type,professor) to (?X,type,person) as the former is more similar to the original triple pattern. A SPARQL query consists of a basic graph pattern which in turn consists of triple patterns. Therefore, the score associated with an answer to a SPARQL query is computed by aggregating the scores of relaxed triple patterns. Each triple pattern consists of a subject, predicate and object parts, and each of them can be potentially relaxed. Their aggregated score gives the score of the triple pattern.

Similarity between Nodes. In a triple pattern t_1, if the subject/object node belongs to class c_1 in the RDFS ontology and is relaxated to class c_2 using the ontology we use the idea of Least Common Ancestor to compute the similarity of the two triple patterns. The Least Common Ancestor denotes the depth of the common ancestor superclass of the two classes from the root in the RDFS ontology.

$$score(c_1, c_2) = \frac{2 * Depth(LCA(c_1, c_2))}{depth(c_1) + depth(c_2)}$$

Similarity between Predicates. In a triple pattern t_1, if the predicate is of type p_1 in the RDFS ontology and is relaxed to type p_2 using the ontology we use the idea of Least Common Ancestor to compute the similarity of the two triple patterns similar to that done for subject/object nodes. The Least Common Ancestor denotes the depth of the common ancestor superproperty of the two types of properties from the root in the RDFS ontology.

$$score(p_1, p_2) = \frac{2 * Depth(LCA(p_1, p_2))}{depth(p_1) + depth(p_2)}$$

Similarity between Triple Patterns. If the triple pattern t_1-$(s_1, p1, o1)$ is relaxed to t_2-(s_2, p_2, o_2) we aggregate the similarity scores of the triple pattern constituents to compute the overall similarity score of relaxed triple pattern.

$$similarity(t_1, t_2) = score(s_1, s_2) + score(p_1, p_2) + score(o_1, o_2)$$

Score of an Answer. The bindings of the relaxed SPARQL queries form the answers to the original SPARQL query. Since the original query is relaxed in a number of ways we need a measure to rank the relevant answers to ensure the quality of results. Thus, we define the score of each relevant answer as the similarity of its corresponding relaxed SPARQL query from which it is produced to the original SPARQL query. The similarity between the two queries is obtained by combining the similarities of the triple patterns in them. Suppose the answer A is obtained from query $Q'(t'_1, t'_2, t'_3...t'_n)$ which was formed after the original query $Q(t_1, t_2, t_3....t_n)$ was relaxed.

$$score(A) = \sum_{i=1}^{n} similarity(t_i, t'_i)$$

4   Query Processing Algorithms

We present an approach to produce approximate answers by generating relaxed SPARQL queries from the original SPARQL query using the RDFS ontologies. It works by assigning scores to the relaxed queries based on the semantic similarity to the original query. Following which, the relaxed queries are executed in the descending order of their semantic similarity scores. The trust values of the triples constituting the results are taken into account and along with the relaxation scores are used to order the results so that trustworthy approximate results appear higher. However, the SPARQL queries generated have many query conditions in common. Therefore, the sequential execution approach of all the queries involves needlessly fetching the same data repeatedly. Later we will present an optimized trust-based query processing algorithm where relaxed queries are generated and answered taking into account trust scores on-the-fly during the query execution resulting in significant performance benefits.

Algorithm 1 describes the compile-time approach to produce approximate answers. Lines 2-8 denote the steps taken to generate multiple relaxed queries. The relaxation procedure is described as a graph, called a relaxation graph here. The algorithm begins by putting the given query as a root in the relaxation graph. Then each triple pattern in the query is relaxed one-by-one and the new query produced as a result is inserted as a child node of the query node in the relaxation graph that led to it being produced. Each triple pattern relaxation is accompanied by computing its relaxation score and this score is attached to its corresponding relaxed query. This process is repeated till all possible relaxed queries are generated. Lines 11-20 execute the relaxed queries produced earlier. To generate ranked approximate results, the relaxed queries are executed in the descending order of their semantic similarity scores with the original query. The relaxed query with the maximum score is executed first following which the next query to be executed is chosen with the highest score amongst its children and so on. As the results are generated for each relaxed query, we compute its aggregated trust score as described in the previous section. Then we calculate the weighted mean of the aggregated trust score and the semantic similarity score of its query with the original query, and use this weighted mean score to obtain a final ranked list of answers. For this purpose, a user-defined variable w_1 ,taking a value from [0,1], is introduced which controls the weight of the semantic similarity score in the final score.

Algorithm 1. Compile-Time Approach

    Input   : :Query Q
    Output: :Approximate answers
1 $relaxationGraph = \phi$
2 Insert Q as root in $relaxationGraph$
3 **while** $Q \neq \phi$ **do**
4   **foreach** *Triple* t_i *in* Q **do**
5   Relax t_i to t_i'
6   compute the relaxScore of relaxation
7   Insert Q_i' as a succeeding node of Q in $relaxationGraph$
8   **end**
9     $Q \Leftarrow Q_{siblingNode}$ or $Q_{succeedingNode}$
10 **end**
11 $Result = \phi$
12 $Candidates = \phi$
13 Insert Q's succeeding nodes from $relaxationGraph$ into $Candidates$.
14 **while** $Candidates > 0$ **do**
15   Select Q_i with maximum relaxScore from $Candidates$
16   Insert Q_i succeeding nodes $relaxationGraph$ into $Candidates$
17    $R \Leftarrow Execute(Q_i)$
18   Compute aggregated $trustScore(R_t)$ according to the user-defined trust function $finalScore = (w_1 * relaxScore + (1 - w_1) * trustScore(R_t))$
19   Add to the sorted list $Result = Result \cup R$ based on $finalScore$ values
20   Add Q_i to processed
21   Remove Q_i from $Candidates$
22 **end**
23 Return $Result$

Figure 3 describes the execution of two queries of Figure 2. The two queries are generated from the query of Figure 1 as described by lines 2-8 in algorithm 1. The top box in Figure 3 shows the execution of the first query in Figure 2 and similarly the bottom box for the other query. As we can see, many of the URIs dereferenced are the same in both the cases. For both of them, the query execution takes place by first dereferencing the university's URI to retrieve its RDF graph. As the triples are processed, they are assigned trust values according to the trust function defined by the information consumer. Then the details of its publications publ1,publ2 and publ3 followed by its authors, John,Peter and Mary, are fetched. The existing approach repeats this process twice for each of the relaxed query when instead we can fetch the shared information once and then perform the relaxation. This motivates us to integrate the relaxation process and the trust assignment process with the execution of the query itself

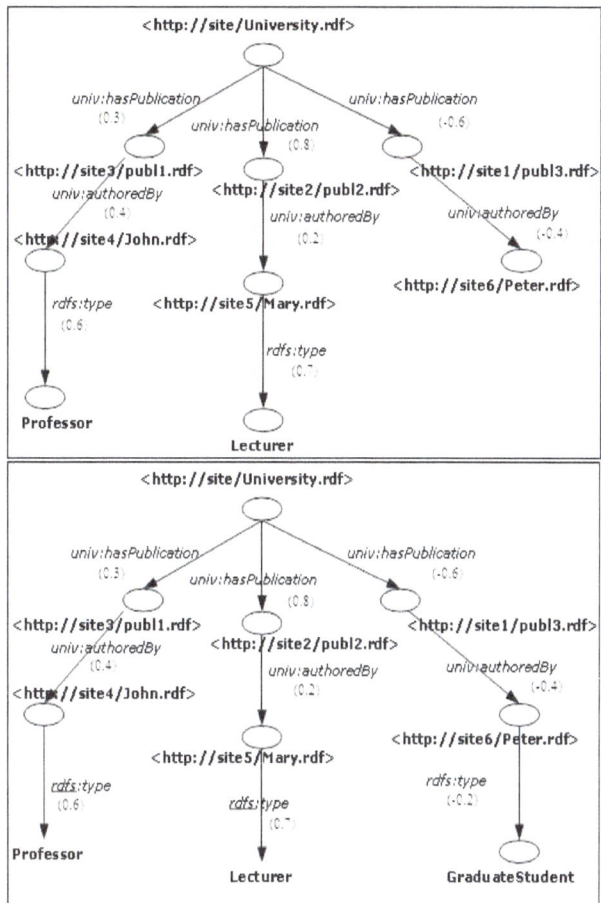

Fig. 3. Execution with compile-time approach

and is described in algorithm 2. The trust scores for the triples are shown in the Figure next to the predicate names. If we consider the aggregated trust score to be the minimum score of all the triples in the result, the trust scores for the results become (0.3),(0.2),(-0.6) respectively. These trust scores are combined with the semantic similarity score of its query with the original query to obtain a final weighted score based on which results are ordered.

Algorithm 2 describes the proposed run-time approach for trust-based optimized approximate answering. Lines 3-20 repeat for each query predicate in the given query. It begins with the seed, fetching its RDF graph. Then the presence of the query predicate is checked for in the fetched RDF graph. If it is present, the relaxation score for the predicate in the graph is given the maximum value of 1.0. Predicates belonging to different namespaces are assumed to be mapped in accordance with the linked data principles. Otherwise, using the metrics described in the earlier section the semantic similarity score for each predicate in

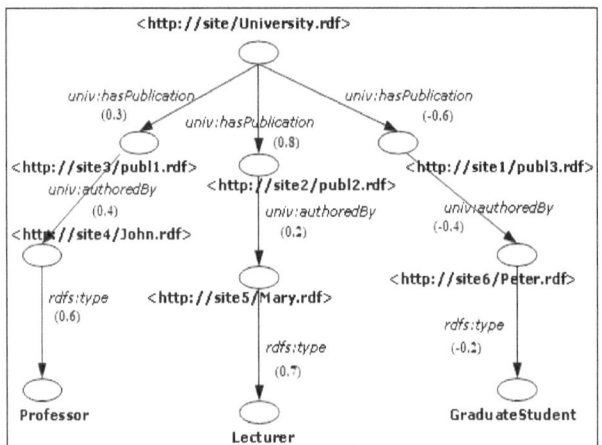

Fig. 4. Execution with run-time approach

the RDF graph compared with the associated predicate in the original query is computed. The similarity scores are combined with the trust values of the triples according to the user defined weight w_1 to obtain an overall score. This overall score is used to sort the predicates in the descending order. The query execution proceeds by updating the seed with the object URIs of the predicates, which are then dereferenced to retrieve their graphs. Further similarities and overall scores are computed and this process is repeated till a set of leaf values are produced. The path from the root to the leaf values in lines 21-23 along the relaxed predicates gives the approximate answers.

Figure 4 shows the query execution with the proposed trust-based run-time approach for the query in Figure 1. The query execution takes place by fetching the university's details, the details of its publications and their authors just once. They are assigned trust values as the RDF data is being fetched according to the trust function defined by the information consumer. Once the publication's authors details have been retrieved the third predicate checking whether the person is of type professor can be relaxed to check for all people in the university like lecturers and graduate students. Its relaxation score is combined with its trust score as described earlier to arrive at a final score which is used to rank the results. Thus in effect the relaxation mechanism has been delayed to be performed on-the-fly at run-time along with the trust assignment mechanism to produce trustworthy approximate results and by doing so the shared data is not fetched repeatedly which results in significant performance benefits.

5 Optimizations

The query processing described in the last section works by performing trust-based relaxations on-the-fly during query execution. This approach serves well

---

Algorithm 2. Run-Time Approach

   Input : :Query Q
   Output: :Approximate answers
1 let γ be the threshold
2 $seed =$ intial set of URIs
3 **foreach** $queryPredicate_k$ in Q **do**
4   Compute trust scores for the triples **while** $seed \neq \phi$ **do**
5   **foreach** $seed_i$ **do**
6   Dereference $seed_i$ and retrieve its RDF graph R
7   Remove $seed_i$ from $seed$
8   **foreach** $predicate$ p_j in R $with$ $subject$ $seed_i$ **do**
9   **if** p_j $matches$ the $corresponding$ $query$ $predicate$
                  $queryPredicate_k$ **then**
10   relaxScore(p_j) = 1
11   **if** $p_{j_{object}}$ is $bound$ in the $Query$ **then**
12   compute relaxScore($p_{j_{object}}$) with $queryPredicate_{k_{object}}$
13   **else**
14   compute the relaxScore(p_j) with $queryPredicate_k$
15             $finalScore_{p_j} = w_1*$relaxScore(p_j) + (1-w_1)*trustScore(p_j)
16   Sort all p_j in the descending order of their finalScores.
17   **foreach** p_j **do**
18   **if** $relaxScore(p_j)$ γ **then**
19   **if** $p_{j_{object}}$ is not $bound$ **then**
20                  $seed \Leftarrow seed \cup p_{j_{object}}$

21 **foreach** $seed_i$ in $seed$ **do**
22   Retrieve the path p from $seed_i$ to root
23   Return p as the approximate answer

---

to optimize the query but there are opportunities that arise during query execution that can be exploited to further optimize the query. To do so the vocabulary(RDFS/OWL) describing the resources which gives the domains and ranges of various predicates as well as the subclass/superclass hierarchy details of all classes is considered. The idea of using the vocabularies to restrict query execution to certain classes of data that produce results has been discussed in [12]. We make use of vocabularies here to prune the relaxation steps which will not lead to results. We also maintain minimum trust score thresholds to ensure quality of results efficiently by determining and pruning unreliable RDF data at an early stage in query execution. There are two cases that arise

Case1: If a predicate p is replaced by p' with the subsequent predicate q, and that $range(p')\cap domain(q)$ is $\emptyset$ the current relaxation of p is pruned as it will not produce results. We also check if the trust score determined by the user-defined trust function is above the minimum threshold α, if it is not we move on to the next relaxation. There may be a situation when the subsequent predicate q is relaxed to q' and $range(p') \cap domain(q')$ is not $\emptyset$ in which case some results are missed. Therefore, a minimum threshold for the score of relaxation is maintained, and if the intersection of $range(p') \cap domain(q)$ is $\emptyset$ the relaxation is pruned only if the score is below the threshold.

Case2: If an object o is replaced by o' with the subsequent predicate q, and that $o' \cap domain(q)$ is $\emptyset$ the current relaxation can be pruned as it will not produce results. We also check if the trust score determined by the user-defined trust function is above the minimum threshold α, if it is not we move on to the next relaxation. There may be a situation when the subsequent predicate q is relaxed to q' and $o' \cap domain(q')$ is not $\emptyset$ in which some results are missed. Therefore, a minimum threshold is maintained for the score of relaxation, and if the intersection of $o' \cap domain(q)$ is $\emptyset$ the relaxation is pruned only if the score is below the threshold.

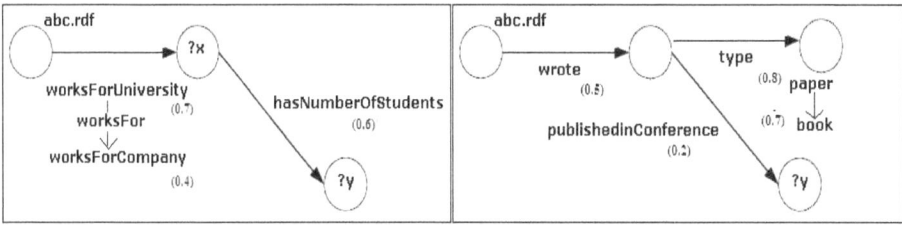

Fig. 5. Optimizations Examples

Figure 5 illustrates the two cases. The first Figure shows the query during whose execution the predicate "worksForUniversity" is relaxed to "worksFor". If there is a predicate "worksForCompany" in the retrieved RDF graph of the entity and as it is a subproperty of "worksFor" the query condition is relaxed to "worksForCompany". But the domain of the predicate succeeding it, that is "hasNumberOfStudents", is the class of universities whereas the range of the predicate "worksinCompany" is the class of Companies whose intersection is $\emptyset$. Thus this relaxation is pruned. Trust score is also computed for the new predicate "worksForCompany" and if it is found to be lesser than a predefined threshold, we move on to the next relaxation. But there is a possibility that the relaxation of the next predicate is from "hasNumberofStudents" to "hasnumberofEmployees". In which case the domain of the new relaxed predicate is the class of companies whose intersection with the range of earlier relaxed predicate is again the class of companies. Hence, if the first relaxation had not been discarded, results could have been produced. To handle this situation, the score of relaxation is taken into account. If the score is above a certain predefined threshold, the relaxation is allowed and the query execution proceeds as usual. The next figure shows the query during whose execution the object node "paper" is relaxed to the class of "books". However, the next predicate "publishedinConference" has the class of papers as it domain. Hence, the relaxation to class of books produces a $\emptyset$ set and can be pruned.

Algorithm 3. Optimizations

Input : :Query Q
Output: :Decision on whether to continue with current approximation
1 let t denote the triple being handled, which is approximated to $t^{'}$
2 let q be the predicate succeeding t
3 let γ be the threshold score of approximation
4 let $alpha$ be the threshold score of trust
5 **if** *predicate p is relaxed to $p^{'}$* **then**
6   Compute $trustScore(t^{'})$ using the user-defined trust function **if** $trustScore(t^{'}) < \alpha$ **then**
7 | try different relaxation of p
8   **end**
9   **if** $range(p^{'}) \cap domain(q) == \emptyset$ **then**
10   **if** $score(t) < \gamma$ **then**
11 | try different relaxation of p
12   **end**
13   **end**
14 **end**
15 **if** *object node o is relaxed to $o^{'}$* **then**
16   Compute $trustScore(t^{'})$ using the user-defined trust function **if** $trustScore(t^{'}) < \alpha$ **then**
17 | try different relaxation of p
18   **end**
19   **if** $o^{'} \cap domain(q) == \emptyset$ **then**
20   **if** $score(t) < \gamma$ **then**
21 | try different relaxation of o
22   **end**
23   **end**
24 **end**

---

6   Experiments

The experiments were conducted on a Pentium 4 machine running windows XP with 1 GB main memory. All the programs were written in Java. The synthetic data used for the simulations was generated with the LUBM benchmark data generator [10]. The LUBM benchmark is basically an university ontology that describes all the constituents of a university like its faculty,courses,students etc. The synthetic data is represented as a web of linked data with 200,890 nodes denoting entities and 500,595 edges denoting the relationships between them.

The efficacy of the proposed trust-based run-time approach was demonstrated by executing a set of queries in Figure 6 on the simulated web of linked data of a university and comparing the results with the compile-time approach. Each of the query below can be relaxed in a number of ways and the compile-time approach generates multiple relaxed queries and sorts and executes them. Their execution considers the trust scores of retrieved RDF data in addition to the similarity scores to rank the results. Whereas in contrast, the proposed trust-based run-time approach integrates the process of relaxation and the trust assignment process with the query execution to produce approximate answers. The trust values are assigned to the triples at random for the purposed of this paper and the experiments were conducted for different values of weight w_1. w_1 as described earlier determines the weight of the relaxation score in the final score that is obtained by combining the relaxation score with the trust score of the triple. This final score is then used to rank the results. The time taken to execute the query is proportional to the number of URIs resolved to fetch their RDF descriptions during the course of query execution. Therefore, this paper uses the reduction in the number of URIs fetched as a metric to judge the results.

Q_1:(?x,type,TeachingAssitant)(?x,teachingAssistantOf,http://www.Department0.University0.C ourse3)(?x,mastersDegreeFrom,http://www.Department0.University0.edu)

Q_2:(?x,teacherOf,?z)(?x,ub,worksFor,University0)(?x,type,AssistantProfessor)

Q_3 (?x,advisor,?y)(?y,type,AssistantProfessor)(?y,researchInterest',Research12')(?y,worksFor,h ttp://www.University0.edu)

Q_4 (?x,advisor,?y)(?y,type,Professor)(?y,worksFor, http://www.University0.edu)

Q_5:(?x,type,JournalArticle)(?y,publicationAuthor,?x) (?y,type,Professor)

Fig. 6. Queries

Query 1 searches for the teaching assistants of a particular course who have a masters degree from a particular university. Approximate answers are generated by relaxing the constraints step by step on the teaching assistant that is the teaching assistant can handle any course and have a master's degree from any university. Query 2 searches for assistant professors who teach a graduate course. Approximate answers are produced by relaxing the conditions in steps to look for all faculty who teach any course. Query 3 looks for assistant professor advisors who have a particular research interest. The query is again relaxed in steps by searching for all the people in the university who have any research interest. Query 4 searches for advisors who are professors and work for a particular university. Approximate answers are produced by looking for advisors who can be any type of faculty and who work for any university. Query 5 searches for professors who have authored a journal article. Approximate answers are produced step by step by looking for all persons including graduate students who have authored any type of paper. As shown in Figure 7, our proposed trust-based run-time approach achieves a significant improvement in performance of upto 90%.

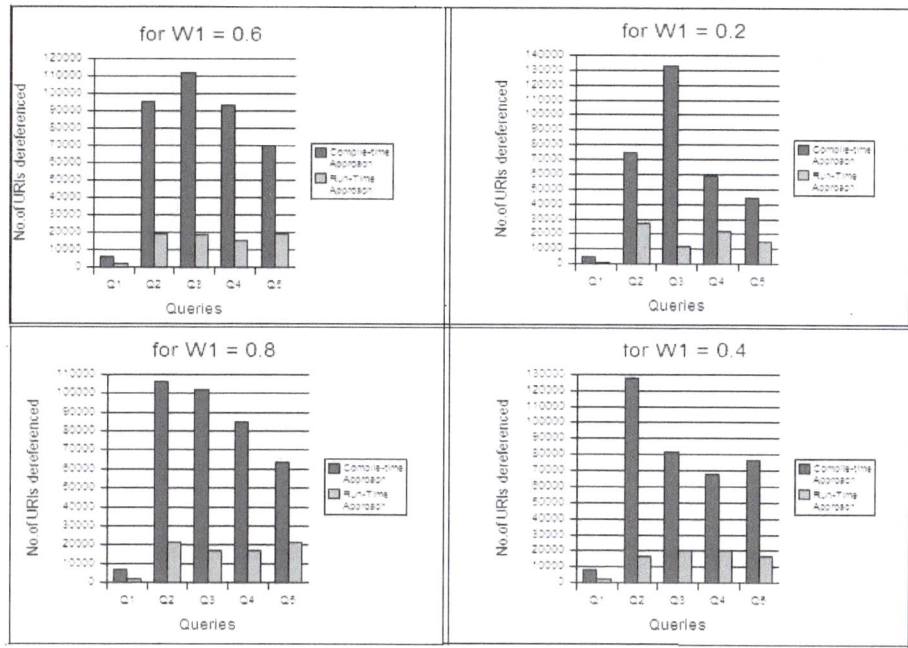

Fig. 7. Results

7   Conclusions

The paper presented an approach towards allowing trust-based approximate querying of the web of linked data using the SPARQL query language. The proposed trust-based run-time approach produces approximate answers by relaxing the query conditions on-the-fly during query execution using the ontologies available on the web of linked data taking into consideration the trust values of RDF data, in contrast with existing compile-time approach which generates multiple relaxed queries and executes them to produce approximate answers. The advantage of proposed trust-based run-time approach is that it is able to avoid fetching the shared data between the relaxed queries repeatedly, which results in significant performance benefits as shown in the experiments. We also presented a trust model which assigned trust scores to RDF data according to the user-defined trust functions which was used along with similarity scores to ensure that trustworthy results were given prominence.

References

1. Bizer, C., Heath, T., Berners-Lee, T.: Linked data – the story so far. International Journal on Semantic Web and Information Systems 5(3), 1–22 (2009)
2. Tummarello, G., Delbru, R., Oren, E.: Sindice.com: Weaving the open linked data, pp. 552–565 (2008)
3. Quilitz, B., Leser, U.: Querying Distributed RDF Data Sources with SPARQL. In: Bechhofer, S., Hauswirth, M., Hoffmann, J., Koubarakis, M. (eds.) ESWC 2008. LNCS, vol. 5021, pp. 524–538. Springer, Heidelberg (2008)
4. Hartig, O., Bizer, C., Freytag, J.-C.: Executing SPARQL Queries over the Web of Linked Data. In: Bernstein, A., Karger, D.R., Heath, T., Feigenbaum, L., Maynard, D., Motta, E., Thirunarayan, K. (eds.) ISWC 2009. LNCS, vol. 5823, pp. 293–309. Springer, Heidelberg (2009)
5. Huang, H., Liu, C., Zhou, X.: Computing Relaxed Answers on RDF Databases. In: Bailey, J., Maier, D., Schewe, K.-D., Thalheim, B., Wang, X.S. (eds.) WISE 2008. LNCS, vol. 5175, pp. 163–175. Springer, Heidelberg (2008)
6. Reddy, K.B., Kumar, P.S.: Efficient approximate SPARQL querying of the web of linked data. In: Proceedings of the ISWC Workshop on Uncertainity Reasoning over the Semantic Web, URSW 2010. CEUR-WS (2010)
7. Hartig, O.: Querying Trust in RDF Data with tSPARQL. In: Aroyo, L., Traverso, P., Ciravegna, F., Cimiano, P., Heath, T., Hyvönen, E., Mizoguchi, R., Oren, E., Sabou, M., Simperl, E. (eds.) ESWC 2009. LNCS, vol. 5554, pp. 5–20. Springer, Heidelberg (2009)
8. Hartig, O.: Provenance information in the web of data. In: LDOW 2009, Madrid, Spain, April 20 (2009)
9. Zaihrayeu, I., da Silva, P.P., McGuinness, D.L.: IWTrust: Improving User Trust in Answers from the Web. In: Herrmann, P., Issarny, V., Shiu, S. (eds.) iTrust 2005. LNCS, vol. 3477, pp. 384–392. Springer, Heidelberg (2005)
10. Gil, Y., Ratnakar, V.: Trusting Information Sources One Citizen at a Time. In: Horrocks, I., Hendler, J. (eds.) ISWC 2002. LNCS, vol. 2342, pp. 162–176. Springer, Heidelberg (2002)
11. Golbeck, J.: Generating Predictive Movie Recommendations from Trust in Social Networks. In: Stølen, K., Winsborough, W.H., Martinelli, F., Massacci, F. (eds.) iTrust 2006. LNCS, vol. 3986, pp. 93–104. Springer, Heidelberg (2006)
12. Reddy, K.B., Kumar, P.S.: Optimizing SPARQL queries over the web of linked data. In: Proceedings of the VLDB Workshop on Semantic Data Management, SemData 2010. CEUR-WS (2010)
13. Guo, Y., Pan, Z., Heflin, J.: LUBM: A benchmark for OWL knowledge base systems. J. Web Sem. 3(2-3), 158–182 (2005)

Author Index